This book provides an introduction, with applications, to three interconnected mathematical topics:

- zeta functions in their rich variety: those of Riemann, Hurwitz, Barnes, Epstein, Selberg, and Ruelle, plus graph zeta functions;
- modular forms: Eisenstein series, Hecke and Dirichlet L-functions, Ramanujan's tau function, and cusp forms;
- vertex operator algebras: correlation functions, quasimodular forms, modular invariance, rationality, and some current research topics including higher-genus conformal field theory.

Applications of the material to physics are presented, including Kaluza–Klein extra-dimensional gravity, bosonic string calculations, a Cardy formula for black hole entropy, Patterson–Selberg zeta function expressions of one-loop quantum field and gravity partition functions, Casimir energy calculations, atomic Schrödinger operators, Bose–Einstein condensation, heat kernel asymptotics, random matrices, quantum chaos, elliptic and theta function solutions of Einstein's equations, a soliton–black hole connection in two-dimensional gravity, and conformal field theory.

Mathematical Sciences Research Institute
Publications

57

A Window into Zeta
and Modular Physics

Mathematical Sciences Research Institute Publications

1	Freed/Uhlenbeck: *Instantons and Four-Manifolds*, second edition
2	Chern (ed.): *Seminar on Nonlinear Partial Differential Equations*
3	Lepowsky/Mandelstam/Singer (eds.): *Vertex Operators in Mathematics and Physics*
4	Kac (ed.): *Infinite Dimensional Groups with Applications*
5	Blackadar: *K-Theory for Operator Algebras*, second edition
6	Moore (ed.): *Group Representations, Ergodic Theory, Operator Algebras, and Mathematical Physics*
7	Chorin/Majda (eds.): *Wave Motion: Theory, Modelling, and Computation*
8	Gersten (ed.): *Essays in Group Theory*
9	Moore/Schochet: *Global Analysis on Foliated Spaces*, second edition
10–11	Drasin/Earle/Gehring/Kra/Marden (eds.): *Holomorphic Functions and Moduli*
12–13	Ni/Peletier/Serrin (eds.): *Nonlinear Diffusion Equations and Their Equilibrium States*
14	Goodman/de la Harpe/Jones: *Coxeter Graphs and Towers of Algebras*
15	Hochster/Huneke/Sally (eds.): *Commutative Algebra*
16	Ihara/Ribet/Serre (eds.): *Galois Groups over* $\mathbb{Q}$
17	Concus/Finn/Hoffman (eds.): *Geometric Analysis and Computer Graphics*
18	Bryant/Chern/Gardner/Goldschmidt/Griffiths: *Exterior Differential Systems*
19	Alperin (ed.): *Arboreal Group Theory*
20	Dazord/Weinstein (eds.): *Symplectic Geometry, Groupoids, and Integrable Systems*
21	Moschovakis (ed.): *Logic from Computer Science*
22	Ratiu (ed.): *The Geometry of Hamiltonian Systems*
23	Baumslag/Miller (eds.): *Algorithms and Classification in Combinatorial Group Theory*
24	Montgomery/Small (eds.): *Noncommutative Rings*
25	Akbulut/King: *Topology of Real Algebraic Sets*
26	Judah/Just/Woodin (eds.): *Set Theory of the Continuum*
27	Carlsson/Cohen/Hsiang/Jones (eds.): *Algebraic Topology and Its Applications*
28	Clemens/Kollár (eds.): *Current Topics in Complex Algebraic Geometry*
29	Nowakowski (ed.): *Games of No Chance*
30	Grove/Petersen (eds.): *Comparison Geometry*
31	Levy (ed.): *Flavors of Geometry*
32	Cecil/Chern (eds.): *Tight and Taut Submanifolds*
33	Axler/McCarthy/Sarason (eds.): *Holomorphic Spaces*
34	Ball/Milman (eds.): *Convex Geometric Analysis*
35	Levy (ed.): *The Eightfold Way*
36	Gavosto/Krantz/McCallum (eds.): *Contemporary Issues in Mathematics Education*
37	Schneider/Siu (eds.): *Several Complex Variables*
38	Billera/Björner/Green/Simion/Stanley (eds.): *New Perspectives in Geometric Combinatorics*
39	Haskell/Pillay/Steinhorn (eds.): *Model Theory, Algebra, and Geometry*
40	Bleher/Its (eds.): *Random Matrix Models and Their Applications*
41	Schneps (ed.): *Galois Groups and Fundamental Groups*
42	Nowakowski (ed.): *More Games of No Chance*
43	Montgomery/Schneider (eds.): *New Directions in Hopf Algebras*
44	Buhler/Stevenhagen (eds.): *Algorithmic Number Theory: Lattices, Number Fields, Curves and Cryptography*
45	Jensen/Ledet/Yui: *Generic Polynomials: Constructive Aspects of the Inverse Galois Problem*
46	Rockmore/Healy (eds.): *Modern Signal Processing*
47	Uhlmann (ed.): *Inside Out: Inverse Problems and Applications*
48	Gross/Kotiuga: *Electromagnetic Theory and Computation: A Topological Approach*
49	Darmon/Zhang (eds.): *Heegner Points and Rankin L-Series*
50	Bao/Bryant/Chern/Shen (eds.): *A Sampler of Riemann–Finsler Geometry*
51	Avramov/Green/Huneke/Smith/Sturmfels (eds.): *Trends in Commutative Algebra*
52	Goodman/Pach/Welzl (eds.): *Combinatorial and Computational Geometry*
53	Schoenfeld (ed.): *Assessing Mathematical Proficiency*
54	Hasselblatt (ed.): *Dynamics, Ergodic Theory, and Geometry*
55	Pinsky/Birnir (eds.): *Probability, Geometry and Integrable Systems*
56	Albert/Nowakowski (eds.): *Games of No Chance 3*
57	Kirsten/Williams (eds.): *A Window into Zeta and Modular Physics*

Volumes 1–4, 6–8, and 10–27 are published by Springer-Verlag

A Window into Zeta and Modular Physics

Edited by

Klaus Kirsten
Baylor University

Floyd L. Williams
University of Massachusetts, Amherst

Klaus Kirsten
Department of Mathematics
Baylor University
Waco, TX 76798
United States
Klaus_Kirsten@baylor.edu

Floyd L. Williams
Department of Mathematics and Statistics
University of Massachusetts
Amherst, MA 01003-9305
United States
williams@math.umass.edu

Silvio Levy (*Series Editor*)
Mathematical Sciences Research Institute
Berkeley, CA 94720
levy@msri.org

The Mathematical Sciences Research Institute wishes to acknowledge support by the National Science Foundation and the *Pacific Journal of Mathematics* for the publication of this series.

CAMBRIDGE UNIVERSITY PRESS
Cambridge, New York, Melbourne, Madrid, Cape Town,
Singapore, São Paulo, Delhi, Mexico City

Cambridge University Press
32 Avenue of the Americas, New York NY 10013-2473, USA

Published in the United States of America by Cambridge University Press, New York

www.cambridge.org
Information on this title: www.cambridge.org/9781107633933

© Mathematical Sciences Research Institute 2010

This publication is in copyright. Subject to statutory exception
and to the provisions of relevant collective licensing agreements,
no reproduction of any part may take place without the written
permission of Cambridge University Press.

First published 2010
First paperback edition 2013

A catalogue record for this publication is available from the British Library

ISBN 978-0-521-19930-8 Hardback
ISBN 978-1-107-63393-3 Paperback

Cambridge University Press has no responsibility for the persistence or accuracy of URLs for external or third-party internet websites referred to in this publication, and does not guarantee that any content on such websites is, or will remain, accurate or appropriate.

Contents

Introduction   1
    KLAUS KIRSTEN AND FLOYD L. WILLIAMS

Introductory Lectures

Lectures on zeta functions, L-functions and modular forms with some physical applications   7
    FLOYD L. WILLIAMS

Basic zeta functions and some applications in physics   101
    KLAUS KIRSTEN

Zeta functions and chaos   145
    AUDREY TERRAS

Vertex operators and modular forms   183
    GEOFFREY MASON AND MICHAEL TUITE

Research Lectures

Applications of elliptic and theta functions to FRLW cosmology with cosmological constant   279
    JENNIE D'AMBROISE

Integrable systems and 2D gravitation: How a soliton illuminates a black hole   295
    SHABNAM BEHESHTI

Functional determinants in higher dimensions using contour integrals   307
    KLAUS KIRSTEN

The role of the Patterson–Selberg zeta function of a hyperbolic cylinder in three-dimensional gravity with a negative cosmological constant   329
    FLOYD L. WILLIAMS

Introduction

Some exciting, bold new cooperative explorations of various interconnections between traditional domains of "pure" mathematics and exotic new developments in theoretical physics have continued to emerge in recent years. The beautiful interlacing of theory and application, and cross-discipline interaction has led, as is usual, to notable, fruitful, and bonus outcomes.

These interconnections range from topology, algebraic geometry, modular forms, Eisenstein series, zeta functions, vertex operators, and knot theory to gauge theory, strings and branes, quantum fields, cosmology, general relativity, and Bose–Einstein condensation. They are broad enough in scope to present the average reader with not only a measure of enchantment but with some mild bewilderment as well. A new journal, *Communications in Number Theory and Physics*, has recently been launched to follow and facilitate interactions and dynamics between these two disciplines, for example. Various books that are now available, in addition to an array of conference and workshop activity, accent this fortunate merger of mathematics and physical theory and assist greatly in bridging the divide, although in some cases the themes are pitched at a level more suitable for advanced readers and researchers.

In an attempt to further bridge the divide, at least in some modest way for students and non-experts, and to provide a window into this adventurous arena of intertwining ideas, a graduate workshop entitled "A Window into Zeta and Modular Physics" was presented at MSRI during the period from 16 to 27 June 2008. The workshop consisted of daily expository lectures, speakers' seminar lectures (where the material was more technical and represented their research, but which at the same time did connect to and enlarge on the daily lectures), and four special student lectures. Given the excellent preparation and presentation of the lectures, it was proposed that it would be of further benefit to the students (especially given their enthusiastic response), and to the mathematical physics community in general, if the lectures were eventually molded in some form as a book. Thus the present volume, with the workshop title and similar style, evolved. Besides the editors, there were three workshop speakers: Geoffrey Mason, Audrey Terras, and Michael Tuite. The student speakers were Jennie

D'Ambroise, Shabnam Beheshti, Savan Kharel, and Paul Nelson. Among these four, Jennie and Shabnam (who have since earned their PhDs) accepted the invitation to have their lectures appear here.

The volume consists of two parts. Part I contains basic, expository lectures, except that it was found convenient to also include some seminar material in the combined presentations of Geoff and Michael, and thus to make those one long set of lectures. Part II consists of speakers' seminar lectures, where we have included the lectures of Jennie and Shabnam.

A more specific description of Part I is as follows. Lectures of Floyd Williams cover topics such as the Riemann zeta function (proof of its meromorphic continuation and functional equation), Euler products (in particular, of Hecke and Dirichlet L-functions), holomorphic modular forms, Dedekind's eta function, the quasimodular form G_2, the Rademacher–Zuckerman formula for the Fourier coefficients of forms of negative weight, and non-holomorphic Eisenstein series, with some physical applications including gravity in extra dimensions, finite temperature zeta functions, the zeta regularization of Casimir energy, a determinant (or path integral) computation in Bosonic string theory, and an abstract Cardy formula for black hole entropy, based on the asymptotic behavior of Fourier coefficients of modular forms of zero weight. Several appendices are included. These provide a proof of the Poisson summation formula, the Jacobi inversion formula, the Fourier expansion of a holomorphic periodic function, etc., and thus they render further completeness of the material. The lectures, in general, provide some background material for some of the other lectures, with some overlap.

Further applications of zeta functions — those of Hurwitz, Barnes, and Epstein, and also spectral zeta functions — are presented in the lectures of Klaus Kirsten. Integral representations of these functions are established and their association with eigenvalue problems for partial differential operators is discussed by way of concrete computational examples, where various types of boundary conditions are imposed. For the atomic Schrödinger operator corresponding to a harmonic oscillator potential in three dimensions, for example, the spectral zeta function is shown to be given in terms of the Barnes zeta function.

Applications of zeta functions to other areas such as the Casimir effect and Bose–Einstein condensation are also discussed in the Kirsten lectures, along with some provocative, motivating questions: *Can one hear the shape of a drum? What does the Casimir effect know about a boundary? What does a Bose gas know about its container?* Regarding the first of these questions — which was posed in 1966 by M. Kac, but may even go back to H. Weyl — some heat kernels and their small-time asymptotics and Minakshisundaram–Pleijel coefficients are presented by way of examples, and by way of a general result for Laplace-type differential operators acting on smooth sections of a vector bundle over a smooth compact Riemannian manifold with or without boundary. Bose–Einstein condensation is the final topic treated, where zeta functions continue to play a relevant role. A Bose–Einstein condensate is the ground state (lowest energy state) assumed by interacting bosons under the influence of some external trapping potential. This condensation phenomenon, which was predicted in 1924 by S. Bose and A. Einstein, is an example of phase transition at the quantum level. It took some 70 years, however, for its first experimental verification, in some 1995 vapor studies of rubidium and sodium.

The subject matter of zeta functions (of Riemann, Selberg, Ruelle, and Ihara) and quantum chaos is taken up in the lectures of Audrey Terras, where emphasis is laid on the Ihara zeta function of a finite graph, and its connections to chaos and random matrix theory. The Selberg zeta function is a zeta function attached to a compact Riemann surface (say of genus at least two), and gives rise to a duality between lengths of closed geodesics and the spectrum of the Laplace–Beltrami operator — closed geodesics being regarded as Selberg "primes". This zeta function can be attached, more generally, to compact quotients of hyperbolic spaces, or in fact to the quotient of a rank 1, noncompact symmetric space modulo a cofinite volume discrete group of isometries. The Ruelle zeta function is a dynamical systems zeta function, whose motivation goes back to work of M. Artin and B. Mazur on the zeta function of a projective algebraic variety over a finite field, where the Frobenius map is replaced by a suitable diffeomorphism of a smooth compact manifold. It is shown in the lectures that the Ihara zeta function is generalized by the Ruelle zeta function, the former function being the graph version of the Selberg zeta function where the duality is replaced by that between closed paths and the spectrum of an *adjacency matrix.*

Some particular topics include edge and path zeta functions, Ihara determinant formulas, a graph prime number theorem, a Riemann hypothesis regarding the poles of the Ihara zeta function (which for a regular connected graph is true precisely when the graph is *Ramanujan* — a condition on the eigenvalues of the adjacency matrix, roughly speaking), the Alon conjectures for regular and irregular graphs (which refer to Riemann hypotheses for these graphs), and some closing material with a focus on the quantum chaos question that concerns a connection between the poles of Ihara zeta and eigenvalues of a random matrix. Here a comparison is made between the spacing of poles and the spacing of the eigenvalues of a certain (non-symmetric) edge adjacency matrix.

Quantum chaos, which is not always precisely defined, is described sometimes as the statistics of eigenvalues (or energy levels) of particular non-classical systems. For example, E. Wigner, in the 1950s, first introduced the notion of the statistical distribution of energy levels of heavy atomic nuclei. From this work there evolved the *Wigner surmise* (which Audrey also discusses) for the probability density for finding two adjacent eigenvalues with a given spacing — a density that differs markedly from a Poisson density.

Part I concludes with an extended set of lectures by Geoff Mason and Michael Tuite that cover basic, introductory material on vertex operator algebras (VOAs) and modular forms, as well as some research material indicative of their presentations during the workshop speakers' seminar.

VOAs, also known as chiral algebras to physicists, special cases of which include W-algebras, were formally defined by R. Borcherds. They figure prominently in many areas such as finite group theory (in particular in regards to "monstrous moonshine"), representations of affine Kac–Moody Lie algebras, knot theory, quantum groups, geometric Langlands theory, etc., and in fact they provide for a mathematical, axiomatic formulation of two-dimensional conformal field theory (CFT). Various CFT manipulations, that sometimes are in want of rigor, are neatly handled, algebraically, in the VOA world. In the so-called operator product expansion for *primary* fields, for example (as postulated by A. Belavin, A. Polyakov, and A. Zamolodchikov), the fundamental requirement of associativity encodes neatly in VOA structure — which in the lectures amounts to the Jacobi identity (page 188). This identity also embodies the key notion of locality. Locality relaxes the requirement of commutativity of fields, which if imposed would be too strong a physical condition.

String theory for bosons can be regarded as a CFT, and it therefore has natural ties to VOAs. For example, to an even, positive definite lattice L is attached a VOA (pages 221–224), which corresponds to a bosonic string compactified on the torus defined by L. Other naturally constructed VOAs are Heisenberg VOAs (that correspond to a free boson), Virasoro VOAs, and VOAs attached to a Lie algebra g equipped with a symmetric, non-degenerate, invariant bilinear form, from whence is attached to g an affine Kac–Moody Lie algebra. The moonshine module also has a VOA structure (due to I. Frenkel, J. Lepowsky, and J. Meurman), which is discussed on page 233.

Modular forms and elliptic functions arise naturally in VOA theory by way of VOA characters (or partition functions) and, more generally, correlation functions. For *nonrational* VOAs, however, these functions might only be quasi-modular. But the character of the (nonrational) Heisenberg VOA, for example, is a modular form (of weight $-\frac{1}{2}$) as it is the reciprocal of the Dedekind eta function. A characteristic property of rational VOAs is that they have only finitely many inequivalent, irreducible representations. In particular, a distinguished class of rational VOAs (called Virasoro *minimal models*) is provided by certain Virasoro quotient modules of zero conformal weight and a central charge $c = 1 - 6(p-q)^2/pq$ parametrized by a pair of coprime integers p, q greater than 1. The (finite) number of irreducible representations here is $(p-1)(q-1)$. $p = 4$ and $q = 3$, for example, give rise to a central charge $\frac{1}{2}$, which at the physical level corresponds to the Ising model in statistical mechanics, whereas $p = 5$, $q = 4$ corresponds to the tricritical Ising model with central charge $\frac{7}{10}$, and $p = 6$, $q = 5$ corresponds to the three-states Potts model with central charge $\frac{4}{5}$.

In addition to one-point correlation functions, Geoff and Michael also consider two-point correlation functions that have relevance to their research on higher genus CFT. These functions are shown to be elliptic. Zhu recursion formulas (for correlation functions), Zhu's finiteness condition (referred to as C_2-cofiniteness in the lectures), the important issue of modular invariance, and other important matters are given careful attention in the latter part of the paper, with a discussion of current research areas. A deep, initial result of Y. Zhu is that for rational VOAs subject to his finiteness condition, the complete (finite) set of inequivalent, irreducible characters enjoys the beautiful modular property of being holomorphic functions on the upper half-plane that span a subspace invariant under the action of the modular group $SL(2, \mathbb{Z})$. At the Lie-algebra representation theory level, apart from VOA theory, modular properties of characters have profound implications — which is the case for Weyl–Kac characters of affine $SU(2)$, for example, where modularity is exploited in the study of BTZ black hole entropy corrections.

Seminar lectures comprise Part II of the volume (with the exception of those of Geoff and Michael, which, as mentioned, appear in Part I). We describe them briefly. Jennie's lecture deals with elliptic and theta function solutions of Einstein's gravitational field equations in the FLRW model. A soliton–black hole connection in two-dimensional gravity is considered in Shabnam's lecture. Floyd discusses the role of the Patterson–Selberg zeta function in three-dimensional gravity. Klaus points out how contour integration methods lead to closed formulas for functional determinants in low and high dimensions.

The editors express sincere thanks and appreciation to their workshop colleagues for their valued contributions, and to MSRI Director Dr. Robert Bryant and Associate Director Dr. Kathleen O'Hara for the invitation to present the workshop and the kind hospitality we were accorded. We are also appreciative of the gifted workshop students that we were fortunate to interact with. Our sincere thanks extend moreover to the MSRI Book Series Editor, Dr. Silvio Levy, for his helpful guidance during the preparation of the volume.

Klaus Kirsten
Floyd L. Williams
August 2009

Lectures on zeta functions, L-functions and modular forms with some physical applications

FLOYD L. WILLIAMS

Introduction

We present nine lectures that are introductory and foundational in nature. The basic inspiration comes from the Riemann zeta function, which is the starting point. Along the way there are sprinkled some connections of the material to physics. The asymptotics of Fourier coefficients of zero weight modular forms, for example, are considered in regards to black hole entropy. Thus we have some interests also connected with Einstein's general relativity. References are listed that cover much more material, of course, than what is attempted here.

Although his papers were few in number during his brief life, which was cut short by tuberculosis, Georg Friedrich Bernhard Riemann (1826–1866) ranks prominently among the most outstanding mathematicians of the nineteenth century. In particular, Riemann published only one paper on number theory [32]: "Über die Anzahl der Primzahlen unter einer gegebenen Grösse", that is, "On the number of primes less than a given magnitude". In this short paper prepared for Riemann's election to the Berlin Academy of Sciences, he presented a study of the distribution of primes based on complex variables methods. There the now famous *Riemann zeta function*

$$\zeta(s) \stackrel{\text{def}}{=} \sum_{n=1}^{\infty} \frac{1}{n^s}, \tag{0.1}$$

defined for $\operatorname{Re} s > 1$, appears along with its analytic continuation to the full complex plane $\mathbb{C}$, and a proof of a functional equation (FE) that relates the values $\zeta(s)$ and $\zeta(1-s)$. The FE in fact was conjectured by Leonhard Euler, who also obtained in 1737 (over 120 years before Riemann) an *Euler product* representation

$$\zeta(s) = \prod_{p>0} \frac{1}{1-p^{-s}} \quad (\operatorname{Re} s > 1) \tag{0.2}$$

of $\zeta(s)$ where the product is taken over the primes p. Moreover, Riemann introduced in that seminal paper a query, now called the *Riemann Hypothesis* (RH), which to date has defied resolution by the best mathematical minds. Namely, as we shall see, $\zeta(s)$ vanishes at the values $s = -2n$, where $n = 1, 2, 3, \ldots$; these are called the *trivial* zeros of $\zeta(s)$. The RH is the (yet unproved) statement that if s is a zero of ζ that is *not* trivial, the real part of s must have the value $\frac{1}{2}$!

Regarding Riemann's analytic approach to the study of the distribution of primes, we mention that his main goal was to set up a framework to facilitate a proof of the *prime number theorem* (which was also conjectured by Gauss) which states that if $\pi(x)$ is the number of primes $\leq x$, for $x \in \mathbb{R}$ a real number, then $\pi(x)$ behaves asymptotically (as $x \to \infty$) as $x/\log x$. That is, one has (precisely) that

$$\lim_{x \to \infty} \frac{\pi(x)}{x/\log x} = 1, \tag{0.3}$$

which was independently proved by Jacques Hadamard and Charles de la Vallée-Poussin in 1896. A key role in the proof of the monumental result (0.3) is the fact that at least all nontrivial zeros of $\zeta(s)$ reside in the interior of the *critical strip* $0 \leq \operatorname{Re} s \leq 1$.

Riemann's deep contributions extend to the realm of physics as well - Riemannian geometry, for example, being the perfect vehicle for the formulation of Einstein's gravitational field equations of general relativity. Inspired by the definition (0.1), or by the Euler product in (0.2), one can construct various other zeta functions (as is done in this volume) with a range of applications to physics. A particular zeta function that we shall consider later will bear a particular relation to a particular solution of the Einstein field equations — namely a *black hole* solution; see my Speaker's Lecture.

There are quite many ways nowadays to find the analytic continuation and FE of $\zeta(s)$. We shall basically follow Riemann's method. For the reader's benefit, we collect some standard background material in various appendices. Thus, to a large extent, we shall attempt to provide details and completeness of the material, although at some points (later for example, in the lecture on modular forms) the goal will be to present a general picture of results, with some (but not all) proofs.

Special thanks are extended to Jennie D'Ambroise for her competent and thoughtful preparation of all my lectures presented in this volume.

CONTENTS

Introduction	7
1. Analytic continuation and functional equation of the Riemann zeta function	9
2. Special values of zeta	17
3. An Euler product expansion	21
4. Modular forms: the movie	30
5. Dirichlet L-functions	46
6. Radiation density integral, free energy, and a finite-temperature zeta function	50
7. Zeta regularization, spectral zeta functions, Eisenstein series, and Casimir energy	57
8. Epstein zeta meets gravity in extra dimensions	66
9. Modular forms of nonpositive weight, the entropy of a zero weight form, and an abstract Cardy formula	70
Appendix	78
References	98

Lecture 1. Analytic continuation and functional equation of the Riemann zeta function

Since $|1/n^s| = 1/n^{\operatorname{Re} s}$, the series in (0.1) converges absolutely for $\operatorname{Re} s > 1$. Moreover, by the Weierstrass M-test, for any $\delta > 0$ one has uniform convergence of that series on the strip

$$S_\delta \stackrel{\text{def}}{=} \{s \in \mathbb{C} \mid \operatorname{Re} s > 1 + \delta\},$$

since $|1/n^s| = 1/n^{\operatorname{Re} s} < 1/n^{1+\delta}$ on S_δ, with

$$\sum_{n=1}^{\infty} \frac{1}{n^{1+\delta}} < \infty.$$

Since any compact subset of the domain $S_0 \stackrel{\text{def}}{=} \{s \in \mathbb{C} \mid \operatorname{Re} s > 1\}$ is contained in some S_δ, the series, in particular, converges absolutely and uniformly on compact subsets of S_0. By Weierstrass's general theorem we can conclude that the Riemann zeta function $\zeta(s)$ in (0.1) is holomorphic on S_0 (since the terms $1/n^s$ are holomorphic in s) and that termwise differentiation is permitted: for $\operatorname{Re} s > 1$

$$\zeta'(s) = -\sum_{n=1}^{\infty} \frac{\log n}{n^s}. \tag{1.1}$$

We wish to analytically continue $\zeta(s)$ to the full complex plane. For that purpose, we begin by considering the world's simplest *theta function* $\theta(t)$, defined for $t > 0$:

$$\theta(t) \overset{\text{def}}{=} \sum_{n \in \mathbb{Z}} e^{-\pi n^2 t} = 1 + 2 \sum_{n=1}^{\infty} e^{-\pi n^2 t} \qquad (1.2)$$

where $\mathbb{Z}$ denotes the ring of integers. It enjoys the remarkable property that its values at t and t inverse (i.e. $1/t$) are related:

$$\theta(t) = \frac{\theta(1/t)}{\sqrt{t}}. \qquad (1.3)$$

The very simple formula (1.3), which however requires some work to prove, is called the *Jacobi inversion formula*. We set up a proof of it in Appendix C, based on the *Poisson Summation Formula* proved in Appendix C. One can of course define more complicated theta functions, even in the context of higher-dimensional spaces, and prove analogous Jacobi inversion formulas.

For $s \in \mathbb{C}$ define

$$J(s) \overset{\text{def}}{=} \int_1^{\infty} \frac{\theta(t) - 1}{2} t^s \, dt. \qquad (1.4)$$

By Appendix A, $J(s)$ is an entire function of s, whose derivative can be obtained, in fact, by differentiation under the integral sign. One can obtain both the analytic continuation and the functional equation of $\zeta(s)$ by introducing the sum

$$I(s) \overset{\text{def}}{=} \sum_{n=1}^{\infty} \int_0^{\infty} (\pi n^2)^{-s} e^{-t} t^{s-1} \, dt, \qquad (1.5)$$

which we will see is well-defined for $\operatorname{Re} s > \frac{1}{2}$, and by computing it in different ways, based on the inversion formula (1.3). Recalling that the gamma function $\Gamma(s)$ is given for $\operatorname{Re} s > 0$ by

$$\Gamma(s) \overset{\text{def}}{=} \int_0^{\infty} e^{-t} t^{s-1} \, dt \qquad (1.6)$$

we clearly have

$$I(s) \overset{\text{def}}{=} \pi^{-s} \left(\sum_{n=1}^{\infty} \frac{1}{n^{2s}} \right) \Gamma(s) = \pi^{-s} \zeta(2s) \Gamma(s), \qquad (1.7)$$

so that $I(s)$ is well-defined for $\operatorname{Re} 2s > 1$: $\operatorname{Re} s > \frac{1}{2}$. On the other hand, by the change of variables $u = t/\pi n^2$ we transform the integral in (1.5) to obtain

$$I(s) = \sum_{n=1}^{\infty} \int_0^{\infty} e^{-\pi n^2 t} t^{s-1} \, dt.$$

We can interchange the summation and integration here by noting that

$$\sum_{n=1}^{\infty} \int_0^{\infty} \left|e^{-\pi n^2 t} t^{s-1}\right| dt = \sum_{n=1}^{\infty} \int_0^{\infty} e^{-\pi n^2 t} t^{\operatorname{Re} s - 1} dt = I(\operatorname{Re} s) < \infty$$

for $\operatorname{Re} s > \frac{1}{2}$; thus

$$I(s) = \int_0^{\infty} \sum_{n=1}^{\infty} e^{-\pi n^2 t} t^{s-1} dt = \int_0^{\infty} \frac{\theta(t) - 1}{2} t^{s-1} dt$$

$$= \int_0^1 \frac{\theta(t) - 1}{2} t^{s-1} dt + \int_1^{\infty} \frac{\theta(t) - 1}{2} t^{s-1} dt, \qquad (1.8)$$

by (1.2). Here

$$\int_0^1 t^{s-1} dt = \lim_{\varepsilon \to 0^+} \int_{\varepsilon}^1 t^{s-1} dt = \frac{1}{s} \qquad (1.9)$$

for $\operatorname{Re} s > 0$. In particular (1.9) holds for $\operatorname{Re} s > \frac{1}{2}$, and we have

$$\int_0^1 \frac{\theta(t) - 1}{2} t^{s-1} dt = \frac{1}{2} \int_0^1 \theta(t) t^{s-1} dt - \frac{1}{2s}. \qquad (1.10)$$

By the change of variables $u = 1/t$, coupled with the Jacobi inversion formula (1.3), we get

$$\int_0^1 \theta(t) t^{s-1} dt = \int_1^{\infty} \theta\left(\frac{1}{t}\right) t^{-1-s} dt = \int_1^{\infty} \theta(t) t^{\frac{1}{2}} t^{-1-s} dt$$

$$= \int_1^{\infty} (\theta(t) - 1) t^{-\frac{1}{2} - s} dt + \int_1^{\infty} t^{-\frac{1}{2} - s} dt$$

$$= \int_1^{\infty} (\theta(t) - 1) t^{-\frac{1}{2} - s} dt + \int_0^1 u^{-\frac{3}{2} + s = (s - \frac{1}{2}) - 1} du$$

$$= \int_1^{\infty} (\theta(t) - 1) t^{-\frac{1}{2} - s} dt + \frac{1}{s - \frac{1}{2}},$$

where we have used (1.9) again for $\operatorname{Re} s > \frac{1}{2}$. Together with equations (1.8) and (1.10), this gives

$$I(s) = \frac{1}{2} \int_1^{\infty} (\theta(t) - 1) t^{-\frac{1}{2} - s} dt + \frac{1}{2(s - \frac{1}{2})} - \frac{1}{2s} + \int_1^{\infty} \frac{\theta(t) - 1}{2} t^{s-1} dt$$

$$= \int_1^{\infty} \frac{\theta(t) - 1}{2} \left(t^{s-1} + t^{-\frac{1}{2} - s}\right) dt + \frac{1}{2s - 1} - \frac{1}{2s},$$

which with equation (1.7) gives

$$\pi^{-s}\zeta(2s)\Gamma(s) = \int_1^\infty \frac{\theta(t)-1}{2}\left(t^{s-1}+t^{-\frac{1}{2}-s}\right)dt + \frac{1}{2s-1} - \frac{1}{2s}, \quad (1.11)$$

for $\operatorname{Re} s > \frac{1}{2}$. Finally, in (1.11) replace s by $s/2$, to obtain

$$\pi^{-s/2}\zeta(s)\Gamma\left(\frac{s}{2}\right) = \int_1^\infty \frac{\theta(t)-1}{2}\left(t^{\frac{s}{2}-1}+t^{-\frac{s}{2}-\frac{1}{2}}\right)dt + \frac{1}{s-1} - \frac{1}{s} \quad (1.12)$$

for $\operatorname{Re} s > 1$. Since $z\Gamma(z) = \Gamma(z+1)$, we have $\Gamma\left(\frac{s}{2}\right)s = 2\Gamma\left(\frac{s}{2}+1\right)$, which proves:

THEOREM 1.13. *For* $\operatorname{Re} s > 1$ *we can write*

$$\zeta(s) = \frac{\pi^{\frac{s}{2}}}{\Gamma\left(\frac{s}{2}\right)}\int_1^\infty \frac{\theta(t)-1}{2}\left(t^{\frac{s}{2}-1}+t^{-\frac{s}{2}-1}\right)dt + \frac{\pi^{\frac{s}{2}}}{\Gamma\left(\frac{s}{2}\right)(s-1)} - \frac{\pi^{\frac{s}{2}}}{2\Gamma\left(\frac{s}{2}+1\right)}.$$

The integral $\int_1^\infty$ in this equality is an entire function of s, since, by (1.4), it equals $J\left(\frac{s}{2}-1\right) + J\left(-\frac{s}{2}-1\right)$. Also, since $1/\Gamma(s)$ is an entire function of s, it follows that the right-hand side of the equality in Theorem 1.13 provides for the analytic continuation of $\zeta(s)$ to the full complex plane, where it is observed that $\zeta(s)$ has only one singularity: $s = 1$ is a *simple pole*.

The fact that $\Gamma\left(\frac{1}{2}\right) = \pi^{1/2}$ allows one to compute the corresponding residue:

$$\lim_{s\to 1}(s-1)\zeta(s) = \lim_{s\to 1}\frac{\pi^{\frac{s}{2}}}{\Gamma\left(\frac{s}{2}\right)} = \frac{\pi^{\frac{1}{2}}}{\Gamma\left(\frac{1}{2}\right)} = 1.$$

An equation that relates the values $\zeta(s)$ and $\zeta(1-s)$, called a *functional equation*, easily follows from the preceding discussion. In fact define

$$X_R(s) \stackrel{\text{def}}{=} \pi^{-s/2}\zeta(s)\Gamma\left(\frac{s}{2}\right) \quad (1.14)$$

for $\operatorname{Re} s > 1$ and note that the right-hand side of equation (1.12) (which provides for the analytic continuation of $X_R(s)$ as a meromorphic function whose simple poles are at $s = 0$ and $s = 1$) is *unchanged* if s there is replaced by $1-s$:

THEOREM 1.15 (THE FUNCTIONAL EQUATION FOR $\zeta(s)$). *Let* $X_R(s)$ *be given by (1.14) and analytically continued by the right-hand side of the (1.12). Then* $X_R(s) = X_R(1-s)$ *for* $s \neq 0, 1$.

One can write the functional equation as

$$\zeta(1-s) = \frac{\pi^{-\frac{s}{2}}\Gamma\left(\frac{s}{2}\right)\zeta(s)}{\pi^{-\left(\frac{1-s}{2}\right)}\Gamma\left(\frac{1-s}{2}\right)} = \frac{\pi^{-s+\frac{1}{2}}\Gamma\left(\frac{s}{2}\right)\zeta(s)}{\Gamma\left(\frac{1-s}{2}\right)} \quad (1.16)$$

for $s \neq 0, 1$, multiply the right-hand side here by $1 = -\left(\frac{s-1}{2}\right)/\left(\frac{1-s}{2}\right)$, use the identity $\left(\frac{1-s}{2}\right)\Gamma\left(\frac{1-s}{2}\right) = \Gamma\left(\frac{3-s}{2}\right)$, and thus also write

$$\zeta(1-s) = -\frac{\pi^{-s+\frac{1}{2}}\Gamma\left(\frac{s}{2}\right)(s-1)\zeta(s)}{2\Gamma\left(\frac{3-s}{2}\right)}, \quad (1.17)$$

an equation that will be useful later when we compute $\zeta'(0)$.

For the computation of $\zeta'(0)$ we make use of the following result, which is of independent interest. $[x]$ denotes the largest integer that does not exceed $x \in \mathbb{R}$.

THEOREM 1.18. *For* $\operatorname{Re} s > 1$,

$$\begin{aligned}\zeta(s) &= \frac{1}{s-1} + \frac{1}{2} + s \int_1^\infty \frac{\left([x] - x + \frac{1}{2}\right) dx}{x^{s+1}} \\ &= \frac{1}{s-1} + 1 + s \int_1^\infty \frac{[x] - x}{x^{s+1}} dx.\end{aligned} \quad (1.19)$$

That these two expressions for $\zeta(s)$ are equal follows from the equality $\int_1^\infty \frac{dx}{x^{s+1}} = \frac{1}{s}$ for $\operatorname{Re} s > 0$; this with the inequalities $0 \leq x - [x] < 1$ allows one to deduce that the improper integrals there converge absolutely for $\operatorname{Re} s > 0$. We base the proof of Theorem 1.18 on a general observation:

LEMMA 1.20. *Let $\phi(x)$ be continuously differentiable on a closed interval $[a, b]$. Then, for $c \in \mathbb{R}$,*

$$\int_a^b \left(x - c - \tfrac{1}{2}\right)\phi'(x)\, dx = \left(b - c - \tfrac{1}{2}\right)\phi(b) - \left(a - c - \tfrac{1}{2}\right)\phi(a) - \int_a^b \phi(x)\, dx.$$

In particular for $[a, b] = [n, n+1]$*, with $n \in \mathbb{Z}$ one gets*

$$\int_n^{n+1} \left(x - [x] - \tfrac{1}{2}\right)\phi'(x)\, dx = \frac{\phi(n+1) + \phi(n)}{2} - \int_n^{n+1} \phi(x)\, dx.$$

PROOF. The first assertion is a direct consequence of integration by parts. Using it, one obtains for the choice $c = n$ the second assertion: $\int_n^{n+1} [x]\phi'(x)\, dx = \int_n^{n+1} n\phi'(x)\, dx$ (since $[x] = n$ for $n \leq x < n+1$); hence

$$\begin{aligned}\int_n^{n+1} &\left(x - [x] - \tfrac{1}{2}\right)\phi'(x)\, dx \\ &= \int_n^{n+1} \left(x - n - \tfrac{1}{2}\right)\phi'(x)\, dx \\ &\doteq \left(n + 1 - n - \tfrac{1}{2}\right)\phi(n+1) - \left(n - n - \tfrac{1}{2}\right)\phi(n) - \int_n^{n+1} \phi(x)\, dx \\ &= \tfrac{1}{2}\phi(n+1) + \tfrac{1}{2}\phi(n) - \int_n^{n+1} \phi(x)\, dx,\end{aligned}$$

$\square$

As a first application of the lemma, note that for integers m_2, m_1 with $m_2 > m_1$,

$$\sum_{n=m_1}^{m_2} [\phi(n+1) + \phi(n)]$$

$$= \sum_{n=m_1}^{m_2} \phi(n+1) + \sum_{n=m_1}^{m_2} \phi(n)$$

$$= \phi(m_1+1) + \phi(m_1+2) + \cdots + \phi(m_2+1) + \phi(m_1) + \phi(m_1+1) + \cdots + \phi(m_2)$$

$$= \phi(m_2+1) + \phi(m_1) + 2 \sum_{n=m_1+1}^{m_2} \phi(n).$$

Also $\sum_{n=m_1}^{m_2} \int_n^{n+1} = \int_{m_1}^{m_2+1}$. Therefore

$$\frac{\phi(m_2+1) + \phi(m_1)}{2} + \sum_{n=m_1+1}^{m_2} \phi(n) = \frac{1}{2} \sum_{n=m_1}^{m_2} [\phi(n+1) + \phi(n)]$$

$$= \sum_{n=m_1}^{m_2} \int_n^{n+1} (x - [x] - \tfrac{1}{2})\phi'(x)\, dx + \sum_{n=m_1}^{m_2} \int_n^{n+1} \phi(x)\, dx$$

(by Lemma 1.20), which equals $\int_{m_1}^{m_2+1} (x - [x] - \tfrac{1}{2})\phi'(x)\, dx + \int_{m_1}^{m_2+1} \phi(x)\, dx$. Thus

$$\sum_{n=m_1+1}^{m_2} \phi(n) = \frac{-\phi(m_2+1) - \phi(m_1)}{2}$$

$$+ \int_{m_1}^{m_2+1} \phi(x)\, dx + \int_{m_1}^{m_2+1} (x - [x] - \tfrac{1}{2})\phi'(x)\, dx \quad (1.21)$$

for $\phi(x)$ continuously differentiable on $[m_1, m_2 + 1]$. Now choose $m_1 = 1$ and

$$\phi(x) \stackrel{\text{def}}{=} x^{-s}$$

for $x > 0$, $\operatorname{Re} s > 1$. Then $\int_1^\infty \frac{dx}{x^s} = \frac{1}{s-1}$. Also $\phi(m_2+1) = (m_2+1)^{-s} \to 0$ as $m_2 \to \infty$, since $\operatorname{Re} s > 0$. Thus in (1.21) let $m_2 \to \infty$:

$$\sum_{n=2}^\infty \frac{1}{n^s} = -\tfrac{1}{2} + \frac{1}{s-1} + \int_1^\infty (x - [x] - \tfrac{1}{2})(-s x^{-s-1})\, dx.$$

That is, for $\operatorname{Re} s > 1$ we have

$$\zeta(s) = 1 + \sum_{n=2}^\infty \frac{1}{n^s} = \frac{1}{2} + \frac{1}{s-1} + s \int_1^\infty \frac{([x] - x + \tfrac{1}{2})}{x^{s+1}}\, dx,$$

which proves Theorem 1.18.

We turn to the second integral in equation (1.19), which we denote by

$$f(s) \stackrel{\text{def}}{=} \int_1^\infty \frac{([x]-x)}{x^{s+1}} dx$$

for $\operatorname{Re} s > 0$. We can write $f(s) = \lim_{n\to\infty} \int_1^n \frac{([x]-x)}{x^{s+1}} dx$, where

$$\int_1^n \frac{([x]-x)}{x^{s+1}} dx = \sum_{j=1}^{n-1} \int_j^{j+1} \frac{([x]-x)}{x^{s+1}} dx = \sum_{j=1}^{n-1} \int_j^{j+1} \frac{j-x}{x^{s+1}} dx, \quad (1.22)$$

since $[x] = j$ for $j \leq x < j+1$. That is, $f(s) = \sum_{j=1}^\infty a_j(s)$ where

$$a_j(s) \stackrel{\text{def}}{=} \int_j^{j+1} \frac{j-x}{x^{s+1}} dx = \frac{j}{s}\left(\frac{1}{j^s} - \frac{1}{(j+1)^s}\right) - \frac{1}{s-1}\left(\frac{1}{j^{s-1}} - \frac{1}{(j+1)^{s-1}}\right)$$

for $s \neq 0, 1$, and where for the second term here $s = 1$ is a removable singularity:

$$\lim_{s \to 1} (s-1)\frac{1}{s-1}\left(\frac{1}{j^{s-1}} - \frac{1}{(j+1)^{s-1}}\right) = 0.$$

Similarly, for the first term $s = 0$ is a removable singularity. That is, the $a_j(s)$ are entire functions. In particular each $a_j(s)$ is holomorphic on the domain $D^+ \stackrel{\text{def}}{=} \{s \in \mathbb{C} \mid \operatorname{Re} s > 0\}$. At the same time, for $\sigma := \operatorname{Re} s > 0$ we have

$$|a_j(s)| \leq \int_j^{j+1} \frac{dx}{x^{\sigma+1}} = \frac{1}{\sigma}\left(\frac{1}{j^\sigma} - \frac{1}{(j+1)^\sigma}\right)$$

(where the inequality comes from $|j-x| = x-j \leq 1$ for $j \leq x \leq j+1$); moreover

$$\sum_{j=1}^n \left(\frac{1}{j^\sigma} - \frac{1}{(j+1)^\sigma}\right) = 1 - \frac{1}{(n+1)^\sigma} \Rightarrow \sum_{j=1}^\infty \left(\frac{1}{j^\sigma} - \frac{1}{(j+1)^\sigma}\right) = 1$$

(i.e. $1/(n+1)^\sigma \to 0$ as $n \to \infty$ for $\sigma > 0$). Hence, by the M-test, $\sum_{j=1}^\infty a_j(s)$ converges absolutely and uniformly on D^+ (and in particular on compact subsets of D^+). $f(s)$ is therefore holomorphic on D^+, by the Weierstrass theorem. Of course, in equation (1.19),

$$s \int_1^\infty \frac{([x] - x + \frac{1}{2})}{x^{s+1}} dx = sf(s) + \frac{1}{2}$$

is also a holomorphic function of s on D^+.

We have deduced:

COROLLARY 1.23. *Let*

$$f(s) \stackrel{\text{def}}{=} \int_1^\infty \frac{([x]-\frac{1}{2})}{x^{s+1}}dx.$$

Then $f(s)$ is well-defined for $\operatorname{Re} s > 0$ and is a holomorphic function on the domain $D^+ \stackrel{\text{def}}{=} \{s \in \mathbb{C} \mid \operatorname{Re} s > 0\}$. For $\operatorname{Re} s > 1$ one has (by Theorem 1.18)

$$\zeta(s) = \frac{1}{s-1} + 1 + sf(s). \qquad (1.24)$$

From this we see that $\zeta(s)$ admits an analytic continuation to D^+. Its only singularity there is a simple pole at $s = 1$ with residue $\lim_{s\to 1}(s-1)\zeta(s) = 1$, as before.

This result is obviously weaker than Theorem 1.13. However, as a further application we show that

$$\lim_{s \to 1} \left(\zeta(s) - \frac{1}{s-1} \right) = \gamma \qquad (1.25)$$

where

$$\gamma \stackrel{\text{def}}{=} \lim_{n\to\infty} \left(1 + \frac{1}{2} + \frac{1}{3} + \cdots + \frac{1}{n} - \log n \right) \qquad (1.26)$$

is the *Euler–Mascheroni constant*; $\gamma \simeq 0.577215665$. By the continuity (in particular) of $f(s)$ at $s = 1$, $f(1) = \lim_{s \to 1} f(s)$. That is, by (1.24), we have

$$\lim_{s \to 1} \left(\zeta(s) - \frac{1}{s-1} \right) = \lim_{s \to 1} (1 + sf(s))$$

$$= 1 + f(1) \stackrel{(1.22)}{=} 1 + \lim_{n\to\infty} \sum_{j=1}^{n-1} \int_j^{j+1} \frac{j-x}{x^2}dx$$

$$= 1 + \lim_{n\to\infty} \sum_{j=1}^{n-1} \left(\frac{1}{j+1} - (\log(j+1) - \log j) \right)$$

$$= 1 + \lim_{n\to\infty} \left(\sum_{j=1}^{n-1} \frac{1}{j+1} - \sum_{j=1}^{n-1} (\log(j+1) - \log j) \right)$$

$$= 1 + \lim_{n\to\infty} \left(\sum_{j=1}^{n-1} \frac{1}{j+1} - \log n \right)$$

$$= 1 + \lim_{n\to\infty} \left(-1 + \left(1 + \frac{1}{2} + \frac{1}{3} + \cdots + \frac{1}{n}\right) - \log n \right)$$

$$= \gamma,$$

as desired.

Since $s = 1$ is a simple pole with residue 1, $\zeta(s)$ has a Laurent expansion

$$\zeta(s) = \frac{1}{s-1} + \gamma_0 + \sum_{k=1}^{\infty} \gamma_k (s-1)^k \qquad (1.27)$$

on a deleted neighborhood of 1. By equation (1.25), $\gamma_0 = \gamma$. One can show that, in fact, for $k = 0, 1, 2, 3, \ldots$

$$\gamma_k = \frac{(-1)^k}{k!} \lim_{n \to \infty} \left(\sum_{l=1}^{n} \frac{(\log l)^k}{l} - \frac{(\log n)^{k+1}}{k+1} \right), \qquad (1.28)$$

a result we will not need (except for the case $k = 0$ already proved) and thus which we will not bother to prove.

The inversion formula (1.3), which was instrumental in the approach above to the analytic continuation and FE of $\zeta(s)$, provides for a function $F(t)$, $t > 0$, that is invariant under the transformation $t \to 1/t$. Namely, let $F(t) \stackrel{\text{def}}{=} t^{1/4} \theta(t)$. Then (1.3) is equivalent to statement that $F(1/t) = F(t)$, for $t > 0$.

Lecture 2. Special values of zeta

In 1736, L. Euler discovered the celebrated special values result

$$\zeta(2n) = \frac{(-1)^{n+1}(2\pi)^{2n} B_{2n}}{2(2n)!} \qquad (2.1)$$

for $n = 1, 2, 3, \ldots$, where B_j is the j-th *Bernoulli number*, defined by

$$\frac{z}{e^z - 1} = \sum_{j=0}^{\infty} \frac{B_j}{j!} z^j,$$

for $|z| < 2\pi$, which is the Taylor expansion about $z = 0$ of the holomorphic function $h(z) \stackrel{\text{def}}{=} z/(e^z - 1)$, which is defined to be 1 at $z = 0$. Since $e^z - 1$ vanishes if and only if $z = 2\pi i n$, for $n \in \mathbb{Z}$, the restriction $|z| < 2\pi$ means that the denominator $e^z - 1$ vanishes only for $z = 0$. The B_j were computed by Euler up to $j = 30$. Here are the first few values:

B_0	B_1	B_2	B_3	B_4	B_5	B_6	B_7	B_8	B_9	B_{10}	B_{11}	B_{12}	B_{13}
1	$-\frac{1}{2}$	$\frac{1}{6}$	0	$-\frac{1}{30}$	0	$\frac{1}{42}$	0	$-\frac{1}{30}$	0	$\frac{5}{66}$	0	$-\frac{691}{2730}$	0

(2.2)

In general, $B_{\text{odd}>1} = 0$. To see this let $H(z) \stackrel{\text{def}}{=} h(z) + z/2$ for $|z| < 2\pi$, which we claim is an *even* function. Namely, for $z \neq 0$ the sum $z/(e^z - 1) + z/(e^{-z} - 1)$ equals $= -z$ by simplification:

$$H(-z) = \frac{-z}{e^{-z} - 1} - \frac{z}{2} = \frac{z}{e^z - 1} + z - \frac{z}{2} = H(z).$$

Then

$$\frac{z}{2} + B_0 + B_1 z + \sum_{j=1}^{\infty} \frac{B_{2j}}{(2j)!} z^{2j} + \sum_{j=1}^{\infty} \frac{B_{2j+1}}{(2j+1)!} z^{2j+1}$$

$$= \frac{z}{2} + \sum_{j=0}^{\infty} \frac{B_j}{j!} z^j = H(z) = H(-z)$$

$$= -\frac{z}{2} + B_0 + B_1(-z) + \sum_{j=1}^{\infty} \frac{B_{2j}}{(2j)!} (-z)^{2j} + \sum_{j=1}^{\infty} \frac{B_{2j+1}}{(2j+1)!} (-z)^{2j+1},$$

which implies

$$0 = (1 + 2B_1)z + 2 \sum_{j=1}^{\infty} \frac{B_{2j+1}}{(2j+1)!} z^{2j+1},$$

and consequently $B_1 = -\frac{1}{2}$ and $B_{2j+1} = 0$ for $j \geq 1$, as claimed. By formula (2.1) (in particular)

$$\zeta(2) = \sum_{n=1}^{\infty} \frac{1}{n^2} = \frac{\pi^2}{6}, \quad \zeta(4) = \sum_{n=1}^{\infty} \frac{1}{n^4} = \frac{\pi^4}{90}, \quad \zeta(6) = \sum_{n=1}^{\infty} \frac{1}{n^6} = \frac{\pi^6}{945}, \quad (2.3)$$

the first formula, $\sum_{n=1}^{\infty} 1/n^2 = \pi^2/6$, being well-known apart from knowledge of the zeta function $\zeta(s)$. We provide a proof of (2.1) based on the summation formula

$$\sum_{n=1}^{\infty} \frac{1}{n^2 + a^2} = \frac{\pi}{2a} \coth \pi a - \frac{1}{2a^2} \qquad (2.4)$$

for $a > 0$; see Appendix E on page 92. Before doing so, however, we note some other special values of zeta.

As we have noted, $1/\Gamma(s)$ is an entire function of s. It has zeros at the points $s = 0, -1, -2, -3, -4, \ldots$. By Theorem 1.13 and the remarks that follow its statement we therefore see that for $n = 1, 2, 3, 4, \ldots$,

$$\zeta(-2n) = \frac{-\pi^{-n}}{2\Gamma(-n+1)} = 0, \quad \zeta(0) = \frac{-1}{2\Gamma(1)} = -\frac{1}{2}. \qquad (2.5)$$

Thus, as mentioned in the Introduction, $\zeta(s)$ vanishes at the real points $s = -2, -4, -6, -8, \ldots$, called the *trivial* zeros of $\zeta(s)$. The value $\zeta(0)$ is nonzero — it equals $-\frac{1}{2}$ by (2.5). Later we shall check that

$$\zeta'(0) = \zeta(0) \log 2\pi = -\tfrac{1}{2} \log 2\pi. \qquad (2.6)$$

Turning to the proof of (2.1), we take $0 < t < 2\pi$ and choose $a = \dfrac{t}{2\pi}$ in (2.4), obtaining successively

$$\frac{\pi^2}{t}\coth\frac{t}{2} - \frac{2\pi^2}{t^2} = 4\pi^2 \sum_{n=1}^{\infty} \frac{1}{t^2 + 4\pi^2 n^2},$$

$$\frac{1}{2}\coth\frac{t}{2} - \frac{1}{t} = 2t \sum_{n=1}^{\infty} \frac{1}{t^2 + 4\pi^2 n^2},$$

$$\frac{1}{e^t - 1} + \frac{1}{2} = \frac{2 + e^t - 1}{2(e^t - 1)} = \frac{e^{-t/2} + e^{t/2}}{2(e^{t/2} - e^{-t/2})}$$

$$= \frac{1}{2}\frac{\cosh(t/2)}{\sinh(t/2)} = \frac{1}{2}\coth\frac{t}{2} = \frac{1}{t} + 2t\sum_{n=1}^{\infty}\frac{1}{t^2 + 4\pi^2 n^2},$$

$$\frac{t}{e^t - 1} + \frac{t}{2} = 1 + 2t^2 \sum_{n=1}^{\infty} \frac{1}{t^2 + 4\pi^2 n^2}. \tag{2.7}$$

Since $B_0 = 1$ and $B_1 = -\frac{1}{2}$ (see (2.2)), and since $B_{2k+1} = 0$ for $k \geq 1$, we can write

$$\frac{t}{e^t - 1} \stackrel{\text{def}}{=} \sum_{k=0}^{\infty} \frac{B_k}{k!} t^k = 1 - \frac{t}{2} + \sum_{k=1}^{\infty} \frac{B_{2k}}{(2k)!} t^{2k},$$

and (2.7) becomes

$$\sum_{k=1}^{\infty} \frac{B_{2k}}{(2k)!} t^{2k} = 2t^2 \sum_{n=1}^{\infty} \frac{1}{t^2 + 4\pi^2 n^2}. \tag{2.8}$$

For $0 < t < 2\pi$, we can use the convergent geometric series

$$\sum_{k=0}^{\infty} \left(\frac{-t^2}{4\pi^2 n^2}\right)^k = \frac{1}{1 + \frac{t^2}{4\pi^2 n^2}} = \frac{4\pi^2 n^2}{t^2 + 4\pi^2 n^2}, \tag{2.9}$$

to rewrite (2.8) as

$$\sum_{k=1}^{\infty} \frac{B_{2k}}{(2k)!} t^{2k} = 2t^2 \sum_{n=1}^{\infty} \sum_{k=0}^{\infty} \frac{1}{4\pi^2 n^2} \left(\frac{-t^2}{4\pi^2 n^2}\right)^k \tag{2.10}$$

$$= 2t^2 \sum_{n=1}^{\infty} \sum_{k=1}^{\infty} \frac{1}{4\pi^2 n^2} \left(\frac{-t^2}{4\pi^2 n^2}\right)^{k-1}.$$

The point is to commute the summations on n and k in this equation. Now

$$\sum_{k=1}^{\infty}\sum_{n=1}^{\infty}\left|\frac{1}{4\pi^2 n^2}\left(-\frac{t^2}{4\pi^2 n^2}\right)^{k-1}\right| = \sum_{k=1}^{\infty}\frac{t^{2(k-1)}}{(4\pi^2)^k}\sum_{n=1}^{\infty}\frac{1}{n^{2k}} \leq \sum_{k=1}^{\infty}\frac{t^{2(k-1)}}{(4\pi^2)^k}\sum_{n=1}^{\infty}\frac{1}{n^2}$$

which is finite since $\sum_{n=1}^{\infty} 1/n^2 = \zeta(2) < \infty$ and $\sum_{n=1}^{\infty} t^{2(k-1)}/(4\pi^2)^k < \infty$, by the ratio test (again for $0 < t < 2\pi$). Commutation of the summation is therefore justified:

$$\sum_{k=1}^{\infty} \frac{B_{2k} t^{2k}}{(2k)!} = 2t^2 \sum_{k=1}^{\infty} \frac{(-t^2)^{k-1}}{4\pi^2 (4\pi^2)^{k-1}} \sum_{n=1}^{\infty} \frac{1}{n^{2k}} = \sum_{k=1}^{\infty} \frac{2(-1)^{k-1}}{(4\pi^2)^k} \zeta(2k) t^{2k}$$

on $(0, 2\pi)$. By equating coefficients, we obtain

$$\frac{B_{2k}}{(2k)!} = \frac{2(-1)^{k-1} \zeta(2k)}{(4\pi^2)^k} \quad \text{for } k \geq 1,$$

which proves Euler's formula (2.1).

Next we turn to a proof of equation (2.6). We start with an easy consequence of the quotient and product rules for differentiation.

LEMMA 2.11 (LOGARITHMIC DIFFERENTIATION WITHOUT LOGS). *If*

$$F(s) = \frac{\phi_1(s) \phi_2(s) \phi_3(s)}{\phi_4(s)},$$

on some neighborhood of $s_0 \in \mathbb{C}$, where the $\phi_j(s)$ are nonvanishing holomorphic functions there, then

$$\frac{F'(s_0)}{F(s_0)} = \frac{\phi_1'(s_0)}{\phi_1(s_0)} + \frac{\phi_2'(s_0)}{\phi_2(s_0)} + \frac{\phi_3'(s_0)}{\phi_3(s_0)} - \frac{\phi_4'(s_0)}{\phi_4(s_0)}.$$

Now choose $\phi_1(s) \stackrel{\text{def}}{=} \pi^{\frac{1}{2}-s}$, $\phi_2(s) \stackrel{\text{def}}{=} \Gamma(\frac{s}{2})$, $\phi_4(s) = 2\Gamma(\frac{3-s}{2})$, say on a small neighborhood of $s = 1$. For the choice of $\phi_3(s)$, we write $\zeta(s) = g(s)/(s-1)$ on a neighborhood N of $s = 1$, for $s \neq 1$, where $g(s)$ is holomorphic on N and $g(1) = 1$. This can be done since $s = 1$ is a simple pole of $\zeta(s)$ with residue $= 1$; for example, see equation (1.27). Assume $0 \notin N$ and take $\phi_3(s) \stackrel{\text{def}}{=} g(s)$ on N. By equation (1.17), $-\zeta(1-s) = \phi_1(s)\phi_2(s)\phi_3(s)/\phi_4(s)$ near $s = 1$, so that by Lemma 2.11 and introducing the function $\psi(s) \stackrel{\text{def}}{=} \Gamma'(s)/\Gamma(s)$, we obtain

$$\frac{\zeta'(1-s)}{-\zeta(1-s)}\bigg|_{s=1}$$
$$= \frac{\pi^{\frac{1}{2}-s}(-\log \pi)}{\pi^{\frac{1}{2}-s}}\bigg|_{s=1} + \psi\left(\frac{s}{2}\right)\frac{1}{2}\bigg|_{s=1} + \frac{g'(s)}{g(s)}\bigg|_{s=1} - \psi\left(\frac{3-s}{2}\right)\left(-\frac{1}{2}\right)\bigg|_{s=1}. \quad (2.12)$$

If γ is the Euler–Mascheroni constant of (1.26), the facts $-\psi(1) = \gamma$ and $\psi(\frac{1}{2}) = -\gamma - 2\log 2$ are known to prevail, which reduces equation (2.12) to

$$\zeta'(0) = \zeta(0)\left(\log \pi + \frac{\gamma}{2} + \log 2 - g'(1) + \frac{\gamma}{2}\right)$$
$$= -\tfrac{1}{2}(\log \pi + \gamma + \log 2 - g'(1)),$$

since $g(1) = 1$ and $\zeta(0) = -\frac{1}{2}$; see (2.5). But $g'(1) = \gamma$, as we will see in a minute; hence we have reached the conclusion that $\zeta'(0) = -\frac{1}{2}\log 2\pi$, which is (2.6). There remains to check that $g'(1) = \gamma$. We have

$$g'(1) \stackrel{\text{def}}{=} \lim_{s \to 1} \frac{g(s) - 1}{s - 1},$$

again since $g(1) = 1$; this in turn equals $\lim_{s \to 1} \left(\zeta(s) - \frac{1}{s-1}\right) = \gamma$, by equation (1.25).

To obtain further special values of zeta we appeal to the special values formula

$$\Gamma\left(\tfrac{1}{2} - n\right) = \frac{(-1)^n \sqrt{\pi}\, 2^{2n} n!}{(2n)!} \tag{2.13}$$

for the gamma function, where $n = 1, 2, 3, 4, \ldots$. This we couple with (2.1) and the functional equation (1.16) to show that

$$\zeta(-1) = -\frac{1}{12} \quad \text{and} \quad \zeta(1 - 2n) = -\frac{B_{2n}}{2n} \quad \text{for } n = 1, 2, 3, 4, \ldots. \tag{2.14}$$

Namely, $\zeta(1-2n) = \pi^{-2n+\frac{1}{2}} \Gamma(n) \zeta(2n) / \Gamma\left(\tfrac{1}{2}-n\right)$, by (1.16); this in turn equals

$$\frac{\pi^{-2n+\frac{1}{2}}(n-1)!\,\zeta(2n)(2n)!}{(-1)^n \sqrt{\pi}\, 2^{2n} n!},$$

by (2.13); whence (2.1) gives

$$\zeta(1-2n) = \frac{(-1)^n (2\pi)^{-2n} \zeta(2n)(2n)!}{n} = -\frac{B_{2n}}{2n}.$$

Taking $n = 1$ gives $\zeta(-1) = -\frac{B_2}{2} = -\frac{1}{12}$, by (2.2), which confirms (2.14).

Lecture 3. An Euler product expansion

For a function $f(n)$ defined on the set $\mathbb{Z}^+ = \{1, 2, 3, \ldots\}$ of positive integers one has a corresponding zeta function or *Dirichlet series*

$$\phi_f(s) \stackrel{\text{def}}{=} \sum_{n=1}^{\infty} \frac{f(n)}{n^s},$$

defined generically for $\operatorname{Re} s$ sufficiently large. If $f(n) = 1$ for all $n \in \mathbb{Z}^+$, for example, then for $\operatorname{Re} s > 1$, $\phi_f(s)$ is of course just the Riemann zeta function $\zeta(s)$, which according to equation (0.2) of the Introduction has an Euler product expansion $\zeta(s) = \prod_{p \in P} \frac{1}{1-p^{-s}}$ over the primes P in $\mathbb{Z}^+$. It is natural to inquire whether, more generally, there are conditions that permit an analogous Euler product expansion of a given Dirichlet series $\phi_f(s)$. Very pleasantly, there is an affirmative result when, for example, the $f(n)$ are *Fourier coefficients* (see Theorem 4.32, where the n-th Fourier coefficient there is denoted by a_n) of certain types of *modular forms*, due to a beautiful theory of E. Hecke. Also see equations (3.20), (3.21) below. Rather than delving directly into that theory at this point we shall instead set up an *abstract* condition for a product expansion. The goal is to show that under suitable conditions on $f(n)$ of course the desired expansion assumes the form

$$\phi_f(s) \stackrel{\text{def}}{=} \sum_{n=1}^{\infty} \frac{f(n)}{n^s} = \frac{f(1)}{\prod_{p \in P}(1 + \alpha(p)p^{-2s} - f(p)p^{-s})} \quad (3.1)$$

for some function $\alpha(p)$ on P; see Theorem 3.17 below. Here we would want to have, in particular, that $f(1) \neq 0$. Before proceeding toward a precise statement and proof of equation (3.1), we note that (again) if $f(n) = 1$ for all $n \in \mathbb{Z}^+$, for example, then for the choice $\alpha(p) = 0$ for all $p \in P$, equation (3.1) reduces to the classical Euler product expansion of equation (0.2).

Given $f : \mathbb{Z}^+ \to \mathbb{R}$ or $\mathbb{C}$, and $\alpha : P \to \mathbb{R}$ or $\mathbb{C}$, we assume the following abstract *multiplicative* condition:

$$f(n)f(p) = \begin{cases} f(np) & \text{if } p \nmid n, \\ f(np) + \alpha(p)f\left(\frac{n}{p}\right) & \text{if } p \mid n, \end{cases} \quad (3.2)$$

for $(n, p) \in \mathbb{Z}^+ \times P$; here $p \mid n$ means that p divides n and $p \nmid n$ means the opposite. Given condition (3.2) we observe first that if $f(1) = 0$ then f vanishes identically, the proof being as follows. For a prime $p \in P$, (3.2) requires that $f(1)f(p) = f(p)$, since $p \nmid 1$; that is, $f(p) = 0$. If $n \in \mathbb{Z}^+$ with $n \geq 2$, there exists $p \in P$ such that $p \mid n$, say $ap = n$, $a \in \mathbb{Z}^+$. Proceed inductively. If $p \nmid a$, $f(a)f(p) = f(ap) = f(n)$, by (3.2), so $f(n) = 0$, as $f(p) = 0$. If $p \mid a$, we have $0 = f(a)f(p)$ (again as $f(p) = 0$), and this equals $f(ap) + \alpha(p)f(a/p) = f(n) + \alpha(p)f(a/p)$, where $1 < p \leq a$ (so $1 \leq a/p < a = n/p < n$). Thus $f(a/p) = 0$, by induction, so $f(n) = 0$, which completes the induction. Thus we see that if $f \not\equiv 0$ then $f(1) \neq 0$.

As in Appendix D (page 88) we set, $m, n \in \mathbb{Z}^+$,

$$d(m, n) \stackrel{\text{def}}{=} \begin{cases} 1 & \text{if } m \mid n, \\ 0 & \text{if } m \nmid n. \end{cases}$$

Fix a finite set of distinct primes $S = \{p_1, p_2, \ldots, p_l\} \subset P$ and define $g(n) = g_S(n)$ on $\mathbb{Z}^+$ by

$$g(n) = f(n) \prod_{j=1}^{l}(1 - d(p_j, n)). \quad (3.3)$$

Fix $p \in P - \{p_1, p_2, \ldots, p_l\}$. Then the next observation is that, for $n \in \mathbb{Z}^+$,

$$g(n)f(p) = \begin{cases} g(np) & \text{if } p \nmid n, \\ g(np) + \alpha(p)g\left(\frac{n}{p}\right) & \text{if } p \mid n \end{cases}, \quad (3.4)$$

which compares with equation (3.2).

PROOF. If $p_j \mid n$ then of course $p_j \mid pn$. If $p_j \nmid n$ then $p_j \nmid pn$; for if $p_j \mid pn$ then $p \mid n$ since p_j, p are relatively prime, given that $p \neq$ each p_j. Thus

$$d(p_j, n) = d(p_j, pn) \quad \text{for } n \in \mathbb{Z}^+, 1 \leq j \leq l. \quad (3.5)$$

Similarly suppose $p \mid n$, say $bp = n$, with $b \in \mathbb{Z}^+$. If $p_j \nmid n/p$ then $p_j \nmid n$; for otherwise $p_j \mid n = bp$ again with p_j, p relatively prime, implying that $p_j \mid b = n/p$. Thus we similarly have

$$d(p_j, n/p) = d(p_j, n) \quad \text{for } n \in \mathbb{Z}^+, 1 \leq j \leq l, \text{ such that } p \mid n. \quad (3.6)$$

Now if $n \in \mathbb{Z}^+$ is such that $p \nmid n$, then

$$g(n)f(p) \stackrel{(3.3)}{=} f(n)f(p) \prod_{j=1}^{l} (1 - d(p_j, n)) \stackrel{(3.2)}{=} f(np) \prod_{j=1}^{l} (1 - d(p_j, pn))$$
$$\stackrel{(3.3)}{=} g(np).$$

On the other hand, if $p \mid n$, then

$$g(n)f(p) \stackrel{(3.3)}{=} f(n)f(p) \prod_{j=1}^{l} (1 - d(p_j, n))$$

$$\stackrel{(3.2)}{=} \left(f(np) + \alpha(p)f\left(\tfrac{n}{p}\right)\right) \prod_{j=1}^{l} (1 - d(p_j, n))$$

$$\stackrel{(3.5)}{\underset{(3.6)}{=}} f(np) \prod_{j=1}^{l} (1 - d(p_j, pn)) + \alpha(p)f\left(\tfrac{n}{p}\right) \prod_{j=1}^{l} \left(1 - d\left(p_j, \tfrac{n}{p}\right)\right)$$

$$\stackrel{(3.3)}{=} g(np) + \alpha(p)g\left(\tfrac{n}{p}\right),$$

which proves (3.4). $\square$

Let $\phi_h(s) = \sum_{n=1}^{\infty} h(n)/n^s$ be a Dirichlet series that converges absolutely, say at some fixed point $s_0 \in \mathbb{C}$. Fix $p \in P$ and some complex number $\lambda(p)$ corresponding to p such that

$$h(n)\lambda(p) = \begin{cases} h(np) & \text{if } p \nmid n, \\ h(np) + \alpha(p)h\left(\frac{n}{p}\right) & \text{if } p \mid n, \end{cases} \quad (3.7)$$

for $n \in \mathbb{Z}^+$. Then

$$\phi_h(s_0)\left(1+\alpha(p)p^{-2s_0}-\lambda(p)p^{-s_0}\right) = \sum_{n=1}^{\infty} \frac{(1-d(p,n))h(n)}{n^{s_0}}. \tag{3.8}$$

PROOF. Define

$$a_n \stackrel{\text{def}}{=} \begin{cases} \dfrac{\alpha(p)h\left(\frac{n}{p}\right)}{(pn)^{s_0}} & \text{if } p \mid n, \\ 0 & \text{if } p \nmid n, \end{cases}$$

for $n \in \mathbb{Z}^+$. Since $p \mid pn$, we have

$$a_{pn} = \frac{\alpha(p)h\left(\frac{pn}{p}\right)}{(ppn)^{s_0}} = \alpha(p)p^{-2s_0}\frac{h(n)}{n^{s_0}},$$

which shows that $\sum_{n=1}^{\infty} a_{pn}$ converges. The Scholium of Appendix D (page 91) then implies that the series $\sum_{n=1}^{\infty} d(p,n)a_n$ converges, and one has

$$\sum_{n=1}^{\infty} a_{pn} = \sum_{n=1}^{\infty} d(p,n)a_n, \tag{3.9}$$

where the left-hand side here is $\alpha(p)p^{-2s_0}\phi_h(s_0)$. Since both $d(p,n), a_n = 0$ if $p \nmid n$, $d(p,n)a_n = a_n$, which is also clear if $p \mid n$. On the other hand, if $p \mid n$,

$$a_n \stackrel{\text{def}}{=} \alpha(p)h\left(\tfrac{n}{p}\right)(pn)^{-s_0} = \left(h(n)\lambda(p) - h(np)\right)(pn)^{-s_0} \tag{3.10}$$

by equation (3.7). Equation (3.10) also holds by (3.7) in case $p \nmid n$, for then both sides are zero. That is, (3.10) holds for all $n \geq 1$ and equation (3.9) reduces to the statement

$$\alpha(p)p^{-2s_0}\phi_h(s_0) = \sum_{n=1}^{\infty}[h(n)\lambda(p) - h(np)](pn)^{-s_0}. \tag{3.11}$$

We apply the Scholium a second time, where this time we define $a_n \stackrel{\text{def}}{=} h(n)/n^{s_0}$; since $|d(p,n)a_n| \leq |a_n|$, the sum $\sum_{n=1}^{\infty} d(p,n)a_n$ converges. By the Scholium, $\sum_{n=1}^{\infty} a_{pn}$ converges and $\sum_{n=1}^{\infty} a_{pn} = \sum_{n=1}^{\infty} d(p,n)a_n$; that is,

$$\sum_{n=1}^{\infty} h(pn)(pn)^{-s_0} = \sum_{n=1}^{\infty} d(p,n)h(n)n^{-s_0},$$

which one plugs into (3.11), to obtain $\alpha(p)p^{-2s_0}\phi_h(s_0) = \lambda(p)p^{-s_0}\phi_h(s_0) - \sum_{n=1}^{\infty} d(p,n)h(n)n^{-s_0}$. This proves equation (3.8).

The proof of the main result does involve various moving parts, and it is a bit lengthy as we have chosen to supply full details. We see, however, that the proof is elementary. One further basic ingredient is needed. Again let $\{p_1, \ldots, p_l\}$ by a fixed, finite set of distinct primes in P. With f, α subject to the multiplicative condition (3.2), we assume that $\phi_f(s_0)$ converges absolutely where $s_0 \in \mathbb{C}$ is some fixed number. For $n \geq 1$, we have $0 \leq \prod_{j=1}^{l}(1 - d(p_j, n)) \leq 1$; therefore the series $\sum_{n=1}^{\infty} (\prod_{j=1}^{l}(1 - d(p_j, n))) f(n)/n^{s_0}$ converges absolutely. We now show by induction on l that

$$\prod_{j=1}^{l}\left(1+\alpha(p_j)p_j^{-2s_0} - f(p_j)p_j^{-s_0}\right)\phi_f(s_0)$$
$$= \sum_{n=1}^{\infty}\left(\prod_{j=1}^{l}(1-d(p_j,n))\right)\frac{f(n)}{n^{s_0}}. \quad (3.12)$$

For $l = 1$, the claim follows by (3.8) with $p = p_1$, $h(n) = f(n)$, $\lambda(p) = f(p)$. Proceeding inductively, we consider a set $\{p_1, p_2, \ldots, p_l, p_{l+1}\}$ of $l+1$ distinct primes in P. Then

$$\prod_{j=1}^{l+1}\left(1+\alpha(p_j)p_j^{-2s_0} - f(p_j)p_j^{-s_0}\right)\phi_f(s_0)$$
$$= \left(1+\alpha(p_{l+1})p_{l+1}^{-2s_0} - f(p_{l+1})p_{l+1}^{-s_0}\right)\prod_{j=1}^{l}\left(1+\alpha(p_j)p_j^{-2s_0} - f(p_j)p_j^{-s_0}\right)\phi_f(s_0)$$
$$= \left(1+\alpha(p_{l+1})p_{l+1}^{-2s_0} - f(p_{l+1})p_{l+1}^{-s_0}\right)\sum_{n=1}^{\infty}\left(\prod_{j=1}^{l}(1-d(p_j,n))\right)\frac{f(n)}{n^{s_0}}$$
$$= \left(1+\alpha(p_{l+1})p_{l+1}^{-2s_0} - f(p_{l+1})p_{l+1}^{-s_0}\right)\sum_{n=1}^{\infty}\frac{g(n)}{n^{s_0}}, \quad (3.13)$$

where the second equality follows by induction and the last one by definition (3.3). We noted, just above (3.12), that $\sum_{n=1}^{\infty}(\prod_{j=1}^{l}(1-d(p_j,n))) f(n)/n^{s_0}$ converges absolutely; that is, $\sum_{n=1}^{\infty} g(n)/n^{s_0}$ converges absolutely. Thus we choose $h(n) = g(n)$, $p = p_{l+1}$, $\lambda(p) = f(p)$. Condition (3.7) is then a consequence of equation (3.4), since $p_{l+1} \in P - \{p_1, p_2, \ldots, p_l\}$, and one is therefore able to apply formula (3.8) again:

$$\left(1+\alpha(p_{l+1})p_{l+1}^{-2s_0} - f(p_{l+1})p_{l+1}^{-s_0}\right)\phi_g(s_0)$$
$$= \sum_{n=1}^{\infty}\frac{(1-d(p_{l+1},n))g(n)}{n^{s_0}}$$
$$\stackrel{(3.3)}{=} \sum_{n=1}^{\infty}\frac{(1-d(p_{l+1},n))}{n^{s_0}}\prod_{j=1}^{l}(1-d(p_j,n))f(n)$$
$$= \sum_{n=1}^{\infty}\prod_{j=1}^{l+1}(1-d(p_j,n))\frac{f(n)}{n^{s_0}} \quad (3.14)$$

which, together with equation (3.13), allows one to complete the induction, and thus the proof of the claim (3.12). $\square$

Now let p_l be the l-th positive prime, so $p_1 = 2, p_2 = 3, p_3 = 5, p_4 = 7, \ldots$. We can write the right-hand side of (3.12) as

$$f(1) + \sum_{n=2}^{\infty} \left(\prod_{j=1}^{l} (1 - d(p_j, n)) \right) \frac{f(n)}{n^{s_0}},$$

since no p_j divides 1. We show that if $2 \leq n \leq l$ then

$$\prod_{j=1}^{l} (1 - d(p_j, n)) = 0. \tag{3.15}$$

Namely, for $n \geq 2$ choose $q \in P$ such that $q \mid n$. If no p_j divides n, $1 \leq j \leq l$, then $q \neq p_1, \ldots, p_l$ (since $q \mid n$); hence $q \geq p_{l+1}$ (since p_l is the l-th prime) and so $q \geq l+1$. But this is impossible since $n \leq l$ (by hypothesis) and $q \leq n$ (since $q \mid n$). This contradiction proves that some p_j divides n, that is, $1 = d(p_j, n)$, which gives (3.15).

It follows that

$$\sum_{n=1}^{\infty} \left(\prod_{j=1}^{l} (1 - d(p_j, n)) \right) \frac{f(n)}{n^{s_0}} - f(1) = \sum_{n=l+1}^{\infty} \left(\prod_{j=1}^{l} (1 - d(p_j, n)) \right) \frac{f(n)}{n^{s_0}},$$

where (using again that $0 \leq \prod_{j=1}^{l}(1 - d(p_j, n)) \leq 1$) we have

$$\left| \sum_{n=l+1}^{\infty} \left(\prod_{j=1}^{l} (1 - d(p_j, n)) \right) \frac{f(n)}{n^{s_0}} \right| \leq \sum_{n=l+1}^{\infty} \left| \frac{f(n)}{n^{s_0}} \right| = \sum_{n=1}^{\infty} \left| \frac{f(n)}{n^{s_0}} \right| - \sum_{n=1}^{l} \left| \frac{f(n)}{n^{s_0}} \right|.$$

But this difference tends to 0 as $l \to \infty$. That is, by equation (3.12), the limit

$$\prod_{p \in P} (1 + \alpha(p) p^{-2s_0} - f(p) p^{-s_0}) \phi_f(s_0)$$

$$\stackrel{\text{def}}{=} \lim_{l \to \infty} \prod_{j=1}^{l} (1 + \alpha(p_j) p_j^{-2s_0} - f(p_j) p^{-s_0}) \phi_f(s_0) \tag{3.16}$$

exists, where $p_j = $ the j-th positive prime, and it equals $f(1)$.

We have therefore finally reached the main theorem.

THEOREM 3.17. (As before, $\mathbb{Z}^+ \stackrel{\text{def}}{=} \{1, 2, 3, \ldots\}$ and P denotes the set of positive primes.) Let $f : \mathbb{Z} \to \mathbb{R}$ or $\mathbb{C}$ and $\alpha : P \to \mathbb{R}$ or $\mathbb{C}$ be functions where f is not identically zero and where f is subject to the multiplicative condition (3.2). Then $f(1) \neq 0$. Let $D \subset \mathbb{C}$ be some subset on which the corresponding Dirichlet series $\phi_f(s) = \sum_{n=1}^{\infty} f(n)/n^s$ converges absolutely. Then on D,

$$\prod_{p \in P} (1 + \alpha(p) p^{-2s} - f(p) p^{-s}) \phi_f(s) = f(1)$$

(see equation (3.16)). *In particular both* $\prod_{p \in P}(1+\alpha(p)p^{-2s} - f(p)p^{-s})$ *and* $\phi_f(s)$ *are nonzero on D and*

$$\phi_f(s) = \frac{f(1)}{\prod_{p \in P}(1+\alpha(p)p^{-2s} - f(p)p^{-s})} \qquad (3.18)$$

on D (which is equation (3.1)).

We remark (again) that the proof of Theorem 3.17 is entirely elementary, if a bit long-winded; it only requires a few basic facts about primes, and a weak version of the fundamental theorem of arithmetic — namely that an integer $n \geq 2$ is divisible by a prime.

As a simple example of Theorem 3.17, suppose $f : \mathbb{Z}^+ \to \mathbb{R}$ or $\mathbb{C}$ is not identically zero, and is *completely multiplicative*: $f(nm) = f(n)f(m)$ for all $n, m \in \mathbb{Z}^+$. Assume also that $s \in \mathbb{C}$ is such that $\sum_{n=1}^{\infty} f(n)/n^s$ converges absolutely. Then

$$\sum_{n=1}^{\infty} \frac{f(n)}{n^s} = \frac{1}{\prod_{p \in P}\left(1 - \frac{f(p)}{p^s}\right)}. \qquad (3.19)$$

To see this, first we note *directly* that $f(1) \neq 0$. In fact since $f \not\equiv 0$, choose $n \in \mathbb{Z}^+$ such that $f(n) \neq 0$. Then $f(n) = f(n \cdot 1) = f(n)f(1)$, so $f(1) = 1$. Also f satisfies condition (3.2) for the choice $\alpha = 0$ (that is, $\alpha(p) = 0$ for all $p \in P$). Equation (3.19) therefore follows by (3.18). In Lecture 5, we apply (3.19) to *Dirichlet L-functions*.

Before concluding this lecture, we feel some obligation to explain the pivotal, abstract multiplicative condition (3.2). This will involve, however, some facts regarding modular forms that will be discussed in the next lecture, Lecture 4. Thus suppose that $f(z)$ is a holomorphic modular form of weight $k = 4, 6, 8, 10, \ldots$, with Fourier expansion

$$f(z) = \sum_{n=0}^{\infty} a_n e^{2\pi i n z} \qquad (3.20)$$

on the upper half-plane π^+. Then there is naturally attached to $f(z)$ a Dirichlet series

$$\phi_f(s) = \sum_{n=1}^{\infty} \frac{a_n}{n^s}, \qquad (3.21)$$

called a *Hecke L-function*, which is known to converge absolutely for $\operatorname{Re} s > k$, and which is holomorphic on this domain. Actually, if $f(z)$ is a *cusp form* (i.e., $a_0 = 0$) then $\phi_f(s)$ is holomorphic on the domain $\operatorname{Re} s > 1 + \frac{k}{2}$. As in the case of the Riemann zeta function, Hecke theory provides for the meromorphic continuation of $\phi_f(s)$ to the full complex plane, and for an appropriate functional equation for $\phi_f(s)$.

For each positive integer $n = 1, 2, 3, \ldots$, there is an operator $T(n)$ (called a *Hecke operator*) on the space of modular forms of weight k given by

$$(T(n)f)(z) = n^{k-1} \sum_{\substack{d>0 \\ d\mid n}} \sum_{a \in \mathbb{Z}/d\mathbb{Z}} f\left(\frac{nz + da}{d^2}\right) d^{-k}. \tag{3.22}$$

Here the inner sum is over a complete set of representatives a in $\mathbb{Z}$ for the cosets $\mathbb{Z}/d\mathbb{Z}$. We shall be interested in the case when $f(z)$ is an eigenfunction of all Hecke operators: $T(n)f = \lambda(n)f$ for all $n \geq 1$, where $f \neq 0$ and $\lambda(n) \in \mathbb{C}$. We assume also that $a_1 = 1$, in which case $f(z)$ is called a *normalized simultaneous eigenform*. For such an eigenform it is known from the theory of Hecke operators that the Fourier coefficients and eigenvalues coincide for $n \geq 1$: $a_n = \lambda(n)$ for $n \geq 1$. Moreover the Fourier coefficients satisfy the "multiplicative" condition

$$a_{n_1} a_{n_2} = \sum_{\substack{d>0 \\ d\mid n_1, d\mid n_2}} d^{k-1} a_{n_1 n_2/d^2} \tag{3.23}$$

for $n_1, n_2 \geq 1$. In particular for a prime $p \in P$ and an integer $n \geq 1$, condition (3.23) clearly reduces to the simpler condition

$$a_n a_p = \begin{cases} a_{np} & \text{if } p \nmid n, \\ a_{np} + p^{k-1} a_{np/p^2} & \text{if } p \mid n, \end{cases} \tag{3.24}$$

which is the origin of condition (3.2), where we see that in the present context we have $f(n) = a_n$ and $\alpha(p) = p^{k-1}$; $f(n)$ here is the function $f : \mathbb{Z} \to \mathbb{C}$ of condition (3.2), of course, and is not the eigenform $f(z)$. By Theorem 3.17, therefore, the following strong result is obtained.

THEOREM 3.25 (EULER PRODUCT FOR HECKE L-FUNCTIONS). *Let $f(z)$ be a normalized simultaneous eigenform of weight k (as desribed above), and let $\phi_f(s)$ be its corresponding Hecke L-function given by definition (3.21) for $\operatorname{Re} s > k$. Then, for $\operatorname{Re} s > k$,*

$$\phi_f(s) = \frac{1}{\prod_{p \in P}(1 + p^{k-1-2s} - a_p p^{-s})}, \tag{3.26}$$

where a_p is the p-th Fourier coefficient $f(z)$; see equation (3.20). If, moreover, $f(z)$ is a cusp form (i.e., the 0-th Fourier coefficient a_0 of $f(z)$ vanishes), then for $\operatorname{Re} s > 1 + \frac{k}{2}$, $\phi_f(s)$ converges (in fact absolutely) and formula (3.26) holds.

The following example is important, though no proofs (which are quite involved) are supplied. If

$$\eta(z) \stackrel{\text{def}}{=} e^{\pi i z/12} \prod_{n=1}^{\infty}(1 - e^{2\pi i n z}) \tag{3.27}$$

is the *Dedekind eta function* on π^+, then the *Ramanujan tau function* $\tau(n)$ on $\mathbb{Z}^+$ is defined by the Fourier expansion

$$\eta(z)^{24} = \sum_{n=1}^{\infty} \tau(n) e^{2\pi i n z}, \tag{3.28}$$

which is an example of equation (3.20), where in fact $\eta(z)^{24}$ is a normalized simultaneous eigenform of weight $k = 12$. It turns out, remarkably, that every $\tau(n)$ is real and is in fact an integer. For example, $\tau(1) = 1$, $\tau(2) = -24$, $\tau(3) = 252$, $\tau(4) = -1472$, $\tau(5) = 4830$. Note also that, since the sum in (3.27) starts at $n = 1$, $\eta(z)^{24}$ is a cusp form. By Theorem 3.25 we get:

COROLLARY 3.29. *For* $\operatorname{Re} s > 1 + \frac{k}{2} = 7$,

$$\sum_{n=1}^{\infty} \frac{\tau(n)}{n^s} = \frac{1}{\prod_{p \in P}(1 + p^{11-2s} - \tau(p) p^{-s})}. \tag{3.30}$$

The Euler product formula (3.30) was actually proved first by L. Mordell (in 1917, before E. Hecke) although it was claimed earlier to be true by S. Ramanujan.

Since we have introduced the Dedekind eta function $\eta(z)$ in (3.27), we check, as a final point, that it is indeed holomorphic on π^+. For

$$a_n(z) \stackrel{\text{def}}{=} -e^{2\pi i n z} \quad \text{and} \quad G(z) \stackrel{\text{def}}{=} \prod_{n=1}^{\infty}(1 + a_n(z)),$$

write $\eta(z) = e^{\pi i z/12} G(z)$. The product $G(z)$ converges absolutely on π^+ since $\sum_{n=1}^{\infty} |a_n(z)| = \sum_{n=1}^{\infty} e^{-2\pi n y}$ (for $z = x + iy$, $x, y \in \mathbb{R}$, $y > 0$) is a convergent geometric series as $e^{-2\pi y} < 1$. We note also that $a_n(z) \neq -1$ since (again) $|a_n(z)| = e^{-2\pi n y} < 1$ for $n \geq 1$. If $K \subset \pi^+$ is any compact subset, then the continuous function $\operatorname{Im} z$ on K has a positive lower bound B: $\operatorname{Im} z \geq B > 0$ for every $z \in K$. Hence

$$|a_n(z)| = e^{-2\pi n \operatorname{Im} z} \leq e^{-2\pi n B} \quad \text{on } K,$$

where $\sum_{n=1}^{\infty} e^{-2\pi n B}$ is a convergent geometric series as $e^{-2\pi B} < 1$ for $B > 0$. Therefore the series $\sum_{n=1}^{\infty} a_n(z)$ converges uniformly on compact subsets of π^+ (by the M-test), which means that the product $G(z)$ converges uniformly on compact subsets of π^+. That is, $G(z)$ is holomorphic on π^+ (as the $a_n(z)$ are holomorphic on π^+), and therefore $\eta(z)$ is holomorphic on π^+.

Lecture 4. Modular forms: the movie

In the previous lecture we proved an Euler product formula for Hecke L-functions, in Theorem 3.25 which followed as a concrete application of Theorem 3.17. That involved, in part, some notions/results deferred to the present lecture for further discussion. Here the attempt is to provide a brief, kaleidoscopic tour of the modular universe, whose space is Lobatchevsky–Poincáre hyperbolic space, the upper half-plane π^+, and whose galaxies of stars are modular forms. As no universe would be complete without zeta functions, Hecke L-functions play that role. In particular we gain, in transit, an enhanced appreciation of Theorem 3.25.

There are many fine texts and expositions on modular forms. These obviously venture much further than our modest attempt here which is designed to serve more or less as a limited introduction and reader's guide. We recommend, for example, the books of Audrey Terras [35], portions of chapter three, (also note her lectures in this volume) and Tom Apostol [2], as supplements.

We begin the story by considering a holomorphic function $f(z)$ on π^+ that satisfies the periodicity condition $f(z+1) = f(z)$. By the remarks following Theorem B.7 of the Appendix (page 84), $f(z)$ admits a Fourier expansion (or q-expansion)

$$f(z) = \sum_{n\in\mathbb{Z}} a_n q(z)^n = \sum_{n\in\mathbb{Z}} a_n e^{2\pi i n z} \tag{4.1}$$

on π^+, where $q(z) \stackrel{\text{def}}{=} e^{2\pi i z}$, and where the a_n are given by formula (B.6). We say that $f(z)$ is *holomorphic at infinity* if $a_n = 0$ for every $n \leq -1$:

$$f(z) = \sum_{n=0}^{\infty} a_n e^{2\pi i n z} \tag{4.2}$$

on π^+. Let $G = \mathrm{SL}(2, \mathbb{R})$ denote the group of 2×2 real matrices $g = \begin{bmatrix} a & b \\ c & d \end{bmatrix}$ with determinant $= 1$, and let $\Gamma = \mathrm{SL}(2, \mathbb{Z}) \subset G$ denote the subgroup of elements $\gamma = \begin{bmatrix} a & b \\ c & d \end{bmatrix}$ with $a, b, c, d \in \mathbb{Z}$. The standard linear fractional action of G on π^+, given by

$$g \cdot z \stackrel{\text{def}}{=} \frac{az+b}{cz+d} \in \pi^+ \quad \text{for } (g, z) \in G \times \pi^+ \tag{4.3}$$

restricts to any subgroup of G, and in particular it restricts to Γ.

A (holomorphic) *modular form of weight* $k \in \mathbb{Z}$, $k \geq 0$, with respect to Γ, is a holomorphic function $f(z)$ on π^+ that satisfies the following two conditions:

(M1) $f(\gamma \cdot z) = (cz+d)^k f(z)$ for $\gamma = \begin{bmatrix} a & b \\ c & d \end{bmatrix} \in \Gamma$, $z \in \pi^+$.

(M2) $f(z)$ is holomorphic at infinity.

Here we note that for the case of $\gamma = T \stackrel{\text{def}}{=} \begin{bmatrix} 1 & 1 \\ 0 & 1 \end{bmatrix} \in \Gamma$, we have $\gamma \cdot z = z + 1$ by (4.3). Then $f(z+1) = f(z)$ by (M1), which means that condition (M2) is well-defined, and therefore $f(z)$ satisfies equation (4.2), which justifies the statement of equation (3.20) of Lecture 3. One can also consider *weak* modular forms, where the assumption that $a_n = 0$ for every $n \leq -1$ is relaxed to allow *finitely many* negative Fourier coefficients to be nonzero. One can consider, moreover, modular forms with respect to various subgroups of Γ. There are two other quick notes to make. First, if $\gamma = -1 \stackrel{\text{def}}{=} \begin{bmatrix} -1 & 0 \\ 0 & -1 \end{bmatrix} \in \Gamma$, then by (4.3) and (M1) we must have $f(z) = (-1)^k f(z)$ which means that $f(z) \equiv 0$ if k is odd. For this reason we always assume that k is *even*. Secondly, condition (M1) is equivalent to the following two conditions:

(M1)′ $f(z+1) = f(z)$, and

(M1)″ $f(-1/z) = z^k f(z)$ for $z \in \pi^+$.

For we have already noted that (M1) $\Rightarrow$ (M1)′, by the choice $\gamma = T$. Also choose $\gamma = S \stackrel{\text{def}}{=} \begin{bmatrix} 0 & -1 \\ 1 & 0 \end{bmatrix} \in \Gamma$. Then by (4.3) and (M1), condition (M1)″ follows. Conversely, the conditions (M1)′ and (M1)″ together, for $k \geq 0$ even, imply condition (M1) since the two elements $T, S \in \Gamma$ generate Γ; a proof of this is provided in Appendix F (page 96).

Basic examples of modular forms are provided by the *holomorphic Eisenstein series* $G_k(z)$, which serve in fact as building blocks for other modular forms:

$$G_k(z) \stackrel{\text{def}}{=} \sum_{(m,n) \in \mathbb{Z} \times \mathbb{Z} - \{(0,0)\}} \frac{1}{(m+nz)^k} \quad (4.4)$$

for $z \in \pi^+$, $k = 4, 6, 8, 10, 12, \ldots$. The issue of absolute or uniform convergence of these series rests mainly on the next observation, whose proof goes back to Chris Henley [2]. Given $A, \delta > 0$ let

$$S_{A,\delta} \stackrel{\text{def}}{=} \{(x, y) \in \mathbb{R}^2 \mid |x| \leq A, y \geq \delta\} \quad (4.5)$$

be the region $\subset \pi^+$, as illustrated:

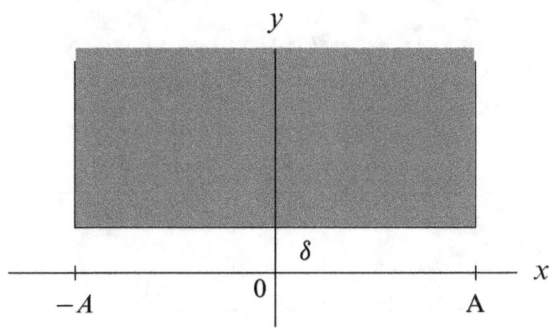

LEMMA 4.6. *There is a constant $K = K(A, \delta) > 0$, depending only on A and δ, such that for any $(x, y) \in S_{A,\delta}$ and $(a, b) \in \mathbb{R}^2$ with $b \neq 0$ the inequality*

$$\frac{(a+bx)^2 + b^2 y^2}{a^2 + b^2} \geq K \tag{4.7}$$

holds. In fact one can take

$$K \stackrel{\text{def}}{=} \frac{\delta^2}{1 + (A+\delta)^2}. \tag{4.8}$$

PROOF. Given $(x, y) \in S_{A,\delta}$ and $(a, b) \in \mathbb{R}^2$ with $b \neq 0$, let $q \stackrel{\text{def}}{=} a/b$. Then (4.7) amounts to

$$\frac{(q+x)^2 + y^2}{1 + q^2} \geq K.$$

Two cases are considered. First, if $|q| \leq A + \delta$, then $1 + q^2 \leq 1 + (A+\delta)^2$, so

$$\frac{1}{1+q^2} \geq \frac{1}{1+(A+\delta)^2}.$$

Also $(q+x)^2 + y^2 \geq y^2 \geq \delta^2$ (since $y \geq \delta$ for $(x, y) \in S_{A,\delta}$). Therefore

$$\frac{(q+x)^2 + y^2}{1+q^2} \geq \frac{\delta^2}{1+(A+\delta)^2} = K,$$

with K as in (4.8).

If instead $|q| > A + \delta$, we have $1/|q| < 1/(A+\delta)$, so $-|x|/q \geq -|x|/(A+\delta)$. Use the triangular inequality and the fact that $|x| \leq A$ for $(x, y) \in S_{A,\delta}$ to write

$$\left|1 + \frac{x}{q}\right| \geq 1 - \left|\frac{x}{q}\right| \geq 1 - \frac{|x|}{A+\delta} \geq 1 - \frac{A}{A+\delta} = \frac{\delta}{A+\delta}.$$

That is, $|q + x| \geq |q|\frac{\delta}{A+\delta}$, which implies $(q+x)^2 \geq \frac{q^2 \delta^2}{(A+\delta)^2}$, or again

$$\frac{(q+x)^2 + y^2}{1+q^2} \geq \frac{(q+x)^2}{1+q^2} \geq \frac{\delta^2}{(A+\delta)^2} \frac{q^2}{1+q^2}. \tag{4.9}$$

On the other hand, $f(x) \stackrel{\text{def}}{=} x^2/(1+x^2)$ is a strictly increasing function on $(0, \infty)$ since $f'(x) = 2x/((1+x^2)^2)$ is positive for $x > 0$. Thus, since $|q| > A + \delta$, we have

$$\frac{q^2}{1+q^2} = f(|q|) > f(A+\delta) = \frac{(A+\delta)^2}{1+(A+\delta)^2},$$

which leads to

$$\frac{(q+x)^2 + y^2}{1+q^2} \geq \frac{\delta^2}{(A+\delta)^2} \frac{(A+\delta)^2}{1+(A+\delta)^2}$$

by the second inequality in (4.9). But the right-hand side is again the constant K of (4.8). This concludes the proof. □

Now suppose $b = 0$, but $a \neq 0$. Then $\frac{(a+bx)^2 + b^2y^2}{a^2+b^2} = \frac{a^2}{a^2} = 1 > \frac{1}{2}$, which by Lemma 4.6 says that

$$\frac{(a+bx)^2 + b^2y^2}{a^2+b^2} \geq K_1 \stackrel{\text{def}}{=} \min(\tfrac{1}{2}, K)$$

for $(x, y) \in S_{A,\delta}$, $(a, b) \in \mathbb{R} \times \mathbb{R} - \{(0, 0)\}$. Hence if $z \in S_{A,\delta}$, say $z = x + iy$, and $(m, n) \in \mathbb{Z} \times \mathbb{Z} - \{(0, 0)\}$, we get $|m+nz|^2 = (m+nx)^2 + n^2y^2 \geq (m^2+n^2)K_1 = K_1|m+ni|^2$. This implies, for $\alpha \geq 0$, that $|m+nz|^\alpha \geq K_1^{\alpha/2}|m+ni|^\alpha$, or

$$\frac{1}{|m+nz|^\alpha} \leq \frac{1}{K_1^{\alpha/2}|m+ni|^\alpha}. \tag{4.10}$$

Moreover — setting for convenience $\mathbb{Z}_*^2 \stackrel{\text{def}}{=} \mathbb{Z} \times \mathbb{Z} - \{(0, 0)\}$ — we know from results in Appendix G that $\sum_{(m,n)\in\mathbb{Z}_*^2} 1/|m+ni|^\alpha$ converges for $\alpha > 2$. This shows that $G_k(z)$ converges absolutely and uniformly on every $S_{A,\delta}$ for $k > 2$, which (since k is even) is why we take $k = 4, 6, 8, 10, 12, \ldots$ in (4.4). In particular, since any compact subset of π^+ is contained in some $S_{A,\delta}$, the holomorphicity of $G_k(z)$ on π^+ is established.

Since the map $(m, n) \mapsto (m-n, n)$ is a bijection of $\mathbb{Z}_*^2$, we have

$$G_k(z+1) = \sum_{(m,n)\in\mathbb{Z}_*^2} \frac{1}{(m+n+nz)^k} = \sum_{(m,n)\in\mathbb{Z}_*^2} \frac{1}{(m-n+n+nz)^k} = G_k(z).$$

Similarly,

$$G_k\left(-\frac{1}{z}\right) = \sum_{(m,n)\in\mathbb{Z}_*^2} \frac{1}{(m-n/z)^k} = z^k \sum_{(m,n)\in\mathbb{Z}_*^2} \frac{1}{(mz-n)^k} = z^k \sum_{(m,n)\in\mathbb{Z}_*^2} \frac{1}{(nz+m)^k},$$

since the map $(m, n) \mapsto (n, -m)$ is a bijection of $\mathbb{Z}_*^2$. This shows that $G_k(z)$ satisfies the conditions (M1)' and (M1)''.

To complete the argument that the $G_k(z)$, for k even ≥ 4, are modular forms of weight k, we must check condition (M2). Although this could be done more directly, we take the route whereby the Fourier coefficients of $G_k(z)$ are actually computed explicitly. For this, consider the function

$$\phi_k(z) \stackrel{\text{def}}{=} \sum_{m\in\mathbb{Z}} \frac{1}{(z+m)^k} \tag{4.11}$$

on π^+ for $k \in \mathbb{Z}$, $k \geq 2$. The inequality (4.10) gives

$$\frac{1}{|m+z|^k} \leq \frac{1}{K_1^{k/2}|m+i|^k} = \frac{1}{K_1^{k/2}(m^2+1)^{k/2}} \qquad (4.12)$$

for $z \in S_{A,\delta}$. Since

$$\sum_{m \in \mathbb{Z}} \frac{1}{(m^2+1)^{k/2}} = 1 + 2\sum_{m=1}^{\infty} \frac{1}{(m^2+1)^{k/2}} \leq 1 + 2\sum_{m=1}^{\infty} \frac{1}{m^{2(k/2)} = m^k} < \infty$$

for $k > 1$, we see that $\phi_k(z)$ converges absolutely and uniformly on every $S_{A,\delta}$, and is therefore a holomorphic function on π^+ such that

$$\phi_k(z+1) = \sum_{m \in \mathbb{Z}} \frac{1}{(z+m+1)^k} = \sum_{m \in \mathbb{Z}} \frac{1}{(z+m)^k} = \phi_k(z).$$

Thus (again) there is a Fourier expansion

$$\phi_k(z) = \sum_{n \in \mathbb{Z}} a_n(k) e^{2\pi i n z} \qquad (4.13)$$

on π^+, where by formula (B.6) of page 84 (with the choice $b_1 = 0$, $b_2 = \infty$)

$$a_n(k) = \int_0^1 \phi_k(t+ib) e^{-2\pi i n(t+ib)} \, dt \qquad (4.14)$$

for $n \in \mathbb{Z}$, $b > 0$.

PROPOSITION 4.15. *In the Fourier expansion* (4.13), $a_n(k) = 0$ *for* $n \leq 0$ *and* $a_n(k) = (-2\pi i)^k n^{k-1}/(k-1)!$ *for* $n \geq 1$. *Therefore*

$$\phi_k(z) = \frac{(-2\pi i)^k}{(k-1)!} \sum_{n=1}^{\infty} n^{k-1} e^{2\pi i n z}$$

is the Fourier expansion of the function $\phi_k(z)$ *of* (4.11) *on* π^+. *Here* $k \in \mathbb{Z}$, $k \geq 2$ *as in* (4.11).

PROOF. For fixed $n \in \mathbb{Z}$ and $b > 0$, define $h_m(t) \stackrel{\text{def}}{=} e^{-2\pi i n t}/(t+ib+m)^k$ on $[0,1]$, for $m \in \mathbb{Z}$. Since $(t,b) \in S_{1,b}$ for $t \in [0,1]$ according to (4.5), the inequality in (4.12) gives

$$|h_m(t)| \leq \frac{1}{K_1^{k/2}(m^2+1)^{k/2}}.$$

As we have seen, $\sum_{m\in\mathbb{Z}} 1/(m^2+1)^{k/2} < \infty$ for $k > 1$, so $\sum_{m\in\mathbb{Z}} h_m(t)$ converges uniformly on $[0, 1]$. By (4.11) and (4.14), therefore, we see that

$$a_n(k) = \int_0^1 \sum_{m\in\mathbb{Z}} h_m(t) e^{2\pi nb} \, dt = e^{2\pi nb} \sum_{m\in\mathbb{Z}} \int_0^1 h_m(t) \, dt$$

$$= e^{2\pi nb} \sum_{m\in\mathbb{Z}} \int_0^1 \frac{e^{-2\pi int}}{(t+ib+m)^k} \, dt. \qquad (4.16)$$

By the change of variables $x = t + m$, we get

$$\int_0^1 \frac{e^{-2\pi int}}{(t+ib+m)^k} \, dt = \int_m^{m+1} \frac{e^{-2\pi in(x-m)}}{(x+ib)^k} \, dx = \int_m^{m+1} \frac{e^{-2\pi inx}}{(x+ib)^k} \, dx,$$

so

$$\sum_{m\in\mathbb{Z}} \int_0^1 \frac{e^{-2\pi int}}{(t+ib+m)^k} \, dt$$

$$= \sum_{m\in\mathbb{Z}} \int_m^{m+1} \frac{e^{-2\pi inx}}{(x+ib)^k} \, dx$$

$$= \sum_{m=0}^{\infty} \int_m^{m+1} \frac{e^{-2\pi inx}}{(x+ib)^k} \, dx + \sum_{m=1}^{\infty} \int_{-m}^{-m+1} \frac{e^{-2\pi inx}}{(x+ib)^k} \, dx$$

$$= \int_0^{\infty} \frac{e^{-2\pi inx}}{(x+ib)^k} \, dx + \int_{-\infty}^0 \frac{e^{-2\pi inx}}{(x+ib)^k} = \int_{-\infty}^{\infty} \frac{e^{-2\pi inx}}{(x+ib)^k} \, dx. \qquad (4.17)$$

(Note that the integrals on the last line are finite for $k > 1$, since

$$\left| \frac{e^{-2\pi inx}}{(x+ib)^k} \right| = \frac{1}{(x^2+b^2)^{k/2}} \qquad (4.18)$$

and the map $x \mapsto 1/(x^2+b^2)^{\sigma}$ lies in $L^1(\mathbb{R}, dx)$ for $2\sigma > 1$.) From (4.16) and (4.17), we have

$$a_n(k) = e^{2\pi nb} \int_{-\infty}^{\infty} \frac{e^{-2\pi inx}}{(x+ib)^k} \, dx. \qquad (4.19)$$

In particular,

$$|a_n(k)| \le e^{2\pi nb} \int_{\mathbb{R}} \frac{dx}{b^k \left(\left(\frac{x}{b}\right)^2 + 1\right)^{k/2}},$$

by (4.18). Setting $t = x/b$, we can rewrite the right-hand side as

$$\frac{e^{2\pi nb}}{b^k} \int_{\mathbb{R}} \frac{b \, dt}{(t^2+1)^{k/2}} = \frac{e^{2\pi nb} c_k}{b^{k-1}}, \qquad (4.20)$$

where $c_k \overset{\text{def}}{=} \int_{\mathbb{R}} dt/(t^2+1)^{k/2} < \infty$ for $k > 1$. Since $b > 0$ is arbitrary, we let $b \to \infty$. For $n = 0$ or for $n < 0$, we see by the inequality (4.20) that $a_n(k) = 0$. Also for $n = 1, 2, 3, 4, \ldots$ and $k > 1$, it is known that

$$\int_{\mathbb{R}} \frac{e^{-2\pi i n x}}{(x+ib)^k} dx = e^{-2\pi n b} \frac{(-2\pi i)^k}{(k-1)!} n^{k-1}; \qquad (4.21)$$

see the Remark below. The proof of Proposition 4.15 is therefore completed by way of equation (4.19). □

REMARK. Equation (4.21) follows from a contour integral evaluation:

$$\int_{-\infty+ib}^{\infty+ib} \frac{e^{-2\pi i \mu z}}{z^k} dz = \frac{(2\pi)^k \mu^{k-1} e^{-k\pi i/2}}{\Gamma(k)} \qquad (4.22)$$

where $\mu, b > 0$, $k > 1$. The left-hand side here is

$$\int_{-\infty}^{\infty} \frac{e^{-2\pi i \mu(x+ib)}}{(x+ib)^k} dx = e^{2\pi \mu b} \int_{-\infty}^{\infty} \frac{e^{-2\pi i \mu x}}{(x+ib)^k} dx.$$

Thus we can write $\int_{-\infty}^{\infty} e^{-2\pi i \mu x} dx/(x+ib)^k = e^{-2\pi \mu b}(-2\pi i)^k \mu^{k-1}/(k-1)!$ for $\mu, b > 0$ and $k > 1$ an integer. The choice $\mu = n$ ($n = 1, 2, 3, 4, \ldots$) gives (4.21).

For $n \in \mathbb{Z}$, define

$$\psi_n(z) \overset{\text{def}}{=} \phi_k(nz) = \sum_{m \in \mathbb{Z}} \frac{1}{(nz+m)^k}$$

(see (4.11) for the last equality) on π^+. In definition (4.4), the $n = 0$ contribution to the sum is $\sum_{m \in \mathbb{Z}-\{0\}} 1/m^k = \sum_{m=1}^{\infty} 1/m^k + \sum_{m=1}^{\infty} 1/(-m)^k = 2\sum_{m=1}^{\infty} 1/m^k$ (since k is even) $= 2\zeta(k)$. Thus we can write $G_k(z) = 2\zeta(k) + \sum_{n \in \mathbb{Z}-\{0\}} \sum_{m \in \mathbb{Z}} 1/(m+nz)^k = 2\zeta(k) + \sum_{n=1}^{\infty} \psi_n(z) + \sum_{n=1}^{\infty} \psi_{-n}(z)$. But $\psi_{-n}(z) = \psi_n(z)$, again because k is even (the easy verification is left to the reader). Therefore

$$G_k(z) = 2\zeta(k) + 2\sum_{n=1}^{\infty} \psi_n(z) = 2\zeta(k) + 2\sum_{n=1}^{\infty} \phi_k(nz)$$

$$= 2\zeta(k) + \frac{2(2\pi i)^k}{(k-1)!} \sum_{n=1}^{\infty} \sum_{m=1}^{\infty} m^{k-1} e^{2\pi i m n z}$$

(by Proposition 4.15, for k even), leading to

$$G_k(z) = 2\zeta(k) + \frac{2(2\pi i)^k}{(k-1)!} \sum_{n=1}^{\infty} \sigma_{k-1}(n) e^{2\pi i n z},$$

by formula (D.8) of Appendix D. This proves:

THEOREM 4.23. *The holomorphic Eisenstein series* $G_k(z)$, $k = 4, 6, 8, 10, 12,$
..., *defined in* (4.4) *satisfy conditions* (M1)′, (M1)″, *and are holomorphic at infinity. In fact*, $G_k(z)$ *has Fourier expansion* (4.2), *where* $a_0 = 2\zeta(k)$ *and*

$$a_n = \frac{2(2\pi i)^k}{(k-1)!}\sigma_{k-1}(n) \stackrel{\text{def}}{=} \frac{2(2\pi i)^k}{(k-1)!} \sum_{\substack{d>0 \\ d\mid n}} d^{k-1}$$

for $n \geq 1$. *The* $G_k(z)$ *are therefore modular forms of weight* k.

Since k is even in Theorem 4.23, formula (2.1) applies to $\zeta(k)$.

As mentioned, a modular form is a *cusp form* if its initial Fourier coefficient a_0 in equation (4.2) vanishes. By Theorem 4.23 the $G_k(z)$, for example, are *not* cusp forms since $\zeta(k) \neq 0$ for k even, $k \geq 4$. In fact we know (by Theorem 3.17) that since $\zeta(s)$ is given by an Euler product, it is nonvanishing for $\operatorname{Re} s > 1$.

We return now to the discussion of Hecke L-functions, where we begin with results on estimates of Fourier coefficients of modular forms. The $G_k(z)$ already provide the example of how the general estimate looks. This involves only an estimate of the divisor function $\sigma_\nu(n)$ for $\nu > 1$, where we first note that $d > 0$ runs through the divisors of $n \in \mathbb{Z}$, $n \geq 1$, as does n/d:

$$\sigma_\nu(n) \stackrel{\text{def}}{=} \sum_{\substack{0<d \\ d\mid n}} d^\nu = \sum_{\substack{0<d \\ d\mid n}} \left(\frac{n}{d}\right)^\nu = n^\nu \sum_{\substack{0<d \\ d\mid n}} \frac{1}{d^\nu} \leq n^\nu \sum_{0<d\in\mathbb{Z}} \frac{1}{d^\nu} = n^\nu \zeta(\nu).$$

Therefore for a_n the n-th Fourier coefficient of $G_k(z)$, Theorem 4.23 gives, for $n \geq 1$, $|a_n| \leq 2(2\pi)^k/(k-1)!\zeta(k-1)n^{k-1} = C(k)n^{k-1}$, where we have set

$$C(k) \stackrel{\text{def}}{=} \frac{2(2\pi)^k}{(k-1)!}\zeta(k-1).$$

In general:

THEOREM 4.24. *For a modular form* $f(z)$ *of weight* $k = 4, 6, 8, 10, \ldots$, *with Fourier expansion given by equation* (4.2), *there is a constant* $C(f, k) > 0$, *depending only on* f *and* k, *such that* $|a_n| < C(f,k)n^{k-1}$ *for* $n \geq 1$. *If* $f(z)$ *is a cusp form*, $C(f, k) > 0$ *can be chosen so that* $|a_n| < C(f,k)n^{k/2}$ *for* $n \geq 1$.

The idea of the proof is to first establish the inequality

$$|a_n| < C(f,k)n^{k/2}, \quad n \geq 1, \qquad (4.25)$$

for a cusp form $f(z)$ by estimating $f(z)$ on a *fundamental domain* $F \subset \pi^+$ for the action of Γ on π^+ (given by restriction of the action of G in equation

(4.3) to Γ), whence an estimate of $f(z)$ on π^+ is readily obtained. Then a_n is estimated from the formula

$$a_n = \int_0^1 f(t+ib)e^{-2\pi i n(t+ib)}\,dt \qquad (4.26)$$

for any $b > 0$; compare formula (4.14). By these arguments one can discover, in fact, that a modular form $f(z)$ of weight k *is a cusp form* if and only if there is a constant $M(f, k) > 0$, depending only on f and k, such that

$$|f(z)| < M(f, k)(\operatorname{Im} z)^{-k/2} \qquad (4.27)$$

on π^+. Once (4.25) is established for a cusp form, the weaker result $|a_n| < C(f, k)n^{k-1}$, $n \geq 1$, for an arbitrary modular form $f(z)$ of weight k follows from the fact that it holds for $G_k(z)$ (as shown above), and the fact that $f(z)$ differs from a cusp form (where one can apply the inequality (4.25)) by a constant multiple of $G_k(z)$. In fact, write $f(z) = a_0 + \sum_{n=1}^\infty a_n e^{2\pi i n z}$, $G_k(z) = b_0 + \sum_{n=1}^\infty b_n e^{2\pi i n z}$ (by equation (4.2)) where $b_0 = 2\zeta(k)$, for example (by Theorem 4.23). Then $f_0(z) \stackrel{\text{def}}{=} f(z) - (a_0/b_0)G_k(z)$ is a modular form of weight k, with Fourier expansion

$$f_0(z) = a_0 + \sum_{n=1}^\infty a_n e^{2\pi i n z} - \frac{a_0}{b_0} b_0 - \sum_{n=1}^\infty \frac{a_0}{b_0} b_n e^{2\pi i n z}$$

$$= \sum_{n=1}^\infty \left(a_n - \frac{a_0 b_n}{b_0}\right) e^{2\pi i n z},$$

which shows that $f_0(z)$ is a cusp form such that

$$f(z) = f_0(z) + \frac{a_0}{b_0} G_k(z). \qquad (4.28)$$

To complete our sketch of the proof of Theorem 4.24, details of which can be found in section 6.15 of [2], for example, we should add further remarks regarding F. By definition, a fundamental domain for the action of Γ on π^+ is an open set $F \subset \pi^+$ such that (F1) no two distinct points of F lie in the same Γ-orbit: if $z_1, z_2 \in F$ with $z_1 \neq z_2$, then there is no $\gamma \in \Gamma$ such that $z_1 = \gamma \cdot z_2$. We also require condition: (F2) given $z \in \pi^+$, there exists some $\gamma \in \Gamma$ such that $\gamma \cdot z \in \bar{F}$ (= the closure of F). The standard fundamental domain, as is well-known, is given by

$$F \stackrel{\text{def}}{=} \{z \in \pi^+ \mid |z| > 1, |\operatorname{Re} z| < \tfrac{1}{2}\}, \qquad (4.29)$$

and is shown at the the top of the next page.

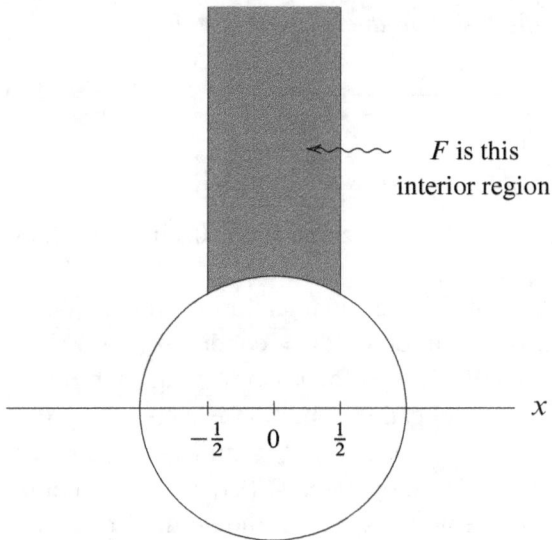

If $M_k(\Gamma)$ and $S_k(\Gamma)$ denote the space of modular forms and cusp forms of weight $k = 4, 6, 8, 10, \ldots$, respectively, there is the $\mathbb{C}$-vector space direct sum decomposition

$$M_k(\Gamma) = S_k(\Gamma) \oplus \mathbb{C} G_k, \tag{4.30}$$

by (4.28). The sum in (4.30) is indeed direct since (as we have seen) $G_k(z)$ is not a cusp form.

If $f \in M_k(\Gamma)$ with Fourier expansion (4.2), the corresponding Hecke L-function $L(s; f)$ is given by definition (3.21):

$$L(s; f) = \phi_f(s) \stackrel{\text{def}}{=} \sum_{n=1}^{\infty} \frac{a_n}{n^s}. \tag{4.31}$$

By Theorem 4.24 this series converges absolutely for Re $s > k$, and for Re $s > 1 + \frac{k}{2}$ if $f \in S_k(\Gamma)$. On these respective domains $L(s; f)$ is holomorphic in s (by an argument similar to that for the Riemann zeta function), as we have asserted in Lecture 3. Since Theorem 3.25 is based on equation (3.24), which is based on equation (3.23), our proof of it actually shows the following reformulation:

THEOREM 4.32. *Suppose the Fourier coefficients a_n of $f \in M_k(\Gamma)$ satisfy the multiplicative condition* (3.23), *with at least one a_n nonzero*:

$$a_{n_1} a_{n_2} = \sum_{\substack{d > 0 \\ d \mid n_1, \, d \mid n_2}} d^{k-1} \frac{a_{n_1 n_2}}{d^2} \tag{4.33}$$

for $n_1, n_2 \geq 1$. Then $L(s; f)$ has the Euler product representation

$$L(s; f) = \prod_{p \in P} \frac{1}{1 + p^{k-1-2s} - a_p p^{-s}} \tag{4.34}$$

for $\operatorname{Re} s > k$. If $f \in S_k(\Gamma)$, equation (4.34) holds for $\operatorname{Re} s > 1 + k/2$.

Here, we only need to note that $a_1 = 1$. For if some $a_n \neq 0$, then by (4.33) $a_n a_1 = a_{n \cdot 1}/1^2 = a_n$ and so $a_1 = 1$.

Theorem 4.32 raises the question of finding modular forms whose Fourier coefficients satisfy the multiplicative condition (3.23) = condition (4.33). This question was answered by Hecke (in 1937), who found, in fact, all such forms. As was observed in Lecture 3, the multiplicative condition is satisfied by normalized simultaneous eigenforms: nonzero forms $f(z)$ with $a_1 = 1$, that are simultaneous eigenfunctions of all the Hecke operators $T(n), n \geq 1$; see definition (3.22). More concretely, among the non-cusp forms the normalized simultaneous eigenforms turn out to be the forms

$$f(z) = \frac{(k-1)!}{2(2\pi i)^k} G_k(z),$$

where indeed, by Theorem 4.23, $a_1 = \sigma_{k-1}(1) = 1$.

We mention that the Hecke operators $\{T(n)\}_{n \geq 1}$ map the space $M_k(\Gamma)$ to itself, and also map the space $S_k(\Gamma)$ to itself. For $f \in M_k(\Gamma)$ with Fourier expansion $f(z) = \sum_{n=0}^{\infty} a_n e^{2\pi i n z}$ on π^+, as in (4.2), $(T(n)f)(z)$ has Fourier expansion

$$(T(n)f)(z) = \sum_{m=0}^{\infty} a_m^{(n)} e^{2\pi i m z}$$

on π^+, where $a_0^{(n)} = a_0 \sigma_{k-1}(n)$ and $a_m^{(n)} = \sum_{0 < d, \, d | n, \, d | m} d^{k-1} a_{\frac{n m}{d \, d}}$ for $m \geq 1$ — a result that leads to condition (4.33); in particular $a_1^{(n)} = a_n$.

Besides his striking observations regarding the connection between modular forms $f(z)$ and their associated Dirichlet series $L(s; f)$, that we have briefly discussed so far, Enrich Hecke also obtained the analytic continuation and functional equation of $L(s; f)$, which we now describe. Hecke showed that

$$(2\pi)^{-s} L(s; f) \Gamma(s)$$
$$= \int_1^{\infty} (f(it) - a_0)(t^{s-1} + i^k t^{k-s-1}) \, dt + a_0 \left(\frac{i^k}{s-k} - \frac{1}{s} \right) \tag{4.35}$$

for $\operatorname{Re} s > k$. Here the integral

$$J_f(s) \stackrel{\text{def}}{=} \int_1^{\infty} (f(it) - a_0) t^{s-1} \, dt \tag{4.36}$$

is an entire function of s. Notice that equation (4.35) is similar in form to (1.12). Again since $1/\Gamma(s)$ is an entire function, and since $(1/s)(1/\Gamma(s)) = 1/\Gamma(s+1)$, we can write equation (4.35) as

$$L(s;f) = \frac{(2\pi)^s}{\Gamma(s)}\left(J_f(s) + i^k J_f(k-s)\right) + \frac{(2\pi)^s}{\Gamma(s)}\frac{a_0 i^k}{(s-k)} - \frac{(2\pi)^s a_0}{\Gamma(s+1)} \qquad (4.37)$$

for $\operatorname{Re} s > k$. If $f(z)$ is a cusp form, we see that $L(s;f)$ extends to an *entire* function by way of the first term in (4.37). In general, we see that $L(s;f)$ extends meromorphically to $\mathbb{C}$, with a single (simple) pole at $s = k$ with residue $(2\pi)^k a_0 i^k/\Gamma(k) = (2\pi)^k a_0 i^k/(k-1)!$. Also by equation (4.37), for $k - s \neq k$ (i.e., $s \neq 0$), we have

$$i^k (2\pi)^{s-k} \Gamma(k-s) L(k-s;f)$$
$$= i^k (2\pi)^{s-k} \Gamma(k-s) \frac{(2\pi)^{k-s}}{\Gamma(k-s)}\left(J_f(k-s) + i^k J_f(s) + \frac{a_0 i^k}{-s} - \frac{a_0 \Gamma(k-s)}{\Gamma(k-s+1)}\right)$$
$$= i^k J_f(k-s) + J_f(s) + \frac{a_0}{-s} - \frac{i^k a_0}{k-s}$$

(again since $\Gamma(w+1) = w\Gamma(w)$, and since $i^{2k} = 1$ for k even); the right-hand side in turn equals $(2\pi)^{-s}\Gamma(s)L(s;f)$. That is,

$$(2\pi)^{-s}\Gamma(s)L(s;f) = i^k (2\pi)^{s-k}\Gamma(k-s)L(k-s;f) \qquad (4.38)$$

for $s \neq 0$, which is the functional equation for $L(s;f)$, which compares with the functional equation for the Riemann zeta function; see [17; 18].

The Eisenstein series $G_k(z)$ can be used as building blocks to construct other modular forms. It is known that any modular form $f(z)$ is, in fact, a finite sum of the form $f(z) = \sum_{n,m\geq 0} c_{nm} G_4(z)^n G_6(z)^m$ for suitable complex numbers c_{nm}. Of particular interest are the *discriminant form*

$$\Delta(z) \stackrel{\text{def}}{=} (60 G_4(z))^3 - 27(140 G_6(z))^2 \qquad (4.39)$$

and the *modular invariant*

$$J(z) \stackrel{\text{def}}{=} (60 G_4(z))^3 / \Delta(z) \qquad (4.40)$$

which is well-defined since it is true that $\Delta(z)$ never vanishes on π^+. $\Delta(z)$ is a modular form of weight 12, since if $f_1(z), f_2(z)$ are modular forms of weight k_1, k_2, then $f_1(z) f_2(z)$ is a modular form of weight $k_1 + k_2$. Similarly $J(z)$ is a weak modular form of weight $k = 0$: $J(\gamma, z) = J(z)$ for $\gamma \in \Gamma$, $z \in \pi^+$. The form $J(z)$ was initially constructed by R. Dedekind in 1877, and by F. Klein in 1878. Associated with it is the equally important *modular j-invariant*

$$j(z) \stackrel{\text{def}}{=} 1728 J(z). \qquad (4.41)$$

$\Delta(z)$ is connected with the Dedekind eta function $\eta(z)$ (see definition (3.27)) by the *Jacobi identity*

$$\Delta(z) = (2\pi)^{12}\eta(z)^{24}, \tag{4.42}$$

which with equation (3.28) shows that $\Delta(z)$ has Fourier expansion

$$\Delta(z) = (2\pi)^{12} \sum_{n=1}^{\infty} \tau(n)e^{2\pi i n z}, \tag{4.43}$$

where $\tau(n)$ is the Ramanujan tau function, and which in particular shows that $\Delta(z)$ is a cusp form: $\Delta(z) \in S_{12}(\Gamma)$, which can also be proved directly by definition (4.39) and Theorem 4.23, for $k = 4, 6$. There are no nonzero cusp forms of weight < 12.

Note that by Theorem 4.24, there is a constant $C > 0$ such that $|\tau(n)| < Cn^6$ for $n \geq 1$. However P. Deligne proved the *Ramanujan conjecture* $|\tau(n)| \leq \sigma_0(n)n^{11/2}$ for $n \geq 1$, where $\sigma_0(n)$ is the number of positive divisors of n. As we remarked in Lecture 3, the $\tau(n)$ (remarkably) are all integers. This can be proved using definition (4.39) and Theorem 4.23, for $k = 4, 6$. It is also true that, thanks to the factor 1728 in definition (4.41), *all of the Fourier coefficients of the modular j-invariant are integers*:

$$j(z) = 1e^{-2\pi i z} + \sum_{n=0}^{\infty} a_n e^{2\pi i n z} \tag{4.44}$$

with each $a_n \in \mathbb{Z}$:

$a_0 = 744$, $a_1 = 196{,}884$, $a_2 = 21{,}493{,}760$, $a_3 = 864{,}299{,}970$,

$$a_4 = 20{,}245{,}856{,}256, \ a_5 = 333{,}202{,}640{,}600, \ \ldots \tag{4.45}$$

An application of the modular invariant $j(z)$, and of the values in (4.45), to three-dimensional gravity with a negative cosmological constant will be given in my Speaker's Lecture; see especially equation (5-8) on page 343 and the subsequent discussion.

In the definition (4.4) of the holomorphic Eisenstein series $G_k(z)$, one cannot take $k = 2$ for convergence reasons. However, Theorem 4.23 provides a suggestion of how one might proceed to construct a series $G_2(z)$. Namely, take $k = 2$ there and thus define

$$G_2(z) \stackrel{\text{def}}{=} 2\zeta(2) + 2(2\pi i)^2 \sum_{n=1}^{\infty} \sigma(n)e^{2\pi i n z} = \frac{\pi^2}{3} - 8\pi^2 \sum_{n=1}^{\infty} \sigma(n)e^{2\pi i n z} \tag{4.46}$$

on π^+, where $\sigma(n) \stackrel{\text{def}}{=} \sigma_1(n) \stackrel{\text{def}}{=} \sum_{0<d,\ d|n} d$. Note that the series on the right does converge on π^+ and, in fact, the convergence is absolute: since $\sigma(n) \leq \sum_{d=1}^{n} d = \frac{1}{2}n(n+1)$ we have $|\sigma(n)e^{2\pi i n z}| \leq \frac{1}{2}n(n+1)e^{-2\pi n \operatorname{Im} z}$, and convergence is assured by the ratio test. Given any compact subset $K \subset \pi^+$, a positive lower bound B for the continuous function $\operatorname{Im} z$ on K exists: we have

Im $z \geq B > 0$ on K, so $|\sigma(n)e^{2\pi i n z}| \leq \frac{1}{2}n(n+1)e^{-2\pi n B}$ on K, and again $\sum_{n=1}^{\infty} n(n+1)e^{-2\pi n B} < \infty$ for $B > 0$. By the M-test, $\sum_{n=1}^{\infty} \sigma(n)e^{2\pi i n z}$ converges uniformly on K, which (by the Weierstrass theorem) means that $G_2(z)$ is a holomorphic function on π^+.

Another expression for $G_2(z)$ is

$$G_2(z) = 2\zeta(2) + \sum_{n \in \mathbb{Z}-\{0\}} \sum_{m \in \mathbb{Z}} \frac{1}{(m+nz)^2}. \tag{4.47}$$

To check this, start by taking $k = 2$ in definition (4.11) and in Proposition 4.15:

$$\frac{1}{z^2} + \sum_{m \in \mathbb{Z}-\{0\}} \frac{1}{(z+m)^2} = \phi_2(z) = (-2\pi i)^2 \sum_{k=1}^{\infty} k e^{2\pi i k z}. \tag{4.48}$$

Replace z by nz in (4.48) and sum on n from 1 to ∞:

$$\frac{1}{z^2}\zeta(2) + \sum_{n=1}^{\infty} \sum_{m \in \mathbb{Z}-\{0\}} \frac{1}{(nz+m)^2} = (-2\pi i)^2 \sum_{n=1}^{\infty} \sum_{k=1}^{\infty} k e^{2\pi i k n z}. \tag{4.49}$$

By (D.6) (see page 90), we obtain

$$\sigma(n) = \sum_{k=1}^{\infty} d(k,n)k. \tag{4.50}$$

For $a_n \stackrel{\text{def}}{=} e^{2\pi i n z}$, $n \geq 1$, $z \in \pi^+$, and for $k \geq 1$ fixed the series $\sum_{n=1}^{\infty} d(k,n)a_n$ clearly converges absolutely, since Im $z > 0$ and $0 \leq d(k,n) \leq 1$. Then the series $\sum_{n=1}^{\infty} a_{kn}$ converges and equals $\sum_{n=1}^{\infty} d(k,n)a_n$, by the Scholium of Appendix D (page 91):

$$\sum_{n=1}^{\infty} e^{2\pi i k n z} = \sum_{n=1}^{\infty} d(k,n) e^{2\pi i n z} \tag{4.51}$$

which gives, by equation (4.50)

$$\sum_{n=1}^{\infty} \sigma(n)e^{2\pi i n z} = \sum_{n=1}^{\infty} \sum_{k=1}^{\infty} d(k,n) e^{2\pi i n z} = \sum_{k=1}^{\infty} k \sum_{n=1}^{\infty} d(k,n) e^{2\pi i n z}$$

$$= \sum_{k=1}^{\infty} k \sum_{n=1}^{\infty} e^{2\pi i k n z} = \sum_{n=1}^{\infty} \sum_{k=1}^{\infty} k e^{2\pi i k n z}, \tag{4.52}$$

which in turn allows for the expression

$$\frac{2}{z^2}\zeta(2) + 2\sum_{n=1}^{\infty} \sum_{m \in \mathbb{Z}-\{0\}} \frac{1}{(nz+m)^2} = 2(2\pi i)^2 \sum_{n=1}^{\infty} \sigma(n) e^{2\pi i n z}$$

$$= G_2(z) - 2\zeta(2) \tag{4.53}$$

by equation (4.49), provided the commutations of the summations over k, n in (4.52) are legal. But for $y = \operatorname{Im} z$,

$$\sum_{n=1}^{\infty} \sum_{k=1}^{\infty} |d(k,n) k e^{2\pi i n z}| = \sum_{n=1}^{\infty} \sum_{k=1}^{\infty} d(k,n) k e^{-2\pi n y} = \sum_{n=1}^{\infty} \sigma(n) e^{-2\pi n y}$$

by (4.50), and this equals $\sum_{n=1}^{\infty} |\sigma(n) e^{2\pi i n z}|$, which is finite, as we have seen. This justifies the first commutation. Similarly,

$$\sum_{k=1}^{\infty} \sum_{n=1}^{\infty} |k e^{2\pi i k n z}| = \sum_{k=1}^{\infty} k \sum_{n=1}^{\infty} (e^{-2\pi k y})^n = \sum_{k=1}^{\infty} k \left(\frac{e^{-2\pi k y}}{1 - e^{-2\pi k y}} \right) = \sum_{k=1}^{\infty} \frac{k}{e^{2\pi k y} - 1}$$

is finite by the integral test:

$$\int_1^{\infty} \frac{t\, dt}{e^{2\pi y t} - 1} = \frac{1}{(2\pi y)^2} \int_{2\pi y}^{\infty} \frac{u\, du}{e^u - 1} < \infty, \tag{4.54}$$

as we shall see later by Theorem 6.1, for example. This justifies the second commutation in equation (4.52). Since for $n \geq 1$

$$\sum_{m \in \mathbb{Z} - \{0\}} \frac{1}{(m - nz)^2} = \sum_{m \in \mathbb{Z} - \{0\}} \frac{1}{(-m - nz)^2} = \sum_{m \in \mathbb{Z} - \{0\}} \frac{1}{(m + nz)^2}, \tag{4.55}$$

the double sum on the right-hand side of (4.47) can be written as

$$\sum_{n \in \mathbb{Z} - \{0\}} \left(\frac{1}{(nz)^2} + \sum_{m \in \mathbb{Z} - \{0\}} \frac{1}{(m + nz)^2} \right)$$

$$= 2 \sum_{n=1}^{\infty} \frac{1}{n^2 z^2} + \sum_{n=1}^{\infty} \sum_{m \in \mathbb{Z} - \{0\}} \frac{1}{(m + nz)^2} + \sum_{n=1}^{\infty} \sum_{m \in \mathbb{Z} - \{0\}} \frac{1}{(m - nz)^2}$$

$$= 2 \frac{\zeta(2)}{z^2} + 2 \sum_{n=1}^{\infty} \sum_{m \in \mathbb{Z} - \{0\}} \frac{1}{(m + nz)^2}, \tag{4.56}$$

which is the left-hand side of (4.53). That is, $G_2(z) - 2\zeta(2)$ equals the double sum on the right-hand side of (4.47), proving (4.47).

Using equation (4.47) one can eventually show that $G_2(z)$ satisfies the rule

$$G_2\left(-\frac{1}{z}\right) = z^2 G_2(z) - 2\pi i z. \tag{4.57}$$

Because of the term $-2\pi i z$ in (4.57), $G_2(z)$ is *not* a modular form of weight 2. That is, condition (M1)″ above is not satisfied, although condition (M1)′ is: $G_2(z+1) = G_2(z)$ by definition (4.46). Equation (4.57) also follows by a transformation property of the Dedekind eta function, whose logarithmic derivative turns out to be a constant multiple of $G_2(z)$. Thus we indicate now an alternative derivation of the rule (4.57).

On the domain $D \stackrel{\text{def}}{=} \{w \in \mathbb{C} \mid |w| < 1\}$, the holomorphic function $1 + w$ is nonvanishing and it therefore has a holomorphic logarithm $g(w)$ that can be chosen so as to vanish at $w = 0$: $e^{g(w)} \stackrel{\text{def}}{=} 1 + w$ on D. In fact $g(w) = -\sum_{n=1}^{\infty} (-w)^n / n$. Again for $q(z) \stackrel{\text{def}}{=} e^{2\pi i z}$, $z \in \pi^+$, consider the n-th partial sum $s_n(z) \stackrel{\text{def}}{=} \sum_{k=1}^{n} g(-q(z)^k)$, which is well-defined, because $q(z) \in D$ for $z \in \pi^+$ implies $-q(z)^k \in D$ for $k > 0$. We claim that the series

$$\psi(z) \stackrel{\text{def}}{=} -\sum_{k=1}^{\infty} g(-q(z)^k) = -\lim_{n \to \infty} s_n(z) \tag{4.58}$$

converges. We have $\psi(z) = \sum_{k=1}^{\infty} \sum_{n=1}^{\infty} (q(z)^k)^n / n$, where for $y = \operatorname{Im} z$ (again)

$$\sum_{n=1}^{\infty} \sum_{k=1}^{\infty} \left| \frac{(q(z)^k)^n}{n} \right| = \sum_{n=1}^{\infty} \frac{1}{n} \sum_{k=1}^{\infty} (e^{-2\pi n y})^k$$

$$= \sum_{n=1}^{\infty} \frac{1}{n} \left(\frac{e^{-2\pi n y}}{1 - e^{-2\pi n y}} \right) = \sum_{n=1}^{\infty} \frac{1}{n} \left(\frac{1}{e^{2\pi n y} - 1} \right),$$

which is finite by the integral test; compare with (4.54), for example. This allows us to write $\sum_{n=1}^{\infty} \sum_{k=1}^{\infty} (q(z)^k)^n / n = \sum_{k=1}^{\infty} \sum_{n=1}^{\infty} (q(z)^k)^n / n$, and shows the finiteness of these series. Hence $\psi(z)$ is finite. Now $\prod_{n=1}^{\infty} (1 - q(z)^n) = \lim_{n \to \infty} \prod_{k=1}^{n} (1 - q(z)^k) = \lim_{n \to \infty} \prod_{k=1}^{n} e^{g(-q(z)^k)}$, by the definition of $g(w)$, and this equals $e^{-\psi(z)}$ by (4.58). That is, for the Dedekind eta function

$$\eta(z) = e^{\pi i z / 12} \prod_{n=1}^{\infty} (1 - q(z)^n) \tag{4.59}$$

on π^+ defined in (3.27) we see that

$$\eta(z) = e^{\pi i z / 12 - \psi(z)}. \tag{4.60}$$

Differentiation of the equation $e^{g(w)} \stackrel{\text{def}}{=} 1 + w$ gives $g'(w) = 1/e^{g(w)} = 1/(1+w)$ (of course), which with termwise differentiation of (4.58) (whose justification we skip) gives

$$\psi'(z) = -\sum_{k=1}^{\infty} g'(-q(z)^k)(-kq(z)^{k-1} q'(z)) = 2\pi i \sum_{k=1}^{\infty} \frac{k q(z)^k}{1 - q(z)^k}$$

$$= 2\pi i \sum_{k=1}^{\infty} k \sum_{n=1}^{\infty} (q(z)^k)^n = 2\pi i \sum_{n=1}^{\infty} \sigma(n) e^{2\pi i n z},$$

by equation (4.52). Therefore, by (4.60), we obtain

$$\eta'(z) = \eta(z)\left(\frac{\pi i}{12} - 2\pi i \sum_{n=1}^{\infty} \sigma(n)e^{2\pi i n z}\right),$$

or

$$\frac{\eta'(z)}{\eta(z)} = -\frac{G_2(z)}{4\pi i} \qquad (4.61)$$

by definition (4.46).

Now $\eta(z)$ satisfies the known transformation rule

$$\eta\left(-\frac{1}{z}\right) = e^{-i\pi/4}\sqrt{z}\,\eta(z),$$

where we take $\arg z \in (-\pi, \pi)$. Differentiation gives

$$\frac{\eta'\left(-\frac{1}{z}\right)\frac{1}{z^2}}{\eta\left(-\frac{1}{z}\right)} = \frac{e^{-i\pi/4}\left(\sqrt{z}\,\eta'(z) + \frac{\sqrt{z}}{2z}\eta(z)\right)}{\eta\left(-\frac{1}{z}\right)} = \frac{\eta'(z)}{\eta(z)} + \frac{1}{2z} \qquad (4.62)$$

which by equation (4.61) says that $-G_2\left(-\frac{1}{z}\right)/4\pi i z^2 = -G_2(z)/4\pi i + \frac{1}{2z}$. This is immediately seen to imply the transformation rule (4.57).

In the lectures of Geoff Mason and Michael Tuite the particular normalization $G_k(z)/(2\pi i)^k$ of the Eisenstein series is considered, which they denote by $E_k(z)$. In particular, by definition (4.46),

$$E_2(z) = -\frac{1}{12} + 2\sum_{n=1}^{\infty} \sigma(n)e^{2\pi i n z}.$$

However, other normalized Eisenstein series appear in the literature that also might be denoted by $E_k(z)$. For example, in [20] there is the normalization (and notation) $E_k(z) \stackrel{\text{def}}{=} G_k(z)/2\zeta(k)$.

Lecture 5. Dirichlet L-functions

Equation (3.19) has an application to *Dirichlet L-functions*, which we now consider. To construct such a function, we need first a *character* χ modulo m, where $m > 0$ is a fixed integer. This is defined as follows. Let U_m denote the group of *units* in the commutative ring $\mathbb{Z}_{(m)} \stackrel{\text{def}}{=} \mathbb{Z}/m\mathbb{Z}$. Thus if $\bar{n} = n + m\mathbb{Z}$ denotes the coset of $n \in \mathbb{Z}$ in $\mathbb{Z}_{(m)}$, we have $\bar{n} \in U_m \iff \exists \bar{a} \in \mathbb{Z}_{(m)}$ such that $\bar{a}\bar{n} = \bar{1}$. One knows of course that $\bar{n} \in U_m \iff (n, m) = 1$ (i.e. n and m are relatively prime). A character modulo m is then (by definition) a group homomorphism $\chi : U_m \to \mathbb{C}^* \stackrel{\text{def}}{=} \mathbb{C} - \{0\}$. For our purpose, however, there is an equivalent way of thinking about characters modulo m. In fact, given χ, define $\chi_{\mathbb{Z}} : \mathbb{Z} \to \mathbb{C}$ by

$$\chi_{\mathbb{Z}}(n) \stackrel{\text{def}}{=} \begin{cases} \chi(\bar{n}) & \text{if } (n, m) = 1, \\ 0 & \text{otherwise,} \end{cases} \quad (5.1)$$

for $n \in \mathbb{Z}$. For $n, n_1, n_2 \in \mathbb{Z}$, $\chi_{\mathbb{Z}}$ satisfies:

(D1) $\chi_{\mathbb{Z}}(n) = 0 \iff (n, m) \neq 1$;

(D2) $\chi_{\mathbb{Z}}(n_1) = \chi_{\mathbb{Z}}(n_2)$ when $\bar{n}_1 = \bar{n}_2$ in $\mathbb{Z}_{(m)}$;

(D3) $\chi_{\mathbb{Z}}(n_1 n_2) = \chi_{\mathbb{Z}}(n_1)\chi_{\mathbb{Z}}(n_2)$ when $(n_1, m) = 1$ and $(n_2, m) = 1$.

Conversely, suppose $\chi_0 : \mathbb{Z} \to \mathbb{C}$ is a function that satisfies the three conditions (D1), (D2), and (D3). Define $\chi : U_m \to \mathbb{C}$ by $\chi(\bar{n}) = \chi_0(n)$ for $n \in \mathbb{Z}$ such that $(n, m) = 1$. The character χ is well-defined by (D2), and $\chi : U_m \to \mathbb{C}^*$ by (D1). By (D3), $\chi(ab) = \chi(a)\chi(b)$ for $a, b \in U_m$, so we see that χ is a character modulo m. Moreover the induced map $(\chi_0)_{\mathbb{Z}} : \mathbb{Z} \to \mathbb{C}$ given by definition (5.1) coincides with χ_0.

Note that $\chi_{\mathbb{Z}}$ is completely multiplicative:

(D4) $\chi_{\mathbb{Z}}(n_1 n_2) = \chi_{\mathbb{Z}}(n_1)\chi_{\mathbb{Z}}(n_2)$ for all $n_1, n_2 \in \mathbb{Z}$.

For if either $(n_1, m) \neq 1$ or $(n_2, m) \neq 1$, then $(n_1 n_2, m) \neq 1$, so that by (D1) both $\chi_{\mathbb{Z}}(n_1)\chi_{\mathbb{Z}}(n_2)$ and $\chi_{\mathbb{Z}}(n_1 n_2)$ are zero. If both $(n_1, m) = 1$ and $(n_2, m) = 1$, then already $\chi_{\mathbb{Z}}(n_1 n_2) = \chi_{\mathbb{Z}}(n_1)\chi_{\mathbb{Z}}(n_2)$ by (D3).

Note also that since $\chi(a) \in \mathbb{C}^*$ for every $a \in U_m$ (that is, $\chi(a) \neq 0$), we have $0 \neq \chi(\bar{1}) = \chi(\bar{1}\bar{1}) = \chi(\bar{1})\chi(\bar{1})$ by (D4), so $\chi(\bar{1}) = 1$. Moreover since $(1, m) = 1$, $\chi_{\mathbb{Z}}(1) = \chi(\bar{1})$, by (5.1), which in turn equals 1.

One final property of $\chi_{\mathbb{Z}}$ that we need is:

(D5) $|\chi(a)| = 1$ for all $a \in U_m$; hence $|\chi_{\mathbb{Z}}(n)| \leq 1$ for all $n \in \mathbb{Z}$.

The proof of (D5) makes use of a little theorem in group theory which says that if G is a finite group with $|G|$ elements, then $a^{|G|} = 1$ for every $a \in G$. Now, given $a \in U_m$, we can write (as just seen) $1 = \chi(\bar{1}) = \chi(a^{|U_m|}) = \chi(a)^{|U_m|}$ (since χ is a group homomorphism), which shows that $\chi(a)$ is a $|U_m|$-th root of unity: $|\chi(a)| = 1$ for all $a \in U_m$. Hence $|\chi_{\mathbb{Z}}(n)| \leq 1$ for all $n \in \mathbb{Z}$, by definition (5.1).

Given a character χ modulo m, it follows that we can form the zeta function, or Dirichlet series

$$L(s, \chi) \stackrel{\text{def}}{=} \sum_{n=1}^{\infty} \frac{\chi_{\mathbb{Z}}(n)}{n^s} = \sum_{(n,m)=1} \frac{\chi(\bar{n})}{n^s}, \quad (5.2)$$

called a *Dirichlet L-function*, which converges for $\operatorname{Re} s > 1$, by (D5). $L(s, \chi)$ is holomorphic on the domain $\operatorname{Re} s > 1$, by the same argument given for the Riemann zeta function $\zeta(s)$. Since $\chi_{\mathbb{Z}} \not\equiv 0$ ($\chi_{\mathbb{Z}}(1) = 1$), and since $\chi_{\mathbb{Z}}$ is completely multiplicative, formula (3.19) implies:

THEOREM 5.3 (EULER PRODUCT FOR DIRICHLET L-FUNCTIONS). *Assume* $\operatorname{Re} s > 1$. *Then*

$$L(s, \chi) = \frac{1}{\prod_{p \in P}\left(1 - \chi_{\mathbb{Z}}(p)p^{-s}\right)} = \frac{1}{\prod_{\substack{p \in P \\ p \nmid m}}\left(1 - \chi_{\mathbb{Z}}(p)p^{-s}\right)}. \qquad (5.4)$$

The second statement of equality follows by (D1), since for a prime $p \in P$, saying that $(p, m) \neq 1$ is the same as saying that $p \mid m$.

As an example, define $\chi_0 : \mathbb{Z} \to \mathbb{C}$ by

$$\chi_0(n) = \begin{cases} 1 & \text{if } (n, m) = 1, \\ 0 & \text{otherwise}, \end{cases}$$

for $n \in \mathbb{Z}$. Then χ_0 satisfies (D1). If $n_1, n_2, l \in \mathbb{Z}$ such that $n_1 = n_2 + lm$ (i.e., $\bar{n}_1 = \bar{n}_2$), then $(n_1, m) = 1 \iff (n_2, m) = 1$, so χ_0 satisfies (D2). If $(n_1, m) = 1$ and $(n_2, m) = 1$, then $(n_1 n_2, m) = 1$, so χ_0 also satisfies (D3), and χ_0 therefore defines a Dirichlet character modulo m. We call χ_0 (or the induced character $U_m \to \mathbb{C}^*$) the *principal character* modulo m. Again since p is a prime, we see by equation (5.4) that for $\operatorname{Re} s > 1$

$$L(s, \chi_0) = \frac{1}{\prod_{\substack{p \in P \\ p \nmid m}}(1 - p^{-s})}. \qquad (5.5)$$

Then for $\operatorname{Re} s > 1$

$$L(s, \chi_0) \frac{1}{\prod_{\substack{p \in P \\ p \mid m}}(1 - p^{-s})} = \frac{1}{\prod_{p \in P}(1 - p^{-s})} = \zeta(s) \qquad (5.6)$$

by formula (0.2). That is,

$$L(s, \chi_0) = \zeta(s) \prod_{\substack{p \in P \\ p \mid m}}(1 - p^{-s}) \qquad (5.7)$$

for $\operatorname{Re} s > 1$.

If $\chi_0 : U_m \to \mathbb{C}^*$ also denotes the character modulo m induced by $\chi_0 : \mathbb{Z} \to \mathbb{C}$, then $\chi_0(\bar{n}) = 1$ for every $\bar{n} \in U_m$ (by (5.1)) since $(n, m) = 1$.

As another simple example, take $m = 5$: $Z_{(5)} = \{\bar{0}, \bar{1}, \bar{2}, \bar{3}, \bar{4}\}$, and it is easily checked that $U_5 = \{\bar{1}, \bar{2}, \bar{3}, \bar{4}\}$. Moreover the equations $\chi(\bar{1}) \stackrel{\text{def}}{=} \chi(\bar{4}) \stackrel{\text{def}}{=} 1$, $\chi(\bar{2}) \stackrel{\text{def}}{=} \chi(\bar{3}) \stackrel{\text{def}}{=} -1$ define a character $\chi^{(5)} = \chi : U_5 \to \mathbb{C}^*$ modulo 5. The induced map $\chi_{\mathbb{Z}} : \mathbb{Z} \to \mathbb{C}$ in definition (5.1) is given by $\chi_{\mathbb{Z}}(1) = 1$, $\chi_{\mathbb{Z}}(2) = -1$, $\chi_{\mathbb{Z}}(3) = -1$, $\chi_{\mathbb{Z}}(4) = 1$, $\chi_{\mathbb{Z}}(5) = 0$ (since $(5, 5) \neq 1$), $\chi_{\mathbb{Z}}(6) = 1$, $\chi_{\mathbb{Z}}(7) = -1$, $\chi_{\mathbb{Z}}(8) = -1$, $\chi_{\mathbb{Z}}(9) = 1, \ldots$ (since $\bar{6} = \bar{1}, \bar{7} = \bar{2}, \bar{8} = \bar{3}, \bar{9} = \bar{4}$, with $(n, 5) = 1$

for $n = 6, 7, 8, 9$). The corresponding Dirichlet L-function is therefore given, for $\operatorname{Re} s > 1$, by

$$L(s, \chi^{(5)}) = 1 - \frac{1}{2^s} - \frac{1}{3^s} + \frac{1}{4^s} + \frac{1}{6^s} - \frac{1}{7^s} - \frac{1}{8^s} + \frac{1}{9^s} \pm \cdots,$$

Next take $m = 8$: $Z_{(8)} = \{\bar{0}, \bar{1}, \bar{2}, \bar{3}, \bar{4}, \bar{5}, \bar{6}, \bar{7}\}$, $U_8 = \{\bar{1}, \bar{3}, \bar{5}, \bar{7}\}$. The map $\chi^{(8)} = \chi : U_8 \to \mathbb{C}^*$ given by $\chi(\bar{1}) \stackrel{\text{def}}{=} \chi(\bar{7}) \stackrel{\text{def}}{=} 1$, $\chi(\bar{3}) \stackrel{\text{def}}{=} \chi(\bar{5}) \stackrel{\text{def}}{=} -1$ is a character modulo 8, with $\chi_\mathbb{Z} : \mathbb{Z} \to \mathbb{C}$ in definition (5.1) given by $\chi_\mathbb{Z}(1) = 1$, $\chi_\mathbb{Z}(2) = 0$, $\chi_\mathbb{Z}(3) = -1$, $\chi_\mathbb{Z}(4) = 0$, $\chi_\mathbb{Z}(5) = -1$, $\chi_\mathbb{Z}(6) = 0$, $\chi_\mathbb{Z}(7) = 1$, $\chi_\mathbb{Z}(8) = 0$, $\chi_\mathbb{Z}(9) = 1$, $\chi_\mathbb{Z}(10) = 0$, $\chi_\mathbb{Z}(11) = -1$, $\chi_\mathbb{Z}(12) = 0$, $\chi_\mathbb{Z}(13) = -1$, $\chi_\mathbb{Z}(14) = 0$, $\chi_\mathbb{Z}(15) = 1$, $\chi_\mathbb{Z}(16) = 0$, $\chi_\mathbb{Z}(17) = 1, \ldots$. Then

$$L(s, \chi^{(8)}) = 1 - \frac{1}{3^s} - \frac{1}{5^s} + \frac{1}{7^s} + \frac{1}{9^s} - \frac{1}{11^s} - \frac{1}{13^s} + \frac{1}{15^s} + \frac{1}{17^s} \pm \cdots.$$

From formula (5.7) it follows that the L-function $L(s, \chi_0)$ admits a meromorphic continuation to the full complex plane, with $s = 1$ as its only singularity — a simple pole with residue $\prod_{p \in P, \, p \mid m}(1 - \frac{1}{p})$. If $\chi \neq \chi_0$ it is known that $L(s, \chi)$ at least extends to $\operatorname{Re} s > 0$ and, moreover, that $L(1, \chi) \neq 0$. For example, for the characters $\chi^{(5)}$, $\chi^{(8)}$ modulo 5 and 8, respectively, constructed in the previous examples, one has

$$L(1, \chi^{(5)}) = \frac{1}{\sqrt{5}} \log\left(\frac{3 + \sqrt{5}}{2}\right), \quad L(1, \chi^{(8)}) = \frac{1}{\sqrt{8}} \log(3 + 2\sqrt{2}).$$

If $\chi \neq \chi_0$ is a *primitive* character modulo m, a notion that we shall define presently, then $L(s, \chi)$ does continue meromorphically to $\mathbb{C}$, and it has a decent functional equation.

First we define the notion of an imprimitive character. Suppose $k > 0$ is a divisor of m. Then there is a natural (well-defined) map $q : \mathbb{Z}/m\mathbb{Z} \to \mathbb{Z}/k\mathbb{Z}$, given by $q(n + m\mathbb{Z}) \stackrel{\text{def}}{=} n + k\mathbb{Z}$ for $n \in \mathbb{Z}$. If $(n, m) = 1$ then $(n, k) = 1$ since $k \mid m$. Therefore the restriction $q^* \stackrel{\text{def}}{=} q|_{U_m}$ maps U_m to U_k, and is a homomorphism between these two groups. Let $\psi^{(k)} : U_k \to \mathbb{C}^*$ be a character modulo k. By definition, $\psi^{(k)}$ is a homomorphism and hence so is $\psi^{(k)} \circ q^* : U_m \to \mathbb{C}^*$. That is, given a positive divisor k of m we have an induced character $\chi \stackrel{\text{def}}{=} \psi^{(k)} \circ q^*$ modulo m. Characters χ modulo m that are induced this way, say for $k \neq m$, are called *imprimitive*. χ is called a *primitive* character if it is not imprimitive, in which case m is also called the *conductor* of χ. Thus for a primitive character χ modulo m, the L-function $L(s, \chi)$ satisfies a theory similar to (but a bit more complicated than) that of the Riemann zeta function $\zeta(s)$.

In the Introduction we referred to the prime number theorem, expressed in equation (0.3), as a *monumental result*, and we noted quite briefly the role of $\zeta(s)$ in its proof. Similarly, the study of the L-functions $L(s, \chi)$ leads to a monumental result regarding primes in an arithmetic progression. Namely, in 1837 Dirichlet proved that there are *infinitely many primes* in any arithmetic progression $n, n+m, n+2m, n+3m, \ldots$, where n, m are positive, relatively prime integers — a key aspect of the proof being the fact (pointed out earlier) that $L(1, X) \neq 0$ if $X \neq X_0$. Dirichlet's proof relates, moreover, $L(1, X)$ to a Gaussian *class number* — an invariant in the study of binary quadratic forms. One can obtain, also, a *prime number theorem for arithmetic progressions* (from the *Siegel–Walfisz theorem*), where the counting function $\pi(x)$ in (0.3) is replaced by the function $\pi(x; m, n) \stackrel{\text{def}}{=}$ the number of primes $p \leq x$, with $p \equiv n \pmod{m}$, for n, m relatively prime. One can also formulate and prove a prime number theorem for *graphs*. This is discussed in section 3.3 of the lectures of Audrey Terras.

Lecture 6. Radiation density integral, free energy, and a finite-temperature zeta function

Theorems 1.13 and 1.18 provide for integral representations of $\zeta(s)$, for $\operatorname{Re} s > 1$, that serve as starting points for its analytic continuation. The following, nice integral representation also serves as a starting point. We apply it to compute Planck's radiation density integral. We also consider a free energy – zeta function connection.

THEOREM 6.1. *For* $\operatorname{Re} s > 1$

$$\zeta(s) = \frac{1}{\Gamma(s)} \int_0^\infty \frac{t^{s-1}\,dt}{e^t - 1}. \tag{6.2}$$

We can regard the integral on the right as the sum

$$\int_0^1 \frac{t^{s-1}\,dt}{e^t - 1} + \int_1^\infty \frac{t^{s-1}\,dt}{e^t - 1},$$

where the second integral converges absolutely for all $s \in \mathbb{C}$, *and the first integral, understood as the limit* $\lim_{\alpha \to 0+} \int_\alpha^1 t^{s-1}\,dt/(e^t - 1)$, *exists for* $\operatorname{Re} s > 1$.

The proof of Theorem 6.1 is developed in two stages. First, for $\alpha, \beta, a \in \mathbb{R}$ and $s \in \mathbb{C}$, with $\alpha < \beta$ and $a > 0$, write $\int_\alpha^\beta e^{-at} t^{s-1}\,dt = a^{-s} \int_{a\alpha}^{a\beta} e^{-v} v^{s-1}\,dv$, by the change of variables $v = at$. In particular,

$$\int_1^\beta e^{-at} t^{s-1}\,dt = a^{-s} \int_a^{a\beta} e^{-t} t^{s-1}\,dt \quad \text{for } \beta > 1, \tag{6.3}$$

$$\int_\alpha^1 e^{-at} t^{s-1}\,dt = a^{-s} \int_{a\alpha}^a e^{-t} t^{s-1}\,dt \quad \text{for } \alpha < 1. \tag{6.4}$$

LECTURES ON ZETA FUNCTIONS, L-FUNCTIONS AND MODULAR FORMS

For $s \in \mathbb{C}, m > 0$, consider the integral

$$I_m(s) \stackrel{\text{def}}{=} \int_0^\infty \frac{t^{s-1}e^{-mt}e^{-t}}{1-e^{-t}} \, dt = \int_0^\infty \frac{t^{s-1}e^{-mt}}{e^t - 1} \, dt \qquad (6.5)$$

which we check does converge for $\operatorname{Re} s > 1$. For $t > 0$, we have $e^t > 1 + t$, so $1/(e^t - 1) < 1/t$, and hence

$$\left| \frac{t^{s-1}e^{-mt}}{e^t - 1} \right| < \frac{t^{\sigma-1}e^{-mt}}{t} = t^{\sigma-2}e^{-mt} \qquad (6.6)$$

for $\sigma = \operatorname{Re} s$. By (6.3), $\int_1^\beta t^{\sigma-2}e^{-mt} \, dt = m^{-(\sigma-1)} \int_m^{m\beta} e^{-t}t^{\sigma-2} \, dt$ for $\beta > 1$. Let $\beta \to \infty$: then $\int_1^\infty t^{\sigma-2}e^{-mt} \, dt$ exists and

$$\int_1^\infty t^{\sigma-2}e^{-mt} \, dt = \frac{1}{m^{\sigma-1}} \int_m^\infty e^{-t}t^{\sigma-2} \, dt.$$

In view of (6.6), therefore, $\int_1^\infty \frac{t^{s-1}e^{-mt}}{e^t - 1} \, dt$ converges absolutely for every $s \in \mathbb{C}$ and $m > 0$, and

$$\left| \int_1^\infty \frac{t^{s-1}e^{-mt}}{e^t - 1} \, dt \right| \leq \int_1^\infty \left| \frac{t^{s-1}e^{-mt}}{e^t - 1} \right| \, dt \leq \frac{1}{m^{\sigma-1}} \int_m^\infty e^{-t}t^{\sigma-2} \, dt. \qquad (6.7)$$

By the change of variables $v = 1/t$ for $t > 0$,

$$\int_\alpha^1 \frac{t^{s-1}e^{-mt}}{e^t - 1} \, dt = \int_1^{1/\alpha} \frac{\left(\frac{1}{v}\right)^{s-1} e^{-m(1/v)}}{(e^{1/v} - 1)v^2} \, dv \qquad (6.8)$$

for $0 < \alpha < 1$. Here, by the inequality in (6.6), we can write

$$\left| \left(\frac{1}{v}\right)^{s-1} \frac{e^{-m(1/v)}}{(e^{1/v} - 1)v^2} \right| < \left(\frac{1}{v}\right)^{\sigma-2} \frac{e^{-m/v}}{v^2} = v^{-\sigma}e^{-m/v} \leq v^{-\sigma}. \qquad (6.9)$$

But $\int_1^{1/\alpha} v^{-\sigma} \, dv = \frac{(1/\alpha)^{1-\sigma} - 1}{1 - \sigma}$, so $\lim_{\alpha \to 0^+} \int_1^{1/\alpha} v^{-\sigma} \, dv = \frac{1}{\sigma - 1}$ for $\sigma > 1$, i.e.,

$$\int_1^\infty \frac{\left(\frac{1}{v}\right)^{s-1} e^{-m(1/v)}}{(e^{1/v} - 1)v^2} \, dv$$

converges absolutely for $\operatorname{Re} s > 1$ and

$$\left| \int_1^\infty \frac{\left(\frac{1}{v}\right)^{s-1} e^{-m(1/v)}}{(e^{1/v} - 1)v^2} \, dv \right| \leq \int_1^\infty \left| \frac{\left(\frac{1}{v}\right)^{s-1} e^{-m(1/v)}}{(e^{1/v} - 1)v^2} \right| \, dv \leq \int_1^\infty v^{-\sigma}e^{-m/v} \, dv,$$

by the inequality in (6.9). The right-hand side equals $\lim_{\beta\to\infty}\int_1^\beta v^{-\sigma}e^{-m/v}\,dv$, or, by the change of variables $t=1/v$,

$$\lim_{\beta\to\infty}\int_{1/\beta}^1 t^{\sigma-2}e^{-mt}\,dt = \lim_{\beta\to\infty}m^{-(\sigma-1)}\int_{m/\beta}^m e^{-t}t^{\sigma-2}\,dt$$

$$= \frac{1}{m^{\sigma-1}}\int_0^m e^{-t}t^{\sigma-2}\,dt.$$

That is, by (6.8), $\lim_{\alpha\to 0^+}\int_\alpha^1 \dfrac{t^{s-1}e^{-mt}}{e^t-1}\,dt$ exists for $\mathrm{Re}\,s>1$, and, with $\sigma=\mathrm{Re}\,s$,

$$\left|\lim_{\alpha\to 0^+}\int_\alpha^1 \frac{t^{s-1}e^{-mt}}{e^t-1}\,dt\right| \le \frac{1}{m^{\sigma-1}}\int_0^m e^{-t}t^{\sigma-2}\,dt. \qquad (6.10)$$

We have therefore checked that the integral $I_m(s)$ defined by (6.5) converges for $\mathrm{Re}\,s>1$ (and in fact the portion $\int_1^\infty t^{s-1}e^{-mt}/(e^t-1)\,dt$ converges absolutely for all $s\in\mathbb{C}$), and that, moreover, for $\sigma=\mathrm{Re}\,s$, we have

$$\left|\int_0^1 \frac{t^{s-1}e^{-mt}}{e^t-1}\,dt\right| \le \frac{1}{m^{\sigma-1}}\int_0^m e^{-t}t^{\sigma-2}\,dt,$$

$$\left|\int_1^\infty \frac{t^{s-1}e^{-mt}}{e^t-1}\,dt\right| \le \frac{1}{m^{\sigma-1}}\int_m^\infty e^{-t}t^{\sigma-2}\,dt,$$

by the inequalities (6.7) and (6.10). This says that

$$|I_m(s)| \le \frac{1}{m^{\sigma-1}}\left(\int_0^m e^{-t}t^{\sigma-2}\,dt + \int_m^\infty e^{-t}t^{\sigma-2}\,dt\right)$$

$$= \frac{1}{m^{\sigma-1}}\int_0^\infty e^{-t}t^{\sigma-2}\,dt = \frac{1}{m^{\sigma-1}}\Gamma(\sigma-1),$$

by definition (1.6). Since $m^{\sigma-1}\to 0$ as $m\to\infty$ for $\sigma>1$, we see that

$$\lim_{m\to\infty}I_m(s)=0 \qquad (6.11)$$

for $\mathrm{Re}\,s>1$!

We move now to the second stage of the proof of Theorem 6.1, which is quite brief. Fix integers n,m with $1\le n\le m$ and $s\in\mathbb{C}$ with $\mathrm{Re}\,s>1$, and set $v=nt$. Again by (1.6), $\int_0^\infty e^{-nt}t^{s-1}\,dt = n^{-s}\int_0^\infty e^{-v}v^{s-1}\,dv = \Gamma(s)/n^s$, so

$$\Gamma(s)\sum_{n=1}^m \frac{1}{n^s} = \int_0^\infty t^{s-1}\left(\sum_{n=1}^m e^{-nt}\right)dt = \int_0^\infty t^{s-1}\frac{e^{-t}(1-e^{-mt})}{1-e^{-t}}\,dt$$

$$= \int_0^\infty \frac{t^{s-1}e^{-t}}{1-e^{-t}}\,dt + I_m(s).$$

(Here the the last equality comes from (6.5) and the last but one from the formula for the partial sum of the geometric series.) Therefore

$$\Gamma(s) \sum_{n=1}^{m} \frac{1}{n^s} = \int_0^\infty \frac{t^{s-1}}{e^t - 1} dt + I_m(s). \tag{6.12}$$

We check the existence of the integral on the right-hand side for $\operatorname{Re} s > 1$: write

$$\frac{t^{s-1}}{e^t - 1} = t^{s-1} e^{-t} + \frac{t^{s-1} e^{-t}}{e^t - 1}$$

for $t > 0$; the integral $\int_0^\infty t^{s-1} e^{-t} dt = \Gamma(s)$ converges for $\operatorname{Re} s > 0$ and the integral of the last term, which equals $I_1(s)$ by (6.5), converges for $\operatorname{Re} s > 1$, as established in the first stage of the proof.

Let $m \to \infty$ in equation (6.12): then $\Gamma(s)\zeta(s) = \int_0^\infty \frac{t^{s-1} dt}{e^t - 1}$ by (6.11) for $\operatorname{Re} s > 1$, concluding the proof of Theorem 6.1.

COROLLARY 6.13. *For $a > 0$, $\operatorname{Re} s > 1$,*

$$\int_0^\infty \frac{t^{s-1} dt}{e^{at} - 1} = \frac{\Gamma(s)\zeta(s)}{a^s}. \tag{6.14}$$

In particular, for $a > 0$, $n = 1, 2, 3, 4, \ldots$

$$\int_0^\infty \frac{t^{2n-1}}{e^{at} - 1} dt = \frac{(-1)^{n+1} (2\pi/a)^{2n} B_{2n}}{4n}. \tag{6.15}$$

PROOF. This follows from (6.2) once the obvious change of variables $v = at$ is executed. By formula (6.14),

$$\int_0^\infty \frac{t^{2n-1}}{e^{at} - 1} dt = \frac{\Gamma(2n)\zeta(2n)}{a^{2n}}$$

for $n \in \mathbb{Z}$, $n \geq 1$. Since $\Gamma(2n) = (2n - 1)!$ (because $\Gamma(m) = (m - 1)!$ for $m \in \mathbb{Z}, m \geq 1$), one can now appeal to formula (2.1) to conclude the proof of equation (6.15). $\square$

As an application of formula (6.15) we shall compute Planck's *radiation density integral*. But first we provide some background.

On 14 December 1900, a paper written by Max Karl Ernst Ludwig Planck and entitled "On the theory of the energy distribution law of the normal spectrum" was presented to the German Physical Society. That date is considered to be the birthday of quantum mechanics, as that paper set forth for the first time the hypothesis that the energy of emitted radiation is *quantized*. Namely, the energy cannot assume arbitrary values but only integral multiples $0, h\nu, 2h\nu, 3h\nu, \ldots$ of the basic energy value $E = h\nu$, where ν is the frequency of the radiation and h is what is now called *Planck's constant*. We borrow a quotation from Hermann

Weyl's notable book [37]: "The magic formula $E = h\nu$ from which the whole of quantum theory is developed, establishes a universal relation between the frequency ν of an oscillatory process and the energy E associated with such a process". The quantization of energy has profound consequences regarding the structure of matter.

Planck was led to his startling hypothesis while searching for a theoretical justification for his newly proposed formula for the energy density of thermal (or "blackbody") radiation. Lord Rayleigh had proposed earlier that year a theoretical explanation for the experimental observation that the rate of energy emission $f(\nu; T)$ by a body at temperature T in the form of electromagnetic radiation of frequency ν grows, under certain conditions, with the square of ν, and the *total* energy emitted grows with the fourth power of T. In the quantitative form derived by James Jeans a few years later, Rayleigh's formula reads

$$f(\nu; T) = \frac{8\pi\nu^2}{c^3}kT, \tag{6.16}$$

where c is the speed of light and k is Boltzmann's constant. As ν grows, however, this formula was known to fail. Wilhelm Wien had already proposed, in 1896, the empirically more accurate formula

$$f(\nu; T) = a\nu^3 e^{-b\nu/T}. \tag{6.17}$$

Unlike the Rayleigh–Jeans formula (6.16), Wien's avoids the "ultraviolet catastrophe". (This colorful name was coined later by Paul Ehrenfest for the notion that a functional form for $f(\nu; T)$ might yield an *infinite value* for the total energy, $\int_0^\infty f(\nu; T)d\nu = \infty$—"ultraviolet" because the divergence sets in at high frequencies.) However, the lack of a theoretical explanation for Wien's law, and its wrong prediction for the asymptotic limit at low frequencies — proportional to ν^3 rather than ν^2 — made it unsatisfactory as well.

By October 1900 Planck had come up with a formula that had the right asymptotic behavior in both directions and was soon found to be very accurate:

$$f(\nu; T) = \frac{8\pi\nu^2}{c^3} \frac{h\nu}{e^{h\nu/kT} - 1}, \tag{6.18}$$

where the new constant h was introduced. In his December paper, already mentioned, he provides a justification for this formula using the earlier notions of electromagnetic oscillators and statistical-mechanical entropy, but invoking the additional assumption that the energy of the oscillators is restricted to multiples of $E_\nu = h\nu$. This then is the genesis of quantization.

Note that given the approximation $e^x \simeq 1 + x$ for a very small value of x, one has the low frequency approximation $e^{h\nu/kT} - 1 \simeq h\nu/kT$, which when used in formula (6.18) gives $f(\nu; T) \simeq 8\pi\nu^2 kT/c^3$ — the Rayleigh–Jeans result (6.16), as expected by our previous remarks.

We now check that in contrast to an infinite total energy value implied by formula (6.16), integration over the full frequency spectrum via Planck's law (6.18) does yield a *finite* value. The result is:

PROPOSITION 6.19. *Planck's radiation density integral*

$$I(T) \stackrel{\text{def}}{=} \int_0^\infty f(\nu; T) d\nu = \frac{8\pi h}{c^3} \int_0^\infty \frac{\nu^3 d\nu}{e^{h\nu/kT} - 1}$$

(see (6.18)) *has the finite value* $8\pi^5 k^4 T^4 / 15 c^3 h^3$.

The proof is quite immediate. In formula (6.15) choose $n = 2$, $a = h/kT$; then

$$I(T) = \frac{8\pi h}{c^3}(-1)^3 \left(\frac{2\pi kT}{h}\right)^4 \frac{B_4}{8}.$$

Since $B_4 = -1/30$ by (2.2), the desired value of $I(T)$ is achieved.

The radiation energy density $f(\nu; T)$ in (6.18) is related to the thermodynamics of the quantized harmonic oscillator; namely, it is related to the thermodynamic *internal energy* $U(T)$. We mention this because $U(T)$, in turn, is related to the *Helmholtz free energy* $F(T)$ which has, in fact, a zeta function connection. A quick sketch of this mix of ideas is as follows, where proofs and details can be found in my book on quantum mechanics [42] (along with some historic remarks).

One of the most basic, elementary facts of quantum mechanics is that the quantized harmonic oscillator of frequency ν has the sequence

$$\{E_n \stackrel{\text{def}}{=} (n + \tfrac{1}{2})h\nu\}_{n=0}^\infty$$

as its energy levels. The corresponding *partition function* $Z(T)$ is given by

$$Z(T) \stackrel{\text{def}}{=} \sum_{n=0}^\infty \exp\left(-\frac{E_n}{kT}\right), \tag{6.20}$$

where again T denotes temperature and k denotes Boltzmann's constant. This sum is easily computed:

$$Z(T) = \sum_{n=0}^\infty e^{-h\nu/2kT} \left(e^{-h\nu/kT}\right)^n = e^{-h\nu/2kT} \frac{1}{1 - e^{-h\nu/kT}};$$

i.e.,

$$Z(T) = \frac{1}{2 \sinh \dfrac{h\nu}{2kT}}. \tag{6.21}$$

The importance of the partition function $Z(T)$ is that from it one derives basic thermodynamic quantities such as

the Hemholtz free energy $\quad F(T) \stackrel{\text{def}}{=} -kT \log Z(T)$,

the entropy $\quad S(T) \stackrel{\text{def}}{=} -\partial F / \partial T$,

the internal energy $\quad U(T) \stackrel{\text{def}}{=} F(T) + TS(T) = F(T) - T\dfrac{\partial F}{\partial T}$.

Using (6.21), one computes that by these definitions

$$F(T) = \frac{h\nu}{2} + kT \log\left(1 - \exp\left(-\frac{h\nu}{kT}\right)\right),$$

$$S(T) = k\left(\frac{h\nu/kT}{\exp(h\nu/kT) - 1} - \log\left(1 - \exp\left(-\frac{h\nu}{kT}\right)\right)\right), \quad (6.22)$$

$$U(T) = \frac{h\nu}{2} + \frac{h\nu}{\exp(h\nu/kT) - 1},$$

which means that the factor $h\nu/(e^{h\nu/kT} - 1)$ of $f(\nu; T)$ in equation (6.18) differs from the internal energy $U(T)$ *exactly* by the quantity $E_0 = \frac{1}{2}h\nu$, which is the *ground state* energy (also called the *zero-point* energy) of the quantized harmonic oscillator (since $E_n = (n + \frac{1}{2})h\nu$). However, our main interest is in setting up a free energy – zeta function connection. Here's how it goes.

For convenience let $\beta \stackrel{\text{def}}{=} 1/(kT)$ denote the inverse temperature. Form the *finite temperature zeta function*

$$\zeta(s; T) \stackrel{\text{def}}{=} \sum_{n \in \mathbb{Z}} \frac{1}{(4\pi^2 n^2 + h^2 \nu^2 \beta^2)^s}, \quad (6.23)$$

which turns out to be well-defined and holomorphic for $\operatorname{Re} s > \frac{1}{2}$. In [42] we show that $\zeta(s; T)$ has a meromorphic continuation to $\operatorname{Re} s < 1$ given by

$$\zeta(s; T) = \frac{\Gamma(s - \frac{1}{2})}{\sqrt{4\pi}\,\Gamma(s) a^{s-1/2}} + 2(\sqrt{a})^{1-2a} \frac{\sin \pi s}{\pi} \int_1^\infty \frac{(x^2 - 1)^{-s}}{\exp(\sqrt{a}x) - 1} dx \quad (6.24)$$

for $a \stackrel{\text{def}}{=} h^2 \nu^2 \beta^2$. Moreover $\zeta(s; T)$ is holomorphic at $s = 0$ and

$$\zeta'(0; T) = -\sqrt{a} - 2\log(1 - \exp(-\sqrt{a})); \quad (6.25)$$

see Theorem 14.4 and Corollary 14.2 of [42]. By definition, $\sqrt{a} = h\nu\beta = h\nu/kT$. Therefore by formula (6.25) (which does require some work to derive from (6.24)), and by the first formula in (6.22), one discovers that

$$F(T) = -\frac{kT}{2} \zeta'(0; T), \quad (6.26)$$

which is the free energy – zeta function connection.

We will meet the zero-point energy $E_0 = \frac{1}{2}h\nu$ again in the next lecture regarding the discussion of Casimir energy. In chapter 16 of [42] another finite temperature zeta function is set up in the context of Kaluza–Klein space-times with spatial sector $\mathbb{R}^m \times \Gamma\backslash G/K$, where Γ is a discrete group of isometries of the rank 1 symmetric space G/K; here K is a maximal compact subgroup of the semisimple Lie group G. In this broad context a partition function $Z(T)$ and free energy-zeta function connection still exist.

Lecture 7. Zeta regularization, spectral zeta functions, Eisenstein series, and Casimir energy

Zeta regularization is a powerful, elegant procedure that allows one to assign to a manifestly infinite quantitiy a *finite* value by providing it a special value zeta interpretation. Such a procedure is therefore of enormous importance in physics, for example, where infinities are prolific. As a simple example, we consider the sum $S = 1 + 2 + 3 + 4 + \cdots = \sum_{n=1}^{\infty} n$, which is obviously infinite. This sum arises naturally in string theory — in the discussion of *transverse Virasoro operators*, for example. A *string* (which replaces the notion of a *particle* in quantum theory, at the Plancktian scale 10^{-33} cm) sweeps out a surface called a *world-sheet* as it moves in d-dimensional space-time $\mathbb{R}^d = \mathbb{R}^1 \times \mathbb{R}^{d-1}$ - in contrast to a world-line of a point-particle. For Bosonic string theory (where there are no *fermions*, but only *bosons*) certain Virasoro constraints force d to assume a specific value. Namely, the condition

$$1 = -\left(\frac{d-2}{2}\right) S \tag{7.1}$$

arises, which as we shall see forces the critical dimension $d = 26$, 1 being the value of a certion *normal ordering* constant. In fact, we write $S = \sum_{n=1}^{\infty} 1/n^s$, where $s = -1$, which means that it is natural to reinterpret S as the special zeta value $\zeta(-1)$. Thus we *zeta regularize* the infinite quantity S by assigning to it the value $-\frac{1}{12}$, according to (2.14). Then by condition (7.1), indeed we must have $d = 26$.

Interestingly enough, the "strange" equation

$$1 + 2 + 3 + 4 + \cdots = -\tfrac{1}{12}, \tag{7.2}$$

which we now understand to be perfectly meaningful, appears in a paper of Ramanujan — though he had no knowledge of the zeta function. It was initially dismissed, of course, as ridiculous and meaningless.

As another simple example we consider "$\infty!$", that is, the product $P = 1 \cdot 2 \cdot 3 \cdot 4 \cdots$, which is also infinite. To zeta regularize P, we consider first $\log P = \sum_{n=1}^{\infty} \log n$ (which is still infinite), and we note that since $\zeta'(s) =$

$-\sum_{n=1}^{\infty}(\log n)n^{-s}$ for $\operatorname{Re} s > 1$, by equation (1.1), if we *illegally* take $s = 0$ the *false* result $-\zeta'(0) = \sum_{n=1}^{\infty} \log n = \log P$ follows. However the left-hand side here is well-defined and in fact it has the value $\frac{1}{2}\log 2\pi$ by equation (2.6). The finite value $\frac{1}{2}\log 2\pi$, therefore, is naturally assigned to $\log P$ and, consequently, we define

$$P = \prod_{n=1}^{\infty} n = \infty! \stackrel{\text{def}}{=} e^{-\zeta'(0)} = e^{\frac{1}{2}\log 2\pi} = \sqrt{2\pi}. \qquad (7.3)$$

More complicated products can be regularized in a somewhat similar manner. A typical set-up for this is as follows. One has a compact smooth manifold M with a Riemannian metric g, and therefore a corresponding *Laplace–Beltrami operator* $\Delta = \Delta(g)$ where $-\Delta$ has a discrete spectrum

$$0 = \lambda_0 < \lambda_1 < \lambda_2 < \lambda_3 < \cdots, \quad \lim_{j \to \infty} \lambda_j = \infty. \qquad (7.4)$$

If n_j denotes the (finite) multiplicity of the j-th eigenvalue λ_j of $-\Delta$, then one can form the corresponding *spectral zeta function* (cf. definition (0.1))

$$\zeta_M(s) = \sum_{j=1}^{\infty} \frac{n_j}{\lambda_j^s}, \qquad (7.5)$$

which is well-defined for $\operatorname{Re} s > \frac{1}{2} \dim M$, due to the discovery by H. Weyl of the asymptotic result $\lambda_j \sim j^{2/\dim M}$, as $j \to \infty$. S. Minakshisundaram and A. Pleijel [26] showed that $\zeta_M(s)$ admits a meromorphic continuation to the complex plane and that, in particular, $\zeta_M(s)$ is holomorphic at $s = 0$. Thus $e^{-\zeta_M(0)}$ is well-defined and, as in definition (7.3), we set

$$\det{'} -\Delta = \prod_{j=1}^{\infty} \lambda_j^{n_j} \stackrel{\text{def}}{=} e^{-\zeta'_M(0)}, \qquad (7.6)$$

where the prime $'$ here indicates that the product of eigenvalues (which in finite dimensions corresponds to the determinant of an operator) is taken over the nonzero ones. Indeed, similar to the preceding example with the infinite product $P = \prod_{j=1}^{\infty} j$, the formal, illegal computation

$$\exp\left(-\frac{d}{ds}\sum_{j=1}^{\infty}\frac{n_j}{\lambda_j^s}\bigg|_{s=0}\right) = \exp\left(\sum_{j=1}^{\infty}\frac{n_j}{\lambda_j^s}\log\lambda_j\bigg|_{s=0}\right) = \exp\left(\sum_{j=1}^{\infty} n_j \log\lambda_j\right)$$

$$= \prod_{j=1}^{\infty} e^{n_j \log \lambda_j} = \prod_{j=1}^{\infty} \lambda_j^{n_j} \qquad (7.7)$$

serves as the motivation for definition (7.6). Clearly this definition of determinant makes sense for more general operators (with a discrete spectrum) on other infinite-dimensional spaces. It is useful, moreover, for Laplace-type operators on smooth sections of a vector bundle over M.

The following example is more involved, where M is a complex torus (in fact M is assumed to be the world-sheet of a bosonic string; see Appendix C of [42], for example). For a fixed complex number $\tau = \tau_1 + i\tau_2$ in the upper half-plane ($\tau_2 > 0$) and for the corresponding integral lattice,

$$L_\tau \stackrel{\text{def}}{=} \{a + b\tau \mid a, b \in \mathbb{Z}\}, \qquad M \stackrel{\text{def}}{=} \mathbb{C} \setminus L_\tau. \tag{7.8}$$

In this case it is known that $-\Delta$ has a multiplicity free spectrum (i.e., every $n_j = 1$) given by

$$\left\{ \lambda_{mn} \stackrel{\text{def}}{=} \frac{4\pi^2}{\tau_2^2} |m + n\tau|^2 \right\}_{m,n \in \mathbb{Z}},$$

and consequently the corresponding spectral zeta function of (7.5) is given by

$$\zeta_M(s) = \frac{\tau_2^s}{(4\pi^2)^s} E^*(s, \tau) \tag{7.9}$$

for $\operatorname{Re} s > 1$, where

$$E^*(s, \tau) \stackrel{\text{def}}{=} \sum_{(m,n) \in \mathbb{Z}_*^2} \frac{\tau_2^s}{|m + n\tau|^{2s}} \tag{7.10}$$

(with $\mathbb{Z}_*^2 = \mathbb{Z} \times \mathbb{Z} - \{(0,0)\}$ as before) is a standard *nonholomorphic Eisenstein series*. That is, in contrast to the series $G_k(\tau)$ in definition (4.4), $E^*(s, \tau)$ is not a holomorphic function of τ. As a function of s, it is a standard fact that $E^*(s, \tau)$, which is holomorphic for $\operatorname{Re} s > 1$, admits a meromorphic continuation to the full complex plane, with a simple pole at $s = 1$ as its only singularity. Hence, by equation (7.9), the same assertion holds for $\zeta_M(s)$. By [7; 14; 35; 38], for example, the continuation of $E^*(s, \tau)$ is given by

$$E^*(s, \tau) = 2\zeta(2s)\tau_2^s + 2\zeta(2s - 1)\sqrt{\pi} \frac{\Gamma(s - \frac{1}{2})}{\Gamma(s)} \tau_2^{-s+1}$$
$$+ \frac{4\pi^s}{\Gamma(s)} \tau_2^{1/2} \sum_{m=1}^\infty \sum_{n=1}^\infty e^{-2\pi i mn\tau_1} \left(\frac{n}{m}\right)^{s-\frac{1}{2}} K_{s-\frac{1}{2}}(2\pi mn\tau_2)$$
$$+ \frac{4\pi^s}{\Gamma(s)} \tau_2^{1/2} \sum_{m=1}^\infty \sum_{n=1}^\infty e^{2\pi i mn\tau_1} \left(\frac{n}{m}\right)^{s-\frac{1}{2}} K_{s-\frac{1}{2}}(2\pi mn\tau_2) \tag{7.11}$$

for $\operatorname{Re} s > 1$, where

$$K_\nu(x) \stackrel{\text{def}}{=} \frac{1}{2} \int_0^\infty \exp\left(-\frac{x}{2}\left(t + \frac{1}{t}\right)\right) t^{\nu-1} \, dt \tag{7.12}$$

is the Macdonald–Bessel (or K-Bessel) function for $v \in \mathbb{C}$, $x > 0$. Introducing the divisor function $\sigma_v(n) \overset{\text{def}}{=} \sum_{0<d,\, d|n} d^v$, and using forumla (D.11) on page 92 on the entire functions of s appearing in the last two sums, we can rewrite (7.11) as

$$= 2\zeta(2s)\tau_2^s + 2\zeta(2s-1)\sqrt{\pi}\frac{\Gamma(s-\tfrac{1}{2})}{\Gamma(s)}\tau_2^{-s+1}$$

$$+ \frac{4\pi^s}{\Gamma(s)}\tau_2^{1/2}\sum_{n=1}^{\infty}\sigma_{-2s+1}(n)e^{-2\pi n\tau_1 i}K_{s-\tfrac{1}{2}}(2\pi n\tau_2)n^{s-\tfrac{1}{2}}$$

$$+ \frac{4\pi^s}{\Gamma(s)}\tau_2^{1/2}\sum_{n=1}^{\infty}\sigma_{-2s+1}(n)e^{2\pi n\tau_1 i}K_{s-\tfrac{1}{2}}(2\pi n\tau_2)n^{s-\tfrac{1}{2}}. \quad (7.13)$$

The sum of the first two terms in (7.13) has $s = \tfrac{1}{2}$ as a removable singularity. To see this, note first that since $\zeta(s)$ has residue $= 1$ at $s = 1$,

$$\lim_{s \to \tfrac{1}{2}} (s - \tfrac{1}{2})\zeta(2s)\tau_2^s = \tfrac{1}{2}\lim_{z \to 1}(z-1)\zeta(z)\tau_2^{z/2}$$

(for $z = 2s$), and this equals $\tau_2^{1/2}/2$. We also have $\Gamma(1) = 1$, $\Gamma(\tfrac{1}{2}) = \sqrt{\pi}$, $\zeta(0) = -\tfrac{1}{2}$ (by (2.5)), and $w\Gamma(w) = \Gamma(w+1)$. Thus

$$\left(\tfrac{z-1}{2}\right)\Gamma\left(\tfrac{z-1}{2}\right) = \Gamma\left(\tfrac{z+1}{2}\right);$$

hence

$$\lim_{s \to \tfrac{1}{2}} (s - \tfrac{1}{2})\zeta(2s-1)\sqrt{\pi}\frac{\Gamma(s-\tfrac{1}{2})}{\Gamma(s)}\tau_2^{-s+1}$$

$$= \lim_{z \to 1}\zeta(z-1)\sqrt{\pi}\left(\tfrac{z-1}{2}\right)\frac{\Gamma\left(\tfrac{z-1}{2}\right)}{\Gamma\left(\tfrac{z}{2}\right)}\tau_2^{-z/2+1}$$

$$= \zeta(0)\frac{\sqrt{\pi}}{\Gamma(\tfrac{1}{2})}\lim_{z \to 1}\Gamma\left(\tfrac{z+1}{2}\right)\tau_2^{1/2} = -\tau_2^{1/2}/2.$$

It follows that the limit as $s \to \tfrac{1}{2}$ of $(s - \tfrac{1}{2}) \times$ the first two terms in (7.13) vanishes, as desired. This proves our claim that $s = 1$ is the only singularity of $E^*(s, \tau)$, which arises as a simple pole from the second term, $2\zeta(2s-1) \times \sqrt{\pi}\Gamma(s-\tfrac{1}{2})/\Gamma(s)\tau_2^{-s+1}$, in (7.13), due to the factor $\zeta(2s-1)$. Moreover the residue at $s = 1$ can be easily evaluated setting $z = 2s - 1$:

$$\lim_{s \to 1} (s-1) 2\zeta(2s-1) \sqrt{\pi}\, \Gamma(s-\tfrac{1}{2})/\Gamma(s) \tau_2^{-s+1}$$
$$= \lim_{z \to 1} (z-1)\zeta(z) \sqrt{\pi} \left(\Gamma\left(\tfrac{z}{2}\right)/\Gamma\left(\tfrac{z+1}{2}\right)\right) \tau_2^{-z/2+1/2}$$
$$= \sqrt{\pi}(\sqrt{\pi}/1) = \pi.$$

From (7.13) we also get $E^*(0, \tau) = 2\zeta(2s)\tau_2^s|_{s=0}$:

$$E^*(0, \tau) = -1. \tag{7.14}$$

Moreover, the functional equation

$$\frac{E^*(1-s, \tau)}{\Gamma(s)} = \frac{\pi^{1-2s} E^*(s, \tau)}{\Gamma(1-s)}, \tag{7.15}$$

say for $s \ne 0, 1$, follows since $\sigma_\nu(n)$, $K_\nu(x)$ satisfy the functional equations

$$\sigma_\nu(n) = n^\nu \sigma_{-\nu}(n), \quad K_\nu(x) = K_{-\nu}(x), \tag{7.16}$$

and since $\zeta(s)$ satisfies the functional equation (1.16). The second equation in (7.16) follows by the change of variables $u = 1/t$ in definition (7.12), and the first equation is the formula

$$\sigma_\nu(n) = n^\nu \sum_{\substack{0 < d \\ d \mid n}} \frac{1}{d^\nu},$$

which we checked in remarks following the statement of Theorem 4.23. Namely, $d > 0$ runs through the divisors of n as $\frac{n}{d}$ does. We check equation (7.15) by replacing s by $1-s$ in equation (7.13). By (7.16), $\sigma_{-2(1-s)+1}(n) n^{1-s-\frac{1}{2}} = \sigma_{-2s+1} n^{s-\frac{1}{2}}$ and $K_{1-s-\frac{1}{2}}(2\pi n \tau_2) = K_{s-\frac{1}{2}}(2\pi n \tau_2)$; hence

$$\frac{4\pi^{1-s}}{\Gamma(1-s)} \tau_2^{1/2} \sum_{n=1}^{\infty} \sigma_{-2(1-s)+1}(n) e^{\pm 2\pi n \tau_1 i} K_{1-s-\frac{1}{2}}(2\pi n \tau_2) n^{1-s-\frac{1}{2}} =$$

$$\frac{4\pi^{1-s}}{\Gamma(1-s)} \tau_2^{1/2} \sum_{n=1}^{\infty} \sigma_{-2s+1}(n) e^{\pm 2\pi n \tau_1 i} K_{s-\frac{1}{2}}(2\pi n \tau_2) n^{s-\frac{1}{2}}. \tag{7.17}$$

In equation (1.16) replace s by $2s$ and $2s - 1$ separately, to obtain

$$\zeta(1-2s) = \frac{\pi^{-2s+\frac{1}{2}} \Gamma(s) \zeta(2s)}{\Gamma(\frac{1}{2}-s)},$$

$$\zeta(2-2s) = \zeta(1-(2s-1)) = \pi^{1/2} \pi^{1-2s} \frac{\Gamma(s-\frac{1}{2}) \zeta(2s-1)}{\Gamma(1-s)}$$

say for $2s, 2s - 1 \neq 0, 1$ (i.e., $s = 0, \frac{1}{2}, 1$). This gives

$$2\zeta(2(1-s))\tau_2^{1-s} + 2\zeta(2(1-s)-1)\sqrt{\pi}\frac{\Gamma(1-s-\frac{1}{2})}{\Gamma(1-s)}\tau_2^{-(1-s)+1}$$

$$= 2\sqrt{\pi}\pi^{1-2s}\frac{\Gamma(s-\frac{1}{2})\zeta(2s-1)}{\Gamma(1-s)}\tau_2^{-s+1}$$

$$+ \frac{2\pi^{-2s+\frac{1}{2}}\Gamma(s)\zeta(2s)\sqrt{\pi}\Gamma(\frac{1}{2}-s)}{\Gamma(\frac{1}{2}-s)\Gamma(1-s)}\tau_2^s. \quad (7.18)$$

By (7.13), (7.17), (7.18), we have for $\dfrac{E^*(1-s,\tau)}{\Gamma(s)}$ the value

$$\frac{\pi^{1-2s}}{\Gamma(1-s)}\left(2\sqrt{\pi}\frac{\Gamma(s-\frac{1}{2})}{\Gamma(s)}\zeta(2s-1)\tau_2^{-s+1} + 2\zeta(2s)\tau_2^s\right)$$

$$+ \frac{\pi^{1-2s}}{\Gamma(1-s)}\left(\frac{4\pi^s}{\Gamma(s)}\tau_2^{1/2}\sum_{n=1}^{\infty}\sigma_{-2s+1}(n)e^{-2\pi n\tau_1 i}K_{s-\frac{1}{2}}(2\pi n\tau_2)n^{s-\frac{1}{2}}\right.$$

$$\left. + \frac{4\pi^s}{\Gamma(s)}\tau_2^{1/2}\sum_{n=1}^{\infty}\sigma_{-2s+1}(n)e^{2\pi n\tau_1 i}K_{s-\frac{1}{2}}(2\pi n\tau_2)n^{s-\frac{1}{2}}\right)$$

$$= \frac{\pi^{1-2s}}{\Gamma(1-s)}E^*(s,\tau)$$

which gives (7.15), as desired. (Note that we have stayed away from $s = 0, \frac{1}{2}, 1$; but equation (7.15) clearly holds for $s = \frac{1}{2}$.)

One other result is needed in order to compute $\zeta'_M(0)$:

THEOREM 7.19 (KRONECKER'S FIRST LIMIT FORMULA).

$$\lim_{s \to 1}\left(E^*(s,\tau) - \frac{\pi}{s-1}\right) = 2\pi\left(\gamma - \log 2 - \log \tau_2^{1/2}|\eta(\tau)|^2\right). \quad (7.20)$$

where γ is the Euler–Mascheroni constant in definition (1.26), and where $\eta(\tau)$ is the Dedekind eta function in definition (3.27).

Formula (7.20) compares with the limit result (1.25), though it is a more involved result; see [35].

For $f(z) \stackrel{\text{def}}{=} \Gamma(z)\pi^{2-2z}/\Gamma(2-z)$

$$f'(1) = -2\log\pi + 2\Gamma'(1) = -2\log\pi - 2\gamma \quad (7.21)$$

by the quotient rule. By equations (7.14), (7.15), (7.21) and Theorem 7.19, and the fact that $(1-z)\Gamma(1-z) = \Gamma(2-z)$ (i.e., $w\Gamma(w) = \Gamma(w+1)$), we have, with $z = 1 - s$,

$$\frac{\partial E^*}{\partial s}(0, \tau) \stackrel{\text{def}}{=} \lim_{s \to 0} \frac{E^*(s, \tau) + 1}{s} = \lim_{z \to 1} \frac{E^*(1 - z, \tau) + 1}{1 - z}$$

$$= \lim_{z \to 1} \left(\frac{E^*(1-z,\tau)\Gamma(z)}{\Gamma(z)(1-z)} - \frac{1}{z-1} \right)$$

$$= \lim_{z \to 1} \left(\frac{\Gamma(z)}{(1-z)} \pi^{1-2z} \frac{E^*(z,\tau)}{\Gamma(1-z)} - \frac{1}{z-1} \right)$$

$$= \lim_{z \to 1} \left(\frac{\Gamma(z)\pi^{1-2z}}{\Gamma(2-z)} \left(E^*(z,\tau) - \frac{\pi}{z-1} \right) + \frac{\Gamma(z)\pi^{1-2z}}{\Gamma(2-z)} \frac{\pi}{z-1} - \frac{1}{z-1} \right)$$

$$\stackrel{\text{def}}{=} 2[\gamma - \log 2 - \log \tau_2^{1/2} |\eta(\tau)|^2] + \left[\lim_{z \to 1} \frac{f(z) - f(1)}{z - 1} = f'(1) \right]$$

$$= -2 \log 2 - 2 \log \tau_2^{1/2} |\eta(\tau)|^2 - 2 \log \pi$$

$$= -\log 4\pi^2 \tau_2 |\eta(\tau)|^4,$$

which, with equations (7.9), (7.14) gives (finally)

$$\zeta'_M(0) = -\log \tau_2^2 |\eta(\tau)|^4. \tag{7.22}$$

Hence, by definition (7.6), the regularized determinant is given by

$$\mathrm{det}' -\Delta = \tau_2^2 |\eta(\tau)|^4. \tag{7.23}$$

One is actually interested in the power $(\mathrm{det}' -\Delta)^{-d/2} = \tau_2^{-d} |\eta(\tau)|^{-2d}$, where $d = 26$ is the critical dimension mentioned above. This power represents a *one-loop* contribution to the "sum of embeddings" of the string world sheet (the complex torus in definition (7.8)) into the target space $\mathbb{R}^{26}$.

In the next lecture we shall make use, similarly, of the meromorphic continuation of the generalized Epstein zeta function

$$E(s, m; \vec{a}, \vec{b}) \stackrel{\text{def}}{=} \sum_{\vec{n}=(n_1,n_2,\ldots,n_d) \in \mathbb{Z}^d} (a_1(n_1-b_1)^2 + \cdots + a_d(n_d-b_d)^2 + m^2)^{-s} \tag{7.24}$$

for $\mathrm{Re}\, s > d/2$, $m > 0$, $\vec{a} = (a_1, \ldots, a_d)$, $\vec{b} = (b_1, \ldots, b_d) \in \mathbb{R}^d$, $a_i > 0$. The result is, setting $\mathbb{Z}^d_* = \mathbb{Z}^d - \{0\}$,

$$E(s, m; \vec{a}, \vec{b}) = \frac{\pi^{d/2} \Gamma(s - \frac{d}{2}) m^{d-2s}}{\sqrt{a_1 a_2 \cdots a_d}\, \Gamma(s)} + \frac{2\pi^s m^{d/2-s}}{\sqrt{a_1 a_2 \cdots a_d}\, \Gamma(s)}$$

$$\times \sum_{\vec{n} \in \mathbb{Z}^d_*} e^{2\pi i \sum_{j=1}^d n_j b_j} \left(\sum_{j=1}^d \frac{n_j^2}{a_j} \right)^{\frac{s-d/2}{2}} K_{\frac{d}{2}-s}\left(2\pi m \left(\sum_{j=1}^d \frac{n_j^2}{a_j}\right)^{\frac{1}{2}}\right) \tag{7.25}$$

for Re $s > d/2$. In particular, we shall need the special value $E(\frac{d+1}{2}, m; \vec{a}, \vec{b})$, which we now compute. For $s = \frac{d+1}{2}$ the first term on the right in (7.25) is

$$\frac{\pi^{\frac{d+1}{2}}}{m(\prod_{j=1}^{d} a_j)^{1/2} \Gamma(\frac{d+1}{2})},$$

since $\Gamma(\frac{1}{2}) = \pi^{1/2}$. Also

$$K_{\frac{d}{2}-s}(x) = K_{-\frac{1}{2}}(x) = K_{\frac{1}{2}}(x),$$

by (7.16). This equals $\sqrt{\pi/2x}\, e^{-x}$ for $x > 0$; hence

$$\left(\sum_{j=1}^{d} \frac{n_j^2}{a_j}\right)^{\frac{s-d/2}{2}} K_{\frac{d}{2}-s}\left(2\pi m \left(\sum_{j=1}^{d} \frac{n_j^2}{a_j}\right)^{\frac{1}{2}}\right)$$

$$= \left(\sum_{j=1}^{d} \frac{n_j^2}{a_j}\right)^{1/4} \sqrt{\frac{\pi}{2}} (2\pi m)^{-\frac{1}{2}} \left(\sum_{j=1}^{d} \frac{n_j^2}{a_j}\right)^{-\frac{1}{4}} \exp\left(-2\pi m \left(\sum_{j=1}^{d} \frac{n_j^2}{a_j}\right)^{\frac{1}{2}}\right).$$

Therefore the second term on the right-hand side of (7.25) is

$$\frac{2\pi^{\frac{d+1}{2}} m^{-1/2}}{(\prod_{j=1}^{d} a_j)^{1/2} \Gamma(\frac{d+1}{2})} \sum_{\vec{n} \in \mathbb{Z}_*^d} \frac{1}{2\sqrt{m}} \exp\left(2\pi i \sum_{j=1}^{d} n_j b_j\right) \exp\left(-2\pi m \left(\sum_{j=1}^{d} \frac{n_j^2}{a_j}\right)^{\frac{1}{2}}\right).$$

That is:

$$E\left(\frac{d+1}{2}, m; \vec{a}, \vec{b}\right) = \frac{\pi^{\frac{d+1}{2}}}{m(\prod_{j=1}^{d} a_j)^{1/2} \Gamma(\frac{d+1}{2})}$$

$$+ \frac{\pi^{\frac{d+1}{2}}}{m(\prod_{j=1}^{d} a_j)^{1/2} \Gamma(\frac{d+1}{2})} \sum_{\vec{n} \in \mathbb{Z}_*^d} e^{2\pi i \vec{n} \cdot \vec{b}} \exp\left(-2\pi m \left(\sum_{j=1}^{d} \frac{n_j^2}{a_j}\right)^{\frac{1}{2}}\right)$$

$$= \frac{\pi^{\frac{d+1}{2}}}{m(\prod_{j=1}^{d} a_j)^{1/2} \Gamma(\frac{d+1}{2})} \sum_{\vec{n} \in \mathbb{Z}^d} e^{2\pi i \vec{n} \cdot \vec{b}} \exp\left(-2\pi m \left(\sum_{j=1}^{d} \frac{n_j^2}{a_j}\right)^{\frac{1}{2}}\right). \quad (7.26)$$

As a final example we consider the zeta regularization of Casimir energy, after a few general remarks.

In Lecture 6 it was observed that the sequence $\{E_n \stackrel{\text{def}}{=} (n + \frac{1}{2})h\nu\}$ is the sequence of energy levels of the quantized harmonic oscillator of frequency ν, where h denotes Planck's constant. In particular there exists a nonvanishing ground state energy (also called the *zero-point energy*) given by $E_0 = \frac{1}{2}h\nu$. Zero-point energy is a prevalent notion in physics, from quantum field theory (QFT), where it is also referred to as *vacuum energy*, to cosmology (concerning issues regarding the cosmological constant, for example), and in between. Based on Planck's radiation density formula (6.18), A. Einstein and O. Stern concluded (in 1913) that even at zero absolute temperature, atomic systems maintain an energy of the amount $E_0 = \frac{1}{2}h\nu$. It is quite well experimentally established that a vacuum (empty space) contains a large supply of residual energy (zero-point energy). Vacuum fluctuations is a large scale study. The energy due to vacuum distortion (*Casimir energy*), for example, was considered by H. Casimir and D. Polder in a 1948 ground-breaking study. Here the vacuum energy was modified by the introduction of a pair of uncharged, parallel, conducting metal plates. A striking prediction emerged: the prediction of the existence of a force of a purely quantum mechanical origin — one arising from zero-point energy changes of harmonic oscillators that make up the normal modes of the electromagnetic field. This force, which has now been measured experimentally by M. Spaarnay, S. Lamoreaux, and others, is called the *Casimir force*.

Casimir energy in various contexts has been computed by many Physicists, including some notable calculations by the co-editor Klaus Kirsten. We refer to his book [22] for much more information on this, and on related matters - a book with 424 references. The author has used the *Selberg trace formula* for general compact space forms $\Gamma \backslash G/K$, mentioned in Lecture 6, of rank-one symmetric spaces G/K to compute the Casimir energy in terms of the *Selberg zeta function* [40; 39]. This was done by Kirsten and others in some special cases.

Consider again a compact smooth Riemannian manifold (M, g) with discrete spectrum of its Laplacian $-\Delta(g)$ given by (7.4). In practice, M is the spatial sector of a space-time manifold $\mathbb{R} \times M$ with metric $-dt^2 + g$. Formally, the Casimir energy in this context is given by the infinite quantity

$$E_C = \frac{1}{2}\sum_{j=1}^{\infty} n_j \lambda_j^{1/2}, \qquad (7.27)$$

up to some omitted factors like h. It is quite clear then how to regularize E_C. Namely, consider $\lambda_j^{1/2}$ as $1/\lambda_j^s$ for $s = -\frac{1}{2}$ and therefore assign to E_C the meaning

$$E_C = \tfrac{1}{2}\zeta_M(-1/2), \qquad (7.28)$$

where $\zeta_M(s)$ is the spectral zeta function of definition (7.5), meromorphically continued. If dim M is even, for example, the poles of $\zeta_M(s)$ are simple, finite in number, and can occur only at one of points $s = 1, 2, 3, \ldots, d/2$ (see [26]), in which case E_C in (7.28) is surely a well-defined, finite quantity. However if dim M is odd, $\zeta_M(s)$ will generally have infinitely many simple poles — at the points $s = \frac{1}{2} \dim M - n$, for $0 \leq n \in \mathbb{Z}$. This would include the point $s = -\frac{1}{2}$ if dim $M = 5$ and $n = 3$, for example. Assume therefore that dim M is even.

When M is one of the above compact space forms, for example, then (based on the results in [41]) E_C can be expressed explicitly in terms of the structure of Γ and the spherical harmonic analysis of G/K — and in terms of the Selberg zeta function attached to $\Gamma \backslash G/K$, as just mentioned. Details of this are a bit too technical to mention here; we have already listed some references. We point out only that by our assumptions on M, the corresponding Lie group pairs (G, K) are given by

$$G = \mathrm{SO}_1(m, 1), \quad K = \mathrm{SO}(m), \quad m \geq 2,$$
$$G = \mathrm{SU}(m, 1), \quad K = U(m), \quad m \geq 2,$$
$$G = \mathrm{SP}(m, 1), \quad K = \mathrm{SP}(m) \times \mathrm{SP}(1), \quad m \geq 2,$$
$$G = F_{4(-20)}, \quad K = \mathrm{Spin}(9),$$

where $F_{4(-20)}$ is a real form of the complex Lie group with exceptional Lie algebra F_4 with Dynkin diagram 0—0 = 0—0. More specifically, $F_{4(-20)}$ is the unique real form for which the difference $\dim G/K - \dim K$ assumes the value -20.

In addition to the reference [22], the books [13; 12] are a good source for information on and examples of Casimir energy, and for applications in general of zeta regularization.

Lecture 8. Epstein zeta meets gravity in extra dimensions

We compute the Kaluza–Klein modes of the 4-dimensional gravitational potential V_{4+d} in the presence of d extra dimensions compactified on a d-torus. The result is known of course [3; 21], but we present here an argument based on the special value $E\left(\frac{d+1}{2}, m; \vec{a}, \vec{b}\right)$ computed in equation (7.26) of the generalized Epstein zeta function $E(s, m; \vec{a}, \vec{b})$ defined in (7.24).

G. Nordström in 1914 and T. Kaluza (independently) in 1921 were the first to unify Einstein's 4-dimensional theory of gravity with Maxwell's theory of electromagnetism. They showed that 5-dimensional general relativity contained both theories, but under an assumption that was somewhat artificial - the so-called "cylinder condition" that in essence restricted physicality of the fifth dimension. O. Klein's idea was to compactify that dimension and thus to render a plausible physical basis for the cylinder assumption.

Consider, for example, the fifth dimension (the "extra dimension") compactified on a circle Γ. This means that instead of considering the Einstein gravitational field equations

$$R_{ij}(g) - \frac{g_{ij}}{2} R(g) - \Lambda g_{ij} = -\frac{8\pi G}{c^4} T_{ij} \qquad (8.1)$$

on a 4-dimensional space-time M^4 [8; 11], one considers these equations on the 5-dimensional product $M^4 \times \Gamma$. In (8.1), $g = [g_{ij}]$ is a Riemannian metric (the solution of the Einstein equations) with Ricci tensor $R_{ij}(g)$ and scalar curvature $R(g)$, Λ is a cosmological constant, and T_{ij} is an energy momentum tensor which describes the matter content of space-time — the left-hand side of (8.1) being pure geometry; G is the Newton constant and c is the speed of light. Given the non-observability of the fifth dimension, however, one takes Γ to be extremely small, say with an extremely small radius $R > 0$. Geometrically we have a fiber bundle $M^4 \times \Gamma \to M^4$ with structure group Γ.

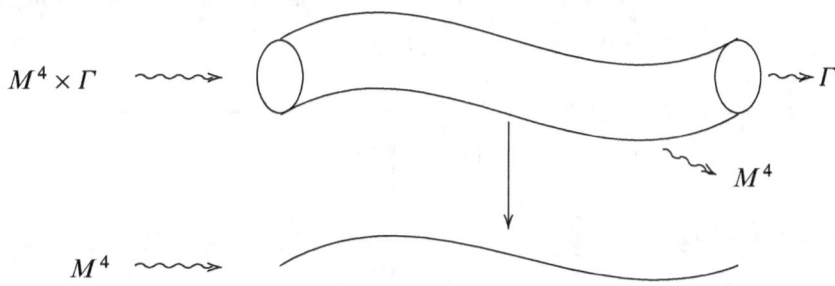

On all "fields" $F(x, \theta) : M^4 \times \mathbb{R} \to \mathbb{C}$ on $M^4 \times \mathbb{R}$ there is imposed, moreover, periodicity in the second variable:

$$F(x, \theta + 2\pi R) = F(x, \theta) \qquad (8.2)$$

for $(x, \theta) \in M^4 \times \mathbb{R}$.

For $n \in \mathbb{Z}$ and $f(x)$ on M^4 fixed, the function $F_{n,f}(x, \theta) \stackrel{\text{def}}{=} f(x) e^{in\theta/R}$ is an example of a field on $M^4 \times \mathbb{R}$ that satisfies equation (8.2). For a general field $F(x, \theta)$, subject to reasonable conditions, and the periodicity condition (8.2), one would have a Fourier series expansion

$$F(x, \theta) = \sum_{n \in \mathbb{Z}} F_{n, f_n} \stackrel{\text{def}}{=} \sum_{n \in \mathbb{Z}} f_n(x) e^{in\theta/R} \qquad (8.3)$$

in which case the functions $f_n(x)$ are called *Kaluza–Klein modes* of $F(x, \theta)$.

Next we consider d extra dimensions compactified on a d-torus

$$\Gamma^d \stackrel{\text{def}}{=} \Gamma_1 \times \cdots \times \Gamma_d,$$

where the Γ_j are circles with extremely small radii $R_j > 0$, and we consider a field $V_{4+d}(x, y, z, x_1, \ldots, x_d)$ on $(\mathbb{R}^3 - \{0\}) \times \mathbb{R}^d$. Thus $\mathbb{R}^3 - \{0\}$ replaces M^4,

Γ^d replaces Γ, and $\vec{x} \stackrel{\text{def}}{=} (x_1, \ldots, x_d) \in \mathbb{R}^d$ replaces θ in the previous discussion. The field is given by

$$V_{4+d}(x, y, z, x_1, \ldots, x_d) \stackrel{\text{def}}{=}$$

$$-MG_{4+d} \sum_{\vec{n}=(n_1,\ldots,n_d) \in \mathbb{Z}^d} \frac{1}{\left(r^2 + \sum_{j=1}^{d}(x_j - 2\pi n_j R_j)^2\right)^{\frac{d+1}{2}}}, \quad (8.4)$$

which is the gravitational potential due to extra dimensions of an object of mass M at a distance $\left(r^2 + \sum_{j=1}^{d} x_j^2\right)^{1/2}$ for $r^2 \stackrel{\text{def}}{=} x^2 + y^2 + z^2$; here G_{4+d} is the $(4+d)$-dimensional Newton constant. Note that, analogously to equation (8.2),

$$V_{4+d}(x, y, z, x_1 + 2\pi R_1, \ldots, x_d + 2\pi R_d)$$

$$\stackrel{\text{def}}{=} -MG_{4+d} \sum_{(n_1,\ldots,n_d) \in \mathbb{Z}^d} \frac{1}{\left(r^2 + \sum_{j=1}^{d}(x_j - 2\pi(n_-1)jR_j)^2\right)^{\frac{d+1}{2}}}$$

$$= -MG_{4+d} \sum_{(n_1,\ldots,n_d) \in \mathbb{Z}^d} \frac{1}{\left(r^2 + \sum_{j=1}^{d}(x_j - 2\pi n_j R_j)^2\right)^{\frac{d+1}{2}}} \quad (8.5)$$

$$= V_{4+d}(x, y, z, x_1, \ldots, x_d),$$

where of course we have used that $n_j - 1$ varies over $\mathbb{Z}$ as n_j does. Thus, analogously to equation (8.3), we look for a Fourier series expansion

$$V_{4+d}(x, y, z, x_1, \ldots, x_d) = \sum_{\vec{n} \in \mathbb{Z}^d} f_{\vec{n}}(x, y, z) \exp\left(i\vec{n} \cdot \overrightarrow{\left(\frac{x_1}{R_1}, \ldots, \frac{x_d}{R_d}\right)}\right), \quad (8.6)$$

where the functions $f_{\vec{n}}(x, y, z)$ on $\mathbb{R}^3 - \{0\}$ would be called the Kaluza–Klein modes of $V_{4+d}(x, y, z, \vec{x} = (x_1, \ldots, x_d))$.

It is easy, in fact, to establish the expansion (8.6) and to compute the modes $f_{\vec{n}}(x, y, z)$ explicitly. For this, define

$$a_j \stackrel{\text{def}}{=} (2\pi R_j)^2 > 0, \qquad b_j \stackrel{\text{def}}{=} \frac{x_j}{2\pi R_j},$$

and note that since $(x_j - 2\pi n_j R_j)^2 = \left(2\pi R_j \left(\frac{x_j}{2\pi R_j} - n_j\right)\right)^2 = a_j(b_j - n_j)^2$ we can write, by definition (8.4),

$$V_{4+d}(x, y, z, \vec{x}) = -MG_{4+d} \sum_{\vec{n} \in \mathbb{Z}^d} \frac{1}{\left(\sum_{j=1}^{d} a_j(n_j - b_j)^2 + r^2\right)^{\frac{d+1}{2}}}$$

$$= -MG_{4+d} E\left(\frac{d+1}{2}, r; \vec{a}, \vec{b}\right), \quad (8.7)$$

LECTURES ON ZETA FUNCTIONS, L-FUNCTIONS AND MODULAR FORMS 69

by definition (7.24). Thus we are in a pleasant position to apply formula (7.26): For

$$\Sigma_d \stackrel{\text{def}}{=} (2\pi)^d \prod_{j=1}^d R_j, \quad \Omega_d \stackrel{\text{def}}{=} \frac{2\pi^{\frac{d+1}{2}}}{\Gamma\left(\frac{d+1}{2}\right)} \tag{8.8}$$

we have $\left(\prod_{j=1}^d a_j\right)^{1/2} = \Sigma_d$, and we see that

$$V_{4+d}(x, y, z, \vec{x}) =$$

$$-\frac{MG_{4+d}\Omega_d}{2r\Sigma_d} \sum_{\vec{n} \in \mathbb{Z}^d} \exp\left(i\vec{n} \cdot \overrightarrow{\left(\frac{x_1}{R_1}, \ldots, \frac{x_d}{R_d}\right)}\right) \exp\left(-r\left(\sum_{j=1}^d \frac{n_j^2}{R_j^2}\right)^{\frac{1}{2}}\right) \tag{8.9}$$

by definition of a_j and b_j, for $r^2 \stackrel{\text{def}}{=} x^2 + y^2 + z^2$. This proves the Fourier series expansion (8.6), where we see that the Kaluza–Klein modes $f_{\vec{n}}(x, y, z)$ are in fact given by

$$f_{\vec{n}}(x, y, z) = -\frac{MG_{4+d}\Omega_d}{2r\Sigma_d} \exp\left(-r\left(\sum_{j=1}^d \frac{n_j^2}{R_j^2}\right)^{\frac{1}{2}}\right) \tag{8.10}$$

for $\vec{n} = (n_1, \ldots, n_d) \in \mathbb{Z}^d, (x, y, z) \in \mathbb{R}^3 - \{0\}$. Since $V_{4+d}(x, y, z, \vec{x})$ is actually real-valued, we write equation (8.9) as

$$V_{4+d}(x, y, z, x_1, \ldots, x_d) =$$

$$-\frac{MG_{4+d}\Omega_d}{2r\Sigma_d} \sum_{\vec{n} \in \mathbb{Z}^d} \exp\left(-r\left(\sum_{j=1}^d \frac{n_j^2}{R_j^2}\right)^{\frac{1}{2}}\right) \cos\left(\vec{n} \cdot \overrightarrow{\left(\frac{x_1}{R_1}, \ldots, \frac{x_d}{R_d}\right)}\right). \tag{8.11}$$

Since $2\pi R_i$ is the length of Γ_i, $\Gamma^d \stackrel{\text{def}}{=} \prod_{i=1}^d \Gamma_i$ has volume $\prod_{i=1}^d 2\pi R_i = (2\pi)^d \prod_{i=1}^d R_i$. That is, Σ_d in definition (8.8) (or in formula (8.11)) is the volume of the compactifying d-torus Γ^d. Similarly Ω_d in (8.8) or in (8.11), one knows, is the surface area of the unit sphere $\{x \in \mathbb{R}^{d+1} \mid \|x\| = 1\}$ in $\mathbb{R}^{d+1}$.

In [21], for example, the choice $x_1 = \cdots = x_d = 0$ is made. Going back to the compactification on a circle, $d = 1$, $R_1 = R$ for example, we can write the sum in (8.11) as

$$1 + \sum_{n \in \mathbb{Z} - \{0\}} e^{-r|n|/R} = 1 + 2 \sum_{n=1}^{\infty} (e^{-r/R})^n$$

$$= 1 + \frac{2e^{-r/R}}{1 - e^{-r/R}} \simeq 1 + 2e^{-r/R} \tag{8.12}$$

for $x_1 = 0$, where we keep in mind that R is extremely small. Thus in (8.12), r/R is extremely large; i.e., $e^{-r/R}$ is extremely small. For

$$K_d \stackrel{\text{def}}{=} MG_{4+d}\Omega_d/2\Sigma_d,$$

we get by (8.11) and (8.12)

$$V_5(x,y,z,x_1) \simeq -\frac{K_1}{r}(1+2e^{-r/R}), \qquad (8.13)$$

which is a correction to the Newtonian potential $V = -K_1/r$ due to an extra dimension.

The approximation (8.13) compares with the general deviations from the Newtonian inverse square law that are known to assume the form

$$V = -\frac{K}{r}(1+\alpha e^{-r/\lambda})$$

for suitable parameters α, λ. Apart from the toroidal compactification that we have considered, other compactifications are important as well [21] — especially Calabi–Yau compactifications. Thus the d-torus Γ^d is replaced by a Calabi–Yau manifold — a compact Kähler manifold whose first Chern class is zero.

Lecture 9. Modular forms of nonpositive weight, the entropy of a zero weight form, and an abstract Cardy formula

A famous formula of John Cardy [9] computes the asymptotic density of states $\rho(L_0)$ (the number of states at level L_0) for a general two-dimensional conformal field theory (CFT): For the holomorphic sector

$$\rho(L_0) = e^{2\pi\sqrt{cL_0/6}}, \qquad (9.1)$$

where the Hilbert space of the theory carries a representation of the Virasoro algebra Vir with generators $\{L_n\}_{n\in\mathbb{Z}}$ and central charge c. Vir has Lie algebra structure given by the usual commutation rule

$$[L_n, L_m] = (n-m)L_{n+m} + \frac{c}{12}n(n^2-1)\delta_{n+m,0} \qquad (9.2)$$

for $n, m \in \mathbb{Z}$. The CFT entropy S is given by

$$S = \log \rho(H_0) = 2\pi\sqrt{\frac{cL_0}{6}}. \qquad (9.3)$$

From the Cardy formula one can derive, for example, the Bekenstein–Hawking formula for BTZ black hole entropy [10]; see also my Speaker's Lecture presented later. More generally, the entropy of black holes in string theory can be derived — the derivation being statistical in nature, and microscopically based [34].

For a CFT on the two-torus with partition function

$$Z(\tau) = trace\, e^{2\pi i(L_0 - \frac{c}{24})\tau} \qquad (9.4)$$

on the upper half-plane π^+ [5], the entropy S can be obtained as follows. Regarding $Z(\tau)$ as a modular form with Fourier expansion

$$Z(\tau) = \sum_{n \geq 0} c_n e^{2\pi i(n-c/24)\tau}, \qquad (9.5)$$

one takes

$$S = \log c_n \qquad (9.6)$$

for large n. In [5], for example, (also see [4]) the Rademacher–Zuckerman exact formula for c_n is applied, where $Z(\tau)$ is assumed to be modular of weight $w = 0$. This is problematic however since the proof of that exact formula works only for modular forms of *negative* weight. In this lecture we indicate how to resolve this contradiction (thanks to some nice work of N. Brisebarre and G. Philibert), and we present what we call an abstract Cardy formula (with logarithmic correction) for holomorphic modular forms of zero weight. In particular we formulate, abstractly, the sub-leading corrections to Bekenstein–Hawking entropy that appear in formula (14) of [5].

The discussion in Lecture 4 was confined to holomorphic modular forms of non-negative integral weight. We consider now forms of *negative* weight $w = -r$ for $r > 0$, where r need not be an integer. The prototypic example will be the function $F_0(z) \stackrel{\text{def}}{=} 1/\eta(z)$, where $\eta(z)$ is the Dedekind eta function defined in (3.27), and where it will turn out that $w = -\frac{1}{2}$. We will use, in fact, the basic properties of $F_0(z)$ to serve as motivation for the general definition of a form of negative weight.

We begin with the *partition function* $p(n)$ on $\mathbb{Z}^+$. For n a positive integer, define $p(n)$ as the number of ways of writing n as an (orderless) sum of positive integers. For example, 3 is expressible as $3 = 1 + 2 = 1 + 1 + 1$, so $p(3) = 3$; $4 = 4 = 1 + 3 = 2 + 2 = 1 + 1 + 2 = 1 + 1 + 1 + 1$, so $p(4) = 5$;

$$5 = 5 = 2 + 3 = 1 + 4 = 1 + 1 + 3 = 1 + 2 + 2 = 1 + 1 + 1 + 2 = 1 + 1 + 1 + 1 + 1,$$

so $p(5) = 7$; similarly $p(2) = 2$, $p(1) = 1$. We set $p(0) \stackrel{\text{def}}{=} 1$. Clearly $p(n)$ grows quite quickly with n. A precise asymptotic formula for $p(n)$ was found by G. Hardy and S. Ramanujan in 1918, and independently by J. Uspensky in 1920:

$$p(n) \sim \frac{e^{\pi \sqrt{2n/3}}}{4n\sqrt{3}} \quad \text{as } n \to \infty \qquad (9.7)$$

(the notation means that the ratio between the two sides of the relation (9.7) tends to 1 as $n \to \infty$). For example, it is known that

$$p(1000) = 24{,}061{,}467{,}864{,}032{,}622{,}473{,}692{,}149{,}727{,}991$$
$$\simeq 2.4061 \times 10^{31}, \qquad (9.8)$$

whereas for $n = 1000$ in (9.7)

$$e^{\pi\sqrt{\frac{2n}{3}}}/4n\sqrt{3} \simeq 2.4402 \times 10^{31}, \tag{9.9}$$

which shows that the asymptotic formula is quite good.

L. Euler found the generating function for $p(n)$. Namely, he showed that

$$\sum_{n=0}^{\infty} p(n)z^n = \frac{1}{\prod_{n=1}^{\infty}(1-z^n)} \tag{9.10}$$

for $z \in \mathbb{C}$ with $|z| < 1$. By this formula and the definition of $\eta(z)$, we see immediately that

$$F_0(z) \stackrel{\text{def}}{=} \frac{1}{\eta(z)} = e^{-\pi i \tau/12} \sum_{n=0}^{\infty} p(n)e^{2\pi i n \tau} \tag{9.11}$$

on π^+.

The following profound result is due to R. Dedekind. To prepare the ground, for $x \in \mathbb{R}$ define

$$((x)) \stackrel{\text{def}}{=} \begin{cases} x - [x] - \frac{1}{2} & \text{if } x \notin \mathbb{Z}, \\ 0 & \text{if } x \in \mathbb{Z}, \end{cases}$$

where, as before, $[x]$ denotes the largest integer not exceeding x.

THEOREM 9.12. *Fix* $\gamma = \begin{bmatrix} a & b \\ c & d \end{bmatrix} \in \Gamma \stackrel{\text{def}}{=} SL(2, \mathbb{Z})$, *with* $c > 0$, *and define*

$$S(\gamma) = \frac{a+d}{12c} - \frac{1}{4} - s(d, c),$$

where $s(d, c)$ *(called a Dedekind sum) is given by*

$$s(d, c) \stackrel{\text{def}}{=} \sum_{\mu \in \mathbb{Z}/c\mathbb{Z}} \left(\!\left(\frac{\mu}{c}\right)\!\right)\left(\!\left(\frac{d\mu}{c}\right)\!\right). \tag{9.13}$$

Then, for $z \in \pi^+$,

$$F_0(\gamma \cdot z) = e^{-i\pi(S(\gamma)+\frac{1}{4})}(-i(cz+d))^{-\frac{1}{2}} F_0(z) \tag{9.14}$$

for $-\pi/2 < \arg(-i(cz+d)) < \pi/2$, *where* $\gamma \cdot z$ *is defined in equation* (4.3).

The sum in definition (9.13) is over a complete set of coset representatives μ in $\mathbb{Z}$. The case $c = 0$ is much less profound; then

$$\gamma = \begin{bmatrix} \pm 1 & b \\ 0 & \pm 1 \end{bmatrix}$$

(since $1 = \det \gamma = ad$), and

$$F_0(z \pm b) = F_0(\gamma \cdot z) = e^{\mp \pi i b/12} F_0(z). \tag{9.15}$$

In particular we can write $F_0(z+1) = e^{-\pi i/12} F_0(z) = e^{-2\pi i/24 + 2\pi i} F_0(z) = e^{2\pi i(1-\frac{1}{24})} F_0(z) = e^{2\pi i \alpha} F_0(z)$ for $\alpha \stackrel{\text{def}}{=} 1 - \frac{1}{24} = \frac{23}{24}$.

In summary, $F_0(z) = 1/\eta(z)$ satisfies the following conditions:

(i) $F_0(z)$ is holomorphic on π^+. (This follows from Lecture 3).)
(ii) $F_0(z+1) = e^{2\pi i \alpha} F_0(z)$ for some real $\alpha \in [0,1)$ (indeed, with $\alpha = \frac{23}{24}$).
(iii) $F_0(\gamma \cdot z) = \varepsilon(a,b,c,d)(-i(cz+d))^{-r} F_0(z)$ for $\gamma = \begin{bmatrix} a & b \\ c & d \end{bmatrix} \in \Gamma$ with $c > 0$, for some $r > 0$, $-\pi/2 < \arg(-i(cz+d)) < \pi/2$, and for a function $\varepsilon(\gamma) = \varepsilon(a,b,c,d)$ on Γ with $|\varepsilon(\gamma)| = 1$ (indeed, for $r = \frac{1}{2}$ and $\varepsilon(a,b,c,d) = \exp(-i\pi(\frac{a+d}{12c} - s(d,c)))$, by Theorem 9.12).
(iv) $F_0(z) = e^{2\pi i \alpha z} \sum_{n=-\mu}^{\infty} a_n e^{2\pi i n z}$ on π^+ for some integer $\mu \geq 1$ (indeed, for $\mu = 1$, $a_n = p(n+1)$ for $n \geq -1$, and $a_n = 0$ for $n \leq -2$, by Euler's formula (9.11)).

Note that by conditions (i) and (ii), the function $f(z) \stackrel{\text{def}}{=} e^{-2\pi i \alpha z} F_0(z)$ is holomorphic on π^+, and it satisfies $f(z+1) = f(z)$. Thus, again by equation (4.1), $f(z)$ has a Fourier expansion $f(z) = \sum_{n \in \mathbb{Z}} a_n e^{2\pi i n z}$ on π^+. That is, conditions (i) and (ii) imply that $F_0(z)$ has a Fourier expansion

$$F_0(z) = e^{2\pi i \alpha z} \sum_{n \in \mathbb{Z}} a_n e^{2\pi i n z}$$

on π^+, and condition (iv) means that we require that $a_{-n} = 0$ for $n > \mu$, for some positive integer μ.

We abstract these properties of $F_0(z)$ and, in general, we define a *modular form of negative weight* $-r$, for $r > 0$, with *multiplier* $\varepsilon : \Gamma \to \{z \in \mathbb{C} \mid |z| = 1\}$ to be a function $F(z)$ on π^+ that satisfies conditions (i), (ii), (iii), and (iv) for some α and μ with $0 \leq \alpha < 1$, $\mu \in \mathbb{Z}$, $\mu \geq 1$. Thus $\eta(z)^{-1}$ is a modular form of weight $-\frac{1}{2}$ and multiplier $\varepsilon(a,b,c,d) = \exp(-i\pi(\frac{a+d}{12c} - s(d,c)))$, with $\alpha = \frac{23}{24}$, $\mu = 1$, and with Fourier coefficients $a_n = p(n+1)$, as we note again.

For modular forms of positive integral weight, there are no general formulas available that explicitly compute their Fourier coefficients — apart from Theorem 4.23 for holomorphic Eisenstein series. For forms of negative weight however, there is a remarkable, explicit (but complicated) formula for their Fourier coefficients, due to H. Rademacher and H. Zuckerman [31]; also see [29; 30].

Before stating this formula we consider some of its ingredients. First, we have the modified Bessel function

$$I_\nu(t) \stackrel{\text{def}}{=} \left(\frac{t}{2}\right)^\nu \sum_{m=0}^{\infty} \frac{(\frac{t}{2})^{2m}}{m!\Gamma(\nu+m+1)} \tag{9.16}$$

for $t > 0$, $\nu \in \mathbb{C}$; the series here converges absolutely by the ratio test. Next, for $k, h \in \mathbb{Z}$ with $k \geq 1$, $h \geq 0$, $(h,k) = 1$, and $h < k$ choose a solution h' of the congruence $hh' \equiv -1 \pmod{k}$. For example, $(h, k) = 1$ means that the equation $xh + yk = 1$ has a solution $(x, y) \in \mathbb{Z} \times \mathbb{Z}$. Then $-xh = -1 + yk$ means that $h' \stackrel{\text{def}}{=} -x$ is a solution. Since $hh' = -1 + lk$ for some $l \in \mathbb{Z}$, we see that $(hh' + 1)/k = l$ is an integer and

$$\det \begin{bmatrix} h' & -(hh'+1)/k \\ k & -h \end{bmatrix} = 1, \quad \text{so} \quad \gamma \stackrel{\text{def}}{=} \begin{bmatrix} h' & -(hh'+1)/k \\ k & -h \end{bmatrix} \in \Gamma.$$

Hence

$$\varepsilon(\gamma) = \varepsilon\left(h', -\frac{hh'+1}{k}, k, -h\right) \tag{9.17}$$

is well-defined. Finally, for $u, v \in \mathbb{C}$ we define the *generalized Kloosterman sum*

$$A_{k,u}(v) = A_k(v, u) \stackrel{\text{def}}{=} \sum_{\substack{0 \le h < k \\ (h,k)=1}} \varepsilon(\gamma)^{-1} e^{-\frac{2\pi i}{k}((u-\alpha)h' + (v+\alpha)h)} \tag{9.18}$$

for $\varepsilon(\gamma)$ given in (9.17), and for $0 \le \alpha < 1$ above. If $k = 1$, for example, then $0 \le h < k$, so $h = 0$, and we can take $h' = 0$:

$$A_{1,\mu}(v) = A_1(v, u) = \varepsilon(0, -1, 1, 0)^{-1} = \varepsilon(0, -1, 1, 0) \stackrel{\text{def}}{=} \varepsilon_0. \tag{9.19}$$

The desired formula expresses the coefficients a_n for $n \ge 0$ in terms of the finitely many coefficients $a_{-\mu}, a_{-\mu+1}, a_{-\mu+2}, \ldots, a_{-2}, a_{-1}$ as follows:

THEOREM 9.20 (H. RADEMACHER AND H. ZUCKERMAN). *Let $F(z)$ be a modular form of negative weight $-r, r > 0$, with multiplier ε, and with Fourier expansion $F(z) = e^{2\pi i \alpha z} \sum_{n=-\mu}^{\infty} a_n e^{2\pi i n z}$ on π^+ given by condition (iv) above, where $0 \le \alpha < 1 \le \mu \in \mathbb{Z}$. Then for $n \ge 0$ with not both $n, \alpha = 0$,*

$$a_n = 2\pi \sum_{j=1}^{\mu} a_{-j} \sum_{k=1}^{\infty} \frac{A_{k,j}(n)}{k} \left(\frac{j-\alpha}{n+\alpha}\right)^{\frac{r+1}{2}} I_{r+1}\left(\frac{4\pi}{k}(j-\alpha)^{1/2}(n+\alpha)^{1/2}\right), \tag{9.21}$$

where $A_{k,j}(n)$ is defined in (9.18) and $I_v(t)$ is the modified Bessel function in (9.16).

Note that for $1 \le j \le \mu$, $j \ge 1 > \alpha \Rightarrow j - \alpha > 0$ in equation (9.21). Also $n + \alpha > 0$ there since $n, \alpha \ge 0$ with not both $n, \alpha = 0$.

Using the asymptotic result

$$\lim_{t \to \infty} \sqrt{2\pi t} I_v(t) e^{-t} = 1 \tag{9.22}$$

for the modified Bessel function $I_v(t)$ in (9.16), and also the trivial estimate $|A_{k,j}(n)| \le k$ that follows from (9.18), one can obtain from the explicit formula

(9.21) the following asymptotic behavior of a_n as $n \to \infty$. Assume that $a_{-\mu} \neq 0$ and define

$$a^\infty(n) \stackrel{\text{def}}{=} \frac{a_{-\mu}}{\sqrt{2}} \varepsilon_0 \frac{(\mu-\alpha)^{\frac{r}{2}+\frac{1}{4}}}{(n+\alpha)^{\frac{r}{2}+\frac{3}{4}}} \exp(4\pi(\mu-\alpha)^{1/2}(n+\alpha)^{1/2}), \qquad (9.23)$$

say for $n \geq 1$, for $\varepsilon_0 \stackrel{\text{def}}{=} \varepsilon(\begin{bmatrix} 0 & -1 \\ 1 & 0 \end{bmatrix})$ in (9.19). Then in [23], for example, it is shown that

$$a_n \sim a^\infty(n) \quad \text{as } n \to \infty \qquad (9.24)$$

which gives the asymptotic behavior of the Fourier coefficients of a modular form $F(z)$ of negative weight $-r$ with Fourier expansion as in the statement of Theorem 9.20. For forms of *zero weight* a quite similar result is given in equation (9.30) below. The asymptotic result (9.7) follows from (9.24) applied to $F_0(z)$, in which case formula (9.21) provides an *exact* formula (due to Rademacher) for $p(n)$ [2; 29; 30].

For $a, b, k \in \mathbb{Z}$ with $k \geq 1$ the *classical Kloosterman sum* $S(a, b; k)$ is defined by

$$S(a, b; k) = \sum_{\substack{h \in \mathbb{Z}/k\mathbb{Z} \\ (h,k)=1}} e^{\frac{2\pi i}{k}(ah + b\bar{h})} \qquad (9.25)$$

where $h\bar{h} \equiv 1 \pmod{k}$. These sums will appear in the next theorem (Theorem 9.27) that is a companion result of Theorem 9.20.

We consider next modular forms $F(z)$ of *weight zero*. That is, $F(z)$ is a holomorphic function on π^+ such that $F(\gamma \cdot z) = F(z)$ for $\gamma \in \Gamma$, and with Fourier expansion

$$F(z) = \sum_{n=-\mu}^{\infty} a_n e^{2\pi i n z} \qquad (9.26)$$

on π^+, for some positive integer μ. In case $F(z)$ is the modular invariant $j(z)$, for example, this expansion is that given in equation (4.44) with $\mu = 1$, in which case the a_n there are computed explicitly by H. Petersson and H. Rademacher [27; 28], independently - by a formula similar in structure to that given in (9.21). For the general case in equation (9.26) the following extension of the Petersson–Rademacher formula is available [6]:

THEOREM 9.27 (N. BRISEBARRE AND G. PHILIBERT). *For a modular form $F(z)$ of weight zero with Fourier expansion given by equation (9.26), its n-th Fourier coefficient a_n is given by*

$$a_n = 2\pi \sum_{j=1}^{\mu} a_{-j} \sqrt{\frac{j}{n}} \sum_{k=1}^{\infty} \frac{S(n,-j;k)}{k} I_1\left(\frac{4\pi\sqrt{nj}}{k}\right) \qquad (9.28)$$

for $n \geq 1$, where $S(n, -j; k)$ is defined in (9.25) and I_1 $(t > 0)$ in (9.16).

M. Knopp's asymptotic argument in [23] also works for a weight zero form (as he shows), provided the trivial estimate $|A_{k,j}(n)| \leq k$ used above is replaced by the less trivial Weil estimate $|S(a, b; k)| \leq C(\varepsilon)(a, b, k)^{1/2} k^{1/2+\varepsilon}$, $\forall \varepsilon > 0$. The conclusion is that if $a_{-\mu} \neq 0$, and if

$$a^{\infty}(n) \stackrel{\text{def}}{=} \frac{a_{-\mu}}{\sqrt{2}} \frac{\mu^{1/4}}{n^{3/4}} e^{4\pi \sqrt{\mu n}}, n \geq 1, \tag{9.29}$$

then

$$a_n \sim a^{\infty}(n): \lim_{n \to \infty} \frac{a_n}{a^{\infty}(n)} = 1. \tag{9.30}$$

We see that, *formally*, definition (9.29) is obtained by taking $\varepsilon_0 = 1, r = 0$, and $\alpha = 0$ in definition (9.23) - in which case formulas (9.21) and (9.28) are also formally the same. Going back to the Fourier expansion of the modular invariant $j(z)$ given in equation (4.44), where $a_{-\mu} = a_{-1} = 1$, we obtain from (9.30) that ([27; 28])

$$a_n \sim \frac{e^{4\pi \sqrt{n}}}{\sqrt{2} n^{3/4}} \text{ as } n \to \infty. \tag{9.31}$$

A stronger result than (9.31), namely that

$$a_n = \frac{e^{4\pi \sqrt{n}}}{\sqrt{2} n^{3/4}} \left(1 - \frac{3}{32\pi \sqrt{n}} + \varepsilon_n\right), |\varepsilon_n| \leq \frac{.055}{n} \tag{9.32}$$

(also due to Brisebarre and Philibert [6]) plays a key role in my study of the asymptotics of the Fourier coefficients of extremal partition functions of certain conformal field theories; see Theorem 5-16 of my Speaker's Lecture (page 345), and the remark that follows it.

Motivated by physical considerations, and by equation (9.5) in particular, we consider a modular form of weight zero with Fourier expansion

$$f(z) = e^{2\pi i \Delta z} \sum_{n \geq 0} c_n e^{2\pi i n z} \tag{9.33}$$

on π^+, where we assume that Δ is a negative integer. Δ corresponds to $-c/24$ in (9.5), say for a positive central charge c; thus $c = 24(-\Delta)$, a case considered in my Speaker's Lecture. $\mu \stackrel{\text{def}}{=} -\Delta$ is a positive integer such that for $a_n \stackrel{\text{def}}{=} c_{n+\mu}$, we have (taking $c_n = 0$ for $n \leq -1$) $a_{-n} = 0$ for $n > \mu$. Moreover, since

$\sum_{n=0}^{\infty} d_{n-\mu} = \sum_{n=-\mu}^{\infty} d_n$, we see that for $d_n \stackrel{\text{def}}{=} a_n e^{2\pi i n z}$ we have

$$\sum_{n=-\mu}^{\infty} a_n e^{2\pi i n z} = \sum_{n=0}^{\infty} a_{n-\mu} e^{2\pi i (n-\mu) z}$$

$$\stackrel{\text{def}}{=} \sum_{n=0}^{\infty} c_n e^{2\pi i (n+\Delta) z} = f(z), \quad (9.34)$$

by (9.33). That is, $f(z)$ has the form (9.26), which means that we can apply formula (9.28), and the asymptotic result (9.30).

Assume that $c_0 \neq 0$, and define

$$c^{\infty}(n) \stackrel{\text{def}}{=} \frac{c_0}{\sqrt{2}} \frac{|\Delta|^{1/4}}{(n+\Delta)^{3/4}} e^{4\pi |\Delta|^{1/2}(n+\Delta)^{1/2}} \quad (9.35)$$

for $n + \Delta \geq 1$. By definition (9.29), for $n - \mu \stackrel{\text{def}}{=} n + \Delta \geq 1$,

$$a^{\infty}(n-\mu) = \frac{a_{-\mu} \mu^{1/4}}{\sqrt{2}(n-\mu)^{3/4}} e^{4\pi \sqrt{\mu(n-\mu)}} \stackrel{\text{def}}{=} c^{\infty}(n),$$

as $a_{-\mu} \stackrel{\text{def}}{=} c_0 \neq 0$. Therefore by (9.30)

$$1 = \lim_{n \to \infty} \frac{a_{n-\mu}}{a^{\infty}(n-\mu)} = \lim_{n \to \infty} \frac{c_n}{c^{\infty}(n)} : c_n \sim c^{\infty}(n) \text{ as } n \to \infty, \quad (9.36)$$

for $c^{\infty}(n)$ defined in (9.35). Thus (9.36) gives the asymptotic behavior of the Fourier coefficients c_n of the modular form $f(z)$ of weight zero in (9.33).

Motivated by equation (9.6), and given the result (9.36) we define *entropy function* $S(n)$ associated to $f(z)$ by

$$S(n) \stackrel{\text{def}}{=} \log c^{\infty}(n) \quad (9.37)$$

for $n + \Delta \geq 1$, in case $c_0 > 0$. Also we set

$$S_0(n) \stackrel{\text{def}}{=} 2\pi \sqrt{4|\Delta|(n+\Delta)}, \quad (9.38)$$

for $n + \Delta \geq 1$. Then for $c = 24(-\Delta) = 24|\Delta|$, as considered above, (i.e., for $4|\Delta| = c/6$) $S_0(n)$ corresponds to the CFT entropy in equation (9.3), where $n + \Delta$ corresponds to the L_0 there. Moreover, by definition (9.37) we obtain

$$S(n) = S_0(n) + \left(\tfrac{1}{4} \log |\Delta| - \tfrac{3}{4} \log(n+\Delta) - \tfrac{1}{2} \log 2 + \log c_0\right), \quad (9.39)$$

which we can regard as an *abstract Cardy formula with logarithmic correction*, given by the four terms parenthesized. Note that, apart from, the term $\log c_0$, equation (9.39) bears an exact resemblance to equation (5.22) of my Speaker's Lecture. We regard $S_0(n)$ in definition (9.38), of course, as an abstract *Bekenstein–Hawking function* associated to the modular form $f(z)$ in (9.33) of zero weight. Equation (9.39) also corresponds to equation (14) of [5].

To close things out, we also apply formula (9.28) to $f(z)$. For $n - \mu \geq 1$,

$$c_n = a_{n-\mu} = 2\pi \sum_{j=1}^{\mu} a_{-j} \sqrt{\frac{j}{n-\mu}} \frac{S(n-\mu, -j; k)}{k} I_1\left(\frac{4\pi \sqrt{(n-\mu)j}}{k}\right). \quad (9.40)$$

Use $\sum_{j=1}^{\mu} d_j = d_\mu + d_{\mu-1} + \ldots + d_2 + d_1 = \sum_{j=0}^{\mu-1} d_{\mu-j}$ and $a_{-(\mu-j)} \stackrel{\text{def}}{=} c_j$ to write equation (9.40) as

$$c_n = 2\pi \sum_{j=0}^{\mu-1} c_j \sqrt{\frac{\mu-j}{n-\mu}} \frac{S(n-\mu, -(\mu-j); k)}{k} I_1\left(\frac{4\pi \sqrt{(n-\mu)(\mu-j)}}{k}\right) \quad (9.41)$$

where $\mu \stackrel{\text{def}}{=} -\Delta$. For $0 \leq j \leq \mu - 1 = -\Delta - 1$, $j \in \mathbb{Z}$, we have $j \geq 0$ and $j + \Delta \leq -1 < 0$. Conversely if $j \geq 0$, $j \in \mathbb{Z}$, and $j + \Delta < 0$, then as $\Delta \in \mathbb{Z}$ we have $j + \Delta \leq -1$, so $0 \leq j \leq -\Delta - 1 = \mu - 1$. Of course $j + \Delta < 0$ also means that $\mu - j = -\Delta - j = |j + \Delta|$. Thus we can write equation (9.41) as

$$c_n = 2\pi \sum_{\substack{j \geq 0 \\ j+\Delta<0}} c_j \sqrt{\frac{|j+\Delta|}{n+\Delta}} \frac{S(n+\Delta, j+\Delta; k)}{k} I_1\left(\frac{4\pi \sqrt{(n+\Delta)|j+\Delta|}}{k}\right) \quad (9.42)$$

for $n + \Delta \, (= n - \mu) \geq 1$:

THEOREM 9.43 (A REFORMULATION OF THEOREM 9.27). *For a modular form $f(z)$ of weight zero with Fourier expansion given by equation (9.33), its n-th Fourier coefficient c_n is given by equation (9.42), for $n + \Delta \geq 1$. Here Δ is assumed to be a negative integer.*

Instead of applying Theorem 9.20 and taking $r = 0$ there, without justification, physicists can now use Theorem 9.43 for a CFT modular invariant partition function, such as that of equation (9.4), and therefore stand on steady mathematical ground.

Appendix

A. Uniform convergence of improper integrals. For the reader's convenience we review the conditions under which an improper integral $f(s) = \int_a^\infty F(t, s)\, dt$ defines a holomorphic function $f(s)$. In particular, a verification of the entirety of the function $J(s)$ in equation (1.4) is provided.

The function $F(t, s)$ is defined on a product $[a, \infty) \times D$ with $D \subset \mathbb{C}$ some open subset, where it is assumed that $\int_a^\infty F(t, s)\, dt$ exists for each $s \in D$ — say $t \mapsto F(t, s)$ is integrable on $[a, b]$ for every $b > a$. Thus $f(s)$ is well-defined on D. By definition, the integral $f(s)$ is *uniformly convergent* on some subset $D_0 \subset D$ if to each $\varepsilon > 0$ there corresponds a number $B(\varepsilon) > a$ such that for $b > B(\varepsilon)$ one has $\left|\int_a^b F(t, s)\, dt - f(s)\right| < \varepsilon$ for all $s \in D_0$. An equivalent definition is given by the following *Cauchy criterion*: $f(s)$ is uniformly convergent on D_0 if and only if to each $\varepsilon > 0$ there corresponds a number $B(\varepsilon) > a$ such that $\left|\int_{b_1}^{b_2} F(t,s)\,dt\right| < \varepsilon$ for all $b_2 > b_1 > B(\varepsilon)$ and all $s \in D_0$. For clearly if $f(s)$ is uniformly convergent on D_0 and $\varepsilon > 0$ is given, we can choose $B(\varepsilon) > a$ such that $\left|\int_a^b F(t,s)\,dt - f(s)\right| < \varepsilon/2$ for $b > B(\varepsilon)$, $s \in D_0$. Then for $b_2 > b_1 > B(\varepsilon)$ and $s \in D_0$, we have

$$\int_{b_1}^{b_2} F(t,s)\,dt = \int_a^{b_2} F(t,s)\,dt - f(s) - \left(\int_a^{b_1} F(t,s)\,dt - f(s)\right),$$

which implies $\left|\int_{b_1}^{b_2} F(t,s)\,dt\right| < \varepsilon/2 + \varepsilon/2 = \varepsilon$. Conversely, assume the alternate condition. Define the sequence $\{f_n(s)\}_{n>a}$ of functions on D_0 by

$$f_n(s) \stackrel{\text{def}}{=} \int_a^n F(t,s)\,dt.$$

Given $\varepsilon > 0$ we can choose, by hypothesis, $B(\varepsilon) > a$ such that $\left|\int_{b_1}^{b_2} F(t,s)\,dt\right| < \varepsilon$ for $b_2 > b_1 > B(\varepsilon)$ and $s \in D_0$. Let $N(\varepsilon)$ be an integer $> B(\varepsilon)$. Then for integers $n > m \geq N(\varepsilon)$ and for all $s \in D_0$ we see that $|f_n(s) - f_m(s)| = \left|\int_m^n F(t,s)\,dt\right| < \varepsilon$. Therefore, by the standard Cauchy criterion, the sequence $\{f_n(s)\}_{n>a}$ converges uniformly on D_0 to a function $g(s)$ on D_0: For any $\varepsilon_1 > 0$, there exists an integer $N(\varepsilon_1) > a$ such that for an integer $n \geq N(\varepsilon_1)$, one has $\varepsilon_1 > |f_n(s) - g(s)| = \left|\int_a^n F(t,s)\,dt - f(s)\right|$ for all $s \in D_0$, since necessarily $g(s) = f(s)$. Now let $\varepsilon > 0$ be given. Again, by hypothesis, we can choose $B(\varepsilon) > a$ such that for $b_2 > b_1 > B(\varepsilon)$ one has that $\left|\int_{b_1}^{b_2} F(t,s)\,dt\right| < \varepsilon/2$ for all $s \in D_0$. Taking the quantity ε_1 considered a few lines above equal to $\varepsilon/2$, we can find an integer $N(\varepsilon_1) > B(\varepsilon_1)$ such that for an integer $n \geq N(\varepsilon_1)$, $\varepsilon/2 = \varepsilon_1 > \left|\int_a^n F(t,s)\,dt - f(s)\right|$ for all $s \in D_0$. Thus suppose $b > N(\varepsilon_1)$. Then, for all $s \in D_0$,

$$\left|\int_a^b F(t,s)\,dt - f(s)\right| = \left|\int_a^{N(\varepsilon_1)} F(t,s)\,dt - f(s) + \int_{N(\varepsilon_1)}^b F(t,s)\,dt\right|$$
$$\leq \frac{\varepsilon}{2} + \frac{\varepsilon}{2} = \varepsilon,$$

where $N(\varepsilon_1) > a$ since $B(\varepsilon_1) > a$, with $N(\varepsilon_1)$ dependent only on ε. This shows that $f(s)$ is uniformly convergent on D_0. The Cauchy criterion is therefore validated.

As an example, we use the Cauchy criterion to prove the following, very useful result:

THEOREM A.1 (WEIERSTRASS M-TEST). *Let $M(t) \geq 0$ be a function on $[a, \infty)$ that is integrable on each $[a, b]$ with $b > a$. Assume also that $I \stackrel{\text{def}}{=}$*

$\int_a^\infty M(t)\,dt$ exists. If $|F(t,s)| \leq M(t)$ on $[a,\infty) \times D_0$, then $f(s) = \int_a^\infty F(t,s)\,dt$ converges uniformly on D_0. Again D_0 is any subset of D.

PROOF. Let $\varepsilon > 0$ be assigned. That $I = \lim_{b \to \infty} \int_a^b M(t)\,dt$ implies there exists a number $B(\varepsilon) > a$ such that $\left|I - \int_a^b M(t)\,dt\right| < \varepsilon/2$ for $b > B(\varepsilon)$. If $b_2 > b_1 > B(\varepsilon)$, then

$$\left|\int_{b_1}^{b_2} M(t)\,dt\right| = \left|\int_a^{b_2} M(t)\,dt - I + I - \int_a^{b_1} M(t)\,dt\right| < \frac{\varepsilon}{2} + \frac{\varepsilon}{2} = \varepsilon;$$

hence $\left|\int_{b_1}^{b_2} F(t,s)\,dt\right| \leq \int_{b_1}^{b_2} |F(t,s)|\,dt \leq \int_{b_1}^{b_2} M(t)\,dt = \left|\int_{b_1}^{b_2} M(t)\,dt\right|$ (since $M(t) \geq 0, b_2 > b_1$). But this is less than ε, for all $s \in D_0$. Theorem A.1 follows, therefore, by the Cauchy criterion. □

The question of the holomorphicity of $f(s)$ is settled by the following theorem.

THEOREM A.2. Again let $F(t,s)$ be defined on $[a,\infty) \times D$ with $D \subset \mathbb{C}$ an open subset. Assume

(i) $F(t,s)$ is continuous on $[a,\infty) \times D$ (in particular for each $s \in D$, $t \mapsto F(t,s)$ is integrable on $[a,b]$ for every $b > a$);
(ii) for every $t \geq a$ fixed, $s \mapsto F(t,s)$ is holomorphic on D;
(iii) for every $s \in D$ fixed, $t \mapsto \partial F(t,s)/\partial s$ is continuous on $[a,\infty)$;
(iv) $f(s) \stackrel{\text{def}}{=} \int_a^\infty F(t,s)\,dt$ converges for every $s \in D$; and
(v) $f(s)$ converges uniformly on compact subsets of D.

Then $f(s)$ is holomorphic on D, and $f'(s) = \int_a^\infty \partial F(t,s)/\partial s\,dt$ for every $s \in D$. Implied here is the existence of the improper integral $\int_a^\infty \partial F(t,s)/\partial s\,dt$ on D.

The idea of the proof is to reduce matters to a situation where the integration $\int_a^\infty$ over an infinite range is replaced by that over a finite range $\int_a^n$, where holomorphicity is known to follow. This is easily done by considering again the sequence $\{f_n(s)\}_{n>a}$ discussed earlier: $f_n(s) \stackrel{\text{def}}{=} \int_a^n F(t,s)\,dt$ on D, which is well-defined by (i). If $K \subset D$ is compact, then given (v), the above argument with D_0 now taken to be K shows exactly (by way of the Cauchy criterion) that $\{f_n(s)\}_{n>a}$ converges uniformly on K (to $f(s)$ by (iv)). On the other hand, by (i), (ii), (iii) we have that (i)′ $F(t,s)$ is continuous on $[a,n] \times D$, (ii)′ for every $t \in [a,n]$ fixed, $s \mapsto F(t,s)$ is holomorphic on D, and (iii)′ for every $s \in D$ fixed, $t \mapsto \partial F(t,s)/\partial s$ is continuous on $[a,n]$; here $a < n \in \mathbb{Z}$. Given (i)′, (ii)′ and (iii)′, it is standard in complex variables texts that $f_n(s) = \int_a^n F(t,s)\,dt$ is holomorphic on D and that $f_n'(s) = \int_a^n \partial F/\partial s(t,s)\,dt$. Since we have noted that the sequence $\{f_n(s)\}_{n>a}$ converges uniformly to $f(s)$ on compact subsets K of D, it follows by the Weierstrass theorem that $f(s)$ is holomorphic on D, and that $f_n'(s) \mapsto f'(s)$ pointwise on D — with uniform convergence on compact subsets of D, in fact. That is, $f'(s) = \lim_{n\to\infty} f'_n(s) = \lim_{n\to\infty} \int_a^n \partial F(t,s)/\partial s\, dt = \int_a^\infty \partial F(t,s)/\partial s\, dt$ on D, which proves Theorem A.2.

As an application, we check that the function $J(s)$ in definition (1.4) is an entire function. First we claim that the function $\theta_0(t) \stackrel{\text{def}}{=} \sum_{n=1}^\infty e^{-\pi n^2 t}$, for $t > 0$, converges uniformly on $[1, \infty)$. This is clear, by the Weierstrass M-test, since for $n, t \geq 1$, we have $n^2 \geq n$, hence $\pi n^2 t \geq \pi n t \geq \pi n$, hence $e^{-\pi n^2 t} \leq e^{-\pi n}$, and moreover $\sum_{n=1}^\infty e^{-\pi n}$ is a convergent geometric series. Therefore $\theta_0(t)$ is continuous on $[1, \infty)$, since the terms $e^{-\pi n^2 t}$ are continuous in t on $[1, \infty)$. By definitions (1.2), (1.4), $J(s) = \int_1^\infty F(t,s)\, dt$ for $F(t,s) \stackrel{\text{def}}{=} \theta_0(t) t^s$ on $[1, \infty) \times \mathbb{C}$, where $F(t,s)$ therefore is also continuous. Again for $n, t \geq 1$, $\pi n^2 t \geq \pi n t$ and also $\pi t \geq \pi$, so $e^{-\pi n^2 t} \leq e^{-\pi n t}$ and $e^{-\pi t} \leq e^{-\pi}$, so $1 - e^{-\pi t} \geq 1 - e^{-\pi}$, so

$$\frac{1}{1 - e^{-\pi t}} \leq \frac{1}{1 - e^{-\pi}} = \frac{e^\pi}{e^\pi - 1} \stackrel{\text{def}}{=} C.$$

That is,

$$\theta_0(t) = \sum_{n=1}^\infty e^{-\pi n^2 t} \leq \sum_{n=1}^\infty e^{-\pi n t} = \sum_{n=1}^\infty (e^{-\pi t})^n = \frac{e^{-\pi t}}{1 - e^{-\pi t}} \leq C e^{-\pi t}$$

for $t \geq 1$, so $|F(t,s)| \leq C e^{-\pi t} t^{\operatorname{Re} s}$ on $[1, \infty) \times \mathbb{C}$, where $\int_1^\infty e^{-bt} t^a\, dt$ converges for $b > 0, a \in \mathbb{R}$. Thus $J(s)$ converges absolutely for every $s \in \mathbb{C}$. We see that conditions (i) and (iv) of Theorem A.2 hold. Conditions (ii) and (iii) certainly hold. To check condition (v), let $K \subset \mathbb{C}$ be any compact subset. The continuous function $s \to \operatorname{Re} s$ on K has an upper bound $\sigma : \operatorname{Re} s \leq \sigma$ on $K \Rightarrow t^{\operatorname{Re} s} \leq t^\sigma$ on $[1, \infty) \times K$ (since $\log t \geq 0$ for $t \geq 1$). That is, on $[1, \infty) \times K$ the estimate $|F(t,s)| \leq C e^{-\pi t} t^\sigma$ holds where $\int_1^\infty e^{-\pi t} t^\sigma\, dt < \infty$, implying that $J(s)$ converges uniformly on K, by Theorem A.1. Therefore $J(s)$ is holomorphic on $\mathbb{C}$ by Theorem A.2.

B. A Fourier expansion (or q-expansion). The function $q(z) \stackrel{\text{def}}{=} e^{2\pi i z}$ is holomorphic and it satisfies the periodicity condition $q(z+1) = q(z)$. Suppose $f(z)$ is an arbitrary holomorphic function defined on an open horizontal strip

$$S_{b_1,b_2} \stackrel{\text{def}}{=} \{z \in \mathbb{C} \mid b_1 < \operatorname{Im} z < b_2\} \tag{B.1}$$

as indicated in the figure at the top of the next page, where $b_1, b_2 \in \mathbb{R}, b_1 < b_2$.

Suppose also that $f(z)$ satisfies the periodicity condition $f(z+1) = f(z)$ on S_{b_1,b_2}; clearly $z \in S_{b_1,b_2}$ implies $z + r \in S_{b_1,b_2}$ for all $r \in \mathbb{R}$. Then $f(z)$ has a *Fourier expansion* (also called a q-*expansion*)

$$f(z) = \sum_{n \in \mathbb{Z}} a_n q(z)^n = \sum_{n \in \mathbb{Z}} a_n e^{2\pi i n z} \tag{B.2}$$

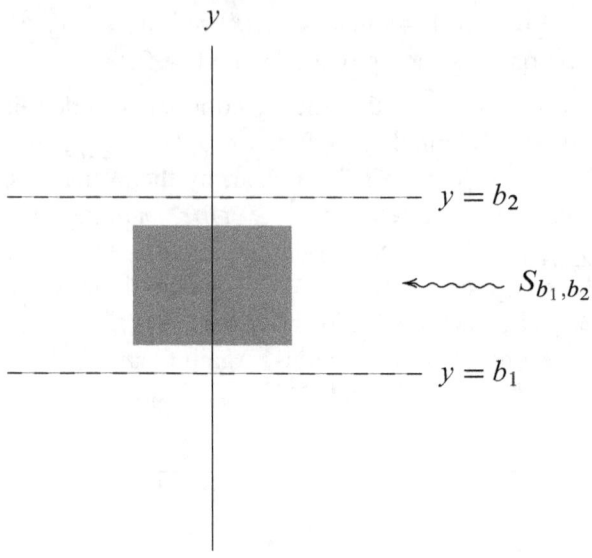

on S_{b_1,b_2}, for suitable coefficients $a_n \in \mathbb{C}$; see Theorem B.7 and equation (B.6) below for an expression of the a_n. The finiteness of b_2 is *not* essential for the validity of equation (B.2). In fact, one of its most useful applications is in case when S_{b_1,b_2} is the upper half plane: $b_1 = 0$, $b_2 = \infty$. The Fourier expansion of $f(z)$ follows from the local invertibility of the function $q(z)$ and the Laurent expansion of the function $(f \circ q^{-1})(z)$. We fill in the details of the proof.

Note first that $q(z)$ is a surjective map of the strip S_{b_1,b_2} onto the annulus $A_{r_1,r_2} \stackrel{\text{def}}{=} \{w \in \mathbb{C} \mid r_1 < |w| < r_2\}$ for $r_1 \stackrel{\text{def}}{=} e^{-2\pi b_2} > 0$, $r_2 \stackrel{\text{def}}{=} e^{-2\pi b_1} > 0$ (see figure below).

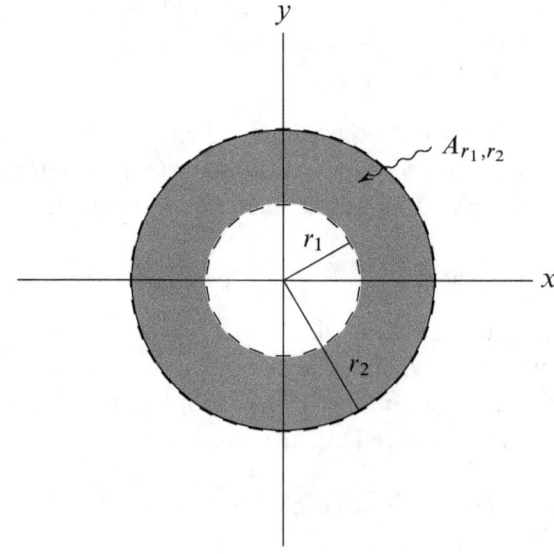

For if $z = x + iy \in S_{b_1,b_2}$, we have $|q(z)| = e^{-2\pi y}$ and $b_1 < y < b_2$, so $e^{-2\pi b_1} > e^{-2\pi y} > e^{-2\pi b_2}$, and $w = q(z) \in A_{r_1,r_2}$. On the other hand, if $w \in A_{r_1,r_2}$ is given choose $t \in \mathbb{R}$ such that $e^{it} = w/|w|$ (since $w \neq 0$), and define $r = \log|w|$. Then one quickly checks that $z \stackrel{\text{def}}{=} t/2\pi + ir/(-2\pi) \in S_{b_1,b_2}$ such that $q(z) = w$, as desired.

From $f(z+1) = f(z)$, it follows by induction that $f(z+n) = f(z)$ for every positive integer n, and therefore for every negative integer n, $f(z) = f(z+n+(-n)) = f(z+n)$; i.e. $f(z+n) = f(z)$ for every $n \in \mathbb{Z}$. Also since $q(z_1) = q(z_2) \iff e^{2\pi i z_1} = e^{2\pi i z_2} \iff e^{2\pi i(z_1-z_2)} = 1 \iff z_1 = z_2 + n$ for some $n \in \mathbb{Z}$, the surjectivitiy of $q(z)$ implies that the equation $F(q(z)) = f(z)$ provides for a well-defined function $F(w)$ on the annulus A_{r_1,r_2}. To check that $F(w)$ is holomorphic, given that $f(z)$ is holomorphic, take any $w_0 \in A_{r_1,r_2}$ and choose $z_0 \in S_{b_1,b_2}$ such that $q(z_0) = w_0$, again by the surjectivity of $q(z)$. Since $q'(z) = 2\pi i e^{2\pi i z}$ implies in particular that $q'(z_0) \neq 0$, one can conclude that $q(z)$ is locally invertible at z_0: there exist $\varepsilon > 0$ and a neighborhood N of z_0, $N \subset S_{b_1,b_2}$, on which q is injective with

$$q(N) = N_\varepsilon(q(z_0)) = N_\varepsilon(w_0) \stackrel{\text{i.e.}}{=} \{w \in \mathbb{C} \mid |w - w_0| < \varepsilon\} \subset A_{r_1,r_2},$$

and with q^{-1} holomorphic on $N_\varepsilon(w_0)$. Thus, on $N_\varepsilon(w_0)$,

$$F(w) = F(q(q^{-1}(w))) = (f \circ q^{-1})(w),$$

which shows that F is holomorphic on $N_\varepsilon(w_0)$ and thus is holomorphic on A_{r_1,r_2}, as $w_0 \in A_{r_1,r_2}$ is arbitrary.

Now $F(w)$ has a Laurent expansion

$$F(w) = \sum_{n=0}^{\infty} \tilde{a}_n w^n + \sum_{m=1}^{\infty} \frac{\tilde{b}_m}{w^m}$$

on the annulus A_{r_1,r_2} where the coefficients $\tilde{a}_n, \tilde{b}_m$ are given by

$$\tilde{a}_n = \frac{1}{2\pi i} \int_\Gamma \frac{F(w)\,dw}{w^{n+1}}, \quad \tilde{b}_m = \frac{1}{2\pi i} \int_\Gamma \frac{F(w)}{w^{-m+1}}\,dw,$$

for any circle Γ in A_{r_1,r_2} that separates the circles $|w| = r_1$, $|w| = r_2$. We choose Γ to be the circle centered at $w = 0$ with radius $R \stackrel{\text{def}}{=} e^{-2\pi b}$, given any b with $b_1 < b < b_2$; $r_1 < R < r_2$. For a continuous function $\psi(w)$ on Γ the change of variables $v(t) = 2\pi t$ on $[0,1]$ permits the expression

$$\int_\Gamma \psi(w)\,dw = \int_0^{2\pi} \psi(Re^{iv}) Rie^{iv}\,dv = 2\pi i \int_0^1 \psi(Re^{2\pi it}) Re^{2\pi it}\,dt$$

$$= 2\pi i \int_0^1 \psi(e^{2\pi i(t+ib)}) e^{2\pi i(t+ib)}\,dt,$$

by definition of R. For the choices $\psi(w) = F(w)/w^{n+1}$, $F(w)/w^{-m+1}$, respectively, one finds that

$$\tilde{a}_n = \int_0^1 F(e^{2\pi i(t+ib)})e^{-2\pi in(t+ib)}\,dt,$$
$$\tilde{b}_m = \int_0^1 F(e^{2\pi i(t+ib)})e^{2\pi im(t+ib)}\,dt,$$
(B.3)

for $n \geq 0$, $m \geq 1$. However $t + ib \in S_{b_1,b_2}$ since $b_1 < b < b_2$ so by definition of $F(w)$ the equations in (B.3) are

$$\tilde{a}_n = \int_0^1 f(t+ib)e^{-2\pi in(t+ib)}\,dt,$$
$$\tilde{b}_m = \int_0^1 f(t+ib)e^{2\pi im(t+ib)}\,dt,$$
(B.4)

for $n \geq 0$, $m \geq 1$, and moreover the Laurent expansion of $F(w)$ has a restatement

$$f(z) = \sum_{n=0}^{\infty} \tilde{a}_n q(z)^n + \sum_{m=1}^{\infty} \frac{\tilde{b}_m}{q(z)^m}$$
(B.5)

on S_{b_1,b_2}. One can codify the preceding formulas by defining

$$a_n \stackrel{\text{def}}{=} \int_0^1 f(t+ib)e^{-2\pi in(t+ib)}\,dt$$
(B.6)

for $n \in \mathbb{Z}$, again for $b_1 < b < b_2$. Then $a_n = \tilde{a}_n$ for $n \geq 0$ and $a_{-n} = \tilde{b}_n$ for $n \geq 1$. By equation (B.5) we have therefore completed the proof of equation (B.2):

THEOREM B.7 (A FOURIER EXPANSION). *Let $f(z)$ be holomorphic on the open strip S_{b_1,b_2} defined in equation (B.1), and assume that $f(z)$ satisfies the periodicity condition $f(z+1) = f(z)$ on S_{b_1,b_2}. Then $f(z)$ has a Fourier expansion on S_{b_1,b_2} given by equation (B.2), where the a_n are given by equation (B.6) for $n \in \mathbb{Z}$, for arbitrary b subject to $b_1 < b < b_2$.*

Theorem B.7 is valid if S_{b_1,b_2} is replaced by the upper half-plane π^+ (with $b_1 = 0$, $b_2 = \infty$), for example, as we have indicated. For clearly the preceding arguments hold for $b_2 = \infty$. Here, in place of the statement that $q : S_{b_1,b_2} \to A_{r_1,r_2}$ is surjective (again for $r_1 \stackrel{\text{def}}{=} e^{-2\pi b_2}$, $r_2 \stackrel{\text{def}}{=} e^{-2\pi b_1}$, $b_2 < \infty$), one simply employs the statement that $q : \pi^+ \to \{w \in \mathbb{C} \mid 0 < |w| < 1\}$ is surjective.

C. Poisson summation and Jacobi inversion.

The Jacobi inversion formula (1.3) can be proved by a special application of the *Poisson summation formula* (PSF). The latter formula, in essence, is the statement

$$\sum_{n \in \mathbb{Z}} f(n) = \sum_{n \in \mathbb{Z}} \hat{f}(n), \tag{C.1}$$

for a suitable class of functions $f(x)$ and a suitable normalization of the Fourier transform $\hat{f}(x)$ of $f(x)$. The purpose here is to prove a slightly more general version of the PSF, which applied in a special case, coupled with a Fourier transform computation, indeed does provide for a proof of equation (1.3).

For a function $h(x)$ on $\mathbb{R}$, the definition

$$\hat{h}(x) \stackrel{\text{def}}{=} \int_{-\infty}^{\infty} h(t) e^{-2\pi i x t} \, dt \tag{C.2}$$

will serve as our normalization of its *Fourier transform*. Here's what we aim to establish:

THEOREM C.3 (POISSON SUMMATION). *Let $f(z)$ be a holomorphic function on an open horizontal strip $S_\delta \stackrel{\text{def}}{=} \{z \in \mathbb{C} \mid -\delta < \operatorname{Im} z < \delta\}, \delta > 0$, say with $f|_\mathbb{R} \in L^1(\mathbb{R}, dx)$. Assume that the series $\sum_{n=0}^{\infty} f(z+n)$, $\sum_{n=1}^{\infty} f(z-n)$ converge uniformly on compact subsets of S_δ. Then for any $z \in S_\delta$*

$$\sum_{n \in \mathbb{Z}} f(z+n) = \sum_{n \in \mathbb{Z}} e^{2\pi i n z} \hat{f}(n). \tag{C.4}$$

In particular for $z = 0$ we obtain equation (C.1).

PROOF. By the Weierstrass theorem, the uniform convergence of the series $\sum_{n=0}^{\infty} f(z+n)$ and $\sum_{n=1}^{\infty} f(z-n)$ on compact subsets of S_δ means that the function

$$F(z) \stackrel{\text{def}}{=} \sum_{n \in \mathbb{Z}} f(z+n) = \sum_{n=0}^{\infty} f(z+n) + \sum_{n=1}^{\infty} f(z-n)$$

on S_δ is holomorphic. $F(z)$ satisfies $F(z+1) = \sum_{n \in \mathbb{Z}} f(z+n+1) = \sum_{n \in \mathbb{Z}} f(z+n) = F(z)$ on S_δ. Therefore, Theorem B.7 of Appendix B is applicable, where the choice $b = 0$ is made ($b_1 = -\delta, b_2 = \delta$): $F(z)$ has a Fourier expansion

$$F(z) = \sum_{n \in \mathbb{Z}} a_n e^{2\pi i n z} \tag{C.5}$$

on S_δ, where $a_n \stackrel{\text{def}}{=} \int_0^1 F(t) e^{-2\pi i n t} \, dt$ for $n \in \mathbb{Z}$. Since $[0, 1] \subset S_\delta$ is compact, for $n \in \mathbb{Z}$ fixed the series $\sum_{l=0}^{\infty} f(t+l) e^{-2\pi i n t}$ and $\sum_{l=1}^{\infty} f(t-l) e^{-2\pi i n t}$ (whose sum is $F(t) e^{-2\pi i n t}$) converge uniformly on $[0, 1]$ (by hypothesis, given of course that $|e^{-2\pi int}| = 1$). Therefore a_n can be obtained by termwise integration; we start by writing

$$a_n = \int_0^1 \left(\sum_{l=0}^{\infty} f(t+l)e^{-2\pi int} + \sum_{l=1}^{\infty} f(t-l)e^{-2\pi int} \right) dt$$

$$= \sum_{l=0}^{\infty} \int_0^1 f(t+l)e^{-2\pi int} dt + \sum_{l=1}^{\infty} \int_0^1 f(t-l)e^{-2\pi int} dt$$

$$= \sum_{l=0}^{\infty} \int_l^{l+1} f(v)e^{-2\pi in(v-l)} dv + \sum_{l=1}^{\infty} \int_{-l}^{-l+1} f(v)e^{-2\pi in(v+l)} dv,$$

by the change of variables $v(t) = t+l$ and $v(t) = t-l$ on $[l, l+1]$, $[-l, -l+1]$, respectively (with $l \in \mathbb{Z}$), This is further equal to $\sum_{l=0}^{\infty} \int_l^{l+1} f(t)e^{-2\pi int} dt + \sum_{l=1}^{\infty} \int_{-l}^{-l+1} f(t)e^{-2\pi int} dt = \int_0^{\infty} f(t)e^{-2\pi int} dt + \int_{-\infty}^0 f(t)e^{-2\pi int} dt = \int_{-\infty}^{\infty} f(t)e^{-2\pi int} dt = \hat{f}(n)$, by definition (C.2). That is, by (C.5), for $z \in S_\delta$ $\sum_{n \in \mathbb{Z}} \hat{f}(n)e^{2\pi inz} = F(z) \stackrel{\text{def}}{=} \sum_{n \in \mathbb{Z}} f(z+n)$, which concludes the proof of Theorem C.3. $\square$

Other proofs of the PSF exist. In contrast to the complex-analytic one just presented, a real-analytic proof (due to Bochner) is given in Chapter 14 of [38], for example, based on *Fejér's Theorem*, which states that the Fourier series of a continuous, 2π-periodic function $\psi(x)$ on $\mathbb{R}$ is *Cesàro summable* to $\psi(x)$.

As an example, choose $f(z) = f_t(z) \stackrel{\text{def}}{=} e^{-\pi z^2 t}$ for $t > 0$ fixed. In this case $f(z)$ is an entire function whose restriction to $\mathbb{R}$ is Lebesgue integrable; the restriction is in fact a Schwartz function. We claim that the series

$$\sum_{n=0}^{\infty} f(z+n) \quad \text{and} \quad \sum_{n=1}^{\infty} f(z-n)$$

converge uniformly on compact subsets K of the plane. Since K is compact the continuous functions $z \mapsto e^{-\pi z^2 t}$ and $z \mapsto \operatorname{Re} z$ on $\mathbb{C}$ are bounded on K: $|e^{-\pi z^2 t}| \leq M_1$, $|\operatorname{Re} z| \leq M_2$ on K for some positive numbers M_1, M_2. Let n_0 be an integer $> 1 + 2M_2$. Then for $n \in \mathbb{Z}$ with $n \geq n_0$ one has $n^2 \geq n(1+2M_2)$, hence $n^2 - 2nM_2 \geq n$, so that $f(z+n) = e^{-\pi z^2 t} e^{-\pi(n^2 + 2nz)t}$ for $z \in K$. But

$$\left| e^{-\pi(n^2 + 2nz)t} \right| = e^{-\pi(n^2 + 2n \operatorname{Re} z)t} \leq e^{-\pi(n^2 - 2nM_2)t} = e^{-\pi nt}$$

and $\left| e^{-\pi z^2 t} \right| \leq M_1$, so $|f(z+n)| \leq M_1 e^{-\pi nt}$ on K, with $\sum_{n=0}^{\infty} M e^{-\pi nt}$ clearly convergent for $t > 0$. Therefore, by the M-test, $\sum_{n=0}^{\infty} f(z+n)$ converges absolutely and uniformly on K. Similarly, for $n \in \mathbb{Z}$, we have $f(z-n) = e^{-\pi z^2 t} e^{-\pi(n^2 - 2nz)t}$, where for $n \geq n_0$ and $z \in K$ again $n^2 - 2nM_2 \geq n$, but where we now use the bound $\operatorname{Re} z \leq M_2$: $n^2 - 2n \operatorname{Re} z \geq n^2 - 2nM_2 \geq n$

$|f(z-n)| \le M_1 e^{-\pi t n}$ on K (for $n \ge n_0$), so $\sum_{n=1}^{\infty} f(z-n)$ converges absolutely and uniformly on K, which shows that $f_t(z) \stackrel{\text{def}}{=} e^{-\pi z^2 t}, t > 0$, satisfies the hypotheses of Theorem C.3. The conclusion

$$\sum_{n \in \mathbb{Z}} e^{-\pi n^2 t} = \sum_{n \in \mathbb{Z}} \hat{f}_t(n) \tag{C.6}$$

is therefore safe, and the left-hand side here is $\theta(t)$ by definition (1.2). One is therefore placed in the pleasant position of computing the Fourier transform

$$\hat{f}_t(x) \stackrel{(C.2)}{=} \int_{-\infty}^{\infty} e^{-\pi y^2 t} e^{-2\pi i x y} \, dy, \tag{C.7}$$

which is a classical computation that we turn to now (for the sake of completeness).

For real numbers a, b, c, t with $a < b$, $t > 0$, note that $e^{-v^2} e^{2\pi i c v / \sqrt{\pi t}} = e^{-v^2} e^{2icv\sqrt{\pi/t}} = e^{-\pi c^2 / t} e^{-(v - ic\sqrt{\pi/t})^2}$. By the change of variables $v(x) = \sqrt{\pi t} x$ on $[a\sqrt{\pi t}, b\sqrt{\pi t}]$, therefore,

$$\int_a^b e^{-\pi x^2 t} e^{2\pi i c x} \, dx = \frac{e^{-\pi c^2 / t}}{\sqrt{\pi t}} \int_{a\sqrt{\pi t}}^{b\sqrt{\pi t}} e^{-(v - ic\sqrt{\pi/t})^2} \, dv. \tag{C.8}$$

Next we show that for $b \in \mathbb{R}$

$$\int_{-\infty}^{\infty} e^{-(x+ib)^2} \, dx = \int_{-\infty}^{\infty} e^{-x^2} \, dx. \tag{C.9}$$

To do this, assume first that $b > 0$ and consider the counterclockwise oriented rectangle C_R of height b and width $2R$: $C_R = C_1 + C_2 + C_3 + C_4$.

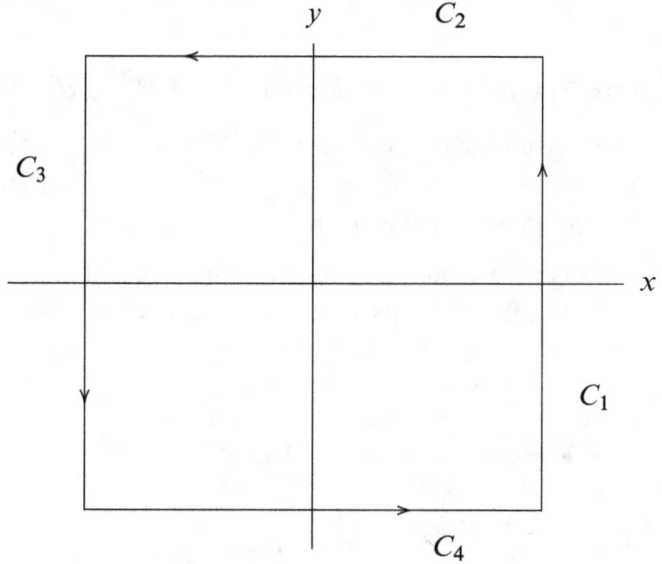

By Cauchy's theorem, $0 = I_R \stackrel{\text{def}}{=} \int_{C_R} e^{-z^2} dz$. Now

$$\int_{C_1} e^{-z^2} dz = \int_{-R}^{R} e^{-x^2} dx,$$
$$\int_{C_2} e^{-z^2} dz = i \int_0^b e^{-(R+ix)^2} dx = ie^{-R^2} \int_0^b e^{-2xRi} e^{x^2} dx,$$
$$\int_{C_3} e^{-z^2} dz = -\int_{-R}^{R} e^{-(x+ib)^2} dx,$$
$$\int_{C_4} e^{-z^2} dz = -\int_{C_2} e^{-z^2} dz$$

Thus $\left| \int_{C_2} e^{-z^2} dz \right| \leq e^{-R^2} \int_0^b e^{x^2} dx$, which tends to 0 as $R \to \infty$. That is, $0 = \lim_{R \to \infty} I_R = \int_{-\infty}^{\infty} e^{-x^2} dx - \int_{-\infty}^{\infty} e^{-(x+ib)^2} dx$, which proves equation (C.9) for $b > 0$. If $b < 0$, write $\int_{-\infty}^{\infty} e^{-(x+ib)^2} dx = \int_{-\infty}^{\infty} e^{-(-x+ib)^2} dx = \int_{-\infty}^{\infty} e^{-(x+i(-b))^2} dx = \int_{-\infty}^{\infty} e^{-x^2} dx$ by the previous case, since $-b > 0$. Thus (C.9) holds for *all* $b \in \mathbb{R}$ (since it clearly holds for $b = 0$). By (C.8) it then follows that

$$\int_{-\infty}^{\infty} e^{-\pi x^2 t} e^{2\pi i c x} dx = \lim_{R \to \infty} \int_{-R}^{R} e^{-\pi x^2 t} e^{2\pi i c x} dx$$
$$= \frac{e^{-\pi c^2/t}}{\sqrt{\pi t}} \lim_{R \to \infty} \int_{-R\sqrt{\pi t}}^{R\sqrt{\pi t}} e^{-(x+i(-c)\sqrt{\pi/t})^2} dx$$
$$= \frac{e^{-\pi c^2/t}}{\sqrt{\pi t}} \int_{-\infty}^{\infty} e^{-x^2} dx = \frac{e^{-\pi c^2/t}}{\sqrt{\pi t}} \sqrt{\pi}. \quad \text{(C.10)}$$

PROPOSITION C.11. *For $c \in \mathbb{R}$ and $t > 0$, we have*

$$\int_{-\infty}^{\infty} e^{-\pi x^2 t} e^{-2\pi i c x} dx = \int_{-\infty}^{\infty} e^{-\pi x^2 t} e^{2\pi i c x} dx = \frac{e^{-\pi c^2/t}}{\sqrt{t}}.$$

Hence equation (C.7) is the statement that $\hat{f}_t(x) = e^{-\pi x^2/t}/\sqrt{t}$.

Having noted that the left-hand side of equation (C.6) is $\theta(t)$, we see that (C.6) (by Proposition C.11) now reads $\theta(t) = \sum_{n \in \mathbb{Z}} e^{-\pi n^2/t}/\sqrt{t} \stackrel{\text{def}}{=} \theta(\frac{1}{t})/\sqrt{t}$, which proves the Jacobi inversion formula (1.3).

D. A divisor lemma and a scholium. The following discussion is taken, nearly word for word, from [38] and thus it has wider applications — for example, applications to the theory of Eisenstein series (see pages 274–276 of that reference). For integers d, n with $d \neq 0$ write $d \mid n$, as usual, if d divides n, and write $d \nmid n$ if d does not divide n. For $n \geq 1$, $v \in \mathbb{C}$ let $\sigma_v(n) \stackrel{\text{def}}{=} \sum_{0 < d, d \mid n} d^v$ denote the *divisor function*, and for $k, n \geq 1$ in $\mathbb{Z}$ let

$$d(k, n) = \begin{cases} 1 & \text{if } k \mid n, \\ 0 & \text{if } k \nmid n. \end{cases}$$

Let $\{a_n\}_{n=1}^{\infty}$ be a sequence of complex numbers such that $a \stackrel{\text{def}}{=} \sum_{n=1}^{\infty} |a_n| < \infty$. We shall prove a lemma to the effect that

$$\sum_{n=1}^{\infty}\left(\sum_{m=1}^{\infty} m^{\nu} a_{mn}\right) = \sum_{n=1}^{\infty} \sigma_{\nu}(n) a_n. \tag{D.1}$$

Before formulating a precise statement of (D.1) it is useful to consider some simple observations. For $m, k \geq 1$ in $\mathbb{Z}$ set

$$s_m^{(k)} \stackrel{\text{def}}{=} \sum_{j=1}^{m} a_{kj}, \qquad t_m^{(k)} \stackrel{\text{def}}{=} \sum_{l=1}^{km} d(k,l) a_l.$$

We first prove by induction that the m-th partial sum $s_m^{(k)}$ equals $t_m^{(k)}$ for every m. For $m = 1$, $s_1^{(k)} = a_k$. On the other hand, $t_1^{(k)} = a_k$ because an integer l in the range $1 \leq l \leq k$ is a multiple of k if and only if $l = k$. Proceeding inductively, one has $s_{m+1}^{(k)} = s_m^{(k)} + a_{k(m+1)} = t_m^{(k)} + a_{k(m+1)}$. On the other hand, $t_{m+1}^{(k)} = t_m^{(k)} + d(k, km+1) a_{km+1} + d(k, km+2) a_{km+2} + \cdots + d(k, km+k) a_{km+k}$. For $l \in \mathbb{Z}$ and $1 \leq l \leq k$, we have $k \mid km+l \iff k \mid l \iff k = l$ (again), so $t_{m+1}^{(k)} = t_m^{(k)} + a_{km+k} = s_{m+1}^{(k)}$, which completes the induction.

Now take $m \to \infty$ in the equality $s_m^{(k)} = t_m^{(k)}$ to conclude that $\sum_{j=1}^{\infty} a_{kj}$ exists and

$$\sum_{j=1}^{\infty} a_{kj} = \sum_{l=1}^{\infty} d(k,l) a_l. \tag{D.2}$$

If $\operatorname{Re} \nu < -1$, then since $|d(k,n) k^{\nu} a_n| \leq |a_n|/k^{-\operatorname{Re} \nu}$, we have

$$\sum_{k=1}^{\infty} |d(k,n) k^{\nu} a_n| \leq |a_n| \zeta(-\operatorname{Re} \nu)$$

(where $\zeta(s) = \sum_{n=1}^{\infty} 1/n^s$, $\operatorname{Re} s > 1$, is the Riemann zeta function) and moreover the iterated series $\sum_{n=1}^{\infty} \left(\sum_{k=1}^{\infty} |d(k,n) k^{\nu} a_n| \right)$ converges:

$$\sum_{n=1}^{\infty} \left(\sum_{k=1}^{\infty} |d(k,n) k^{\nu} a_n| \right) \leq a \zeta(-\operatorname{Re} \nu). \tag{D.3}$$

By elementary facts regarding double series (found in advanced calculus texts) it follows that one can conclude that the double series $\sum_{n=1}^{\infty} \sum_{k=1}^{\infty} d(k,n) k^{\nu} a_n$ converges absolutely, and that

$$\sum_{n=1}^{\infty} \left(\sum_{k=1}^{\infty} d(k,n) k^{\nu} a_n \right) = \sum_{k=1}^{\infty} \left(\sum_{n=1}^{\infty} d(k,n) k^{\nu} a_n \right). \tag{D.4}$$

Similarly for $k \geq 1$ fixed, equation (D.2) (with $\{a_n\}_{n=1}^{\infty}$ replaced by $\{|a_n|\}_{n=1}^{\infty}$) yields $\sum_{m=1}^{\infty} |k^{\nu} a_{km}| = k^{\operatorname{Re} \nu} \sum_{m=1}^{\infty} |a_{km}| = k^{\operatorname{Re} \nu} \sum_{l=1}^{\infty} d(k,l) |a_l|$. That is,

$\sum_{m=1}^{\infty}|k^{\nu}a_{km}|<\infty$ and moreover the iterated series $\sum_{k=1}^{\infty}\left(\sum_{m=1}^{\infty}|k^{\nu}a_{km}|\right)$ converges, as it equals $\sum_{k=1}^{\infty}k^{\operatorname{Re}\nu}\sum_{l=1}^{\infty}d(k,l)|a_l| \leq a\zeta(-\operatorname{Re}\nu)$. Thus, similarly to equation (D.4), one has that the double series $\sum_{k=1}^{\infty}\sum_{m=1}^{\infty}k^{\nu}a_{km}$ converges absolutely, and equality of the corresponding iterated series prevails:

$$\sum_{k=1}^{\infty}\left(\sum_{m=1}^{\infty}k^{\nu}a_{km}\right) = \sum_{m=1}^{\infty}\left(\sum_{k=1}^{\infty}k^{\nu}a_{km}\right). \tag{D.5}$$

Given these observations, we can now state and prove the main lemma regarding the validity of equation (D.1):

DIVISOR LEMMA. *Let $\{a_n\}_{n=1}^{\infty}$ be a sequence of complex numbers such that the series $\sum_{n=1}^{\infty}|a_n|$ converges. Let $\sigma_{\nu}(n) = \sum_{0<d,\,d\,|\,n}d^{\nu}$ be the divisor function, as above, for $\nu \in \mathbb{C}, n \geq 1$ in $\mathbb{Z}$. If $\operatorname{Re}\nu < -1$, then the series $\sum_{n=1}^{\infty}\sigma_{\nu}(n)a_n$ converges absolutely, the iterated series $\sum_{n=1}^{\infty}\left(\sum_{m=1}^{\infty}m^{\nu}a_{mn}\right)$ converges and formula (D.1) holds, i.e.,*

$$\sum_{n=1}^{\infty}\left(\sum_{m=1}^{\infty}m^{\nu}a_{mn}\right) = \sum_{n=1}^{\infty}\sigma_{\nu}(n)a_n.$$

The double series $\sum_{m=1}^{\infty}\sum_{n=1}^{\infty}m^{\nu}a_{mn}$, in fact, converges absolutely and the corresponding iterated series $\sum_{n=1}^{\infty}\left(\sum_{m=1}^{\infty}m^{\nu}a_{mn}\right)$, $\sum_{m=1}^{\infty}\left(\sum_{n=1}^{\infty}m^{\nu}a_{mn}\right)$ coincide.

PROOF. $\sum_{k=1}^{n}d(k,n)k^{\nu} \stackrel{\text{def}}{=} \sum_{1\leq k\leq n,\,k\,|\,n}k^{\nu} \stackrel{\text{def}}{=} \sigma_{\nu}(n)$, where $d(k,n) = 0$ for $k > n$. That is,

$$\sum_{k=1}^{\infty}d(k,n)k^{\nu} = \sigma_{\nu}(n) \tag{D.6}$$

for any $\nu \in \mathbb{C}$. The series $\sum_{n=1}^{\infty}\sigma_{\operatorname{Re}\nu}(n)|a_n|$ is, by (D.6), the iterated series $\sum_{n=1}^{\infty}\left(\sum_{k=1}^{\infty}d(k,n)k^{\operatorname{Re}\nu}|a_n|\right)$, which we have seen converges and is bounded above by $\zeta(-\operatorname{Re}\nu)$ according to (D.3). Clearly $|\sigma_{\nu}(n)| \leq \sigma_{\operatorname{Re}\nu}(n)$, so that also $\sum_{n=1}^{\infty}|\sigma_{\nu}(n)||a_n|$ converges. Again by (D.6), we have $\sum_{n=1}^{\infty}\sigma_{\nu}(n)a_n = \sum_{n=1}^{\infty}\left(\sum_{k=1}^{\infty}d(k,n)k^{\nu}a_n\right)$. Now apply equations (D.4), (D.2), (D.5), successively, to express the latter iterated series as $\sum_{k=1}^{\infty}\left(\sum_{n=1}^{\infty}d(k,n)k^{\nu}a_n\right) = \sum_{k=1}^{\infty}\left(k^{\nu}\sum_{m=1}^{\infty}a_{km}\right) = \sum_{m=1}^{\infty}\left(\sum_{k=1}^{\infty}k^{\nu}a_{km}\right)$, which proves (D.1). We have already seen that the double series $\sum_{m=1}^{\infty}\sum_{n=1}^{\infty}m^{\nu}a_{mn}$ (which equals $\sum_{k=1}^{\infty}\sum_{m=1}^{\infty}k^{\nu}a_{km}$) converges absolutely. By equation (D.5), then one derives the equality of the corresponding iterated series $\sum_{m=1}^{\infty}\left(\sum_{n=1}^{\infty}m^{\nu}a_{mn}\right)$ and $\sum_{n=1}^{\infty}\left(\sum_{m=1}^{\infty}m^{\nu}a_{mn}\right)$. □

Going back to the equality $s_m^{(k)} = t_m^{(k)}$ of the previous page, we actually have the following fact, recorded for future application:

SCHOLIUM. *Given a sequence of complex numbers $\{a_n\}_{n=1}^\infty$ and $k \in \mathbb{Z}$ with $k \geq 1$, the series $\sum_{j=1}^\infty a_{kj}$ converges if and only if the series $\sum_{j=1}^\infty d(k,j)a_j$ converges, in which case these series coincide.*

As an example, we use the Divisor Lemma to prove the next lemma, which is important for Lecture 4.

LEMMA D.7. *Fix $z, k \in \mathbb{C}$ with $\operatorname{Im} z > 0$, $\operatorname{Re} k > 2$. Then the iterated series $\sum_{m=1}^\infty \left(\sum_{n=1}^\infty n^{k-1} e^{2\pi i m n} \right)$ exists, the series $\sum_{n=1}^\infty \sigma_{1-k}(n) n^{k-1} e^{2\pi i n z}$ converges absolutely, and*

$$\sum_{m=1}^\infty \left(\sum_{n=1}^\infty n^{k-1} e^{2\pi i m n} \right) = \sum_{n=1}^\infty \sigma_{1-k}(n) n^{k-1} e^{2\pi i n z} = \sum_{n=1}^\infty \sigma_{k-1}(n) e^{2\pi i n z}. \quad \text{(D.8)}$$

PROOF. The last equality comes from $\sigma_\nu(n) = n^\nu \sigma_{-\nu}(n)$. To show the first, set $a_n \stackrel{\text{def}}{=} n^{k-1} e^{2\pi i n z}$. Then $\sum_{n=1}^\infty |a_n| = \sum_{n=1}^\infty n^{\operatorname{Re} k - 1} e^{-2\pi n \operatorname{Im} z}$ converges by the ratio test, since $\operatorname{Im} z > 0$. Also $m^{1-k} a_{mn} = m^{1-k} (mn)^{k-1} e^{2\pi i m n z} = n^{k-1} e^{2\pi i m n z}$, where $\operatorname{Re} k > 2 \Rightarrow \operatorname{Re}(1-k) < -1$. By the Divisor Lemma (for $\nu = 1-k$), the series $\sum_{n=1}^\infty \sigma_{1-k}(n) a_n$ converges absolutely, the iterated series $\sum_{n=1}^\infty \left(\sum_{m=1}^\infty m^{1-k} a_{mn} \right)$ converges, and $\sum_{n=1}^\infty \left(\sum_{m=1}^\infty m^{1-k} a_{mn} \right) = \sum_{n=1}^\infty \sigma_{1-k}(n) a_n$. Substituting the value of a_n proves the desired equality. □

As another example, consider the sequence $\{a_n\}_{n=1}^\infty$ given by

$$a_n \stackrel{\text{def}}{=} e^{\pm 2\pi n x i} K_{s-\frac{1}{2}}(2\pi n y) n^{s-\frac{1}{2}}$$

for $x, y \in \mathbb{R}$, $y > 0$, $s \in \mathbb{C}$ fixed, where

$$K_\nu(z) \stackrel{\text{def}}{=} \frac{1}{2} \int_0^\infty \exp\left(-\frac{z}{2}\left(t + \frac{1}{t} \right) \right) t^{\nu-1} \, dt \quad \text{(D.9)}$$

is the K-Bessel function for $\operatorname{Re} z > 0$, $\nu \in \mathbb{C}$. To see that $\sum_{n=1}^\infty |a_n| < \infty$, one applies the asymptotic result

$$\lim_{t \to \infty} \sqrt{t} K_\nu(t) e^t = \sqrt{\frac{\pi}{2}}. \quad \text{(D.10)}$$

In particular, choose $M_\nu > 0$ such that $\left| \sqrt{t} K_\nu(t) e^t - \sqrt{\pi/2} \right| < \sqrt{\pi/2}$ for $t > M_\nu$; that is, $|K_\nu(t)| < 2\sqrt{\pi/2t} e^{-t}$ for $t > M_\nu$. Then if $N_{s,y}$ is an integer with $N_{s,y} > M_{s-\frac{1}{2}}/2\pi y$ we see that for $n \geq N_{s,y}$, $2\pi n y > M_{s-\frac{1}{2}}$, so

$$\left| K_{s-\frac{1}{2}}(2\pi n y) \right| < 2 \sqrt{\frac{\pi}{2 \cdot 2\pi n y}} e^{-2\pi n y} = \frac{n^{-1/2}}{\sqrt{y}} e^{-2\pi n y};$$

therefore $|a_n| < (n^{\operatorname{Re} s - 1}/\sqrt{y}) e^{-2\pi n y}$, where $\sum_{n=1}^\infty n^{\operatorname{Re} s - 1} e^{-2\pi n y}$ converges by the ratio test since $y > 0$.

Now assume that $\operatorname{Re} s > 1$ so that $\nu \stackrel{\text{def}}{=} -2s + 1$ satisfies $\operatorname{Re} \nu < -1$. Also $m^\nu a_{mn} \stackrel{\text{def}}{=} e^{\pm 2\pi mnxi} K_{s-\frac{1}{2}}(2\pi mny)\left(\frac{n}{m}\right)^{s-\frac{1}{2}}$ for $m, n \geq 1$. The Divisor Lemma gives

$$\sum_{m=1}^{\infty}\left(\sum_{n=1}^{\infty} e^{\pm 2\pi mnxi} K_{s-\frac{1}{2}}(2\pi mny)\left(\frac{n}{m}\right)^{s-\frac{1}{2}}\right)$$
$$= \sum_{n=1}^{\infty} \sigma_{-2s+1}(n) e^{\pm 2\pi nxi} K_{s-\frac{1}{2}}(2\pi ny) n^{s-\frac{1}{2}}, \quad \text{(D.11)}$$

with absolute convergence of the latter series, and convergence of the iterated series which coincides with the iterated series

$$\sum_{n=1}^{\infty}\left(\sum_{m=1}^{\infty} e^{\pm 2\pi mnxi} K_{s-\frac{1}{2}}(2\pi mny)\left(\frac{n}{m}\right)^{s-\frac{1}{2}}\right).$$

E. Another summation formula and a proof of formula (2.4). In addition to the useful Poisson summation formula

$$\sum_{n \in \mathbb{Z}} f(n) = \sum_{n \in \mathbb{Z}} \hat{f}(n) \quad \text{(E.1)}$$

of Theorem C.3, there are other very useful, well known summation formulas. The one that we consider here assumes the form

$$\sum_{n \in \mathbb{Z}} f(n) = -\text{the sum of residues of } (\pi \cot \pi z) f(z) \text{ at the poles of } f(z), \quad \text{(E.2)}$$

for a suitable class of functions $f(z)$. As we applied formula (E.1) to a specific function (namely the function $f(z) = e^{-\pi z^2 t}$ for $t > 0$ fixed) to prove the Jacobi inversion formula (1.3), we will, similarly, apply formula (E.2) to a specific function (namely the function $f(z) = (z^2 + a^2)^{-1}$ for $a > 0$ fixed) to prove formula (2.4) of Lecture 2. The main observation towards the proof of formula (E.2) is that there is a nice bound for $|\cot \pi z|$ on a square C_N with side contours R_N, L_N and top and bottom contours T_N, B_N, as illustrated on the next page, for a fixed integer $N > 0$. The bound, in fact, is *independent* of N. Namely,

$$|\cot \pi z| \leq \max(1, B) < 2 \quad \text{(E.3)}$$

on C_N, for $B \stackrel{\text{def}}{=} (1 + e^{-\pi})/(1 - e^{-\pi})$. We begin by checking this known result.

For $z = x + iy$, $x, y \in \mathbb{R}$, we have $i\pi z = -\pi y + i\pi x$, and simple manipulations give

$$\cot \pi z = i \frac{e^{i\pi z} + e^{-i\pi z}}{e^{i\pi z} - e^{-i\pi z}} = i \frac{e^{-\pi y} e^{i\pi x} + e^{\pi y} e^{-i\pi x}}{e^{-\pi y} e^{i\pi x} - e^{\pi y} e^{-i\pi x}}. \quad \text{(E.4)}$$

Hence

$$|\cot \pi z| \leq \frac{e^{-\pi y} + e^{\pi y}}{|e^{-\pi y} e^{i\pi x} - e^{\pi y} e^{-i\pi x}|} \leq \frac{e^{-\pi y} + e^{\pi y}}{|e^{-\pi y} - e^{\pi y}|}$$

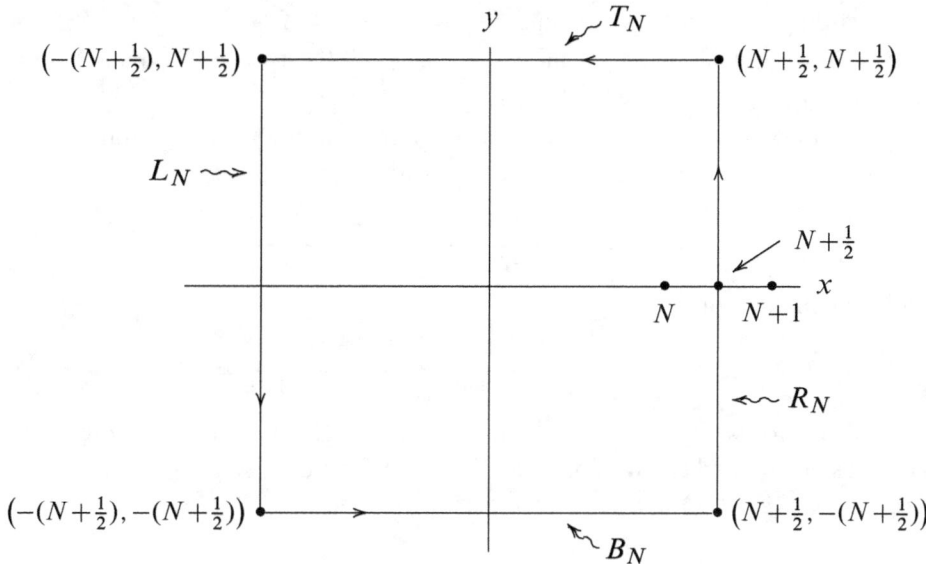

(since $|a-b| \geq ||a|-|b||$ for $a, b \in \mathbb{C}$). Since $|a| \geq \pm a$ for $a \in \mathbb{R}$, this becomes

$$|\cot \pi z| \leq \frac{e^{-\pi y} + e^{\pi y}}{\pm(e^{-\pi y} - e^{\pi y})}. \tag{E.5}$$

Suppose (in general) that $y > \frac{1}{2}$. Then $2\pi y > \pi$, so $e^{-2\pi y} < e^{-\pi}$ and $1 - e^{-2\pi y} > 1 - e^{-\pi}$, which, by the choice of the minus sign in (E.5), allows us to write

$$|\cot \pi z| \leq \left(\frac{e^{-\pi y} + e^{\pi y}}{e^{\pi y} - e^{-\pi y}}\right) \frac{e^{-\pi y}}{e^{-\pi y}} = \frac{e^{-2\pi y} + 1}{1 - e^{-2\pi y}} < \frac{e^{-\pi} + 1}{1 - e^{-\pi}} \stackrel{\text{def}}{=} B,$$

for $y > \frac{1}{2}$. Similarly if $y < -\frac{1}{2}$, then $2\pi y < -\pi$, so $e^{2\pi y} < e^{-\pi}$ and $1 - e^{2\pi y} > 1 - e^{-\pi}$, and by the choice of the plus sign in (E.5) we get

$$|\cot \pi z| \leq \left(\frac{e^{-\pi y} + e^{\pi y}}{e^{-\pi y} - e^{\pi y}}\right) \frac{e^{\pi y}}{e^{\pi y}} = \frac{1 + e^{2\pi y}}{1 - e^{2\pi y}} < \frac{1 + e^{-\pi}}{1 - e^{-\pi}} = B.$$

Thus we see that

$$|\cot \pi z| < B \stackrel{\text{def}}{=} \frac{1 + e^{-\pi}}{1 - e^{-\pi}} \tag{E.6}$$

for $z \in \mathbb{C}$ with either $\operatorname{Im} z > \frac{1}{2}$ or $\operatorname{Im} z < -\frac{1}{2}$. In particular on T_N, $\operatorname{Im} z = N + \frac{1}{2} > \frac{1}{2}$ and on B_N, $\operatorname{Im} z = -(N + \frac{1}{2}) < -\frac{1}{2}$ so by (E.6) the estimate

$$|\cot \pi z| < B \tag{E.7}$$

holds on both the contours T_N and B_N.

Regarding the contours R_N and L_N we have $z = N + \frac{1}{2} + iy$ on R_N and $z = -(N + \frac{1}{2}) + iy$ on L_N. On R_N we have $e^{2\pi i z} = -e^{-2\pi y}$ (since $N \in \mathbb{Z}$) and, similarly, $e^{2\pi i z} = -e^{-2\pi y}$ on L_N.

Now consider the first equation in (E.4) and multiply the fraction by $1 = e^{i\pi z}/e^{i\pi z}$. On both R_N and L_N this leads to

$$\cot \pi z = i \frac{e^{2\pi i z} + 1}{e^{2\pi i z} - 1} = i \frac{-e^{-2\pi y} + 1}{-e^{-2\pi y} - 1},$$

and we conclude that

$$|\cot \pi z| = \frac{|1 - e^{-2\pi \operatorname{Im} z}|}{1 + e^{-2\pi \operatorname{Im} z}} \leq \frac{1 + e^{-2\pi \operatorname{Im} z}}{1 + e^{-2\pi \operatorname{Im} z}} = 1 \quad (E.8)$$

on both contours R_N and L_N. The inequalities (E.7), (E.8) therefore imply (E.3), as desired, where we note that $e^t \geq 1 + t$ for $t \in \mathbb{R}$, so $e^\pi \geq 1 + \pi > 3 \Rightarrow 2(e^\pi - 1) - (e^\pi + 1) = e^\pi - 3 > 0$. That is, indeed

$$2 > \left(\frac{e^\pi + 1}{e^\pi - 1}\right) \frac{e^{-\pi}}{e^{-\pi}} = \frac{1 + e^{-\pi}}{1 - e^{-\pi}} \stackrel{\text{def}}{=} B.$$

We note also that $\sin \pi z = 0 \iff z = n \in \mathbb{Z}$. That is, since (again) $N \in \mathbb{Z}$ we cannot have $\sin \pi z = 0$ on C_N; in particular C_N avoids the poles of $\cot \pi z = \cos \pi z / \sin \pi z$, and $\cot \pi z$ is continuous on C_N.

Consider now a function $f(z)$ subject to the following two conditions:

C1. $f(z)$ is meromorphic on $\mathbb{C}$, with only finitely many poles $z_1, z_2, \ldots, z_k$, none of which is an integer.

C2. There are numbers $M, \rho > 0$ such that $|f(z)| \leq M/|z|^2$ holds for $|z| > \rho$.

Then:

THEOREM E.9. $\lim_{N \to \infty} \sum_{n=-N}^{N} f(n)$ *exists and equals minus the sum of the residues of the function* $f(z) \pi \cot \pi z$ *at the poles* $z_1, z_2, \ldots, z_k$ *of* $f(z)$. *One can replace condition C2, in fact, by the more general condition (E.11) below.*

PROOF. Since the poles z_j are finite in number we can choose N sufficiently large that C_N encloses all of them. The function $\pi \cot \pi z$ has simple poles at the integers (again as $\sin \pi z = 0 \iff z = n \in \mathbb{Z}$) and the residue at $z = n \in \mathbb{Z}$ is immediately calculated to be 1. Therefore the residue of $\phi(z) \stackrel{\text{def}}{=} f(z) \pi \cot \pi z$ at $z = n \in \mathbb{Z}$ is $f(n)$. As none of the z_j are integers (by C1) the poles of $\phi(z)$ within C_N are given precisely by the set $\{z_j, n \mid 1 \leq j \leq k, -N \leq n \leq N, n \in \mathbb{Z}\}$. By the residue theorem, accordingly, we deduce that

$$\frac{1}{2\pi i} \int_{C_N} \phi(z) \, dz =$$

the sum of the residues of $\phi(z)$ at the $z_j + \sum_{n=-N}^{N} f(n)$. (E.10)

Now if
$$\lim_{N\to\infty}\int_{C_N}\phi(z)dz = 0, \tag{E.11}$$
we can let $N \to \infty$ in equation (E.10) and conclude the validity of Theorem E.9 (more generally, without condition C2).

We check that condition (E.11) is implied by condition C2. Since $|z| \geq N + \frac{1}{2}$ on C_N, we have for $N + \frac{1}{2} > \rho$ and z on C_N the bound
$$|f(z)| \leq \frac{M}{|z|^2} \leq \frac{M}{(N+\frac{1}{2})^2}.$$

By the main inequality (E.3), we have $|\pi \cot \pi z| < 2\pi$ on C_N, so $|\phi(z)| = |f(z)\pi \cot \pi z| < 2\pi M/(N+\frac{1}{2})^2$ on C_N. Given that the length of C_N is $4(2(N+\frac{1}{2}))$ we therefore have the following estimate (for $N + \frac{1}{2} > \rho$):
$$\left|\int_{C_N}\phi(z)dz\right| \leq \frac{2\pi M}{(N+\frac{1}{2})^2} 8(N+\frac{1}{2}) = \frac{32\pi M}{2N+1}, \tag{E.12}$$
where we note that $\phi(z)$ is continuous on C_N, because, as seen, $\cot \pi z$ is continuous on C_N. The inequality (E.12) clearly establishes the condition (E.11), by which the proof of Theorem E.9 is concluded. □

As an example of Theorem E.9 we choose
$$f(z) = \frac{1}{z^2 + a^2} = \frac{1}{(z-ai)(z+ai)}$$
for $a > 0$ fixed. Hence f is meromorphic on $\mathbb{C}$ with exactly two simple poles $z_1 \stackrel{\text{def}}{=} ai$, $z_2 \stackrel{\text{def}}{=} -ai$. Suppose, for example, that $|z| > \sqrt{2}a$: Then
$$1 - \frac{a^2}{|z|^2} > \frac{1}{2}, \quad \text{so} \quad \left|1 + \frac{a^2}{z^2}\right| \geq 1 - \frac{a^2}{|z|^2}, \quad \text{so} \quad \left|\frac{z^2}{z^2+a^2}\right| = \frac{1}{\left|1+\frac{a^2}{z^2}\right|} < 2.$$

Therefore $|f(z)| < 2/|z|^2$; that is, $f(z)$ satisfies conditions C1, C2 with $M = 2$, $\rho = \sqrt{2}a$. The residue of $\phi(z) \stackrel{\text{def}}{=} f(z)\pi \cot \pi z$ at z_1 is
$$\lim_{z\to z_1}(z-z_1)\phi(z) = \lim_{z\to ai}\frac{\pi \cot \pi z}{z+ai} = \frac{\pi \cot \pi ai}{2ai} = -\frac{\pi}{2a}\coth \pi a,$$
since $\cos iw = \cosh w$, $\sin iw = i \sinh w$. Similarly, the residue of $\phi(z)$ at z_2 is $-(\pi/2a)\coth \pi a$. As $f(-z) = f(z)$, $\sum_{n=-N}^{N} f(n) = \frac{1}{a^2} + 2\sum_{n=1}^{N}\frac{1}{n^2+a^2}$. Theorem E.9 therefore gives
$$\frac{1}{a^2} + 2\sum_{n=1}^{\infty}\frac{1}{n^2+a^2} = -\left(-\frac{\pi}{2a}\coth \pi a - \frac{\pi}{2a}\coth \pi a\right) = \frac{\pi}{a}\coth \pi a,$$

which proves the summation formula (2.4).

F. Generators of $SL(2, \mathbb{Z})$. Let $\Gamma \stackrel{\text{def}}{=} SL(2, \mathbb{Z})$. To prove that the elements $T \stackrel{\text{def}}{=} \begin{bmatrix} 1 & 1 \\ 0 & 1 \end{bmatrix}$ and $S \stackrel{\text{def}}{=} \begin{bmatrix} 0 & -1 \\ 1 & 0 \end{bmatrix} \in \Gamma$ generate Γ, we start with a lemma.

LEMMA F.1. *Let* $\gamma = \begin{bmatrix} a & b \\ c & d \end{bmatrix} \in \Gamma$ *with* $c \geq 1$. *Then* γ *is a finite product* $\gamma_1 \cdots \gamma_l$, *where each* $\gamma_j \in \Gamma$ *has the form* $\gamma_j = T^{n_j} S^{m_j}$ *for some* $n_j, m_j \in \mathbb{Z}$. *Here for a group element* g *and* $0 > n \in \mathbb{Z}$, *we have set* $g^n \stackrel{\text{def}}{=} (g^{-1})^{-n}$.

The proof is by induction on c. If $c = 1$, then $1 = \det \gamma = ad - bc = ad - b$, so $b = ad - 1$, so

$$\gamma = \begin{bmatrix} a & ad-1 \\ 1 & d \end{bmatrix} = \begin{bmatrix} 1 & a \\ 0 & 1 \end{bmatrix}\begin{bmatrix} 0 & -1 \\ 1 & 0 \end{bmatrix}\begin{bmatrix} 1 & d \\ 0 & 1 \end{bmatrix}\begin{bmatrix} 1 & 0 \\ 0 & 1 \end{bmatrix} = \gamma_1 \gamma_2$$

for $\gamma_1 = \begin{bmatrix} 1 & a \\ 0 & 1 \end{bmatrix}\begin{bmatrix} 0 & -1 \\ 1 & 0 \end{bmatrix} = T^a S^1$, $\gamma_2 = \begin{bmatrix} 1 & d \\ 0 & 1 \end{bmatrix}\begin{bmatrix} 1 & 0 \\ 0 & 1 \end{bmatrix} = T^d S^0$. Proceeding by induction, we use the Euclidean algorithm to write $d = qc + r$ for $q, r \in \mathbb{Z}$ with $0 \leq r < c \geq 2$, say. If $r = 0$, then $1 = \det \gamma = ad - bc = (aq - b)c$, which shows that c is a positive divisor of 1. That is, the contradiction $c = 1$ implies that $r > 0$. Now

$$\gamma T^{-q} = \begin{bmatrix} a & b \\ c & d \end{bmatrix}\begin{bmatrix} 1 & -q \\ 0 & 1 \end{bmatrix} = \begin{bmatrix} a & -aq+b \\ c & -cq+d \end{bmatrix} = \begin{bmatrix} a & -aq+b \\ c & r \end{bmatrix},$$

so $\gamma T^{-q} S = \begin{bmatrix} a & -aq+b \\ c & r \end{bmatrix}\begin{bmatrix} 0 & -1 \\ 1 & 0 \end{bmatrix} = \begin{bmatrix} -aq+b & -a \\ r & -c \end{bmatrix}$, which equals $\gamma_1 \cdots \gamma_l$ by induction (since $1 \leq r < c$), where each γ_j has the form $\gamma_j = T^{n_j} S^{m_j}$ for some $n_j, m_j \in \mathbb{Z}$. Consequently,

$$\gamma = \gamma_1 \cdots \gamma_l S^{-1} T^q = (\gamma_1 \cdots \gamma_{l-1})(T^{n_l} S^{m_l - 1}) T^q S^0,$$

which has the desired form for γ and which therefore completes the induction and the proof of Lemma F.1.

THEOREM F.2. *The elements* T, S *generate* Γ: *Every* $\gamma \in \Gamma$ *is a finite product* $\gamma_1 \cdots \gamma_l$ *where each* $\gamma_j \in \Gamma$ *has the form* $\gamma_j = T^{n_j} S^{m_j}$ *for some* $n_j, m_j \in \mathbb{Z}$.

PROOF. Let $\gamma = \begin{bmatrix} a & b \\ c & d \end{bmatrix} \in \Gamma$ be arbitrary. If $c = 0$, then $1 = \det \gamma = ad$, so $a = d = \pm 1$, so

$$\gamma = \begin{bmatrix} 1 & b \\ 0 & 1 \end{bmatrix} = T^b S^0 \quad \text{or} \quad \gamma = \begin{bmatrix} -1 & b \\ 0 & -1 \end{bmatrix} = \begin{bmatrix} 1 & -b \\ 0 & 1 \end{bmatrix}\begin{bmatrix} -1 & 0 \\ 0 & -1 \end{bmatrix} = T^{-b} S^2.$$

Since the case $c \geq 1$ is already settled by Lemma F.1, there remains only the case $c \leq -1$. Then $\gamma S^2 = \gamma \begin{bmatrix} -1 & 0 \\ 0 & -1 \end{bmatrix} = \begin{bmatrix} -a & -b \\ -c & -d \end{bmatrix} = \gamma_1 \cdots \gamma_l$, by Lemma F.1 (since $-c \geq 1$), where the γ_j have the desired form. Thus $\gamma = \gamma_1 \cdots \gamma_l \cdot (T^0 S^{-2})$, as desired. □

G. Convergence of the sum of $|m+ni|^{-\alpha}$ for $\alpha > 2$. To complete the argument that the Eisenstein series $G_k(z)$ converge absolutely and uniformly on each of the strips $S_{A,\delta}$ in definition (4.5) (for $k = 4, 6, 8, 10, 12, \dots$), we must show, according to the inequality (4.10), that the series

$$S(\alpha) \stackrel{\text{def}}{=} \sum_{(m,n) \in \mathbb{Z}_*^2} \frac{1}{|m+ni|^\alpha}$$

converges for $\alpha > 2$, where $\mathbb{Z}_*^2 = \mathbb{Z} \times \mathbb{Z} - \{(0,0)\}$.

For $n \geq 1, n \in \mathbb{Z}$, let π_n denote the set of *integer* points on the boundary of the square with vertices (n,n), $(-n,n)$, $(-n,-n)$, $(n,-n)$. As an example, π_3 is illustrated here, with $24 = 8 \cdot 3$ points.

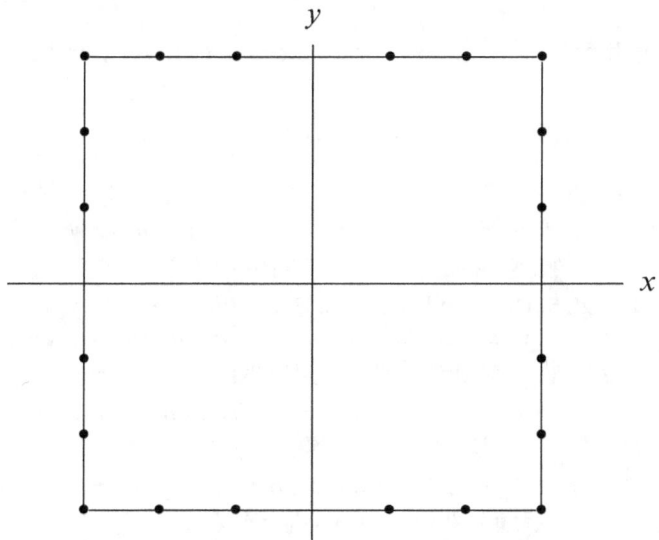

In general π_n has $|\pi_n| \stackrel{(i)}{=} 8n$ points. Also the π_n partition out all of the nonzero integer pairs:

$$\mathbb{Z} \times \mathbb{Z} - \{(0,0)\} = \bigcup_{n=1}^{\infty} \pi_n \qquad (G.1)$$

is a disjoint union.

LEMMA G.2. $\sum_{(a,b) \in \pi_n} \frac{1}{|a+bi|^\alpha} \leq \frac{8}{n^{\alpha-1}}$ *for* $\alpha \geq 0, n \geq 1$.

PROOF. For $(a,b) \in \pi_n$, either $a = \pm n$ or $b = \pm n$, according to whether (a,b) lies on one of the vertical sides of the square (as illustrated above for π_3), or on one of the horizontal sides, respectively. Thus $a^2 + b^2 = $ either $n^2 + b^2$ or

$a^2+n^2 \Rightarrow a^2+b^2 \geq n^2$. That is, for $(a,b) \in \pi_n$ we have $|a+bi|^2 = a^2+b^2 \geq n^2$, so $|a+bi|^\alpha \geq n^\alpha$ (since $\alpha \geq 0$). Inverting and summing we get

$$\sum_{(a,b)\in\pi_n} \frac{1}{|a+bi|^\alpha} \leq \sum_{(a,b)\in\pi_n} \frac{1}{n^\alpha} \leq \frac{8n}{n^\alpha} \quad \text{(by (i))}$$
$$= \frac{8}{n^{\alpha-1}},$$

which proves Lemma G.2. □

Now use (G.1) and Lemma G.2 to write

$$S(\alpha) = \sum_{n=1}^{\infty} \sum_{(a,b)\in\pi_n} \frac{1}{|a+bi|^\alpha} \leq \sum_{n=1}^{\infty} \frac{8}{n^{\alpha-1}} \tag{G.3}$$

for $\alpha \geq 0$, which proves that $S(\alpha) < \infty$ for $\alpha - 1 > 1$, as desired.

References

[1] T. M. Apostol, *Introduction to analytic number theory*, Springer, New York, 1976.

[2] ———, *Modular functions and Dirichlet series in number theory*, second ed., Graduate Texts in Mathematics, no. 41, Springer, New York, 1990.

[3] N. Arkani-Hamed, S. Dimopoulos, and G. Dvali, *Phenomenology, astrophysics, and cosmology of theories with submillimeter dimensions and tev scale quantum gravity*, Phys. Rev. D **59** (1999), art. #086004, See also hep-ph/9807344 on the arXiv.

[4] D. Birmingham, I. Sachs, and S. Sen, *Exact results for the BTZ black hole*, Internat. J. Modern Phys. D **10** (2001), no. 6, 833–857.

[5] D. Birmingham and S. Sen, *Exact black hole entropy bound in conformal field theory*, Phys. Rev. D (3) **63** (2001), no. 4, 047501, 3.

[6] N. Brisebarre and G. Philibert, *Effective lower and upper bounds for the Fourier coefficients of powers of the modular invariant j*, J. Ramanujan Math. Soc. **20** (2005), no. 4, 255–282.

[7] D. Bump, J. W. Cogdell, E. de Shalit, D. Gaitsgory, E. Kowalski, and S. S. Kudla, *An introduction to the Langlands program*, Birkhäuser, Boston, MA, 2003.

[8] J. J. Callahan, *The geometry of spacetime: An introduction to special and general relativity*, Springer, New York, 2000.

[9] J. L. Cardy, *Operator content of two-dimensional conformally invariant theories*, Nuclear Phys. B **270** (1986), no. 2, 186–204.

[10] S. Carlip, *Logarithmic corrections to black hole entropy, from the Cardy formula*, Classical quantum gravity **17** (2000), no. 20, 4175–4186.

[11] S. Carroll, *Spacetime and geometry: An introduction to general relativity*, Addison Wesley, San Francisco, 2004.

[12] E. Elizalde, *Ten physical applications of spectral zeta functions*, Lecture Notes in Physics. New Series: Monographs, no. 35, Springer, Berlin, 1995.

[13] E. Elizalde, S. D. Odintsov, A. Romeo, A. A. Bytsenko, and S. Zerbini, *Zeta regularization techniques with applications*, World Scientific, River Edge, NJ, 1994.

[14] S. S. Gelbart and S. D. Miller, *Riemann's zeta function and beyond*, Bull. Amer. Math. Soc. (N.S.) **41** (2004), no. 1, 59–112.

[15] R. C. Gunning, *Lectures on modular forms*, Annals of Mathematics Studies, no. 48, Princeton University Press, Princeton, N.J., 1962.

[16] S. W. Hawking, *Zeta function regularization of path integrals in curved spacetime*, Comm. Math. Phys. **55** (1977), no. 2, 133–148.

[17] E. Hecke, *Über die Bestimmung Dirichletscher Reihen durch ihre Funktionalgleichung*, Math. Ann. **112** (1936), no. 1, 664–699.

[18] _____, *Über Modulfunktionen und die Dirichletschen Reihen mit Eulerscher Produktentwicklung, I*, Math. Ann. **114** (1937), no. 1, 1–28.

[19] A. Ivić, *The Riemann zeta-function: The theory of the riemann zeta-function with applications*, Wiley, New York, 1985.

[20] H. Iwaniec, *Topics in classical automorphic forms*, Graduate Studies in Mathematics, no. 17, American Mathematical Society, Providence, RI, 1997.

[21] A. Kehagias and K. Sfetsos, *Deviations from the $1/r^2$ Newton law due to extra dimensions*, Phys. Lett. B **472** (2000), no. 1-2, 39–44.

[22] K. Kirsten, *Spectral functions in mathematics and physics*, Chapman and Hall / CRC, 2002.

[23] M. I. Knopp, *Automorphic forms of nonnegative dimension and exponential sums*, Michigan Math. J. **7** (1960), 257–287.

[24] D. Lyon, *The physics of the riemann zeta function*, online lecture, available at http://tinyurl.com/yar2znr.

[25] J. E. Marsden, *Basic complex analysis*, W. H. Freeman and Co., San Francisco, Calif., 1973.

[26] S. Minakshisundaram and Å. Pleijel, *Some properties of the eigenfunctions of the Laplace-operator on Riemannian manifolds*, Canadian J. Math. **1** (1949), 242–256.

[27] H. Petersson, *Über die Entwicklungskoeffizienten der automorphen Formen*, Acta Math. **58** (1932), no. 1, 169–215.

[28] H. Rademacher, *The Fourier coefficients of the modular invariant $J(\tau)$*, Amer. J. Math. **60** (1938), no. 2, 501–512.

[29] _____, *Fourier expansions of modular forms and problems of partition*, Bull. Amer. Math. Soc. **46** (1940), 59–73.

[30] _____, *Topics in analytic number theory*, Grundlehren math. Wissenschaften, vol. Band 169, Springer, New York, 1973.

[31] H. Rademacher and Herbert S. Zuckerman, *On the Fourier coefficients of certain modular forms of positive dimension*, Ann. of Math. (2) **39** (1938), no. 2, 433–462.

[32] B. Riemann, *Über die Anzahl der Primzahlen unter einer gegebenen Grösse*, Mon. Not. Berlin Akad. (1859), 671–680, An English translation by D. Wilkins is available at http://www.maths.tcd.ie/pub/histmath/people/riemann/zeta.

[33] P. Sarnak, *Some applications of modular forms*, Cambridge Tracts in Mathematics, no. 99, Cambridge University Press, Cambridge, 1990.

[34] A. Strominger and C. Vafa, *Microscopic origin of the Bekenstein–Hawking entropy*, Phys. Lett. B **379** (1996), no. 1-4, 99–104.

[35] A. Terras, *Harmonic analysis on symmetric spaces and applications, I*, Springer, New York, 1985.

[36] E. C. Titchmarsh, *The theory of the Riemann zeta-function*, Clarendon Press, Oxford, 1951.

[37] H. Weyl, *The theory of groups and quantum mechanics*, Dover, New York, 1931.

[38] F. L. Williams, *Lectures on the spectrum of $L^2(\Gamma \backslash G)$*, Pitman Research Notes in Mathematics Series, no. 242, Longman, Harlow, 1991.

[39] ———, *The role of Selberg's trace formula in the computation of Casimir energy for certain Clifford–Klein space-times*, African Americans in mathematics (Piscataway, NJ, 1996), DIMACS Ser. Discrete Math. Theoret. Comput. Sci., no. 34, Amer. Math. Soc., Providence, RI, 1997, pp. 69–82.

[40] ———, *Topological Casimir energy for a general class of Clifford–Klein space-times*, J. Math. Phys. **38** (1997), no. 2, 796–808.

[41] ———, *Meromorphic continuation of Minakshisundaram–Pleijel series for semisimple Lie groups*, Pacific J. Math. **182** (1998), no. 1, 137–156.

[42] ———, *Topics in quantum mechanics*, Progress in Mathematical Physics, no. 27, Birkhäuser Boston Inc., Boston, MA, 2003.

FLOYD L. WILLIAMS
DEPARTMENT OF MATHEMATICS AND STATISTICS
LEDERLE GRADUATE RESEARCH TOWER
710 NORTH PLEASANT STREET
UNIVERSITY OF MASSACHUSETTS
AMHERST, MA 01003-9305
UNITED STATES
williams@math.umass.edu

Basic zeta functions and some applications in physics

KLAUS KIRSTEN

1. Introduction

It is the aim of these lectures to introduce some basic zeta functions and their uses in the areas of the Casimir effect and Bose–Einstein condensation. A brief introduction into these areas is given in the respective sections; for recent monographs on these topics see [8; 22; 33; 34; 57; 67; 68; 71; 72]. We will consider exclusively spectral zeta functions, that is, zeta functions arising from the eigenvalue spectrum of suitable differential operators. Applications like those in number theory [3; 4; 23; 79] will not be considered in this contribution.

There is a set of technical tools that are at the very heart of understanding analytical properties of essentially every spectral zeta function. Those tools are introduced in Section 2 using the well-studied examples of the Hurwitz [54], Epstein [38; 39] and Barnes zeta function [5; 6]. In Section 3 it is explained how these different examples can all be thought of as being generated by the same mechanism, namely they all result from eigenvalues of suitable (partial) differential operators. It is this relation with partial differential operators that provides the motivation for analyzing the zeta functions considered in these lectures. Motivations come for example from the questions "Can one hear the shape of a drum?", "What does the Casimir effect know about a boundary?", and "What does a Bose gas know about its container?" The first two questions are considered in detail in Section 4. The last question is examined in Section 5, where we will see how zeta functions can be used to analyze the phenomenon of Bose–Einstein condensation. Section 6 will point towards recent developments for the analysis of spectral zeta functions and their applications.

2. Some basic zeta functions

In this section we will construct analytical continuations of basic zeta functions. From these we will determine the meromorphic structure, residues at singular points and special function values.

2.1. Hurwitz zeta function. We start by considering a generalization of the Riemann zeta function

$$\zeta_R(s) = \sum_{n=1}^{\infty} \frac{1}{n^s}. \tag{2-1}$$

DEFINITION 2.1. Let $s \in \mathbb{C}$ and $0 < a < 1$. Then for $\operatorname{Re} s > 1$ the Hurwitz zeta function is defined by

$$\zeta_H(s, a) = \sum_{n=0}^{\infty} \frac{1}{(n+a)^s}.$$

Clearly, $\zeta_H(s, 1) = \zeta_R(s)$. Results for $a = 1 + b > 1$ follow by observing

$$\zeta_H(s, 1+b) = \sum_{n=0}^{\infty} \frac{1}{(n+1+b)^s} = \zeta_H(s, b) - \frac{1}{b^s}.$$

In order to determine properties of the Hurwitz zeta function, one strategy is to express it in term of 'known' zeta functions like the Riemann zeta function.

THEOREM 2.2. *For $0 < a < 1$ we have*

$$\zeta_H(s, a) = \frac{1}{a^s} + \sum_{k=0}^{\infty} (-1)^k \frac{\Gamma(s+k)}{\Gamma(s)k!} a^k \zeta_R(s+k).$$

PROOF. Note that for $|z| < 1$ we have the binomial expansion

$$(1-z)^{-s} = \sum_{k=0}^{\infty} \frac{\Gamma(s+k)}{\Gamma(s)k!} z^k.$$

So for $\operatorname{Re} s > 1$ we compute

$$\zeta_H(s, a) = \frac{1}{a^s} + \sum_{n=1}^{\infty} \frac{1}{n^s} \frac{1}{(1+\frac{a}{n})^s} = \frac{1}{a^s} + \sum_{n=1}^{\infty} \frac{1}{n^s} \sum_{k=0}^{\infty} (-1)^k \frac{\Gamma(s+k)}{\Gamma(s)k!} \left(\frac{a}{n}\right)^k$$

$$= \frac{1}{a^s} + \sum_{k=0}^{\infty} (-1)^k \frac{\Gamma(s+k)}{\Gamma(s)k!} a^k \sum_{n=1}^{\infty} \frac{1}{n^{s+k}}$$

$$= \frac{1}{a^s} + \sum_{k=0}^{\infty} (-1)^k \frac{\Gamma(s+k)}{\Gamma(s)k!} a^k \zeta_R(s+k). \qquad \square$$

From here it is seen that $s = 1$ is the only pole of $\zeta_H(s, a)$ with Res $\zeta_H(1, a) = 1$.

In determining certain function values of $\zeta_H(s, a)$ the following polynomials will turn out to be useful.

DEFINITION 2.3. For $x \in \mathbb{C}$ we define the Bernoulli polynomials $B_n(x)$ by the equation

$$\frac{ze^{xz}}{e^z - 1} = \sum_{n=0}^{\infty} \frac{B_n(x)}{n!} z^n, \qquad \text{where } |z| < 2\pi. \tag{2-2}$$

Examples are $B_0(x) = 1$ and $B_1(x) = x - \frac{1}{2}$. The numbers $B_n(0)$ are called Bernoulli numbers and are denoted by B_n. Thus

$$\frac{z}{e^z - 1} = \sum_{n=0}^{\infty} \frac{B_n}{n!} z^n, \qquad \text{where } |z| < 2\pi. \tag{2-3}$$

LEMMA 2.4. *The Bernoulli polynomials satisfy*

(1) $\qquad B_n(x) = \sum_{k=0}^{n} \binom{n}{k} B_k x^{n-k}$,

(2) $\qquad B_n(x+1) - B_n(x) = nx^{n-1} \qquad \text{if } n \geq 1$,

(3) $\qquad (-1)^n B_n(-x) = B_n(x) + nx^{n-1}$,

(4) $\qquad B_n(1-x) = (-1)^n B_n(x)$.

EXERCISE 1. Use relations (2-2) and (2-3) to show assertions (1)–(4).

We now establish elementary properties of $\zeta_H(s, a)$.

THEOREM 2.5. *For* Re $s > 1$ *we have*

$$\zeta_H(s, a) = \frac{1}{\Gamma(s)} \int_0^{\infty} t^{s-1} \frac{e^{-at}}{1 - e^{-t}} dt. \tag{2-4}$$

Furthermore, for $k \in \mathbb{N}_0$ *we have*

$$\zeta_H(-k, a) = -\frac{B_{k+1}(a)}{k+1}.$$

PROOF. We use the definition of the gamma function and have

$$\Gamma(s) = \int_0^{\infty} u^{s-1} e^{-u} du = \lambda^s \int_0^{\infty} t^{s-1} e^{-t\lambda} dt. \tag{2-5}$$

This shows the first part of the theorem,

$$\zeta_H(s, a) = \sum_{n=0}^{\infty} \frac{1}{\Gamma(s)} \int_0^{\infty} t^{s-1} e^{-t(n+a)} dt = \frac{1}{\Gamma(s)} \int_0^{\infty} t^{s-1} \sum_{n=0}^{\infty} e^{-t(n+a)} dt$$

$$= \frac{1}{\Gamma(s)} \int_0^{\infty} t^{s-1} \frac{e^{-at}}{1 - e^{-t}} dt.$$

Furthermore we have

$$\zeta_H(s,a) = \frac{1}{\Gamma(s)} \int_0^\infty t^{s-2} \frac{te^{-ta}}{1-e^{-t}} dt$$

$$= \frac{1}{\Gamma(s)} \int_0^1 t^{s-2} \frac{(-t)e^{-ta}}{e^{-t}-1} dt + \frac{1}{\Gamma(s)} \int_1^\infty t^{s-2} \frac{(-t)e^{-ta}}{e^{-t}-1} dt.$$

The integral in the second term is an entire function of s. Given that the gamma function has singularities at $s = -k$, $k \in \mathbb{N}_0$, only the first term can possibly contribute to the properties $\zeta_H(-k,a)$ considered. We continue and write

$$\frac{1}{\Gamma(s)} \int_0^1 t^{s-2} \frac{(-t)e^{-ta}}{e^{-t}-1} dt = \frac{1}{\Gamma(s)} \int_0^1 t^{s-2} \sum_{n=0}^\infty \frac{B_n(a)}{n!} (-t)^n dt$$

$$= \frac{1}{\Gamma(s)} \sum_{n=0}^\infty \frac{B_n(a)}{n!} \frac{(-1)^n}{s+n-1},$$

which provides the analytical continuation of the integral to the complex plane. From here we observe again

$$\operatorname{Res} \zeta_H(1,a) = B_0(a) = 1.$$

Furthermore the second part of the theorem follows:

$$\zeta_H(-k,a) = \lim_{\varepsilon \to 0} \frac{1}{\Gamma(-k+\varepsilon)} \frac{B_{k+1}(a)}{(k+1)!} \frac{(-1)^{k+1}}{\varepsilon}$$

$$= \lim_{\varepsilon \to 0} (-1)^k k! \varepsilon \frac{B_{k+1}(a)}{(k+1)!} \frac{(-1)^{k+1}}{\varepsilon} = -\frac{B_{k+1}(a)}{k+1}. \qquad \square$$

The disadvantage of the representation (2-4) is that it is valid only for $\operatorname{Re} s > 1$. This can be improved by using a complex contour integral representation. The starting point is the following representation for the gamma function [46].

LEMMA 2.6. *For* $z \notin \mathbb{Z}$ *we have*

$$\Gamma(z) = -\frac{1}{2i \sin(\pi z)} \int_C (-t)^{z-1} e^{-t} dt,$$

where the anticlockwise contour C consists of a circle C_3 of radius $\varepsilon < 2\pi$ and straight lines C_1, respectively C_2, just above, respectively just below, the x-axis; see Figure 1.

PROOF. Assume $\operatorname{Re} z > 1$. As the integrand remains bounded along C_3, no contributions will result as $\varepsilon \to 0$. Along C_1 and C_2 we parametrize as given in

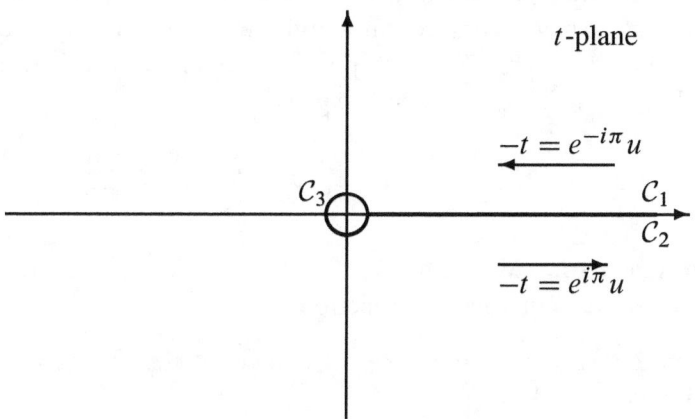

Figure 1. Contour in Lemma 2.6.

Figure 1 and thus for $\operatorname{Re} z > 1$

$$\lim_{\varepsilon \to 0} \int_C (-t)^{z-1} e^{-t} dt = \int_\infty^0 e^{-i\pi(z-1)} u^{z-1} e^{-u} du + \int_0^\infty e^{i\pi(z-1)} u^{z-1} e^{-u} du$$

$$= -\int_0^\infty u^{z-1} e^{-u} \left(e^{i\pi z} - e^{-i\pi z} \right) du$$

$$= -2i \sin(\pi z) \int_0^\infty u^{z-1} e^{-u} du,$$

which implies the assertion by analytical continuation. □

This representation for the gamma function can be used to show the following result for the Hurwitz zeta function.

THEOREM 2.7. *For $s \in \mathbb{C}$, $s \notin \mathbb{N}$, we have*

$$\zeta_H(s, a) = -\frac{\Gamma(1-s)}{2\pi i} \int_C \frac{(-t)^{s-1} e^{-ta}}{1 - e^{-t}} dt,$$

with the contour C given in Figure 1.

PROOF. We follow the previous calculation to note

$$\int_C \frac{(-t)^{s-1} e^{-ta}}{1 - e^{-t}} dt = -2i \sin(\pi s) \int_0^\infty t^{s-1} \frac{e^{-ta}}{1 - e^{-t}} dt,$$

and we use [46]

$$\sin(\pi s) \Gamma(s) = \frac{\pi}{\Gamma(1-s)}$$

to conclude the assertion. □

From here, properties previously given can be easily derived. For $s \in \mathbb{Z}$ the integrand does not have a branch cut and the integral can easily be evaluated using the residue theorem. The only possible singularity enclosed is at $t = 0$ and to read off the residue we use the expansion

$$-(-t)^{s-2}\frac{(-t)e^{-ta}}{e^{-t}-1} = -(-t)^{s-2}\sum_{n=0}^{\infty}\frac{B_n(a)}{n!}(-t)^n.$$

2.2. Barnes zeta function.
The Barnes zeta function is a multidimensional generalization of the Hurwitz zeta function.

DEFINITION 2.8. Let $s \in \mathbb{C}$ with $\operatorname{Re} s > d$ and $c \in \mathbb{R}_+$, $\vec{r} \in \mathbb{R}_+^d$. The Barnes zeta function is defined as

$$\zeta_B(s, c|\vec{r}) = \sum_{\vec{m} \in \mathbb{N}_0^d} \frac{1}{(c + \vec{m} \cdot \vec{r})^s}. \tag{2-6}$$

If $c = 0$ it is understood that the summation ranges over $\vec{m} \neq \vec{0}$.

For $\vec{r} = \vec{1}_d := (1, 1, \ldots, 1, 1)$, the Barnes zeta function can be expanded in terms of the Hurwitz zeta function.

EXAMPLE 2.9. Consider $d = 2$ and $\vec{r} = (1, 1)$. Then

$$\zeta_B(s, c|\vec{1}_2) = \sum_{\vec{m} \in \mathbb{N}_0^2} \frac{1}{(c + m_1 + m_2)^s} = \sum_{k=0}^{\infty} \frac{k+1}{(c+k)^s} = \sum_{k=0}^{\infty} \frac{k+c+1-c}{(c+k)^s}$$
$$= \zeta_H(s-1, c) + (1-c)\zeta_H(s, c).$$

EXAMPLE 2.10. Let $e_k^{(d)}$ be the number of possibilities to write an integer k as a sum over d non-negative integers. We then can write

$$\zeta_B(s, c|\vec{1}_d) = \sum_{\vec{m} \in \mathbb{N}_0^d} \frac{1}{(c + m_1 + \cdots + m_d)^s} = \sum_{k=0}^{\infty} e_k^{(d)} \frac{1}{(c+k)^s}.$$

The coefficient $e_k^{(d)}$ can be determined for example as follows. Consider

$$\frac{1}{(1-x)^d} = \frac{1}{1-x} \cdots \frac{1}{1-x} = \left(\sum_{l_1=0}^{\infty} x^{l_1}\right) \cdots \left(\sum_{l_d=0}^{\infty} x^{l_d}\right)$$
$$= \sum_{l_1=0}^{\infty} \cdots \sum_{l_d=0}^{\infty} x^{l_1 + \cdots + l_d} = \sum_{k=0}^{\infty} e_k^{(d)} x^k.$$

On the other side, using the binomial expansion

$$\frac{1}{(1-x)^d} = \sum_{k=0}^{\infty} \frac{\Gamma(d+k)}{\Gamma(d)k!} x^k = \sum_{k=0}^{\infty} \frac{(d+k-1)!}{(d-1)!k!} x^k = \sum_{k=0}^{\infty} \binom{d+k-1}{d-1} x^k.$$

This shows

$$\zeta_B(s,c|\vec{1}_d) = \sum_{k=0}^{\infty} \binom{d+k-1}{d-1} \frac{1}{(c+k)^s},$$

which, once the dimension d is specified, allows to write the Barnes zeta function as a sum of Hurwitz zeta functions along the lines in Example 2.9.

It is possible to obtain similar formulas for r_i rational numbers [27; 28].

For some properties of the Barnes zeta function the use of complex contour integral representations turns out to be the best strategy.

THEOREM 2.11. *We have the following representations*:

$$\zeta_B(s,c|\vec{r}) = \frac{1}{\Gamma(s)} \int_0^\infty t^{s-1} \frac{e^{-ct}}{\prod_{j=1}^d (1-e^{-r_j t})} dt$$

$$= -\frac{\Gamma(1-s)}{2\pi i} \int_C (-t)^{s-1} \frac{e^{-ct}}{\prod_{j=1}^d (1-e^{-r_j t})} dt,$$

with the contour C given in Figure 1.

EXERCISE 2. Use equation (2-5), and again Lemma 2.6, to prove Theorem 2.11.

The residues of the Barnes zeta function and its values at non-positive integers are best described using generalized Bernoulli polynomials [70].

DEFINITION 2.12. We define the generalized Bernoulli polynomials $B_n^{(d)}(x|\vec{r})$ by the equation

$$\frac{e^{-xt}}{\prod_{j=1}^d (1-e^{-r_j t})} = \frac{(-1)^d}{\prod_{j=1}^d r_j} \sum_{n=0}^{\infty} \frac{(-t)^{n-d}}{n!} B_n^{(d)}(x|\vec{r}).$$

Using Definition 2.12 in Theorem 2.11 one immediately obtains the following properties of the Barnes zeta function.

THEOREM 2.13.

(1) $\operatorname{Res} \zeta_B(z,c|\vec{r}) = \dfrac{(-1)^{d+z}}{(z-1)!(d-z)! \prod_{j=1}^d r_j} B_{d-z}^{(d)}(c|\vec{r}), \qquad z = 1, 2, \ldots, d,$

(2) $\zeta_B(-n, c|\vec{r}) = \dfrac{(-1)^d n!}{(d+n)! \prod_{j=1}^d r_j} B_{d+n}^{(d)}(c|\vec{r}).$

EXERCISE 3. Use the first representation of $\zeta_B(s, c|\vec{r})$ in Theorem 2.11 together with Definition 2.12 to show Theorem 2.13. Follow the steps of the proof in Theorem 2.5.

EXERCISE 4. Use the second representation of $\zeta_B(s, c|\vec{r})$ in Theorem 2.11 together with Definition 2.12 and the residue theorem to show Theorem 2.13.

2.3. Epstein zeta function. We now consider zeta functions associated with sums of squares of integers [38; 39].

DEFINITION 2.14. Let $s \in \mathbb{C}$ with $\operatorname{Re} s > d/2$ and $c \in \mathbb{R}_+, \vec{r} \in \mathbb{R}_+^d$. The Epstein zeta function is defined as

$$\zeta_\mathcal{E}(s, c|\vec{r}) = \sum_{\vec{m} \in \mathbb{Z}^d} \frac{1}{(c + r_1 m_1^2 + r_2 m_2^2 + \cdots + r_d m_d^2)^s}.$$

If $c = 0$ it is understood that the summation ranges over $\vec{m} \neq \vec{0}$.

LEMMA 2.15. *For* $\operatorname{Re} s > d/2$, *we have*

$$\zeta_\mathcal{E}(s, c|\vec{r}) = \frac{1}{\Gamma(s)} \int_0^\infty t^{s-1} \sum_{\vec{m} \in \mathbb{Z}^d} e^{-t(r_1 m_1^2 + \cdots + r_d m_d^2 + c)} dt.$$

PROOF. This follows as before from property (2-5) of the gamma function. □

As we have noted in the proof of Theorem 2.5, it is the small-t behavior of the integrand that determines residues of the zeta function and special function values. The way the integrand is written in Lemma 2.15 this $t \to 0$ behavior is not easily read off. A suitable representation is obtained by using the Poisson resummation [53].

LEMMA 2.16. *Let* $r \in \mathbb{C}$ *with* $\operatorname{Re} r > 0$ *and* $t \in \mathbb{R}_+$, *then*

$$\sum_{l=-\infty}^\infty e^{-trl^2} = \sqrt{\frac{\pi}{tr}} \sum_{l=-\infty}^\infty e^{-\frac{\pi^2}{rt} l^2}.$$

EXERCISE 5. If $F(x)$ is continuous such that

$$\int_{-\infty}^\infty |F(x)| dx < \infty,$$

then we define its Fourier transform by

$$\hat{F}(u) = \int_{-\infty}^\infty F(x) e^{-2\pi i x u} dx.$$

If

$$\int_{-\infty}^\infty |\hat{F}(u)| du < \infty,$$

then we have the Fourier inversion formula

$$F(x) = \int_{-\infty}^{\infty} \hat{F}(u) \, e^{2\pi i x u} \, du.$$

Show the following Theorem: Let $F \in L^1(\mathbb{R})$. Suppose that the series

$$\sum_{n \in \mathbb{Z}} F(n+v)$$

converges absolutely and uniformly in v, and that

$$\sum_{m \in \mathbb{Z}} |\hat{F}(m)| < \infty.$$

Then

$$\sum_{n \in \mathbb{Z}} F(n+v) = \sum_{n \in \mathbb{Z}} \hat{F}(n) e^{2\pi i n v}.$$

Hint: Note that

$$G(v) = \sum_{n \in \mathbb{Z}} F(n+v)$$

is a function of v of period 1.

EXERCISE 6. Apply Exercise 5 with a suitable function $F(x)$ to show the Poisson resummation formula Lemma 2.16.

In Lemma 2.16 it is clearly seen that the only term on the right hand side that is not exponentially damped as $t \to 0$ comes from the $l = 0$ term. Using the resummation formula for all d sums in Lemma 2.15, after resumming the $\vec{m} = \vec{0}$ term contributes

$$\zeta_{\mathcal{E}}^{\vec{0}}(s, c|\vec{r}) = \frac{1}{\Gamma(s)} \int_0^\infty t^{s-1} \frac{\pi^{d/2}}{t^{d/2} \sqrt{r_1 \cdots r_d}} e^{-ct} dt$$

$$= \frac{\pi^{d/2}}{\sqrt{r_1 \cdots r_d} \, \Gamma(s)} \int_0^\infty t^{s-d/2-1} e^{-ct} dt = \frac{\pi^{d/2}}{\sqrt{r_1 \cdots r_d}} \frac{\Gamma\left(s - \frac{d}{2}\right)}{\Gamma(s) c^{s-d/2}}.$$

All other contributions after resummation are exponentially damped as $t \to 0$ and can be given in terms of modified Bessel functions [46].

DEFINITION 2.17. Let $\operatorname{Re} z^2 > 0$. We define the modified Bessel function $K_\nu(z)$ by

$$K_\nu(z) = \frac{1}{2} \left(\frac{z}{2}\right)^\nu \int_0^\infty e^{-t - \frac{z^2}{4t}} t^{-\nu - 1} dt.$$

Performing the resummation in Lemma 2.15 according to Lemma 2.16, with Definition 2.17 one obtains the following representation of the Epstein zeta function valid in the whole complex plane [34; 78].

THEOREM 2.18. *We have*

$$\zeta_{\mathcal{E}}(s,c|\vec{r}) = \frac{\pi^{d/2}}{\sqrt{r_1 \cdots r_d}} \frac{\Gamma\left(s-\frac{d}{2}\right)}{\Gamma(s)} c^{\frac{d}{2}-s} + \frac{2\pi^s c^{\frac{d-2s}{4}}}{\Gamma(s)\sqrt{r_1 \cdots r_d}}$$

$$\times \sum_{\vec{n} \in \mathbb{Z}^d / \{\vec{0}\}} \left(\frac{n_1^2}{r_1} + \cdots + \frac{n_d^2}{r_d}\right)^{\frac{1}{2}\left(s-\frac{d}{2}\right)} K_{\frac{d}{2}-s}\left(2\pi\sqrt{c}\left(\frac{n_1^2}{r_1} + \cdots + \frac{n_d^2}{r_d}\right)^{\frac{1}{2}}\right).$$

EXERCISE 7. Show Theorem 2.18 along the lines indicated.

From Definition 2.17 it is clear that the Bessel function is exponentially damped for large Re z^2. As a result the representation above is numerically very effective as long as the argument of $K_{d/2-s}$ is large. The terms involving the Bessel functions are analytic for all values of s, the first term contains poles. As an immediate consequence of the properties of the gamma function one can show the following properties of the Epstein zeta function.

THEOREM 2.19. *For d even, $\zeta_{\mathcal{E}}(s,c|\vec{r})$ has poles at $s = \frac{d}{2}, \frac{d}{2}-1, \ldots, 1$, whereas for d odd they are located at $s = \frac{d}{2}, \frac{d}{2}-1, \ldots, \frac{1}{2}, -\frac{2l+1}{2}, l \in \mathbb{N}_0$. Furthermore,*

$$\mathrm{Res}\, \zeta_{\mathcal{E}}(j,c|\vec{r}) = \frac{(-1)^{\frac{d}{2}+j} \pi^{\frac{d}{2}} c^{\frac{d}{2}-j}}{\sqrt{r_1 \cdots r_d}\, \Gamma(j)\Gamma\left(\frac{d}{2}-j+1\right)},$$

$$\zeta_{\mathcal{E}}(-p,c|\vec{r}) = \begin{cases} 0 & \text{for } d \text{ odd}, \\ \dfrac{(-1)^{\frac{d}{2}} p! \pi^{\frac{d}{2}} c^{\frac{d}{2}+p}}{\sqrt{r_1 \cdots r_d}\, \Gamma\left(\frac{d}{2}+p+1\right)} & \text{for } d \text{ even}. \end{cases}$$

EXERCISE 8. Use Theorem 2.18 and properties of the gamma function to show Theorem 2.19.

This concludes the list of examples for zeta functions to be considered in what follows. A natural question is what the motivations are to consider these zeta functions. Before we describe a few aspects relating to this question let us mention how all these zeta functions, and many others, result from a common principle.

3. Boundary value problems and associated zeta functions

In this section we explain how the considered zeta functions, and others, are all associated with eigenvalue problems of (partial) differential operators.

EXAMPLE 3.1. Let $M = [0, L]$ be some interval and consider the *Dirichlet* boundary value problem.

$$P\phi_n(x) := -\frac{\partial^2}{\partial x^2}\phi_n(x) = \lambda_n \phi_n(x), \qquad \phi_n(0) = \phi_n(L) = 0.$$

The solutions to the boundary value problem have the general form

$$\phi_n(x) = A \sin(\sqrt{\lambda_n} x) + B \cos(\sqrt{\lambda_n} x).$$

Imposing the Dirichlet boundary condition shows we need

$$\phi_n(0) = B = 0, \qquad \phi_n(L) = A \sin(L\sqrt{\lambda_n}) = 0,$$

which implies

$$\lambda_n = \frac{n^2 \pi^2}{L^2}, \qquad n \in \mathbb{N}.$$

We only need to consider $n \in \mathbb{N}$ because non-positive integers lead to linearly dependent eigenfunctions. The zeta function $\zeta_P(s)$ associated with this boundary value problem is defined to be the sum over all eigenvalues raised to the power $(-s)$, namely

$$\zeta_P(s) = \sum_{n=1}^{\infty} \lambda_n^{-s}, \qquad \operatorname{Re} s > \tfrac{1}{2}.$$

So here the associated zeta function is a multiple of the zeta function of Riemann,

$$\zeta_P(s) = \sum_{n=1}^{\infty} \left(\frac{n\pi}{L}\right)^{-2s} = \left(\frac{L}{\pi}\right)^{2s} \zeta_R(2s).$$

EXAMPLE 3.2. The previous example can be easily generalized to higher dimensions. We consider explicitly two dimensions; for the higher dimensional situation see [1]. Let $M = \{(x, y) | x \in [0, L_1], y \in [0, L_2]\}$. We consider the boundary value problem with Dirichlet boundary conditions on M, that is

$$P\phi_{n,m}(x, y) = \left(-\frac{\partial^2}{\partial x^2} - \frac{\partial^2}{\partial y^2} + c\right) \phi_{n,m}(x, y) = \lambda_{n,m} \phi_{n,m}(x, y),$$
$$\phi_{n,m}(0, y) = \phi_{n,m}(L_1, y) = \phi_{n,m}(x, 0) = \phi_{n,m}(x, L_2) = 0.$$

Using the process of separation of variables, eigenfunctions are seen to be

$$\phi_{n,m}(x, y) = A \sin\left(\frac{n\pi x}{L_1}\right) \sin\left(\frac{m\pi y}{L_2}\right),$$

with the eigenvalues

$$\lambda_{n,m} = \left(\frac{n\pi}{L_1}\right)^2 + \left(\frac{m\pi}{L_2}\right)^2 + c, \qquad n, m \in \mathbb{N}.$$

The associated zeta function therefore is

$$\zeta_P(s) = \sum_{n=1}^{\infty} \sum_{m=1}^{\infty} \left[\left(\frac{n\pi}{L_1}\right)^2 + \left(\frac{m\pi}{L_2}\right)^2 + c \right]^{-s},$$

which can be expressed in terms of the Epstein zeta function given in Definition 2.14 as follows:

$$\zeta_P(s) = \tfrac{1}{4}\zeta_\mathcal{E}\left(s, c \,\Big|\, \left(\left(\frac{\pi}{L_1}\right)^2, \left(\frac{\pi}{L_2}\right)^2\right)\right)$$
$$- \tfrac{1}{4}\zeta_\mathcal{E}\left(s, c \,\Big|\, \left(\frac{\pi}{L_1}\right)^2\right) - \tfrac{1}{4}\zeta_\mathcal{E}\left(s, c \,\Big|\, \left(\frac{\pi}{L_2}\right)^2\right) + \tfrac{1}{4} c^{-s}. \quad (3\text{-}1)$$

EXAMPLE 3.3. Similarly one can consider periodic boundary conditions instead of Dirichlet boundary conditions, this means the manifold M is given by $M = S^1 \times S^1$. In this case the eigenfunctions have to satisfy

$$\phi_{n,m}(0, y) = \phi_{n,m}(L_1, y), \qquad \frac{\partial}{\partial x}\phi_{n,m}(0, y) = \frac{\partial}{\partial x}\phi_{n,m}(L_1, y),$$
$$\phi_{n,m}(x, 0) = \phi_{n,m}(x, L_2), \qquad \frac{\partial}{\partial y}\phi_{n,m}(x, 0) = \frac{\partial}{\partial y}\phi_{n,m}(x, L_2).$$

This shows that

$$\phi_{n,m}(x, y) = A e^{i 2\pi n x / L_1} e^{i 2\pi m y / L_2},$$

which implies for the eigenvalues

$$\lambda_{n,m} = \left(\frac{2\pi n}{L_1}\right)^2 + \left(\frac{2\pi m}{L_2}\right)^2 + c, \qquad (n, m) \in \mathbb{Z}^2.$$

The associated zeta function therefore is

$$\zeta_P(s) = \zeta_\mathcal{E}\left(s, c \,|\, \vec{r}\right), \qquad \vec{r} = \left(\left(\frac{2\pi}{L_1}\right)^2, \left(\frac{2\pi}{L_2}\right)^2\right).$$

Clearly, in d dimensions one finds

$$\zeta_P(s) = \zeta_\mathcal{E}\left(s, c \,|\, \vec{r}\right), \qquad \vec{r} = \left(\left(\frac{2\pi}{L_1}\right)^2, \ldots, \left(\frac{2\pi}{L_d}\right)^2\right).$$

EXAMPLE 3.4. As a final example we consider the Schrödinger equation of atoms in a harmonic oscillator potential. In this case $M = \mathbb{R}^3$, and the eigenvalue equation reads

$$\left(-\frac{\hbar^2}{2m}\Delta + \frac{m}{2}\left(\omega_1 x^2 + \omega_2 y^2 + \omega_3 z^2\right)\right)\phi_{n_1,n_2,n_3}(x,y,z)$$
$$= \lambda_{n_1,n_2,n_3}\phi_{n_1,n_2,n_3}(x,y,z).$$

This differential equation is augmented by the condition that eigenfunctions must be square integrable, $\phi_{n_1,n_2,n_3}(x,y,z) \in \mathcal{L}^2(\mathbb{R}^3)$. As is well known, this gives the eigenvalues

$$\lambda_{n_1,n_2,n_3} = \hbar\omega_1\left(n_1 + \tfrac{1}{2}\right) + \hbar\omega_2\left(n_2 + \tfrac{1}{2}\right) + \hbar\omega_3\left(n_3 + \tfrac{1}{2}\right),$$

for $(n_1, n_2, n_3) \in \mathbb{N}_0^3$. This clearly leads to the Barnes zeta function

$$\zeta_P(s) = \zeta_B(s, c|\vec{r}),$$

where

$$c = \tfrac{1}{2}\hbar(\omega_1 + \omega_2 + \omega_3), \qquad \vec{r} = \hbar(\omega_1, \omega_2, \omega_3).$$

If $M = \mathbb{R}$ is chosen the Hurwitz zeta function results.

The examples above illustrate how the zeta functions considered in Section 2 are all related in a natural way to eigenvalues of specific boundary value problems. In fact, zeta functions in a much more general context are studied in great detail. For our purposes the relevant setting is the setting of Laplace-type operators on a Riemannian manifold M, possibly with a boundary ∂M. Laplace-type means the operator P can be written as

$$P = -g^{jk}\nabla_j^V \nabla_k^V - E,$$

where g^{jk} is the metric of M, ∇^V is the connection on M acting on a smooth vector bundle V over M, and where E is an endomorphism of V. Imposing suitable boundary conditions, eigenvalues λ_n and eigenfunctions ϕ_n do exist,

$$P\phi_n(x) = \lambda_n \phi_n(x),$$

and assuming $\lambda_n > 0$ the zeta function is defined to be

$$\zeta_P(s) = \sum_{n=1}^{\infty} \lambda_n^{-s}$$

for $\operatorname{Re} s$ sufficiently large. If there are modes with $\lambda_n = 0$ those have to be excluded from the sum. Also, if finitely many eigenvalues are negative the zeta function can be defined by choosing nonstandard definitions of the principal value for the argument of complex numbers, but we will not need to consider those cases.

4. Some motivations to consider zeta functions

There are many situations where properties of zeta functions in the above context of Laplace-type operators are needed. In the following we present a few of them, but many more can be found for example in the context of number theory [3; 4; 23; 79] and quantum field theory [8; 14; 15; 16; 26; 30; 31; 33; 41; 42; 57; 74].

4.1. Can one hear the shape of a drum? Let M be a two-dimensional membrane representing a drum with boundary ∂M. The drum is fixed along its boundary. Then possible vibrations of the drum and its fundamental tones are described by the eigenvalue problem

$$-\left(\frac{\partial^2}{\partial x^2} + \frac{\partial^2}{\partial y^2}\right)\phi_n(x,y) = \lambda_n \phi_n(x,y), \qquad \phi_n(x,y)|_{(x,y)\in\partial M} = 0.$$

Here, (x, y) denotes the variables in the plane, the eigenfunctions $\phi_n(x, y)$ describe the amplitude of the vibrations and λ_n its fundamental tones. In 1966 Kac [56] asked if just by listening with a perfect ear, so by knowing all the fundamental tones λ_n, it is possible to hear the shape of the drum. One problem in answering this question is, of course, that in general it will be impossible to write down the eigenvalues λ_n in a closed form and to read off relations with the shape of the drum directly. Instead one has to organize the spectrum intelligently in form of a spectral function to reveal relationships between the eigenvalues and the shape of the drum. In this context a particularly fruitful spectral function is the heat kernel

$$K(t) = \sum_{n=1}^{\infty} e^{-\lambda_n t},$$

which as t tends to zero clearly diverges. Given that some relations between the fundamental tones and properties of the drum are hidden in the $t \to 0$ behavior let us consider this asymptotic behavior very closely. Before we come back to the setting of the drum, let us use a few examples to get an idea what the structure of the $t \to 0$ behavior of the heat kernel is expected to be.

EXAMPLE 4.1. Let $M = S^1$ be the circle with circumference L and let $P = -\partial^2/\partial x^2$. Imposing periodic boundary conditions eigenvalues are

$$\lambda_k = \left(\frac{2\pi k}{L}\right)^2, \qquad k \in \mathbb{Z},$$

and the heat kernel reads

$$K_{S^1}(t) = \sum_{k=-\infty}^{\infty} e^{-(2\pi k/L)^2 t}.$$

From Lemma 2.16 we find the $t \to 0$ behavior

$$K_{S^1}(t) = \frac{1}{\sqrt{4\pi t}} L + \text{(exponentially damped terms)}.$$

With the obvious notation this could be written as

$$K_{S^1}(t) = \frac{1}{\sqrt{4\pi t}} \text{vol } M + \text{(exponentially damped terms)}.$$

EXAMPLE 4.2. The heat kernel for the d-dimensional torus $M = S^1 \times \cdots \times S^1$ with $P = -\Delta$ clearly gives a product of the above and thus

$$K_M(t) = K_{S^1}(t) \times \cdots \times K_{S^1}(t) = \frac{1}{(4\pi t)^{d/2}} \text{vol } M + \text{e.d.t.}$$

EXAMPLE 4.3. To avoid the impression that there is always just one term that is not exponentially damped consider M as above but $P = -\Delta + m^2$. Then

$$K(t) = e^{-m^2 t} K_M(t) = e^{-m^2 t} \left(\frac{1}{(4\pi t)^{d/2}} \text{vol } M + \text{e.d.t.} \right)$$

$$= \frac{1}{(4\pi)^{d/2}} \text{vol } M \sum_{\ell=0}^{\infty} \frac{(-1)^\ell}{\ell!} m^{2\ell} t^{\ell - \frac{d}{2}} + \text{e.d.t.}$$

In fact, the structure of the heat kernel observed in this last example is the structure observed for the general class of Laplace-type operators.

THEOREM 4.4. *Let M be a d-dimensional smooth compact Riemannian manifold without boundary and let*

$$P = -g^{jk} \nabla_j^V \nabla_k^V - E,$$

where g^{jk} is the metric of M, ∇^V is the connection on M acting on a smooth vector bundle V over M, and where E is an endomorphism of V. Then as $t \to 0$,

$$K(t) \sim \sum_{k=0}^{\infty} a_k \, t^{k-d/2}$$

with the so-called heat kernel coefficients a_k.

PROOF. See, e.g., [44]. □

In Example 4.3 one sees that

$$a_k = \frac{1}{(4\pi)^{d/2}} \frac{(-1)^k}{k!} m^{2k} \text{vol } M.$$

In general, the heat kernel coefficients are significantly more complicated and they depend upon the geometry of the manifold M and the endomorphism E [44].

Up to this point we have only considered manifolds without boundary. In order to consider in more detail questions relating to the drum, let us now see what relevant changes in the structure of the small-t heat kernel expansion occur if boundaries are present.

EXAMPLE 4.5. Let $M = [0, L]$ and $P = -\partial^2/\partial x^2$ with Dirichlet boundary conditions imposed. Normalized eigenfunctions are then given by

$$\varphi_\ell(x) = \sqrt{\frac{2}{L}} \sin \frac{\pi \ell x}{L}$$

and the associated eigenvalues are

$$\lambda_\ell = \left(\frac{\pi \ell}{L}\right)^2, \qquad \ell \in \mathbb{N}.$$

Using Lemma 2.16 this time we obtain

$$K(t) = \frac{1}{\sqrt{4\pi t}} \text{vol } M - \tfrac{1}{2} + \text{(exponentially damped terms)}. \qquad (4\text{-}1)$$

Notice that in contrast to previous results we have integer and half-integer powers in t occurring.

EXERCISE 9. There is a more general version of the Poisson resummation formula than the one given in Lemma 2.16, namely

$$\sum_{\ell=-\infty}^{\infty} e^{-t(\ell+c)^2} = \sqrt{\frac{\pi}{t}} \sum_{\ell=-\infty}^{\infty} e^{-\frac{\pi^2}{t}\ell^2 - 2\pi i \ell c}. \qquad (4\text{-}2)$$

Apply Exercise 5 with a suitable function $F(x)$ to show equation (4-2).

EXERCISE 10. Consider the setting described in Example 4.5. The local heat kernel is defined as the solution of the equation

$$\left(\frac{\partial}{\partial t} - \frac{\partial^2}{\partial x^2}\right) K(t, x, y) = 0$$

with the initial condition

$$\lim_{t \to 0} K(t, x, y) = \delta(x, y).$$

In terms of the quantities introduced in Example 4.5 it can be written as

$$K(t, x, y) = \sum_{\ell=1}^{\infty} \varphi_\ell(x) \varphi_\ell(y) e^{-\lambda_\ell t}.$$

Use the resummation (4-2) for $K(t, x, y)$ and the fact that

$$K(t) = \int_0^L K(t, x, x)dx$$

to rediscover the result (4-1).

EXERCISE 11. Let $M = [0, L]$ and

$$P = -\frac{\partial^2}{\partial x^2} + m^2$$

with Dirichlet boundary conditions imposed. Find the small-t asymptotics of the heat kernel.

EXERCISE 12. Let $M = [0, L] \times S^1 \times \cdots \times S^1$ be a d-dimensional manifold and

$$P = -\frac{\partial^2}{\partial x^2} + m^2.$$

Impose Dirichlet boundary conditions on $[0, L]$ and periodic boundary conditions on the circle factors. Find the small-t asymptotics of the heat kernel.

As the examples and exercises above suggest, one has the following result.

THEOREM 4.6. *Let M be a d-dimensional smooth compact Riemannian manifold with smooth boundary and let*

$$P = -g^{jk}\nabla_j^V \nabla_k^V - E,$$

where g^{jk} is the metric of M, ∇^V is the connection on M acting on a smooth vector bundle V over M, and where E is an endomorphism of V. We impose Dirichlet boundary conditions. Then as $t \to 0$,

$$K(t) \sim \sum_{k=0,\frac{1}{2},1,\ldots}^{\infty} a_k\, t^{k-d/2}$$

with the heat kernel coefficients a_k.

PROOF. See, e.g., [44]. □

As for the manifold without boundary case, Theorem 4.4, the heat kernel coefficients depend upon the geometry of the manifold M and the endomorphism E, and in addition on the geometry of the boundary. Note, however, that in contrast to Theorem 4.4 the small-t expansion contains integer and half-integer powers in t.

The same structure of the small-t asymptotics is found for other boundary conditions like Neumann or Robin, see [44], and the coefficients then also depend on the boundary condition chosen. In particular, for Dirichlet boundary conditions one can show the identities

$$a_0 = (4\pi)^{-d/2} \operatorname{vol} M, \qquad a_{1/2} = (4\pi)^{-(d-1)/2} \left(-\tfrac{1}{4}\right) \operatorname{vol} \partial M, \qquad (4\text{-}3)$$

a result going back to McKean and Singer [66]. In the context of the drum, what the formula shows is that by listening with a perfect ear one can indeed hear certain properties like the area of the drum and the circumference of its boundary. But as has been shown by Gordon, Webb and Wolpert [45], one cannot hear all details of the shape.

EXERCISE 13. Use Exercise 12 to verify the general formulas (4-3) for the heat kernel coefficients.

Instead of using the heat kernel coefficients to make the preceding statements, one could equally well have used zeta function properties for equivalent statements. Consider the setting of Theorem 4.6. The associated zeta function is

$$\zeta_P(s) = \sum_{n=1}^{\infty} \lambda_n^{-s},$$

where it follows from Weyl's law [80; 81] that this series is convergent for $\operatorname{Re} s > d/2$. The zeta function is related with the heat kernel by

$$\zeta_P(s) = \frac{1}{\Gamma(s)} \int_0^{\infty} t^{s-1} K(t) \, dt, \qquad (4\text{-}4)$$

where equation (2-5) has been used. This equation allows us to relate residues and function values at certain points with the small-t behavior of the heat kernel. In detail,

$$\operatorname{Res} \zeta_P(z) = \frac{a_{(d/2)-z}}{\Gamma(z)}, \qquad z = \frac{d}{2}, \frac{d-1}{2}, \ldots, \frac{1}{2}, -\frac{2n+1}{2}, \; n \in \mathbb{N}_0, \quad (4\text{-}5)$$

$$\zeta_P(-q) = (-1)^q q!\, a_{\frac{d}{2}+q}, \qquad q \in \mathbb{N}_0. \qquad (4\text{-}6)$$

Keeping in mind the vanishing of the heat kernel coefficients a_k with half-integer index for $\partial M = \varnothing$, see Theorem 4.4, this means for d even the poles are actually located only at $z = d/2, d/2-1, \ldots, 1$. In addition, for d odd we get $\zeta_P(-q) = 0$ for $q \in \mathbb{N}_0$.

EXERCISE 14. Use Theorem 4.6 and proceed along the lines indicated in the proof of Theorem 2.5 to show equations (4-5) and (4-6).

Going back to the setting of the drum properties of the zeta function relate with the geometry of the surface. In particular, from (4-3) and (4-5) one can show the identities

$$\operatorname{Res} \zeta_P(1) = \frac{\operatorname{vol} M}{4\pi}, \qquad \operatorname{Res} \zeta_P(\tfrac{1}{2}) = -\frac{\operatorname{vol} \partial M}{2\pi},$$

and the remarks below equation (4-3) could be repeated.

4.2. What does the Casimir effect know about a boundary? We next consider an application in the context of quantum field theory in finite systems. The importance of this topic lies in the fact that in recent years, progress in many fields has been triggered by the continuing miniaturization of all kinds of technical devices. As the separation between components of various systems tends towards the nanometer range, there is a growing need to understand every possible detail of quantum effects due to the small sizes involved.

Very generally speaking, effects resulting from the finite extension of systems and from their precise form are known as the Casimir effect. In modern technical devices this effect is responsible for up to 10% of the forces encountered in microelectromechanical systems [19; 20]. Casimir forces are of direct practical relevance in nanotechnology where, e.g., sticking of mobile components in micromachines might be caused by them [76]. Instead of fighting the occurrence of the effect in technological devices, the tendency is now to try and take technological advantage of the effect.

Experimental progress in recent years has been impressive and for some configurations allows for a detailed comparison with theoretical predictions. The best tested situations are those of parallel plates [12] and of a plate and a sphere [20; 21; 62; 63; 69]; recently also a plate and a cylinder has been considered [13; 37]. Experimental data and theoretical predictions are in excellent agreement, see, e.g., [8; 25; 61; 64]. This interplay between theory and experiments, and the intriguing technological applications possible, are the main reasons for the heightened interest in this effect in recent years.

In its original form, the effect refers to the situation of two uncharged, parallel, perfectly conducting plates. As predicted by Casimir [17], the plates should attract with a force per unit area, $F(a) \sim 1/a^4$, where a is the distance between the plates. Two decades later Boyer [10] found a repulsive pressure of magnitude $F(R) \sim 1/R^4$ for a perfectly conducting spherical shell of radius R. Up to this day an intuitive understanding of the opposite signs found is lacking. One of the main questions in the context of the Casimir effect therefore is how the occurring forces depend on the geometrical properties of the system considered. Said differently, the question is "What does the Casimir effect know about a boundary?" In the absence of general answers one approach consists in accumulating further knowledge by adding bits of understanding based on specific calculations for specific configurations. Several examples will be provided in this section and we will see the dominant role the zeta functions introduced play. However, before we come to specific settings let us briefly introduce the zeta function regularization of the Casimir energy and force that we will use later.

We will consider the Casimir effect in a quantum field theory of a non-interacting scalar field under *external* conditions. The action in this case is [55]

$$S[\Phi] = -\frac{1}{2} \int_M \Phi(x)\,(\Delta - V(x))\,\Phi(x)\,dx, \qquad (4\text{-}7)$$

describing a scalar field $\Phi(x)$ in the background potential $V(x)$. We assume the Riemannian manifold M to be of the form $M = S^1 \times M_s$, where the circle S^1 of radius β is used to describe finite temperature $T = 1/\beta$ and M_s, in general, is a d-dimensional Riemannian manifold with boundary. For the action (4-7) the corresponding field equations are

$$(\Delta - V(x))\Phi(x) = 0. \qquad (4\text{-}8)$$

If M_s has a boundary ∂M_s, these equations of motion have to be supplemented by boundary conditions on ∂M_s. Along the circle, for a scalar field, periodic boundary conditions are imposed.

Physical properties like the Casimir energy of the system are conveniently described by means of the path-integral functionals

$$Z[V] = \int e^{-S[\Phi]}\,D\Phi, \qquad (4\text{-}9)$$

where we have neglected an infinite normalization constant, and the functional integral is to be taken over all fields satisfying the boundary conditions. Formally, equation (4-9) is easily evaluated to be

$$\Gamma[V] = -\ln Z[V] = \tfrac{1}{2}\ln\det\left((-\Delta + V(x))/\mu^2\right), \qquad (4\text{-}10)$$

where μ is an arbitrary parameter with dimension of a mass to adjust the dimension of the arguments of the logarithm.

EXERCISE 15. In order to motivate equation (4-10) show that for P a positive definite Hermitian $(N \times N)$-matrix one has

$$\int_{\mathbb{R}^n} e^{-(x,Px)/2}(dx) = (\det P)^{-1/2},$$

where

$$(dx) = d^n x\,(2\pi)^{-n/2}.$$

For $P = -\Delta + V(x)$ and interpreting the scalar product (x, Px) as an $L^2(M)$-product, one is led to (4-10) by identifying $D\Phi$ with (dx).

Equation (4-10) is purely formal, because the eigenvalues λ_n of $-\Delta+V(x)$ grow without bound for $n \to \infty$ and thus expression (4-10) needs further explanations.

In order to motivate the basic definition let P be a Hermitian ($N \times N$)-matrix with positive eigenvalues λ_n. Clearly

$$\ln \det P = \sum_{n=1}^{N} \ln \lambda_n = -\frac{d}{ds} \sum_{n=1}^{N} \lambda_n^{-s} \bigg|_{s=0} = -\frac{d}{ds} \zeta_P(s) \bigg|_{s=0},$$

and the determinant of P can be expressed in terms of the zeta function associated with P. This very same definition, namely

$$\ln \det P = -\zeta'_P(0) \tag{4-11}$$

with

$$\zeta_P(s) = \sum_{n=1}^{\infty} \lambda_n^{-s} \tag{4-12}$$

is now applied to differential operators as in (4-10). Here, the series representation is valid for $\operatorname{Re} s$ large enough, and in (4-11) the unique analytical continuation of the series to a neighborhood about $s = 0$ is used.

This definition was first used by the mathematicians Ray and Singer [73] to give a definition of the Reidemeister–Franz torsion. In physics, this regularization scheme took its origin in ambiguities of dimensional regularization when applied to quantum field theory in curved spacetime [29; 51]. For applications beyond the ones presented here see, e.g., [14; 15; 26; 30; 31; 41; 42; 74].

The quantity $\Gamma[V]$ is called the effective action and the argument V indicates the dependence of the effective action on the external fields. The Casimir energy is obtained from the effective action via

$$E = \frac{\partial}{\partial \beta} \Gamma[V] = -\frac{1}{2} \frac{\partial}{\partial \beta} \zeta'_{P/\mu^2}(0). \tag{4-13}$$

Here, we will only consider the zero temperature Casimir energy

$$E_{\text{Cas}} = \lim_{\beta \to \infty} E \tag{4-14}$$

and we will next derive a suitable representation for E_{Cas}. We want to concentrate on the influence of boundary conditions and therefore we set $V(x) = 0$. The relevant operator to be considered therefore is

$$P = -\frac{\partial^2}{\partial \tau^2} - \Delta_s,$$

where $\tau \in S^1$ is the imaginary time and Δ_s is the Laplace operator on M_s. In order to analyze the zeta function associated with P we note that eigenfunctions, respectively eigenvalues, are of the form

$$\phi_{n,j}(\tau, y) = \frac{1}{\beta} e^{2\pi i n \tau/\beta} \varphi_j(y),$$

$$\lambda_{n,j} = \left(\frac{2\pi n}{\beta}\right)^2 + E_j^2, \qquad n \in \mathbb{Z},$$

with

$$-\Delta_s \varphi_j(y) = E_j^2 \varphi_j(y),$$

where $y \in M_s$. For the non-self-interacting case considered here, E_j are the one-particle energy eigenvalues of the system. The relevant zeta function therefore has the structure

$$\zeta_P(s) = \sum_{n=-\infty}^{\infty} \sum_{j=1}^{\infty} \left(\left(\frac{2\pi n}{\beta}\right)^2 + E_j^2\right)^{-s}. \tag{4-15}$$

We repeat the analysis outlined previously, namely we use equation (2-5) and we apply Lemma 2.16 to the n-summation. In this process the zeta function

$$\zeta_{P_s}(s) = \sum_{j=1}^{\infty} E_j^{-2s}$$

and the heat kernel

$$K_{P_s}(t) = \sum_{j=1}^{\infty} e^{-E_j^2 t} \sim \sum_{k=0,\frac{1}{2},1,\ldots}^{\infty} a_k \, t^{k-(d/2)}$$

of the spatial section are the most natural quantities to represent the answer,

$$\zeta_P(s) = \frac{1}{\Gamma(s)} \sum_{n=-\infty}^{\infty} \int_0^{\infty} t^{s-1} e^{-(2\pi n/\beta)^2 t} K_{P_s}(t) \, dt$$

$$= \frac{\beta}{\sqrt{4\pi}} \frac{\Gamma(s-\frac{1}{2})}{\Gamma(s)} \zeta_{P_s}\left(s-\frac{1}{2}\right) + \frac{\beta}{\sqrt{\pi}\,\Gamma(s)} \sum_{n=1}^{\infty} \int_0^{\infty} t^{s-\frac{3}{2}} e^{-\frac{n^2 \beta^2}{4t}} K_{P_s}(t) \, dt.$$

For the Casimir energy we need ($D = d + 1$)

$$\zeta'_{P/\mu^2}(0) = \zeta'_P(0) + \zeta_P(0) \ln \mu^2$$

$$= -\beta \left(FP\, \zeta_{P_s}(-\tfrac{1}{2}) + 2(1 - \ln 2)\, \text{Res}\, \zeta_{P_s}(-\tfrac{1}{2}) - \tfrac{1}{\beta}\zeta_P(0) \ln \mu^2 \right)$$

$$+ \frac{\beta}{\sqrt{\pi}} \sum_{n=1}^{\infty} \int_0^{\infty} t^{-3/2} e^{-\left(\frac{n^2 \beta^2}{4t}\right)} K_{P_s}(t)\, dt$$

$$= -\beta \left(FP\, \zeta_{P_s}(-\tfrac{1}{2}) - \frac{1}{\sqrt{4\pi}} a_{D/2}\bigl((\ln \mu^2) + 2(1 - \ln 2)\bigr) \right)$$

$$+ \frac{\beta}{\sqrt{\pi}} \sum_{n=1}^{\infty} \int_0^{\infty} t^{-3/2} e^{-\frac{n^2 \beta^2}{4t}} K_{P_s}(t)\, dt, \qquad (4\text{-}16)$$

with the finite part FP of the zeta function and where equations (4-5) and (4-6) together with the fact that

$$K_M(t) = K_{S^1}(t)\, K_{P_s}(t)$$

have been used, in particular

$$\text{Res}\, \zeta_{P_s}(-\tfrac{1}{2}) = -\frac{a_{D/2}}{2\sqrt{\pi}}, \qquad \zeta_P(0) = \frac{\beta}{\sqrt{4\pi}} a_{D/2}. \qquad (4\text{-}17)$$

At $T = 0$ we obtain for the Casimir energy, see equations (4-13) and (4-14),

$$E_{\text{Cas}} = \lim_{\beta \to \infty} E = \tfrac{1}{2} FP\, \zeta_{P_s}(-\tfrac{1}{2}) - \frac{1}{2\sqrt{4\pi}} a_{D/2} \ln \tilde{\mu}^2, \qquad (4\text{-}18)$$

with the scale $\tilde{\mu} = (\mu e/2)$. Equation (4-18) implies that as long as $a_{D/2} \neq 0$ the Casimir energy contains a finite ambiguity and renormalization issues need to be discussed. Note from (4-17) that whenever $\zeta_{P_s}(-\tfrac{1}{2})$ is finite no ambiguity exists because $a_{D/2} = 0$. In the specific examples chosen later we will make sure that these ambiguities are absent and therefore a discussion of renormalization will be unnecessary.

In a purely formal calculation one essentially is also led to equation (4-18). As mentioned, in the quantum field theory of a free scalar field the eigenvalues of a Laplacian are the square of the energies of the quantum fluctuations. Writing the Casimir energy as (one-half) the sum over the energy of all quantum fluctuations one has

$$E_{\text{Cas}} = \frac{1}{2} \sum_{k=0}^{\infty} \lambda_k^{1/2}, \qquad (4\text{-}19)$$

and a formal identification "shows" that

$$E_{\text{Cas}} = \tfrac{1}{2}\zeta_{P_s}(-\tfrac{1}{2}). \qquad (4\text{-}20)$$

Clearly, the expression (4-19) is purely formal as the series diverges. However, when $\zeta_{P_s}(-\frac{1}{2})$ turns out to be finite this formal identification yields the correct result. Otherwise, the ambiguities given in (4-18) remain as discussed above.

An alternative discussion leading to definition (4-18) can be found in [7].

As a first example let us consider the configuration of two parallel plates a distance a apart analyzed originally by Casimir [17]. For simplicity we concentrate on a scalar field instead of the electromagnetic field and we impose Dirichlet boundary conditions on the plates. The boundary value problem to be solved therefore is

$$-\Delta u_k(x, y, z) = \lambda_k u_k(x, y, z),$$

with $u_k(0, y, z) = u_k(a, y, z) = 0$.

For the time being, we compactify the (y, z)-directions to a torus with perimeter length R and impose periodic boundary conditions in these directions. Later on, the limit $R \to \infty$ is performed to recover the parallel plate configuration. Using separation of variables one obtains normalized eigenfunctions in the form

$$u_{\ell_1 \ell_2 \ell}(x, y, z) = \sqrt{\frac{2}{aR^2}} \sin \frac{\pi \ell x}{a} e^{i 2\pi \ell_1 y / R} e^{i 2\pi \ell_2 z / R}$$

with eigenvalues

$$\lambda_{\ell_1 \ell_2 \ell} = \left(\frac{2\pi \ell_1}{R}\right)^2 + \left(\frac{2\pi \ell_2}{R}\right)^2 + \left(\frac{\pi \ell}{a}\right)^2, \quad (\ell_1, \ell_2) \in \mathbb{Z}^2, \quad \ell \in \mathbb{N}.$$

This means we have to study the zeta function

$$\zeta(s) = \sum_{(\ell_1, \ell_2) \in \mathbb{Z}^2} \sum_{\ell=1}^{\infty} \left(\left(\frac{2\pi \ell_1}{R}\right)^2 + \left(\frac{2\pi \ell_2}{R}\right)^2 + \left(\frac{\pi \ell}{a}\right)^2\right)^{-s}. \quad (4\text{-}21)$$

As $R \to \infty$ the Riemann sum turns into an integral and we compute using polar coordinates in the (y, z)-plane

$$\zeta(s) = \left(\frac{R}{2\pi}\right)^2 \sum_{\ell=1}^{\infty} \int_{-\infty}^{\infty} \int_{-\infty}^{\infty} \left(k_1^2 + k_2^2 + \left(\frac{\pi \ell}{a}\right)^2\right)^{-s} dk_2 \, dk_1$$

$$= \left(\frac{R}{2\pi}\right)^2 \sum_{\ell=1}^{\infty} 2\pi \int_0^{\infty} k \left(k^2 + \left(\frac{\pi \ell}{a}\right)^2\right)^{-s} dk$$

$$= \frac{R^2}{2\pi} \frac{1}{2(1-s)} \sum_{\ell=1}^{\infty} \left(k^2 + \left(\frac{\pi \ell}{a}\right)^2\right)^{-s+1} \Big|_0^{\infty}$$

$$= -\frac{R^2}{4\pi(1-s)} \sum_{\ell=1}^{\infty} \left(\frac{\pi \ell}{a}\right)^{2(-s+1)} = -\frac{R^2}{4\pi(1-s)} \left(\frac{\pi}{a}\right)^{2-2s} \zeta_R(2s-2).$$

Setting $s = -\frac{1}{2}$ as needed for the Casimir energy we obtain

$$\zeta(-\tfrac{1}{2}) = -\frac{R^2}{4\pi}\frac{2}{3}\left(\frac{\pi}{a}\right)^3 \zeta_R(-3) = -\frac{R^2\pi^2}{720a^3}. \tag{4-22}$$

The resulting Casimir force *per area* is

$$F_{\text{Cas}} = -\frac{\partial}{\partial a}\frac{E_{\text{Cas}}}{R^2} = -\frac{\pi^2}{480a^4}. \tag{4-23}$$

Note that this computation takes into account only those quantum fluctuations from between the plates. But in order to find the force acting on the, say, right plate the contribution from the right to this plate also has to be counted. To find this part we place another plate at the position $x = L$ where at the end we take $L \to \infty$. Following the preceding calculation, we simply have to replace a by $L - a$ to see that the associated zeta function produces

$$\zeta(-\tfrac{1}{2}) = -\frac{R^2\pi^2}{720(L-a)^3}$$

and the contribution to the force on the plate at $x = a$ reads

$$F_{\text{Cas}} = \frac{\pi^2}{480(L-a)^4}.$$

This shows the plate at $x = a$ is always attracted to the closer plate. As $L \to \infty$ it is seen that equation (4-23) also describes the total force on the plate at $x = a$ for the parallel plate configuration.

EXERCISE 16. Consider the Casimir energy that results in the previous discussion when the compactification length R is kept finite. Use Lemma 2.18 to give closed answers for the energy and the resulting force. Can the force change sign depending on a and R?

More realistically plates will have a finite extension. An interesting setting that we are able to analyze with the tools provided are pistons. These have received an increasing amount of interest because they allow the unambiguous prediction of forces [18; 52; 58; 65; 77].

Instead of having parallel plates let us consider a box with side lengths L_1, L_2 and L_3. Although it is possible to find the Casimir force acting on the plate at $x = L_1$ resulting from the interior of the box, the exterior problem has remained unsolved until today. No analytical procedure is known that allows to obtain the Casimir energy or force for the outside of the box. This problem is avoided by adding on another box with side lengths $L - L_1, L_2$ and L_3 such that the wall at $x = L_1$ subdivides the bigger box into two chambers. The wall at $x = L_1$ is assumed to be movable and is called the piston. Each chamber can be dealt with separately and total energies and forces are obtained by adding up the two contributions. Assuming again Dirichlet boundary conditions and starting with the left chamber, the relevant spectrum reads

$$\lambda_{\ell_1\ell_2\ell_3} = \left(\frac{\pi\ell_1}{L_1}\right)^2 + \left(\frac{\pi\ell_2}{L_2}\right)^2 + \left(\frac{\pi\ell_3}{L_3}\right)^2, \quad \ell_1,\ell_2,\ell_3 \in \mathbb{N}, \qquad (4\text{-}24)$$

and the associated zeta function is

$$\zeta(s) = \sum_{\ell_1,\ell_2,\ell_3 \in \mathbb{N}} \left(\left(\frac{\pi\ell_1}{L_1}\right)^2 + \left(\frac{\pi\ell_2}{L_2}\right)^2 + \left(\frac{\pi\ell_3}{L_3}\right)^2\right)^{-s}. \qquad (4\text{-}25)$$

One way to proceed is to rewrite (4-25) in terms of the Epstein zeta function in Definition 2.14.

EXERCISE 17. Use Lemma 2.18 in order to find the Casimir energy for the inside of the box with side lengths L_1, L_2 and L_3 and with Dirichlet boundary conditions imposed.

Instead of using Lemma 2.18 we proceed as follows. We write first

$$\zeta(s) = \frac{1}{2} \sum_{\ell_1=-\infty}^{\infty} \sum_{\ell_2,\ell_3=1}^{\infty} \left(\left(\frac{\pi\ell_1}{L_1}\right)^2 + \left(\frac{\pi\ell_2}{L_2}\right)^2 + \left(\frac{\pi\ell_3}{L_3}\right)^2\right)^{-s}$$
$$- \frac{1}{2} \sum_{\ell_2,\ell_3=1}^{\infty} \left(\left(\frac{\pi\ell_2}{L_2}\right)^2 + \left(\frac{\pi\ell_3}{L_3}\right)^2\right)^{-s}. \qquad (4\text{-}26)$$

This shows that it is convenient to introduce

$$\zeta_C(s) = \sum_{\ell_2,\ell_3=1}^{\infty} \left(\left(\frac{\pi\ell_2}{L_2}\right)^2 + \left(\frac{\pi\ell_3}{L_3}\right)^2\right)^{-s}. \qquad (4\text{-}27)$$

We note that this could be expressed in terms of the Epstein zeta function given in Definition 2.14. However, it will turn out that this is unnecessary.

Also, to simplify the notation let us introduce

$$\mu^2_{\ell_2\ell_3} = \left(\frac{\pi\ell_2}{L_2}\right)^2 + \left(\frac{\pi\ell_3}{L_3}\right)^2.$$

Using equation (2-5) for the first line in (4-26) we continue

$$\zeta(s) = \frac{1}{2\Gamma(s)} \sum_{\ell_1=-\infty}^{\infty} \sum_{\ell_2,\ell_3=1}^{\infty} \int_0^{\infty} t^{s-1} \exp\left(-t\left(\left(\frac{\pi\ell_1}{L_1}\right)^2 + \mu^2_{\ell_2\ell_3}\right)\right) dt - \frac{1}{2}\zeta_C(s).$$

We now apply the Poisson resummation in Lemma 2.16 to the ℓ_1-summation and therefore we get

$$\zeta(s) = \frac{L_1}{2\sqrt{\pi}\,\Gamma(s)} \sum_{\ell_1=-\infty}^{\infty} \sum_{\ell_2,\ell_3=1}^{\infty} \int_0^{\infty} t^{s-3/2} \exp\left(-\frac{L_1^2 \ell_1^2}{t} - t\mu_{\ell_2 \ell_3}^2\right) dt$$

$$-\tfrac{1}{2}\zeta_C(s). \quad (4\text{-}28)$$

The $\ell_1 = 0$ term gives a ζ_C-term, the $\ell_1 \neq 0$ terms are rewritten using (2.17). The outcome reads

$$\zeta(s) = \frac{L_1 \Gamma(s - \tfrac{1}{2})}{2\sqrt{\pi}\,\Gamma(s)} \zeta_C(s - \tfrac{1}{2}) - \tfrac{1}{2}\zeta_C(s)$$

$$+ \frac{2L_1^{s+\tfrac{1}{2}}}{\sqrt{\pi}\,\Gamma(s)} \sum_{\ell_1,\ell_2,\ell_3=1}^{\infty} \left(\frac{\ell_1^2}{\mu_{\ell_2\ell_3}^2}\right)^{\tfrac{1}{2}(s-\tfrac{1}{2})} K_{\tfrac{1}{2}-s}(2L_1 \ell_1 \mu_{\ell_2 \ell_3}). \quad (4\text{-}29)$$

We need the zeta function about $s = -\tfrac{1}{2}$ in order to find the Casimir energy and Casimir force.

Let $s = -\tfrac{1}{2} + \varepsilon$. In order to expand equation (4-29) about $\varepsilon = 0$ we need to know the pole structure of $\zeta_C(s)$. From equation (2.18) it is expected that $\zeta_C(s)$ has at most a first order pole at $s = -\tfrac{1}{2}$ and that it is analytic about $s = -1$. So for now let us simply assume the structure

$$\zeta_C(-\tfrac{1}{2} + \varepsilon) = \frac{1}{\varepsilon}\operatorname{Res}\zeta_C(-\tfrac{1}{2}) + \operatorname{FP}\zeta_C(-\tfrac{1}{2}) + \mathcal{O}(\varepsilon),$$

$$\zeta_C(-1 + \varepsilon) = \zeta_C(-1) + \varepsilon\zeta_C'(-1) + \mathcal{O}(\varepsilon^2),$$

where $\operatorname{Res}\zeta_C(-\tfrac{1}{2})$ and $\operatorname{FP}\zeta_C(-\tfrac{1}{2})$ will be determined later. With this structure assumed, we find

$$\zeta(-\tfrac{1}{2}+\varepsilon) = \frac{1}{\varepsilon}\left(\frac{L_1}{4\pi}\zeta_C(-1) - \tfrac{1}{2}\operatorname{Res}\zeta_C(-\tfrac{1}{2})\right)$$

$$+ \frac{L_1}{4\pi}\left(\zeta_C'(-1) + \zeta_C(-1)(\ln 4 - 1)\right) - \tfrac{1}{2}\operatorname{FP}\zeta_C(-\tfrac{1}{2})$$

$$- \frac{1}{\pi}\sum_{\ell_1,\ell_2,\ell_3=1}^{\infty} \left|\frac{\mu_{\ell_2 \ell_3}}{\ell_1}\right| K_1(2L_1 \ell_1 \mu_{\ell_2\ell_3}). \quad (4\text{-}30)$$

This shows that the Casimir energy for this setting is unambiguously defined only if $\zeta_C(-1) = 0$ and $\operatorname{Res}\zeta_C(-\tfrac{1}{2}) = 0$.

EXERCISE 18. Show the following analytical continuation for $\zeta_C(s)$:

$$\zeta_C(s) = -\frac{1}{2}\left(\frac{L_3}{\pi}\right)^{2s}\zeta_R(2s) + \frac{L_2\Gamma(s-\frac{1}{2})}{2\sqrt{\pi}\,\Gamma(s)}\left(\frac{L_3}{\pi}\right)^{2s-1}\zeta_R(2s-1) \qquad (4\text{-}31)$$

$$+\frac{2L_2^{s+1/2}}{\sqrt{\pi}\,\Gamma(s)}\sum_{\ell_2=1}^{\infty}\sum_{\ell_3=1}^{\infty}\left(\frac{\ell_2 L_3}{\pi \ell_3}\right)^{s-1/2} K_{\frac{1}{2}-s}\left(\frac{2\pi L_2 \ell_2 \ell_3}{L_3}\right).$$

Read off that $\zeta_C(-1) = \operatorname{Res}\zeta_C(-\frac{1}{2}) = 0$.

Using the results from Exercise 18 the Casimir energy, from equation (4-30), can be expressed as

$$E_{\mathrm{Cas}} = \frac{L_1}{8\pi}\zeta_C'(-1) - \tfrac{1}{4}\mathrm{FP}\,\zeta_C(-\tfrac{1}{2}) - \frac{1}{2\pi}\sum_{\ell_1,\ell_2,\ell_3=1}^{\infty}\left|\frac{\mu_{\ell_2\ell_3}}{\ell_1}\right| K_1(2L_1\ell_1\mu_{\ell_2\ell_3}). \qquad (4\text{-}32)$$

EXERCISE 19. Use representation (4-31) to give an explicit representation of the Casimir energy (4-32).

For the force this shows

$$F_{\mathrm{Cas}} = -\frac{1}{8\pi}\zeta_C'(-1) + \frac{1}{2\pi}\sum_{\ell_1,\ell_2,\ell_3=1}^{\infty}\left|\frac{\mu_{\ell_2\ell_3}}{\ell_1}\right|\frac{\partial}{\partial L_1} K_1(2L_1\ell_1\mu_{\ell_2\ell_3}). \qquad (4\text{-}33)$$

EXERCISE 20. Use Definition 2.17 to show that $K_\nu(x)$ is a monotonically decreasing function for $x \in \mathbb{R}_+$.

EXERCISE 21. Determine the sign of $\zeta_C'(-1)$. What is the sign of the Casimir force as $L_1 \to \infty$? What about $L_1 \to 0$?

Remember that the results given describe the contributions from the interior of the box only. The contributions from the right chamber are obtained by replacing L_1 with $L - L_1$. This shows for the right chamber

$$E_{\mathrm{Cas}} = \frac{L-L_1}{8\pi}\zeta_C'(-1) - \tfrac{1}{4}\mathrm{FP}\,\zeta_C\!\left(-\tfrac{1}{2}\right)$$

$$-\frac{1}{2\pi}\sum_{\ell_1,\ell_2,\ell_3=1}^{\infty}\left|\frac{\mu_{\ell_2\ell_3}}{\ell_1}\right| K_1(2(L-L_1)\ell_1\mu_{\ell_2\ell_3}),$$

$$F_{\mathrm{Cas}} = \frac{1}{8\pi}\zeta_C'(-1) + \frac{1}{2\pi}\sum_{\ell_1,\ell_2,\ell_3=1}^{\infty}\left|\frac{\mu_{\ell_2\ell_3}}{\ell_1}\right|\frac{\partial}{\partial L_1} K_1(2(L-L_1)\ell_1\mu_{\ell_2\ell_3}).$$

Adding up, the total force on the piston is

$$F_{\text{Cas}}^{\text{tot}} = \frac{1}{2\pi} \sum_{\ell_1,\ell_2,\ell_3=1}^{\infty} \left|\frac{\mu_{\ell_2\ell_3}}{\ell_1}\right| \frac{\partial}{\partial L_1} K_1(2L_1\ell_1\mu_{\ell_2\ell_3})$$
$$+ \frac{1}{2\pi} \sum_{\ell_1,\ell_2,\ell_3=1}^{\infty} \left|\frac{\mu_{\ell_2\ell_3}}{\ell_1}\right| \frac{\partial}{\partial L_1} K_1(2(L-L_1)\ell_1\mu_{\ell_2\ell_3}). \quad (4\text{-}34)$$

This shows, using the results of Exercise 20, that the piston is always attracted to the closer wall.

Although we have presented the analysis for a piston with rectangular cross-section, our result in fact holds in much greater generality. The fact that we analyzed a rectangular cross-section manifests itself in the spectrum (4-24), namely the part

$$\left(\frac{\pi\ell_2}{L_2}\right)^2 + \left(\frac{\pi\ell_3}{L_3}\right)^2$$

is a direct consequence of it. If instead we had considered an arbitrary cross-section $\mathcal{C}$, the relevant spectrum had the form

$$\lambda_{\ell_1 i} = \left(\frac{\pi\ell_1}{L_1}\right)^2 + \mu_i^2,$$

where, assuming still Dirichlet boundary conditions on the boundary of the cross-section $\mathcal{C}$, μ_i^2 is determined from

$$-\left(\frac{\partial^2}{\partial y^2} + \frac{\partial^2}{\partial z^2}\right)\phi_i(y,z) = \mu_i^2 \phi_i(y,z), \qquad \phi_i(y,z)\Big|_{(y,z)\in\partial\mathcal{C}} = 0.$$

Proceeding in the same way as before, replacing $\mu_{\ell_2\ell_3}$ with μ_i and introducing $\zeta_\mathcal{C}(s)$ as the zeta function for the cross-section,

$$\zeta_\mathcal{C}(s) = \sum_{i=1}^{\infty} \mu_i^{-2s},$$

equation (4-28) remains valid, as well as equations (4-29) and (4-30). So also for an arbitrary cross-section the total force on the piston is described by equation (4-34) with the replacements given and the piston is attracted to the closest wall.

EXERCISE 22. In going from equation (4-28) to (4-29) we used the fact that $\mu_{\ell_2\ell_3}^2 > 0$. Above we used $\mu_i^2 > 0$ which is true because we imposed Dirichlet boundary conditions. Modify the calculation if boundary conditions are chosen (like Neumann boundary conditions) that allow for d_0 zero modes $\mu_i^2 = 0$ [58].

We have presented the piston set-up for three spatial dimensions, but a similar analysis can be performed in the presence of extra dimensions [58]. Once this kind of calculation is fully understood for the electromagnetic field it is hoped that future high-precision measurements of Casimir forces for simple configurations such as parallel plates can serve as a window into properties of the dimensions of the universe that are somewhat hidden from direct observations.

As we have seen for the example of the piston, there are cases where an unambiguous prediction of Casimir forces is possible. Of course the set-up we have chosen was relatively simple and for many other configurations even the sign of Casimir forces is unknown. This is a very active field of research; some references are [8; 36; 43; 67; 68; 75]. Further discussion is provided in the Conclusions.

5. Bose–Einstein condensation of Bose gases in traps

We now turn to applications in statistical mechanics. We have chosen to apply the techniques in a quantum mechanical system described by the Schrödinger equation

$$\left(-\frac{\hbar^2}{2m}\Delta + V(x,y,z)\right)\phi_k(x,y,z) = \lambda_k \phi_k(x,y,z), \tag{5-1}$$

that is we consider a gas of quantum particles of mass m under the influence of the potential $V(x, y, z)$. Specifically, later we will consider in detail the harmonic oscillator potential

$$V(x,y,z) = \frac{m}{2}(\omega_1 x^2 + \omega_2 y^2 + \omega_3 z^2)$$

briefly mentioned in Example 3.4, as well as a gas confined in a finite cavity.

Thermodynamic properties of a *Bose gas*, which is what we shall consider in the following, are described by the (grand canonical) partition sum

$$q = -\sum_{k=0}^{\infty} \ln\left(1 - e^{-\beta(\lambda_k - \mu)}\right), \tag{5-2}$$

where β is the inverse temperature and μ is the chemical potential. We assume the index $k = 0$ labels the unique ground state, that is, the state with smallest energy eigenvalue λ_0. From this partition sum all thermodynamical properties are obtained. For example the particle number is

$$N = \frac{1}{\beta}\frac{\partial q}{\partial \mu}\bigg|_{T,V} = \sum_{k=0}^{\infty} \frac{1}{e^{\beta(\lambda_k - \mu)} - 1}, \tag{5-3}$$

where the notation $(\partial q/\partial \mu|_{T,V})$ indicates that the derivative has to be taken with temperature T and volume V kept fixed. The particle number is the most important quantity for the phenomenon of Bose–Einstein condensation. Although this phenomenon was predicted more than 80 years ago [9; 32] it was only relatively recently experimentally verified [2; 11; 24]. Bose–Einstein condensation is one of the most interesting properties of a system of bosons. Namely, under certain conditions it is possible to have a phase transition at a critical value of the temperature in which all of the bosons can condense into the ground state. In order to understand at which temperature the phenomenon occurs a detailed study of N, or alternatively q, is warranted. This is the subject of this section.

We first note that from the fact that the particle number in each state has to be non-negative it is clear that $\mu < \lambda_0$ has to be imposed. It is seen in (5-2) that as $\beta \to 0$ (high temperature limit) the behavior of q cannot be easily understood. But contour integral techniques together with the zeta function information provided makes the analysis feasible and it will allow for the determination of the critical temperature of the Bose gas.

Let us start by noting that from

$$\ln(1-x) = -\sum_{n=1}^{\infty} \frac{x^n}{n}, \quad \text{for } |x| < 1,$$

the partition sum can be rewritten as

$$q = \sum_{n=1}^{\infty} \sum_{k=0}^{\infty} \frac{1}{n} e^{-\beta(\lambda_k - \mu)n}. \tag{5-4}$$

The $\beta \to 0$ behavior is best found using the following representation of the exponential.

EXERCISE 23. Given that

$$\lim_{|y|\to\infty} |\Gamma(x+iy)| \, e^{\frac{\pi}{2}|y|} \, |y|^{\frac{1}{2}-x} = \sqrt{2\pi}, \quad x, y \in \mathbb{R},$$

and

$$\Gamma(z) = \sqrt{2\pi} e^{(z-\frac{1}{2})\log z - z} (1 + o(1)),$$

as $|z| \to \infty$, show that

$$e^{-a} = \frac{1}{2\pi i} \int_{\sigma - i\infty}^{\sigma + i\infty} a^{-t} \, \Gamma(t) \, dt, \tag{5-5}$$

valid for $\sigma > 0$, $|\arg a| < \frac{\pi}{2} - \delta$, $0 < \delta \leq \pi/2$.

Before we apply this result to the partition sum (5-4) let us use a simple example to show how this formula allows us to determine asymptotic behavior of certain series in a relatively straightforward fashion. From Lemma 2.16 we know that

$$\sum_{\ell=1}^{\infty} e^{-\beta \ell^2} = \frac{1}{2} \sum_{\ell=-\infty}^{\infty} e^{-\beta \ell^2} - \frac{1}{2}$$

$$= \frac{1}{2}\sqrt{\frac{\pi}{\beta}} - \frac{1}{2} + \sqrt{\frac{\pi}{\beta}} \sum_{\ell=1}^{\infty} e^{-\frac{\pi^2}{\beta}\ell^2}. \quad (5\text{-}6)$$

As $\beta \to 0$ it is clear that the series on the left diverges and Lemma 2.16 shows that the leading behavior is described by a $1/\sqrt{\beta}$ term, followed by a constant term, followed by exponentially damped corrections. Let us see how we can easily find the polynomial behavior as $\beta \to 0$ from (5-5). We first write

$$\sum_{\ell=1}^{\infty} e^{-\beta \ell^2} = \sum_{\ell=1}^{\infty} \frac{1}{2\pi i} \int_{\sigma - i\infty}^{\sigma + i\infty} (\beta \ell^2)^{-t} \Gamma(t) \, dt.$$

Here, $\sigma > 0$ is assumed by Exercise 23. However, in order to be allowed to interchange summation and integration we need to impose $\sigma > \frac{1}{2}$ and find

$$\sum_{\ell=1}^{\infty} e^{-\beta \ell^2} = \frac{1}{2\pi i} \int_{\sigma - i\infty}^{\sigma + i\infty} \beta^{-t} \Gamma(t) \sum_{\ell=1}^{\infty} \ell^{-2t} \, dt$$

$$= \frac{1}{2\pi i} \int_{\sigma - i\infty}^{\sigma + i\infty} \beta^{-t} \Gamma(t) \zeta_R(2t) \, dt.$$

In order to find the small-β behavior, the strategy now is to shift the contour to the left. In doing so we cross over poles of the integrand generating polynomial contributions in β. For this example, the right most pole is at $t = \frac{1}{2}$ (pole of the zeta function of Riemann) and the next pole is at $t = 0$ (from the gamma function). Those are all singularities present as $\zeta_R(-2n) = 0$ for $n \in \mathbb{N}$. Therefore,

$$\sum_{\ell=1}^{\infty} e^{-\beta \ell^2} = \beta^{-1/2} \Gamma(\tfrac{1}{2}) \tfrac{1}{2} + \beta^0 \cdot 1 \cdot \zeta_R(0) + \frac{1}{2\pi i} \int_{\tilde{\sigma} - i\infty}^{\tilde{\sigma} + i\infty} \beta^{-t} \Gamma(t) \zeta_R(2t) \, dt$$

$$= \frac{1}{2}\sqrt{\frac{\pi}{\beta}} - \frac{1}{2} + \frac{1}{2\pi i} \int_{\tilde{\sigma} - i\infty}^{\tilde{\sigma} + i\infty} \beta^{-t} \Gamma(t) \zeta_R(2t) \, dt,$$

where $\tilde{\sigma} < 0$ and where contributions from the horizontal lines between $\tilde{\sigma} \pm i\infty$ and $\sigma \pm i\infty$ are neglected. For the remaining contour integral plus the neglected horizontal lines one can actually show that they will produce the exponentially damped terms as given in (5-6). How exactly this actually happens has been described in detail in [35].

EXERCISE 24. Argue how $\sum_{n=1}^{\infty} e^{-\beta n^{\alpha}}$, $\beta > 0, \alpha > 0$, behaves as $\beta \to 0$ by using the procedure above. Determine the leading three terms in the expansion assuming that the contributions from the contour at infinity can be neglected.

EXERCISE 25. Find the leading three terms of the small-β behavior of

$$\sum_{n=1}^{\infty} \log(1 - e^{-\beta n})$$

assuming that the contributions from the contour at infinity can be neglected.

We next apply these ideas to the partition sum (5-4). As a further warmup, for simplicity, let us first set $\mu = 0$. Not specifying λ_k for now and using

$$\zeta(s) = \sum_{k=0}^{\infty} \lambda_k^{-s}$$

for $\mathrm{Re}\, s > M$ large enough to make this series convergent, we write

$$q = \sum_{n=1}^{\infty} \sum_{k=0}^{\infty} \frac{1}{n} e^{-\beta \lambda_k n} = \sum_{n=1}^{\infty} \sum_{k=0}^{\infty} \frac{1}{n} \frac{1}{2\pi i} \int_{\sigma - i\infty}^{\sigma + i\infty} (\beta \lambda_k n)^{-t} \Gamma(t)\, dt$$

$$= \frac{1}{2\pi i} \int_{\sigma - i\infty}^{\sigma + i\infty} \beta^{-t} \Gamma(t) \left(\sum_{n=1}^{\infty} n^{-t-1} \right) \left(\sum_{k=0}^{\infty} \lambda_k^{-t} \right) dt$$

$$= \frac{1}{2\pi i} \int_{\sigma - i\infty}^{\sigma + i\infty} \beta^{-t} \Gamma(t) \zeta_R(t+1) \zeta(t)\, dt.$$

Here $\sigma > M$ is needed for the algebraic manipulations to be allowed. It is clearly seen that the integrand has a double pole at $t = 0$. The right most pole (at M) therefore comes from $\zeta(t)$, and the location of this pole determines the leading $\beta \to 0$ behavior of the partition sum.

For the harmonic oscillator potential, in the notation of Example 3.4, the Barnes zeta function occurs and we have

$$q = \frac{1}{2\pi i} \int_{\sigma - i\infty}^{\sigma + i\infty} \beta^{-t} \Gamma(t) \zeta_R(t+1) \zeta_B(t, c|\vec{r})\, dt. \tag{5-7}$$

The location of the poles and its residues are known for the Barnes zeta function, see Definition 2.12 and Theorem 2.13, in particular one has

$$\mathrm{Res}\, \zeta_B(3, c|\vec{r}) = \frac{1}{2\hbar^3 \Omega^3},$$

where, as is common, the geometric mean of the oscillator frequencies
$$\Omega = (\omega_1 \omega_2 \omega_3)^{1/3}$$
has been used. The leading order of the partition sum therefore is
$$q = \frac{\pi^4}{90} \frac{1}{(\beta \hbar \Omega)^3} + \mathcal{O}(\beta^{-2}).$$

EXERCISE 26. Use Definition 2.12 and Theorem 2.13 to find the subleading order of the small-β expansion of the partition sum q.

EXERCISE 27. Consider the harmonic oscillator potential in d dimensions and find the leading and subleading order of the small-β expansion of the partition sum q.

If instead of considering a Bose gas in a trap we consider the gas in a finite three-dimensional cavity M with boundary ∂M we have to augment the Schrödinger equation (5-1) by boundary conditions. We choose Dirichlet boundary conditions and thus the results for the heat kernel coefficients (4-3) are valid.

From equation (4-5) we also conclude that the rightmost pole of $\zeta(s)$ is located at $s = 3/2$ and that
$$\text{Res } \zeta\left(\tfrac{3}{2}\right) = \frac{a_0}{\Gamma\left(\tfrac{3}{2}\right)} = \frac{\text{vol } M}{4\pi^2};$$
furthermore the next pole is located at $s = 1$. For this case, the leading order of the partition sum therefore is
$$q = \frac{1}{(4\pi\beta)^{3/2}} \zeta_R\left(\tfrac{5}{2}\right) \text{vol } M + \mathcal{O}(\beta^{-1}).$$

One way to read this result is that the Bose gas does know the volume of its container because it can be found from the partition sum. This is completely analogous to the statement for the drum where we used the heat kernel instead of the partition sum.

Subleading orders of the partition sum reveal more information about the cavity, see the following exercise. But as for the drums, the gas does not know all the details of the shape of the cavity because there are different cavities leading to the same eigenvalue spectrum [45]. Those cavities cannot be distinguished by the above analysis.

EXERCISE 28. Consider a Bose gas in a d-dimensional cavity M with boundary ∂M. Use (4-3) and (4-5) to find the leading and subleading order of the small-β expansion of the partition sum q. What does the Bose gas know about its container, meaning what information about the container can be read of from the high-temperature behavior of the partition sum?

In order to examine the phenomenon of Bose–Einstein condensation we have to consider non-vanishing chemical potential. Close to the phase transition, as we will see, more and more particles have to reside in the ground state and the value of the chemical potential will be close to the smallest eigenvalue, which is the 'critical' value for the chemical potential, $\mu_c = \lambda_0$. Near the phase transition, for the expansion to be established, it will turn out advantageous to rewrite $\lambda_k - \mu$ such that the small quantity $\mu_c - \mu$ appears,

$$\lambda_k - \mu = \lambda_k - \mu_c + \mu_c - \mu = \lambda_k - \lambda_0 + \mu_c - \mu.$$

Given the special role of the ground state, we separate off its contribution and write

$$q = q_0 + \sum_{n=1}^{\infty} \sum_{k=1}^{\infty} \frac{1}{n} e^{-\beta n (\lambda_k - \lambda_0)} e^{-\beta n (\mu_c - \mu)}.$$

Note that the k-sum starts with $k = 1$, which means that the ground state is not included in this summation. Employing the representation (5-5) only to the first exponential factor and proceeding as before we obtain

$$q = q_0 + \frac{1}{2\pi i} \int_{\sigma - i\infty}^{\sigma + i\infty} \beta^{-t} \Gamma(t) \mathrm{Li}_{1+t}\!\left(e^{-\beta(\mu_c - \mu)}\right) \zeta_0(t) \, dt, \qquad (5\text{-}8)$$

with the polylogarithm

$$\mathrm{Li}_n(x) = \sum_{\ell=1}^{\infty} \frac{x^\ell}{\ell^n}, \qquad (5\text{-}9)$$

and the spectral zeta function

$$\zeta_0(s) = \sum_{k=1}^{\infty} (\lambda_k - \lambda_0)^{-s}.$$

In order to determine the small-β behavior of expression (5-8) let us discuss the pole structure of the integrand. Given $\mu_c - \mu > 0$, the polylogarithm $\mathrm{Li}_{1+t}(e^{-\beta(\mu_c-\mu)})$ does not generate any poles. Concentrating on the harmonic oscillator, we find

$$\mathrm{Res}\,\zeta_0(3) = \frac{1}{2(\hbar\Omega)^3}, \quad \mathrm{Res}\,\zeta_0(2) = \frac{1}{2\hbar^2}\left(\frac{1}{\omega_1\omega_2} + \frac{1}{\omega_1\omega_3} + \frac{1}{\omega_2\omega_3}\right).$$

Note that $\zeta_0(s)$ is the Barnes zeta function as given in Definition 2.8 with $c = 0$ where we have to exclude $\vec{m} = \vec{0}$ from the summation. However, clearly the residues at $s = 3$ and $s = 2$ can still be obtained from Theorem 2.13 with $c \to 0$ taken.

Shifting the contour to the left we now find

$$q = q_0 + \frac{1}{(\beta\hbar\Omega)^3}\text{Li}_4\left(e^{-\beta(\mu_c-\mu)}\right)$$
$$+ \frac{1}{2(\beta\hbar)^2}\text{Li}_3\left(e^{-\beta(\mu_c-\mu)}\right)\left(\frac{1}{\omega_1\omega_2} + \frac{1}{\omega_1\omega_3} + \frac{1}{\omega_2\omega_3}\right) + \cdots$$

In order to find the particle number N we need the relation for the polylogarithm

$$\frac{\partial \text{Li}_n(x)}{\partial x} = \frac{1}{x}\text{Li}_{n-1}(x),$$

which follows from (5-9). So

$$N = N_0 + \frac{1}{(\beta\hbar\Omega)^3}\text{Li}_3\left(e^{-\beta(\mu_c-\mu)}\right)$$
$$+ \frac{1}{2(\beta\hbar)^2}\text{Li}_2\left(e^{-\beta(\mu_c-\mu)}\right)\left(\frac{1}{\omega_1\omega_2} + \frac{1}{\omega_1\omega_3} + \frac{1}{\omega_2\omega_3}\right) + \cdots$$

EXERCISE 29. Use (5-5) and (5-9) to show

$$\text{Li}_n\left(e^{-x}\right) = \zeta_R(n) - x\zeta_R(n-1) + \cdots$$

valid for $n > 2$. What does the subleading term look like for $n = 2$?

As the critical temperature is approached $\mu \to \mu_c$ and with Exercise 29 the particle number close to the transition temperature becomes

$$N = N_0 + \frac{\zeta_R(3)}{(\beta\hbar\Omega)^3} + \frac{\zeta_R(2)}{2(\beta\hbar)^2}\left(\frac{1}{\omega_1\omega_2} + \frac{1}{\omega_1\omega_3} + \frac{1}{\omega_2\omega_3}\right) + \cdots \quad (5\text{-}10)$$

The second and third terms give the number of particles in the *excited* levels (at high temperature close to the phase transition).

The critical temperature is defined as the temperature where all excited levels are completely filled such that lowering the temperature the ground state population will start to build up. This means the defining equation for the critical temperature $T_c = 1/\beta_c$ in the approximation considered is

$$N = \frac{1}{(\beta_c\hbar\Omega)^3}\zeta_R(3) + \frac{1}{2(\beta_c\hbar)^2}\zeta_R(2)\left(\frac{1}{\omega_1\omega_2} + \frac{1}{\omega_1\omega_3} + \frac{1}{\omega_2\omega_3}\right). \quad (5\text{-}11)$$

Solving for β_c one finds

$$T_c = T_0\left(1 - \frac{\zeta_R(2)}{3\zeta_R(3)^{2/3}}\,\delta\,N^{-1/3}\right).$$

Here, T_0 is the critical temperature in the bulk limit ($N \to \infty$)

$$T_0 = \hbar\Omega\left(\frac{N}{\zeta_R(3)}\right)^{1/3}$$

and

$$\delta = \tfrac{1}{2}\Omega^{2/3}\left(\frac{1}{\omega_1\omega_2} + \frac{1}{\omega_1\omega_3} + \frac{1}{\omega_2\omega_3}\right).$$

Different approaches can be used to obtain the same answers [47; 48; 49; 50].

If only a few thousand particles are used in the experiment the finite-N correction is actually quite important. For example the first successful experiments on Bose–Einstein condensates were done with rubidium [2] at frequencies $\omega_1 = \omega_2 = 240\pi/\sqrt{8}$ s^{-1} and $\omega_3 = 240\pi$s^{-1}. With $N = 2000$ one finds $T_c \sim 31.9$nK $= 0.93\, T_0$ [59], a significant correction compared to the thermodynamic limit.

EXERCISE 30. Consider the Bose gas in a d-dimensional cavity. Find the particle number and the critical temperature along the lines described for the harmonic oscillator. What is the correction to the critical temperature caused by the finite size of the cavity? (For a solution to this problem see [60].)

6. Conclusions

In these lectures some basic zeta functions are introduced and used to analyze the Casimir effect and Bose–Einstein condensation for particular situations. The basic zeta functions considered are the Hurwitz, the Barnes and the Epstein zeta function. Although these zeta functions differ from each other they have one property in common: they are based upon a sequence of numbers that is explicitly known and given in closed form. The analysis of these zeta functions and of the indicated applications in physics is heavily based on this explicit knowledge in that well-known summation formulas are used.

In most cases, however, an explicit knowledge of the eigenvalues of, say, a Laplacian will not be available and an analysis of the associated zeta functions will be more complicated. In recent years a new class of examples where eigenvalues are defined implicitly as solutions to transcendental equations has become accessible. In some detail let us assume that eigenvalues are determined by equations of the form

$$F_\ell(\lambda_{\ell,n}) = 0 \tag{6-1}$$

with ℓ, n suitable indices. For example when trying to find eigenvalues and eigenfunctions of the Laplacian whenever possible one resorts to separation of variables and ℓ and n would be suitable 'quantum numbers' labeling eigenfunctions. To be specific consider a scalar field in a three dimensional ball of radius R with Dirichlet boundary conditions. The eigenvalues λ_k for this situation, with k as a multiindex, are thus determined through

$$-\Delta \phi_k(x) = \lambda_k \phi_k(x), \qquad \phi_k(x)|_{|x|=R} = 0.$$

In terms of spherical coordinates (r, Ω), a complete set of eigenfunctions may be given in the form

$$\phi_{l,m,n}(r, \Omega) = r^{-1/2} J_{l+1/2}(\sqrt{\lambda_{l,n}} r) Y_{l,m}(\Omega),$$

where $Y_{l,m}(\Omega)$ are spherical surface harmonics [40], and J_ν are Bessel functions of the first kind [46]. Eigenvalues of the Laplacian are determined as zeroes of Bessel functions. In particular, for a given angular momentum quantum number l, imposing Dirichlet boundary conditions, eigenvalues $\lambda_{l,n}$ are determined by

$$J_{l+1/2}(\sqrt{\lambda_{l,n}} R) = 0. \tag{6-2}$$

Although some properties of the zeroes of Bessel functions are well understood [46], there is no closed form for them available and we encounter the situation described by (6-1). In order to find properties of the zeta function associated with this kind of boundary value problems the idea is to use the argument principle or Cauchy's residue theorem. For the situation of the ball one writes the zeta function in the form

$$\zeta(s) = \sum_{l=0}^{\infty} (2l+1) \frac{1}{2\pi i} \int_\gamma k^{-2s} \frac{\partial}{\partial k} \ln J_{l+1/2}(kR) \, dk, \tag{6-3}$$

where the contour γ runs counterclockwise and must enclose all solutions of (6-2). The factor $(2l+1)$ represents the degeneracy for each angular momentum l and the summation is over all angular momenta. The integrand has singularities exactly at the eigenvalues and one can show that the residues are one such that the definition of the zeta function is recovered. More generally, in other coordinate systems, one would have, somewhat symbolically,

$$\zeta(s) = \sum_j d_j \frac{1}{2\pi i} \int_\gamma k^{-2s} \frac{\partial}{\partial k} \ln F_j(k) \, dk, \tag{6-4}$$

the task being to construct the analytical continuation of this object. The details of the procedure will depend very much on the properties of the special function F_j that enters, but often all the information needed can be found [57]. Nevertheless, for many separable coordinate systems this program has not been performed but efforts are being made in order to obtain yet unknown precise values for the Casimir energy for various geometries.

Acknowledgements

This work is supported by the National Science Foundation Grant PHY-0757791. Part of the work was done while the author enjoyed the hospitality and partial support of the Department of Physics and Astronomy of the University of Oklahoma. Thanks go in particular to Kimball Milton and his group who made this very pleasant and exciting visit possible.

References

[1] J. Ambjorn and S. Wolfram. Properties of the vacuum, 1: Mechanical and thermodynamic. *Ann. Phys.*, 147:1–32, 1983.

[2] M. H. Anderson, J. R. Ensher, M. R. Matthews, C. E. Wieman, and E. A. Cornell. Observation of Bose–Einstein condensation in a dilute atomic vapor. *Science*, 269:198–201, 1995.

[3] T. M. Apostol. *Introduction to analytic number theory*. Springer Verlag, Berlin, 1976.

[4] T. M. Apostol. *Modular function and Dirichlet series in number theory*. Springer Verlag, Berlin, 1990.

[5] E. W. Barnes. On the asymptotic expansion of integral functions of multiple linear sequence. *Trans. Camb. Philos. Soc.*, 19:426–439, 1903.

[6] E. W. Barnes. On the theory of the multiple gamma function. *Trans. Camb. Philos. Soc.*, 19:374–425, 1903.

[7] S. K. Blau, M. Visser, and A. Wipf. Zeta functions and the Casimir energy. *Nucl. Phys.*, B310:163–180, 1988.

[8] M. Bordag, U. Mohideen, and V. M. Mostepanenko. New developments in the Casimir effect. *Phys. Rept.*, 353:1–205, 2001.

[9] S. N. Bose. Planck's law and light quantum hypothesis. *Z. Phys.*, 26:178–181, 1924.

[10] T. H. Boyer. Quantum electromagnetic zero point energy of a conducting spherical shell and the Casimir model for a charged particle. *Phys. Rev.*, 174:1764–1774, 1968.

[11] C. C. Bradley, C. A. Sackett, J. J. Tollett, and R. G. Hulet. Evidence of Bose–Einstein condensation in an atomic gas with attractive interactions. *Phys. Rev. Lett.*, 75:1687–1690, 1995.

[12] G. Bressi, G. Carugno, R. Onofrio, and G. Ruoso. Measurement of the Casimir force between parallel metallic surfaces. *Phys. Rev. Lett.*, 88:041804, 2002.

[13] M. Brown-Hayes, D. A. R. Dalvit, F. D. Mazzitelli, W. J. Kim, and R. Onofrio. Towards a precision measurement of the Casimir force in a cylinder-plane geometry. *Phys. Rev.*, A72:052102, 2005.

[14] I. L. Buchbinder, S. D. Odintsov, and I. L. Shapiro. *Effective action in quantum gravity*. Hilger, Bristol, 1992.

[15] A. A. Bytsenko, G. Cognola, E. Elizalde, V. Moretti, and S. Zerbini. *Analytic aspects of quantum fields*. World Scientific, London, 2003.

[16] A. A. Bytsenko, G. Cognola, L. Vanzo, and S. Zerbini. Quantum fields and extended objects in space-times with constant curvature spatial section. *Phys. Rept.*, 266:1–126, 1996.

[17] H. B. G. Casimir. On the attraction between two perfectly conducting plates. *Kon. Ned. Akad. Wetensch. Proc.*, 51:793–795, 1948.

[18] R. M. Cavalcanti. Casimir force on a piston. *Phys. Rev.*, D69:065015, 2004.

[19] H. B. Chan, V. A. Aksyuk, R. N. Kleiman, D. J. Bishop, and F. Capasso. Nonlinear micromechanical Casimir oscillator. *Phys. Rev. Lett.*, 87:211801, 2001.

[20] H. B. Chan, V. A. Aksyuk, R. N. Kleiman, D. J. Bishop, and F. Capasso. Quantum mechanical actuation of microelectromechanical systems by the Casimir force. *Science*, 291:1941–1944, 2001.

[21] F. Chen, U. Mohidden, G. L. Klimchitskaya, and V. M. Mostepanenko. Experimental test for the conductivity properties from the Casimir force between metal and semiconductor. *Phys. Rev.*, A74:022103, 2006.

[22] F. Dalvovo, S. Giorgini, L. Pitaevskii, and S. Stringari. Theory of Bose-Einstein condensation in trapped gases. *Rev. Mod. Phys.*, 71:463–512, 1999.

[23] H. Davenport. *Multiplicative number theory*. Springer Verlag, Berlin, 1967.

[24] K. B. Davis, M.-O. Mewes, M. R. Andrews, N. J. van Druten, D. S. Durfee, D. M. Kurn, and W. Ketterle. Bose–Einstein condensation in a gas of sodium atoms. *Phys. Rev. Lett.*, 75:3969–3973, 1995.

[25] R. S. Decca, D. López, E. Fischbach, G. L. Klimchitskaya, D. E. Krause, and V. M. Mostepanenko. Precise comparison of theory and new experiment for the Casimir force leads to stronger constraints on thermal quantum effects and long range interaction. *Ann. Phys.*, 318:37–80, 2005.

[26] A. Dettki and A. Wipf. Finite size effects from general covariance and Weyl anomaly. *Nucl. Phys.*, B377:252–280, 1992.

[27] J. S. Dowker. Effective action in spherical domains. *Commun. Math. Phys.*, 162:633–648, 1994.

[28] J. S. Dowker. Functional determinants on spheres and sectors. *J. Math. Phys.*, 35:4989–4999, 1994.

[29] J. S. Dowker and R. Critchley. Effective Lagrangian and energy momentum tensor in de Sitter space. *Phys. Rev.*, D13:3224–3232, 1976.

[30] G. V. Dunne. Functional determinants in quantum field theory. *J. Phys.*, A41: 304006, 2008.

[31] G. V. Dunne, J. Hur, C. Lee, and H. Min. Precise quark mass dependence of instanton determinant. *Phys. Rev. Lett.*, 94:072001, 2005.

[32] A. Einstein. Quantentheorie des einatomigen idealen Gases. *Sitzungsberichte der Preussichen Akademie der Wissenschaften*, 22:261–267, 1924.

[33] E. Elizalde. *Ten physical applications of spectral zeta functions*. Lecture Notes in Physics m35, Springer, Berlin, 1995.

[34] E. Elizalde, S. D. Odintsov, A. Romeo, A. A. Bytsenko, and S. Zerbini. *Zeta regularization techniques with applications*. World Scientific, Singapore, 1994.

[35] E. Elizalde and A. Romeo. Rigorous extension of the proof of zeta function regularization. *Phys. Rev.*, D40:436–, 1989.

[36] T. Emig, N. Graham, R. L. Jaffe, and M. Kardar. Casimir forces between arbitrary compact objects. *Phys. Rev. Lett.*, 99:170403, 2007.

[37] T. Emig, R. L. Jaffe, M. Kardar, and A. Scardicchio. Casimir interaction between a plate and a cylinder. *Phys. Rev. Lett.*, 96:080403, 2006.

[38] P. Epstein. Zur Theorie allgemeiner Zetafunctionen. *Math. Ann.*, 56:615–644, 1903.

[39] P. Epstein. Zur Theorie allgemeiner Zetafunctionen, II. *Math. Ann.*, 63:205–216, 1907.

[40] A. Erdélyi, W. Magnus, F. Oberhettinger, and F. G. Tricomi. *Higher transcendental functions*. Based on the notes of Harry Bateman, McGraw-Hill, New York, 1955.

[41] G. Esposito. *Quantum gravity, quantum cosmology and Lorentzian geometries*. Lecture Notes in Physics m12, Springer, Berlin, 1994.

[42] G. Esposito. *Dirac operators and spectral geometry*. Cambridge University Press, Cambridge, 1998.

[43] S. A. Fulling, L. Kaplan, K. Kirsten, Z. H. Liu, and K. A. Milton. Vacuum stress and closed paths in rectangles, pistons, and pistols. *J. Phys.*, A42:155402, 2009.

[44] P. B. Gilkey. *Invariance theory, the heat equation and the Atiyah–Singer index theorem*. CRC Press, Boca Raton, 1995.

[45] C. Gordon, D. Webb, and S. Wolpert. One cannot hear the shape of a drum. *Bull. Amer. Math. Soc.*, 27:134–138, 1992.

[46] I. S. Gradshteyn and I. M. Ryzhik. *Table of integrals, series and products*. Academic Press, New York, 1965.

[47] S. Grossmann and M. Holthaus. λ-transition to the Bose-Einstein condensate. *Z. Naturforsch.*, A50:921–930, 1995.

[48] S. Grossmann and M. Holthaus. On Bose-Einstein condensation in harmonic traps. *Phys. Lett.*, A208:188–192, 1995.

[49] H. Haugerud, T. Haugset, and F. Ravndal. A more accurate analysis of Bose-Einstein condensation in harmonic traps. *Phys. Lett.*, A225:18–22, 1997.

[50] T. Haugset, H. Haugerud, and J. O. Anderson. Bose–Einstein condensation in anisotropic harmonic traps. *Phys. Rev.*, A55:2922–2929, 1997.

[51] S. W. Hawking. Zeta function regularization of path integrals in curved space-time. *Commun. Math. Phys.*, 55:133–148, 1977.

[52] M. P. Hertzberg, R. L. Jaffe, M. Kardar, and A. Scardicchio. Attractive Casimir forces in a closed geometry. *Phys. Rev. Lett.*, 95:250402, 2005.

[53] E. Hille. *Analytic function theory*, Vol. 2. Ginn, Boston, 1962.

[54] A. Hurwitz. Einige Eigenschaften der Dirichlet'schen Functionen $f(s) = \sum d_n n^{-s}$ die bei der Bestimmung der Classenzahlen binärer quadratischer Formen auftreten. *Zeitschrift für Math. und Physik*, 27:86–101, 1882.

[55] C. Itzykson and J.-B. Zuber. *Quantum field theory*. McGraw-Hill, New York, 1980.

[56] M. Kac. Can one hear the shape of a drum? *Am. Math. Mon.*, 73:1–23, 1966.

[57] K. Kirsten. *Spectral Functions in Mathematics and Physics*. Chapman&Hall/CRC, Boca Raton, FL, 2002.

[58] K. Kirsten and S. A. Fulling. Kaluza–Klein models as pistons. *Phys. Rev.*, D79: 065019, 2009.

[59] K. Kirsten and D. J. Toms. Bose–Einstein condensation of atomic gases in a general harmonic-oscillator confining potential trap. *Phys. Rev.*, A54:4188–4203, 1996.

[60] K. Kirsten and D. J. Toms. Bose-Einstein condensation in arbitrarily shaped cavities. *Phys. Rev.*, E59:158–167, 1999.

[61] G. L. Klimchitskaya, F. Chen, R. S. Decca, E. Fischbach, D. E. Krause, D. López, U. Mohideen, and V. M. Mostepanenko. Rigorous approach to the comparison between experiment and theory in Casimir force measurements. *J. Phys. A: Math. Gen.*, 39:6485–6493, 2006.

[62] S. K. Lamoreaux. Demonstration of the Casimir force in the 0.6 to 6 micrometers range. *Phys. Rev. Lett.*, 78:5–7, 1997.

[63] S. K. Lamoreaux. A reply to the comment by Astrid Lambrecht and Serge Renaud. *Phys. Rev. Lett.*, 84:5673–5673, 2000.

[64] S. K. Lamoreaux. The Casimir force: background, experiments, and applications. *Rep. Prog. Phys.*, 68:201–236, 2005.

[65] V. Marachevsky. Casimir interaction of two plates inside a cylinder. *Phys. Rev.*, D75:085019, 2007.

[66] H. P. McKean and I. M. Singer. Curvature and eigenvalues of the Laplacian. *J. Diff. Geom.*, 1:43–69, 1967.

[67] K. A. Milton. *The Casimir effect: physical manifestations of zero-point energy*. World Scientific, River Edge, 2001.

[68] K. A. Milton. The Casimir effect: recent controversies and progress. *J. Phys.*, A37:R209–R277, 2004.

[69] U. Mohideen and A. Roy. Precision measurement of the Casimir force from 0.1 to 0.9 μm. *Phys. Rev. Lett.*, 81:4549–4552, 1998.

[70] N. E. Nörlund. Mémoire sur les polynômes de Bernoulli. *Acta Math.*, 43:121–196, 1922.

[71] C. J. Pethick and H. Smith. *Bose–Einstein condensation in dilute gases.* Cambridge University Press, Cambridge, 2002.

[72] L. Pitaevskii and S. Stringari. *Bose–Einstein condensation.* Oxford University Press, Oxford, 2003.

[73] D. B. Ray and I. M. Singer. R-torsion and the Laplacian on Riemannian manifolds. *Advances in Math.*, 7:145–210, 1971.

[74] I. Sachs and A. Wipf. Finite temperature Schwinger model. *Helv. Phys. Acta*, 65:652–678, 1992.

[75] M. Schaden. Dependence of the direction of the Casimir force on the shape of the boundary. *Phys. Rev. Lett.*, 102:060402, 2009.

[76] F. M. Serry, D. Walliser, and G. J. Maclay. The anharmonic Casimir oscillator (ACO): the Casimir effect in a model microelectromechanical system. *J. Microelectromech. Syst.*, 4:193–215, 1995.

[77] L. P. Teo. Finite temperature Casimir effect in spacetime with extra compactified dimensions. *Phys. Lett.*, B672:190–195, 2009.

[78] A. Terras. Bessel series expansion of the Epstein zeta function and the functional equation. *Trans. Amer. Math. Soc.*, 183:477–486, 1973.

[79] E. C. Titchmarsh. *The theory of the Riemann zeta function.* Oxford Science Publications, Oxford, 1951.

[80] H. Weyl. Das asymptotische Verteilungsgesetz der Eigenwerte linearer partieller Differentialgleichungen. *Math. Ann.*, 71:441–479, 1912.

[81] H. Weyl. Das asymptotische Verteilungsgesetz der Eigenschwingungen eines beliebig gestalteten elastischen Körpers. *Rend. Circ. Mat. Palermo*, 39:1–50, 1915.

KLAUS KIRSTEN
DEPARTMENT OF MATHEMATICS
BAYLOR UNIVERSITY
WACO, TX 76798
UNITED STATES
Klaus_Kirsten@baylor.edu

Zeta functions and chaos

AUDREY TERRAS

1. Introduction

This paper is an expanded version of lectures given at MSRI in June of 2008. It provides an introduction to various zeta functions emphasizing zeta functions of a finite graph and connections with random matrix theory and quantum chaos.

For the number theorist, most zeta functions are multiplicative generating functions for something like primes (or prime ideals). The Riemann zeta is the chief example. There are analogous functions arising in other fields such as Selberg's zeta function of a Riemann surface, Ihara's zeta function of a finite connected graph. All of these are introduced in Section 2. We will consider the Riemann hypothesis for the Ihara zeta function and its connection with expander graphs.

Chapter 3 starts with the Ruelle zeta function of a dynamical system, which will be shown to be a generalization of the Ihara zeta. A determinant formula is proved for the Ihara zeta function. Then we prove the graph prime number theorem.

In Section 4 we define two more zeta functions associated to a finite graph: the edge and path zetas. Both are functions of several complex variables. Both are reciprocals of polynomials in several variables, thanks to determinant formulas. We show how to specialize the path zeta to the edge zeta and then the edge zeta to the original Ihara zeta. The Bass proof of Ihara's determinant formula for the Ihara zeta function is given. The edge zeta allows one to consider graphs with weights on the edges. This is of interest for work on quantum graphs. See [Smilansky 2007] or [Horton et al. 2006b].

Lastly we consider what the poles of the Ihara zeta have to do with the eigenvalues of a random matrix. That is the sort of question considered in quantum chaos theory. Physicists have long studied spectra of Schrödinger operators and random matrices thanks to the implications for quantum mechanics where eigenvalues are viewed as energy levels of a system. Number theorists such as A.

Odlyzko have found experimentally that (assuming the Riemann hypothesis) the high zeros of the Riemann zeta function on the line Re(s) = 1/2 have spacings that behave like the eigenvalues of a random Hermitian matrix. Thanks to our two determinant formulas we will see that the Ihara zeta function, for example, has connections with spectra of more that one sort of matrix.

References [Terras 2007] and [Terras 2010] may be helpful for more details on some of these matters. The first is some introductory lectures on quantum chaos given at Park City, Utah in 2002. The second is a draft of a book on zeta functions of graphs.

2. Three zeta functions

2.1. The Riemann zeta function *Riemann's zeta function* for $s \in \mathbb{C}$ with Re(s) > 1 is defined to be

$$\zeta(s) = \sum_{n=1}^{\infty} \frac{1}{n^s} = \prod_{p \text{ prime}} \left(1 - \frac{1}{p^s}\right)^{-1}.$$

In 1859 Riemann extended the definition of zeta to an analytic function in the whole complex plane except for a simple pole at $s = 1$. He also showed that there is a *functional equation*

$$\Lambda(s) = \pi^{-s/2} \Gamma\left(\frac{s}{2}\right) \zeta(s) = \Lambda(1 - s). \tag{2-1}$$

The *Riemann hypothesis*, or *RH*, says that the nonreal zeros of $\zeta(s)$ (equivalently those with $0 < \operatorname{Re} s < 1$) are on the line $\operatorname{Re} s = \frac{1}{2}$. It is equivalent to giving an explicit error term in the prime number theorem stated below. The Riemann hypothesis has been checked to the 10^{13}-th zero as of 12 October 2004, by Xavier Gourdon with the help of Patrick Demichel. See Ed Pegg Jr.'s website for an article called the Ten Trillion Zeta Zeros: http://www.maa.org/editorial/mathgames. Proving (or disproving) the Riemann hypothesis is one of the million-dollar problems on the Clay Mathematics Institute website.

There is a duality between the primes and the zeros of zeta, given analytically through the Hadamard product formula as various sorts of explicit formulas. See [Davenport 1980] and [Murty 2001]. Such results lead to the *prime number theorem* which says

$$\#\{p = \text{prime} \mid p \leq x\} \sim \frac{x}{\log x}, \quad \text{as } x \to \infty.$$

The spacings of high zeros of zeta have been studied by A. Odlyzko; see the page www.dtc.umn.edu/~odlyzko/doc/zeta.htm. He has found that experimentally they look like the spacings of the eigenvalues of random Hermitian matrices (GUE). We will say more about this in the last section. See also [Conrey 2003].

EXERCISE 1. Use Mathematica to make a plot of the Riemann zeta function.
Hint. The function Zeta[s] in Mathematica can be used to compute the Riemann zeta function.

There are many other kinds of zeta function. One is the Dedekind zeta of an algebraic number field F such as $\mathbb{Q}(\sqrt{2}) = \{a+b\sqrt{2} \mid a,b \in \mathbb{Q}\}$, where primes are replaced by prime ideals $\mathfrak{p}$ in the ring of integers O_F (which is $\mathbb{Z}[\sqrt{2}] = \{a+b\sqrt{2} \mid a,b \in \mathbb{Z}\}$, if $F = \mathbb{Q}(\sqrt{2})$). Define the *norm of an ideal* of O_F to be $N\mathfrak{a} = |O_F/\mathfrak{a}|$. Then the *Dedekind zeta function* is defined for Re $s > 1$ by

$$\zeta(s, F) = \prod_{\mathfrak{p}} (1 - N\mathfrak{p}^{-s})^{-1},$$

where the product is over all prime ideals of O_F. The Riemann zeta function is $\zeta(s, \mathbb{Q})$.

Hecke gave the analytic continuation of the Dedekind zeta to all complex s except for a simple pole at $s = 1$. And he found the functional equation relating $\zeta(s, F)$ and $\zeta(1-s, F)$. The value at 0 involves the interesting number h_F=the class number of O_F which measures how far O_F is from having unique factorization into prime numbers rather than prime ideals ($h_{\mathbb{Q}(\sqrt{2})} = 1$). Also appearing in $\zeta(0, F)$ is the regulator which is a determinant of logarithms of units (i.e., elements $u \in O_F$ such that $u^{-1} \in O_F$). For $F = \mathbb{Q}(\sqrt{2})$, the regulator is $\log(1+\sqrt{2})$. The formula is

$$\zeta(0, F) = -\frac{hR}{w}, \qquad (2\text{-}2)$$

where w is the number of roots of unity in F ($w = 2$ for $F = \mathbb{Q}(\sqrt{2})$). One has $\zeta(0, \mathbb{Q}) = -\frac{1}{2}$. See [Stark 1992] for an introduction to this subject meant for physicists.

2.2. The Selberg zeta function. This zeta function is associated to a compact (or finite volume) Riemannian manifold. Assuming M has constant curvature -1, it can be realized as a quotient of the *Poincaré upper half-plane*

$$H = \{x+iy \mid x, y \in \mathbb{R}, \, y > 0\}.$$

The *Poincaré arc length* element is

$$ds^2 = \frac{dx^2 + dy^2}{y^2},$$

which can be shown invariant under fractional linear transformation

$$z \mapsto \frac{az+b}{cz+d}, \quad \text{where } a, b, c, d \in \mathbb{R}, \, ad - bc > 0.$$

It is not hard to see that *geodesics*—curves minimizing the Poincaré arc length—are half-lines and semicircles in H orthogonal to the real axis. Calling these geodesics straight lines creates a model for non-Euclidean geometry since Euclid's fifth postulate fails. There are infinitely many geodesics through a fixed point not meeting a given geodesic.

The *fundamental group* Γ of M acts as a discrete group of distance-preserving transformations. The favorite group of number theorists is the *modular group* $\Gamma = \mathrm{SL}(2,\mathbb{Z})$ of 2×2 matrices of determinant one and integer entries or the quotient $\overline{\Gamma} = \Gamma/\{\pm I\}$. However the Riemann surface $M = \mathrm{SL}(2,\mathbb{Z})\backslash H$ is not compact, although it does have finite volume.

Selberg defined primes in the compact Riemannian manifold $M = \Gamma\backslash H$ to be primitive closed geodesics C in M. Here *primitive* means you only go around the curve once.

Define the *Selberg zeta function,* for $\mathrm{Re}(s)$ sufficiently large, as

$$Z(s) = \prod_{[C]} \prod_{j \geq 1} \left(1 - e^{-(s+j)v(C)}\right).$$

The product is over all primitive closed geodesics C in $M = \Gamma\backslash H$ of Poincaré length $v(C)$. By the Selberg trace formula (which we do not discuss here), there is a duality between the lengths of the primes and the spectrum of the Laplace operator on M. Here

$$\Delta = y^2 \left(\frac{\partial^2}{\partial x^2} + \frac{\partial^2}{\partial y^2} \right).$$

Moreover one can show that the Riemann hypothesis (suitably modified to fit the situation) can be proved for Selberg zeta functions of compact Riemann surfaces.

EXERCISE 2. Show that $Z(s+1)/Z(s)$ has a product formula which is more like that for the Riemann zeta function.

The closed geodesics in $M = \Gamma\backslash H$ correspond to geodesics in H itself. One can show that the endpoints of such geodesics in $\mathbb{R}$ (the real line = the boundary of H) are fixed by hyperbolic elements of Γ; i.e., the matrices $\begin{pmatrix} a & b \\ c & d \end{pmatrix}$ with trace $a+d > 2$. Primitive closed geodesics correspond to hyperbolic elements that generate their own centralizer in Γ.

Some references for this subject are [Selberg 1989] and [Terras 1985].

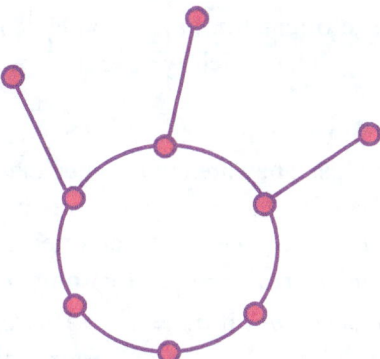

Figure 1. An example of a bad graph for zeta functions.

2.3. The Ihara zeta function.
We will see that the Ihara zeta function of a graph has similar properties to the preceding zetas. A good reference for graph theory is [Biggs 1974].

First we must figure out what primes in graphs are. Recalling what they are for manifolds, we expect that we need to look at closed paths that minimize distance. What is distance? It is the number of oriented edges in a path.

First suppose that X is a finite connected unoriented graph. Thus it is a collection of vertices and edges. Usually we assume the graph is not a cycle or a cycle with hair (i.e., degree 1 vertices). Thus Figure 1 is a bad graph. We do allow our graphs to have loops and multiple edges however.

Let E be the set of unoriented (or undirected) edges of X and V the set of vertices. We orient (or direct) the edges arbitrarily and label them $e_1, e_2, \ldots, e_{|E|}$. An example is shown in Figure 2. Then we label the inverse edges (meaning

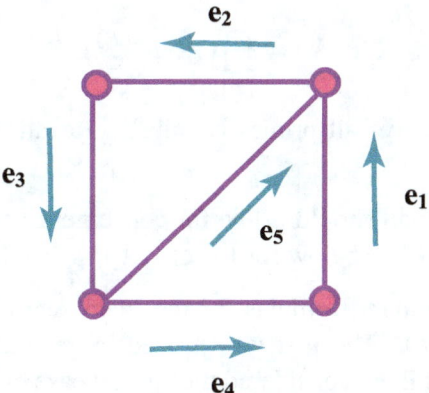

Figure 2. We choose an arbitrary orientation of the edges of a graph. Then we label the inverse edges via $e_{j+|E|} = e_j^{-1}$, for $j = 1, \ldots, 5$.

the edge with the opposite orientation) $e_{j+|E|} = e_j^{-1}$, for $j = 1, \ldots, |E|$. The oriented edges give an alphabet which we use to form words representing the paths in our graph.

Now we can define primes in the graph X. They correspond to closed geodesics in compact manifolds. They are equivalence classes $[C]$ of tailless primitive closed paths C. We define these last adjectives in the next paragraph.

A path or walk $C = a_1 \cdots a_s$, where a_j is an oriented or directed edge of X, is said to have a *backtrack* if $a_{j+1} = a_j^{-1}$, for some $j = 1, \ldots, s-1$. A path $C = a_1 \cdots a_s$ is said to have a *tail* if $a_s = a_1^{-1}$. The *length* of $C = a_1 \cdots a_s$ is $s = \nu(C)$. A *closed path* means the starting vertex is the same as the terminal vertex. The closed path $C = a_1 \cdots a_s$ is called a *primitive or prime path* if it has no backtrack or tail and $C \neq D^f$, for $f > 1$. For the path $C = a_1 \cdots a_s$, the *equivalence class* $[C]$ means

$$[C] = \{a_1 \cdots a_s, a_2 \cdots a_s a_1, \ldots, a_s a_1 \cdots a_{s-1}\}.$$

That is, we call two prime paths *equivalent* if we get one from the other by changing the starting point. A *prime* in the graph X is an equivalence class $[C]$ of prime paths.

Examples of primes in a graph. For the graph in Figure 2, we have primes $[C] = [e_2 e_3 e_5]$, $[D] = [e_1 e_2 e_3 e_4]$, $E = [e_1 e_2 e_3 e_4 e_1 e_{10} e_4]$. Here $e_{10} = e_5^{-1}$ and the lengths of these primes are: $\nu(C) = 3$, $\nu(D) = 4$, $\nu(E) = 7$. We have infinitely many primes since $E_n = [(e_1 e_2 e_3 e_4)^n e_1 e_{10} e_4]$ is prime for all $n \geq 1$. But we don't have unique factorization into primes. The only nonprimes are powers of primes.

DEFINITION 3. The *Ihara zeta function* is defined for $u \in \mathbb{C}$, with $|u|$ sufficiently small by

$$\zeta(u, X) = \prod_{[P]} (1 - u^{\nu(P)})^{-1}, \qquad (2\text{-}3)$$

where the product is over all primes $[P]$ in X. Recall that $\nu(P)$ denotes the length of P.

EXERCISE 4. How small should $|u|$ be for convergence of $\zeta(u, X)$?

Hint. See formula (3-5) below for $\log \zeta(u, X)$.

There are two determinant formulas for the Ihara zeta function (see formulas (2-4) and (3-1) below). The first was proved in general by Bass [1992] and Hashimoto [1989], as Ihara considered the special case of *regular* graphs (those all of whose vertices have the same *degree*; i.e., the same number of oriented edges coming out of the vertex) and in fact was considering p-adic groups and not graphs. Moreover the degree had to be $1 + p^e$, where p is a prime number.

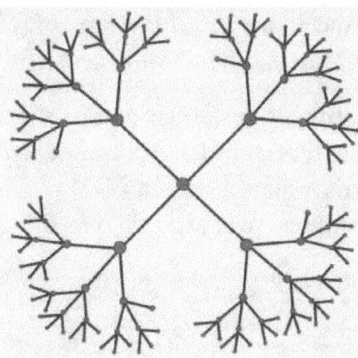

Figure 3. Part of the 4-regular tree. The tree itself is infinite.

The (vertex) *adjacency matrix* A of X is a $|V| \times |V|$) matrix whose i, j entry is the number of directed edges from vertex i to vertex j. The matrix Q is defined to be a diagonal matrix whose j-th diagonal entry is 1 less than the degree of the j-th vertex. If there is a loop at a vertex, it contributes 2 to the degree.

Then we have the *Ihara determinant formula*

$$\zeta(u, X)^{-1} = (1 - u^2)^{r-1} \det(I - Au + Qu^2). \qquad (2\text{-}4)$$

Here r is the rank of the fundamental group of the graph. This is $r = |E| - |V| + 1$. In Section 4 we will give a version of Bass's proof of this formula.

In the case of regular graphs, one can prove the formula using the Selberg trace formula for the graph realized as a quotient $\Gamma \backslash T$, where T is the universal covering tree of the graph and Γ is the fundamental group of the graph. A graph T is a *tree* if it is a connected graph without any closed backtrackless paths. For a tree to be regular, it must be infinite. We will discuss covering graphs in the last section of this paper. For a discussion of the Selberg trace formula on $\Gamma \backslash T$, see the last chapter of [Terras 1999].

Figure 3 shows part of the 4-regular tree T_4. As the tree is infinite, we cannot put the whole thing on a page. It can be identified with the 3-adic quotient $\mathrm{SL}(2, \mathbb{Q}_3) / \mathrm{SL}(2, \mathbb{Z}_3)$. A finite 4-regular graph X is a quotient of T_4 modulo the fundamental group of X.

EXAMPLE 5. The *tetrahedron graph* K_4 is the complete graph on 4 vertices and its zeta function is given by

$$\zeta(u, K_4)^{-1} = (1 - u^2)^2 (1 - u)(1 - 2u)(1 + u + 2u^2)^3.$$

EXAMPLE 6. Let $X = K_4 - e$ be the graph obtained from K_4 by deleting an edge e. See Figure 2. Then

$$\zeta(u, X)^{-1} = (1 - u^2)(1 - u)(1 + u^2)(1 + u + 2u^2)(1 - u^2 - 2u^3).$$

EXERCISE 7. Compute the Ihara zeta functions of your favorite graphs; e.g., the cube, the icosahedron, the buckyball or soccer ball graph.

EXERCISE 8. Obtain a functional equation for the Ihara zeta function of a $(q+1)$-regular graph. It will relate $\zeta(u, X)$ and $\zeta(1/qu, X)$.
Hint. Use the Ihara determinant formula (2-4).
There are various possible answers to this question. One answer is:

$$\Lambda_X(u) = (1-u^2)^{r-1+n/2}(1-q^2u^2)^{n/2}\zeta_X(u) = (-1)^n \Lambda_X\left(\frac{1}{qu}\right).$$

In the special case of a $(q+1)$-regular graph the substitution $u = q^{-s}$ makes the Ihara zeta more like Riemann zeta. That is we set $f(s) = \zeta(q^{-s}, X)$ when X is $(q+1)$-regular. Then the functional equation relates $f(s)$ and $f(1-s)$. See Exercise 8.

The *Riemann hypothesis* for Ihara's zeta function of a $(q+1)$-regular graph says that

$$\zeta(q^{-s}, X) \text{ has no poles with } 0 < \operatorname{Re} s < 1 \text{ unless } \operatorname{Re} s = \tfrac{1}{2}. \quad (2\text{-}5)$$

It turns out (using the Ihara determinant formula again) that the Riemann hypothesis means that the graph is *Ramanujan*; i.e., the nontrivial spectrum of the adjacency matrix of the graph is contained in the spectrum of the adjacency operator on the universal covering tree which is the interval $[-2\sqrt{q}, 2\sqrt{q}]$. This definition was introduced by Lubotzky, Phillips and Sarnak [Lubotzky et al. 1988], who showed that for each fixed degree of the form $p^e + 1$, $p =$ prime, there is a family of Ramanujan graphs X_n with $|V(X_n)| \to \infty$. Ramanujan graphs are of interest to computer scientists because they provide efficient communication networks. The graph is a good expander.

EXERCISE 9. Show that for a $(q+1)$-regular graph the Riemann hypothesis is equivalent to saying that the graph is Ramanujan; i.e. if λ is an eigenvalue of the adjacency matrix A of the graph such that $|\lambda| \neq q+1$, then $|\lambda| \leq 2\sqrt{q}$.
Hint. Use the Ihara determinant formula (2-4).

What is an expander graph? There are 4 ideas.

(1) There is a spectral property of some matrix associated to our finite graph X. Choose one of three matrices:

(a) the (vertex) adjacency matrix A,

(b) the Laplacian $D - A$ or $I - D^{-\frac{1}{2}}AD^{-\frac{1}{2}}$, where D is the diagonal matrix of degrees of vertices, or

(c) the edge adjacency matrix W_1 to be defined in the next section.

Following [Lubotzky 1995], a graph is Ramanujan if the spectrum of the adjacency matrix for X is inside the spectrum of the analogous operator on the universal covering tree of X. One could ask for the analogous property of the other operators such as the Laplacian or the edge adjacency matrix.

(2) X behaves like a random graph in some sense.

(3) Information is passed quickly in the gossip network based on X. The graph has a large expansion constant. This is defined by formula (2-6) below.

(4) The random walker on the graph gets lost FAST.

DEFINITION 10. For sets of vertices S, T of X, define

$E(S, T) = \{e \mid e$ is edge of X with one vertex in S and the other vertex in $T\}$.

DEFINITION 11. If S is a set of vertices of X, we say the *boundary* is $\partial S = E(S, X - S)$.

DEFINITION 12. A graph X with vertex set V and $n = |V|$ has *expansion constant*

$$h(X) = \min_{\substack{S \subset V \\ |S| \leq n/2}} \frac{|\partial S|}{|S|}. \tag{2-6}$$

The expansion constant is an analog of the Cheeger constant for differentiable manifolds. References for these things include [Chung 2007; Hoory et al. 2006; Terras 1999; 2010]. The first of these references gives relations between the expansion constant and the *spectral gap* $\lambda_X = \min\{\lambda_1, 2 - \lambda_{n-1}\}$ if $0 = \lambda_0 \leq \lambda_1 \leq \cdots \leq \lambda_n$ are the eigenvalues of $I - D^{-1/2} A D^{-1/2}$. Fan Chung proves that $2h_X \geq \lambda_X \geq h_X^2/2$. This is an analog of the Cheeger inequality in differential geometry. She also connects these inequalities with webpage search algorithms of the sort used by Google.

The possible locations of poles u of $\zeta(u, X)$ of a $(q + 1)$-regular graph can be found in Figure 4. The poles satisfying the Riemann hypothesis are those on the circle of radius $1/\sqrt{q}$. Any nontrivial pole; i.e., $u \neq \pm 1, \pm 1/q$, which is not on that circle is a non-RH pole. In the $(q + 1)$-regular graph case, $1/q$ is always the closest pole of the Ihara zeta to the origin.

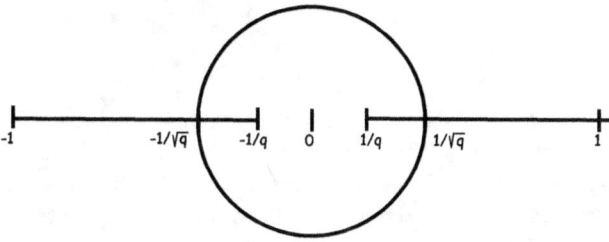

Figure 4. Possible locations of poles of zeta for a $(q + 1)$-regular graph.

EXERCISE 13. Show that Figure 4 correctly locates the position of possible poles of the Ihara zeta function of a $(q+1)$-regular graph.

Hint. Use the Ihara determinant formula (2-4).

The *Alon conjecture* for regular graphs says that the RH is *approximately* true for "most" regular graphs See [Friedman 2008]for a proof. See [Miller and Novikoff 2008] for experiments leading to the conjecture that the percent of regular graphs exactly satisfying the RH approaches 27% as the number of vertices approaches infinity. The argument involves the Tracy–Widom distribution from random matrix theory.

Newland [2005] performed graph analogs of Odlyzko's experiments on the spacings of imaginary parts of zeros of Riemann zeta. See Figure 5 below and Figure 8 on page 173.

An obvious question is: What is the meaning of the RH for irregular graphs? To understand this we need a definition.

DEFINITION 14. $R_X = R$ is the *radius of the largest circle of convergence* of the Ihara zeta function.

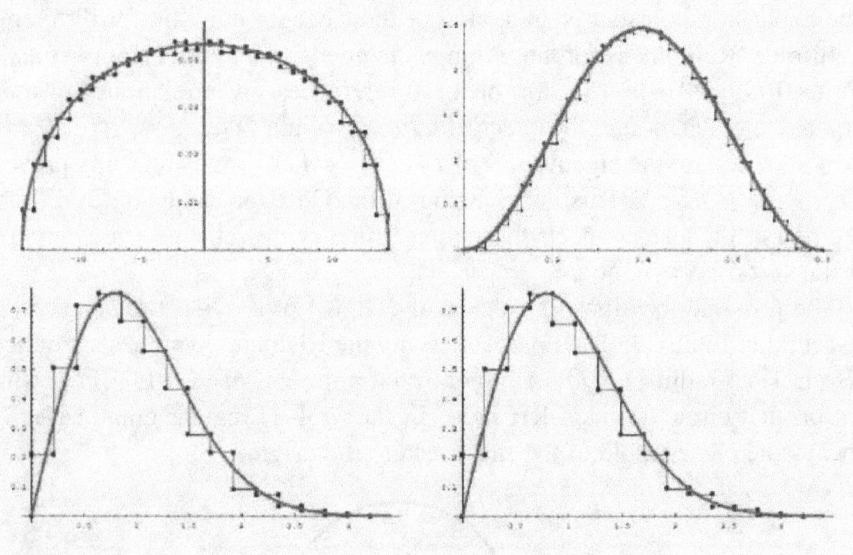

Figure 5. For a pseudo-random regular graph with degree 53 and 2000 vertices, generated by Mathematica, the top row shows the distributions of the eigenvalues of the adjacency matrix on the left and imaginary parts of the Ihara zeta poles on the right. The bottom row contains their respective level spacings. The red line on the bottom left is the Wigner surmise for the GOE $y = ((\pi x/2))e^{-\pi x^2/4}$. From [Newland 2005].

As a power series in the complex variable u, the Ihara zeta function has nonnegative coefficients. Thus, by a classic theorem of Landau, both the series and the product defining $\zeta_X(u)$ will converge absolutely in a circle $|u| < R_X$ with a singularity (pole of order 1 for connected X) at $u = R_X$. See [Apostol 1976, p. 237] for Landau's theorem.

Define the *spectral radius* of a matrix M to be the maximum absolute value of all eigenvalues of M. We will see in the next section that by the Perron–Frobenius theorem in linear algebra (see [Horn and Johnson 1990], for example), $1/R_X$ is the spectral radius of the edge adjacency matrix W_1 which will be defined at the beginning of the next section. To apply the theorem, one must show that the edge adjacency matrix of a graph (under our usual assumptions) satisfies the necessary hypotheses. See [Terras and Stark 2007]. It is interesting to see that the quantity R_X can be viewed from two points of view; complex analysis and linear algebra.

For a $(q+1)$-regular graph $R_X = 1/q$. If the graph is not regular, one sees by experiment that generally there is no functional equation. Thus when we make the change of variables $u = R^s$ in our zeta, the critical strip $0 \leq \operatorname{Re} s \leq 1$ is too large. We should only look at half of it and our Riemann hypothesis becomes:

The graph theory RH for irregular graphs:

$$\zeta(u, X) \text{ is pole free in } R < |u| < \sqrt{R}. \tag{2-7}$$

If the graph is $(q+1)$-regular (by the functional equations), this is equivalent to the Riemann hypothesis stated earlier in formula (2-5).

To investigate this we need to define the edge adjacency matrix W_1, which is found in the next section. We will consider examples in the last section.

EXERCISE 15. Consider the graph $X = K_4 - e$ from Exercise 6. Show that the poles of $\zeta(u, X)$ are not invariant under the map $u \to R/u$. This means there is no functional equation of the sort that occurs for regular graphs in Exercise 8. Do the poles satisfy the Riemann hypothesis? Do they satisfy a weak RH meaning that $\zeta(u, X)$ is pole free in $R < |u| < 1/\sqrt{q}$?

3. Ruelle's zeta function of a dynamical system, a determinant formula, and the graph prime number theorem

3.1. The edge adjacency matrix of a graph and another determinant formula for the Ihara zeta. In this section we consider some zeta functions that arise from those in algebraic geometry. For the Ihara zeta function, we prove a simple determinant formula. We also consider a proof of the graph theory prime number theorem.

DEFINITION 16. The *edge adjacency matrix* W_1 is defined to be the $2|E|\times 2|E|$ matrix with i, j entry 1 if edge i feeds into edge j (meaning that the terminal vertex of edge i is the initial vertex of edge j) provided that edge i is not the inverse of edge j.

We will soon prove a second determinant formula for the Ihara zeta function:

$$\zeta(u, X)^{-1} = \det(I - uW_1). \tag{3-1}$$

COROLLARY 17. *The poles of the Ihara zeta are the reciprocals of the eigenvalues of* W_1.

Recall that R is the radius of convergence of the product defining the Ihara zeta function. By the corollary it is the reciprocal of the spectral radius of W_1 as well as the closest pole of zeta to the origin. It is necessarily positive. See the last chapter of [Horn and Johnson 1990] or [Apostol 1976, p. 237] for Landau's theorem, which implies the same thing.

There are many proofs of formula (3-1). We give the dynamical systems version here. There is another related proof in Section 4.1 and in [Terras 2010].

3.2. Ruelle zeta (aka dynamical systems zeta or Smale zeta).

For the material here see [Ruelle 1994], [Bedford et al. 1991] or [Lagarias 1999]. Ruelle's motivation for his definition came partially from [Artin and Mazur 1965], whose authors were in turn inspired by the zeta function of a projective nonsingular algebraic variety V of dimension n over a finite field k with q elements. If N_m denotes the number of points of V with coordinates in the degree m extension field of k, the *zeta function of a variety V over a finite field* is

$$Z_V(u) = \exp\left(\sum_{m\geq 1} \frac{N_m u^m}{m}\right). \tag{3-2}$$

Example of varieties are given by taking solutions of polynomial equations over finite fields; e.g., $x^2 + y^2 = 1$ and $y^2 = x^3 + ax + b$. You actually have to look at the homogeneous version of the equations in projective space. For more information on these zeta functions, see [Lorenzini 1996, p. 280] or [Rosen 2002].

Let F be the *Frobenius map* taking a point on the variety with coordinates x_i to the point with coordinates x_i^q. Here q is the number of elements in the finite field k. Define $\text{Fix}(F^m) = \{x \in M \mid F^m(x) = x\}$; then $N_m = |\text{Fix}(F^m)|$.

Weil conjectured that zeta satisfies a functional equation relating the values $Z_V(u)$ and $Z_V\left(\frac{1}{q^n u}\right)$. He also conjectured that

$$Z_V(u) = \prod_{j=0}^{2n} P_j(u)^{(-1)^{j+1}},$$

where the P_j are polynomials with zeros of absolute value $q^{-j/2}$. Weil proved the conjectures for the case of curves ($n = 1$). The proof was later simplified. See Rosen [Rosen 2002]. For general n, the Weil conjectures were proved by Deligne. Further, the P_j have a cohomological meaning as $\det\left(1-uF^*|_{H^j(V)}\right)$. Here the Frobenius has induced an action on the ℓ-adic étale cohomology. The case that $n = 1$ is very similar to that of the Ihara zeta function for a $(q+1)$-regular graph.

Artin and Mazur replace the Frobenius of V with a diffeomorphism f of a smooth compact manifold M such that its iterates f^k all have isolated fixed points. The *Artin–Mazur zeta function* is defined by

$$\zeta(u) = \exp\left(\sum_{m\geq 1} \frac{u^m}{m} |\mathrm{Fix}(f^m)|\right). \tag{3-3}$$

The Ruelle zeta function involves a function $f : M \to M$ on a compact manifold M. Assume $\mathrm{Fix}(f^m)$ is finite for all $m \geq 1$. The (first type of) *Ruelle zeta* is defined for a matrix valued function $\varphi : M \to \mathbb{C}^{d \times d}$ by

$$\zeta(u) = \exp\left(\sum_{m\geq 1} \frac{u^m}{m} \sum_{x \in \mathrm{Fix}(f^m)} \mathrm{Tr} \prod_{k=0}^{m-1} \varphi(f^k(x))\right). \tag{3-4}$$

Here we consider only the special case that $d = 1$ and φ is identically 1, when formula (3-4) looks exactly like formula (3-3).

Let I be a finite nonempty set (*our alphabet*). For a graph X, I is the set of directed edges. The transition matrix t is a matrix of zeros and ones with indices in I. In the case of a graph X, t is the 0,1 edge adjacency W_1 from Definition 16.

Since $I^{\mathbb{Z}}$ is compact, the following subset is closed:

$$\Lambda = \left\{(\xi_k)_{k \in \mathbb{Z}} \mid t_{\xi_k \xi_{k+1}} = 1 \text{ for all } k\right\}.$$

In the graph case, $t = W_1$ and $\xi \in \Lambda$ corresponds to a path without backtracking.

A continuous function $\tau : \Lambda \to \Lambda$ such that $\tau(\xi)_k = \xi_{k+1}$ is called a *subshift of finite type*. In the graph case, this shifts the path right, assuming the paths go from left to right.

Then we can find a new formula for the Ihara zeta function which shows that it is a Ruelle zeta. To understand this formula, we need a definition.

DEFINITION 18. $N_m = N_m(X)$ is the *number of closed paths of length m without backtracking and tails in the graph X.*

From Definition 2-3 of the Ihara zeta, we prove in the next paragraph that

$$\log \zeta(u, X) = \sum_{m\geq 1} \frac{N_m}{m} u^m. \tag{3-5}$$

Compare this formula with formula (3-2) defining the zeta function of a projective variety over a finite field.

To prove formula (3-5), take the logarithm of Definition 2-3 where the product is over primes $[P]$ in the graph X:

$$\log \zeta(u, X) = \log \left(\prod_{[P] \text{ prime}} (1 - u^{\nu(P)})^{-1} \right) = -\sum_{[P]} \log(1 - u^{\nu(P)})$$

$$= \sum_{[P]} \sum_{j \geq 1} \frac{1}{j} u^{j\nu(P)} = \sum_{P} \sum_{j \geq 1} \frac{1}{j\nu(P)} u^{j\nu(P)}$$

$$= \sum_{P} \sum_{j \geq 1} \frac{1}{\nu(P^j)} u^{\nu(P^j)}$$

$$= \sum_{C \text{ closed, backtrackless, tailless path}} \frac{1}{\nu(C)} u^{\nu(C)} = \sum_{m \geq 1} \frac{N_m}{m} u^m.$$

Here we have used the power series for $\log(1 - x)$ to see the third equality. Then the fourth equality comes from the fact that there are $\nu(P)$ elements in the equivalence class $[P]$, for any prime $[P]$. The sixth equality is proved using the fact that any closed backtrackless tailless path C in the graph is a power of some prime path P. The last equality comes from Definition 18 of N_m.

If the subshift of finite type τ is as defined above for the graph X, we have

$$|\text{Fix}(\tau^m)| = N_m. \tag{3-6}$$

It follows from this result and formula (3-5) that the Ihara zeta is a special case of the Ruelle zeta.

Next we claim that

$$N_m = \text{Tr}(W_1^m). \tag{3-7}$$

To see this, set $B = W_1$, with entries b_{ef}, for oriented edges e, f. Then

$$\text{Tr}(W_1^m) = \text{Tr}(B^m) = \sum_{e_1, \ldots, e_m} b_{e_1 e_2} b_{e_2 e_3} \cdots b_{e_m e_1},$$

where the sum is over all oriented edges of the graph. The b_{ef} are 0 unless edge e feeds into edge f without backtracking; i.e., the terminal vertex of e is the initial vertex of f and $f \neq e^{-1}$. Thus $b_{e_1 e_2} b_{e_2 e_3} \cdots b_{e_m e_1} = 1$ means that the path $C = e_1 e_2 \cdots e_m$ is closed, backtrackless, tailless of length m.

It follows that:

$$\log \zeta(u, X) = \sum_{m \geq 1} \frac{u^m}{m} \text{Tr}(W_1^m) = \text{Tr}\left(\sum_{m \geq 1} \frac{u^m}{m} W_1^m \right)$$

$$= \text{Tr}\left(\log (I - uW_1)^{-1} \right) = \log \det (I - uW_1)^{-1}.$$

Here we have used formula (3-7) and the continuous linear property of trace. Then we need the power series for the matrix logarithm and the following exercise.

EXERCISE 19. Show that $\exp \mathrm{Tr}(A) = \det(\exp A)$, for any matrix A. To prove this, you need to know that there is a nonsingular matrix B such that $BAB^{-1} = T$ is upper triangular. See your favorite linear algebra book.

This proves formula (3-1) for the Ihara zeta function which says $\zeta(u, X) = \det(I - uW_1)^{-1}$. This is known as the Bowen–Lanford theorem for subshifts of finite type in the context of Ruelle zeta functions.

3.3. Graph prime number theorem. Next we prove the graph prime number theorem. This requires two definitions and a theorem.

DEFINITION 20. The *prime counting function* is

$$\pi(n) = \#\{\text{primes } [P] \mid n = \nu(P) = \text{length of } P\}.$$

DEFINITION 21. The *greatest common divisor of the prime path lengths* is

$$\Delta_X = \gcd\{\nu(P) \mid [P] \text{ prime of } X\}.$$

Kotani and Sunada prove the following theorem. Their proof makes heavy use of the Perron–Frobenius theorem from linear algebra. A proof can also be found in [Terras 2010].

THEOREM 22 [Kotani and Sunada 2000]. *Assume, as usual, that the graph X is connected, has fundamental group of rank $r > 1$, and has no degree 1 vertices.*

(1) *Every pole u of $\zeta_X(u)$ satisfies $R_X \leq |u| \leq 1$, with R_X from Definition 14, and*

$$q^{-1} \leq R_X \leq p^{-1}. \qquad (3\text{-}8)$$

(2) *For a graph X, if $q+1$ is the maximum degree of X and $p+1$ is the minimum degree of X, then every nonreal pole u of $\zeta_X(u)$ satisfies the inequality*

$$q^{-1/2} \leq |u| \leq p^{-1/2}. \qquad (3\text{-}9)$$

(3) *The poles of ζ_X on the circle $|u| = R_X$ have the form $R_X e^{2\pi i a/\Delta_X}$, where $a = 1, \ldots, \Delta_X$. Here Δ_X is from Definition 21.*

EXERCISE 23. Look up [Kotani and Sunada 2000] and figure out their proof of the result that we needed in proving the prime number theorem. Another version of this proof can be found in [Terras 2010].

THEOREM 24 (GRAPH PRIME NUMBER THEOREM). *Assume X satisfies the hypotheses of the preceding theorem. Suppose that R_X is as in Definition 14. If $\pi(m)$ and Δ_X are as in Definitions 20 and 21, then $\pi(m) = 0$ unless Δ_X divides m. If Δ_X divides m, we have*

$$\pi(m) \sim \Delta_X \frac{R_X^{-m}}{m} \quad as\ m \to \infty.$$

PROOF. If N_m is as in Definition 18, then formula (3-5) implies we have

$$u \frac{d}{du} \log \zeta_X(u) = \sum_{m \geq 1} N_m u^m. \tag{3-10}$$

Now observe that the defining formula for the Ihara zeta function can be written as

$$\zeta_X(u) = \prod_{n \geq 1} (1 - u^n)^{-\pi(n)}.$$

Then

$$u \frac{d}{du} \log \zeta_X(u) = \sum_{n \geq 1} \frac{n\pi(n) u^n}{1 - u^n} = \sum_{m \geq 1} \sum_{d \mid m} d\pi(d) u^m.$$

Here the inner sum is over all positive divisors of m. Thus we obtain the *relation between N_m and $\pi(n)$*

$$N_m = \sum_{d \mid m} d\pi(d).$$

This sort of relation occurs frequently in number theory and combinatorics. It is inverted using the *Möbius function* $\mu(n)$ defined by

$$\mu(n) = \begin{cases} 1 & \text{if } n = 1, \\ (-1)^r & \text{if } n = p_1 \cdots p_r \text{ for distinct primes } p_i, \\ 0 & \text{otherwise.} \end{cases}$$

Then, by the *Möbius inversion formula* (which can be found in any elementary number theory book),

$$\pi(m) = \frac{1}{m} \sum_{d \mid m} \mu\left(\frac{m}{d}\right) N_d. \tag{3-11}$$

Next use formula (3-1) to see that

$$u \frac{d}{du} \log \zeta_X(u) = -u \frac{d}{du} \sum_{\lambda \in \text{Spec } W_1} \log(1 - \lambda u) = \sum_{\lambda \in \text{Spec } W_1} \sum_{n \geq 1} (\lambda u)^n.$$

From this, we get the *formula relating N_m and the spectrum of the edge adjacency W_1*:

$$N_m = \sum_{\lambda \in \text{Spec } W_1} \lambda^m. \tag{3-12}$$

The dominant terms in this last sum are those coming from $\lambda \in \text{Spec } W_1$ such that $|\lambda| = R^{-1}$, with $R = R_X$ from Definition 14.

By Theorem 22, the largest absolute value of an eigenvalue λ occurs Δ_X times with these eigenvalues having the form $e^{2\pi i a/\Delta_X} R^{-1}$, where $a = 1, \ldots, \Delta_X$. we see that

$$\pi(n) \sim \frac{1}{n} \sum_{|\lambda| \text{ maximal}} \lambda^n = \frac{R^{-n}}{n} \sum_{a=1}^{\Delta_X} e^{2\pi i a n/\Delta_X}.$$

The orthogonality relations for exponential sums (see [Terras 1999]) which are basic to the theory of the finite Fourier transform say that

$$\sum_{a=1}^{\Delta_X} e^{2\pi i a n/\Delta_X} = \begin{cases} 0 & \text{if } \Delta_X \text{ does not divide } n, \\ \Delta_X & \text{if } \Delta_X \text{ divides } n. \end{cases} \quad (3\text{-}13)$$

The graph prime number theorem follows. $\square$

Note that the Riemann hypothesis gives information on the size of the error term in the prime number theorem.

EXERCISE 25. Fill in the details in the proof of the graph theory prime number theorem. In particular, prove the orthogonality relations for exponential sums which implies formula (3-13) above.

As is the case for the Riemann zeta function, it is clear that the graph theory Riemann hypothesis gives information on the size of the error term in the prime number theorem.

EXAMPLE 26 (TETRAHEDRON OR K_4). We saw that

$$\zeta_{K_4}(u)^{-1} = (1-u^2)^2 (1-u)(1-2u)(1+u+2u^2)^3.$$

From this we find that

$$u \frac{d}{du} \log \zeta_X(u) = \sum_{m \geq 1} N_m u^m$$

$$= 24x^3 + 24x^4 + 96x^6 + 168x^7 + 168x^8 + 528x^9 + \cdots.$$

Then the question becomes what are the corresponding $\pi(n)$? We see that $\pi(3) = N_3/3 = 8$, $\pi(4) = N_4/4 = 6$; $\pi(5) = N_5 = 0$. Then, because $\pi(1) = \pi(2) = 0$, we have

$$N_6 = \sum_{d|5} d\pi(d) = 3\pi(3) + 6\pi(6),$$

which implies $\pi(6) = 12$.

It follows from the fact that there are paths of lengths 3 and 4 that the greatest common divisor of the lengths of the prime paths $\Delta = 1$.

The poles of zeta for K_4 are $\{1, 1, 1, -1, \frac{1}{2}, a, a, a, b, b, b\}$, where

$$a = \frac{1 + \sqrt{-7}}{4}, \quad b = \frac{1 - \sqrt{-7}}{4}.$$

Then $|a| = |b| = 1/\sqrt{2}$. The closest pole of zeta to the origin is $\frac{1}{2}$.

The prime number theorem for K_4 says that

$$\pi(m) \sim \frac{2^m}{m}, \quad \text{as } m \to \infty.$$

EXAMPLE 27 (TETRAHEDRON MINUS AN EDGE, $X = K_4 - e$). We saw that

$$\zeta_X(u)^{-1} = (1 - u^2)(1 - u)(1 + u^2)(1 + u + 2u^2)(1 - u^2 - 2u^3).$$

From this, we have

$$u \frac{d}{du} \log \zeta_X(u) = \sum_{m \geq 1} N_m u^m$$
$$= 12x^3 + 8x^4 + 24x^6 + 28x^7 + 8x^8 + 48x^9 + \cdots.$$

It follows that $\pi(3) = 4$, $\pi(4) = 2$, $\pi(5) = 0$, $\pi(6) = 2$. Again Δ, the gcd of lengths of primes, is equal to 1.

The poles of zeta for $K_4 - e$ are $\{1, 1, -1, i, -i, a, b, \alpha, \beta, \bar{\beta}\}$. Here

$$a = \frac{1 + \sqrt{-7}}{4}, \quad b = \frac{1 - \sqrt{-7}}{4},$$

and $\alpha = R$ is the real root of the cubic, while $\beta, \bar{\beta}$ are the remaining (nonreal) roots of the cubic.

The prime number theorem for $K_4 - e$ becomes, for $1/\alpha \cong 1.5$

$$\pi(m) \sim \frac{\alpha^{-m}}{m}, \quad \text{as } m \to \infty.$$

EXERCISE 28. Compute the Ihara zeta function of your favorite graph and then use formula (3-10) to compute the first 5 nonzero N_m. State the prime number theorem explicitly for this graph. Next do the same computations for the graph with one edge removed.

EXERCISE 29. (a) Show that the radius of convergence of the Ihara zeta function of a $(q+1)$-regular graph is $R = 1/q$.

(b) A graph is a *bipartite graph* if and only if the set of vertices can be partitioned into 2 disjoint sets S, T such that no vertex in S is adjacent to any other vertex in S and no vertex in T is adjacent to any other vertex in T. Assume your graph is non bipartite and prove the prime number theorem using the Ihara determinant formula (2-4).

(c) What happens if the graph is $(q + 1)$-regular graph and bipartite?

EXERCISE 30. List all the zeta functions you can and what they are good for. The website www.maths.ex.ac.uk/~mwatkins lists lots of them.

Now that we have the prime number theorem, we can also produce analogs of the explicit formulas of analytic number theory. That is, we seek an analog of Weil's explicit formula for the Riemann zeta function. In Weil's original work he used the result to formulate an equivalent statement to the Riemann hypothesis. See [Weil 1952].

Our analog of the von Mangoldt function from elementary number theory is N_m. Using formula (3-1), we have

$$u\frac{d}{du}\log\zeta(u, X) = -u\frac{d}{du}\sum_{\lambda \in \text{Spec } W_1} \log(1 - \lambda u)$$

$$= \sum_{\lambda \in \text{Spec } W_1} \frac{\lambda u}{1 - \lambda u} = -\sum_{\rho \text{ pole of } \zeta} \frac{u}{u - \rho}. \quad (3\text{-}14)$$

Then it is not hard to prove the following result following the method of [Murty 2001, p. 109].

PROPOSITION 31 (AN EXPLICIT FORMULA). *Let $0 < a < R$, where R is the radius of convergence of $\zeta(u, X)$. Assume $h(u)$ is meromorphic in the plane and holomorphic outside the circle of center 0 and radius $a - \varepsilon$, for small $\varepsilon > 0$. Assume also that $h(u) = O(|u|^p)$ as $|u| \to \infty$ for some $p < -1$. Also assume that its transform $\widehat{h}_a(n)$ decays rapidly enough for the right hand side of the formula to converge absolutely. Then if N_m is as in Definition 3-10, we have*

$$\sum_{\rho} \rho h(\rho) = \sum_{n \geq 1} N_n \widehat{h}_a(n),$$

where the sum on the left is over the poles of $\zeta(u, X)$ and

$$\widehat{h}_a(n) = \frac{1}{2\pi i} \oint_{|u|=a} u^n h(u)\, du.$$

PROOF. We follow the method of [Murty 2001, p. 109]. Look at

$$\frac{1}{2\pi i} \oint_{|u|=a} \left(u \frac{d}{du} (\log \zeta(u, X)) \right) h(u) \, du.$$

Use Cauchy's integral formula to move the contour over to the circle $|u| = b > 1$. Then let $b \to \infty$. Also use formulas (3-14) and (3-10). Note that $N_n \sim \Delta_X / R_X^m$, as $m \to \infty$. □

Such explicit formulas are basic to work on the pair correlation of complex zeros of zeta (see [Montgomery 1973]). They can also be viewed as an analog of Selberg's trace formula. See [Horton 2007] or [Terras and Wallace 2003] for discussion of Selberg's trace formula for a $q + 1$ regular graph. In these papers various kernels (e.g., Green's, characteristic functions of intervals, heat) were plugged in to the trace formula deducing various things such as McKay's theorem on the distribution of eigenvalues of the adjacency matrix and the Ihara determinant formula for the Ihara zeta. It would be an interesting research project to do the same sort of thing for irregular graphs.

4. Edge and path zeta functions and their determinant formulas; connections with quantum chaos

4.1. Proof of Ihara's determinant formula for Ihara zeta.
Before we give our version of the Bass proof of formulas (2-4) and (3-1) from [Stark and Terras 2000], we define a new graph zeta function with many complex variables. We orient and label the edges of our undirected graph as usual.

DEFINITION 32. The *edge matrix* W for graph X is a $2m \times 2m$ matrix with a, b entry corresponding to the oriented edges a and b. This a, b entry is the complex variable w_{ab} if edge a feeds into edge b (i.e., the terminal vertex of a is the starting vertex of b) and $b \neq a^{-1}$ and the a, b entry is 0 otherwise.

DEFINITION 33. Given a path C in X, which is written as a product of oriented edges $C = a_1 a_2 \cdots a_s$, the *edge norm* of C is

$$N_E(C) = w_{a_1 a_2} w_{a_2 a_3} \cdots w_{a_{s-1} a_s} w_{a_s a_1}.$$

The *edge Ihara zeta function* is

$$\zeta_E(W, X) = \prod_{[P]} (1 - N_E(P))^{-1},$$

where the product is over primes in X. Here assume that all $|w_{ab}|$ are sufficiently small for convergence.

Properties and applications of edge zeta

(1) By the definitions, if you set all nonzero variables in W equal to u, the *edge zeta function specializes to the Ihara zeta function*; i.e,

$$\zeta_E(W, X)|_{\substack{0 \neq w_{ab}=u \\ \text{for all a,b}}} = \zeta(u, X). \tag{4-1}$$

(2) If you cut or delete an edge of a graph, you can compute the edge zeta for the new graph with one less edge by setting all variables equal to 0 if the cut or deleted edge or its inverse appear in a subscript.

(3) The edge zeta allows one to define a zeta function for a weighted or quantum graph. See [Smilansky 2007] or [Horton et al. 2006b].

(4) There is an application of the edge zeta to error correcting codes. See [Koetter et al. 2005].

The following result is a generalization of formula (3-1).

THEOREM 34 (DETERMINANT FORMULA FOR THE EDGE ZETA).

$$\zeta_E(W, X) = \det(I - W)^{-1}.$$

We prove the theorem after giving an example.

EXAMPLE 35 (DUMBBELL GRAPH). Figure 6 shows the labeled picture of the dumbbell graph X. For this graph we find that

$$\zeta_E(W, X)^{-1} = \det \begin{pmatrix} w_{11}-1 & w_{12} & 0 & 0 & 0 & 0 \\ 0 & -1 & w_{23} & 0 & 0 & w_{26} \\ 0 & 0 & w_{33}-1 & 0 & w_{35} & 0 \\ 0 & w_{42} & 0 & w_{44}-1 & 0 & 0 \\ w_{51} & 0 & 0 & w_{54} & -1 & 0 \\ 0 & 0 & 0 & 0 & w_{65} & w_{66}-1 \end{pmatrix}.$$

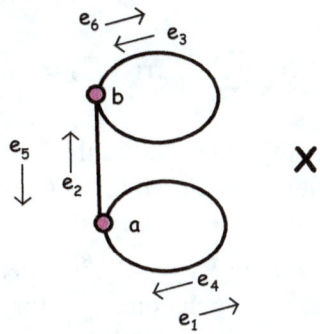

Figure 6. The dumbbell graph.

If we cut or delete the vertical edges which are edges e_2 and e_5, we should specialize all the variables with 2 or 5 in them to be 0. This yields the edge zeta function of the subgraph with the vertical edge removed, and incidentally diagonalizes the matrix W. This also diagonalizes the edge matrix W. Of course the resulting graph consists of 2 disconnected loops. So zeta is the product of two loop zetas.

EXERCISE 36. Do another example computing the edge zeta function of your favorite graph. Then see what happens if you delete an edge.

PROOF OF THEOREM 34. First note that, from the Euler product for the edge zeta function, we have

$$-\log \zeta_E(W, X) = \sum_{[P]} \sum_{j \geq 1} \frac{1}{j} N_E(P)^j.$$

Then, since there are $\nu(P)$ elements in $[P]$, we have

$$-\log \zeta_E(W, X) = \sum_{\substack{m \geq 1 \\ j \geq 1}} \frac{1}{jm} \sum_{\substack{P \\ \nu(P)=m}} N_E(P)^j.$$

It follows that

$$-\log \zeta_E(W, X) = \sum_C \frac{1}{\nu(C)} N_E(C).$$

This comes from the fact that any closed path C without backtracking or tail has the form P^j for a prime path P. Then by the Exercise below, we see that

$$-\log \zeta_E(W, X) = \sum_{m \geq 1} \frac{1}{m} \text{Tr}(W^m).$$

Finally, again using the Exercise below, we see that the right hand side of the preceding formula is $\log \det (I - W)^{-1}$ This proves the theorem. □

EXERCISE 37. Prove that

$$\sum_C \frac{1}{\nu(C)} N_E(C) = \sum_{m \geq 1} \frac{1}{m} \text{Tr}(W^m) = \log \det(I - W)^{-1}.$$

Hints. (1) For the first equality, you need to think about $\text{Tr}(W^m)$ as an m-fold sum of products of w_{ij} in terms of closed paths C of length m just as we did in proving formula (3-7) above.

(2) Exercise 19 says that $\det(\exp(B)) = e^{\text{Tr}(B)}$. Then write $\log((I - W)^{-1}) = B$, using the matrix logarithm (which converges for small w_{ij}), and see that

$$\log \det((I - W)^{-1}) = \text{Tr}(\log(I - W)^{-1}).$$

Theorem 34 gives another proof of formula (3-1) for the Ihara zeta by specializing all the nonzero w_{ij} to be u.

Next we give a version of Bass's proof of the Ihara determinant formula (2-4) using formula (3-1). In what follows, n is the number of vertices of X and m is the number of unoriented edges of X.

First define some matrices. Set $J = \begin{pmatrix} 0 & I_m \\ I_m & 0 \end{pmatrix}$. Then define the $n \times 2m$ start matrix S and the $n \times 2m$ terminal matrix T by setting

$$s_{ve} = \begin{cases} 1 & \text{if } v \text{ is the starting vertex of the oriented edge } e, \\ 0 & \text{otherwise,} \end{cases}$$

and

$$t_{ve} = \begin{cases} 1 & \text{if } v \text{ is the terminal vertex of the oriented edge } e, \\ 0 & \text{otherwise.} \end{cases}$$

PROPOSITION 38 (SOME MATRIX IDENTITIES). *Using these definitions, the following formulas hold. We write ^tM for the transpose of the matrix M.*

(1) $SJ = T$ and $TJ = S$.
(2) *If A is the adjacency matrix of X and $Q + I_n$ is the diagonal matrix whose jth diagonal entry is the degree of the jth vertex of X, then $A = S\,^tT$ and $Q + I_n = S\,^tS = T\,^tT$.*
(3) *The 0,1 edge adjacency W_1 from Definition 16 satisfies $W_1 + J = {}^tTS$.*

PROOF. (1) This comes from the fact that the starting (terminal) vertex of edge e_j is the terminal (starting) vertex of edge $e_{j+|E|}$, according to our edge numbering system.

(2) Consider

$$(S\,^tT)_{ab} = \sum_e s_{ae} t_{be}.$$

The right hand side is the number of oriented edges e such that a is the initial vertex and b is the terminal vertex of e, which is the a,b entry of A. Note that $A_{a,a} = 2*$number of loops at vertex a. Similar arguments prove the second formula.

(3) We have

$$({}^tTS)_{ef} = \sum_v t_{ve} s_{vf}.$$

The sum is 1 if and only if edge e feeds into edge f, even if $f = e^{-1}$. □

BASS'S PROOF OF THE GENERALIZED IHARA DETERMINANT FORMULA (2-4). In the following identity all matrices are $(n+2m) \times (n+2m)$, where the first block is $n \times n$, if n is the number of vertices of X and m is the number of unoriented edges of X. Use the preceding proposition to see that

$$\begin{pmatrix} I_n & 0 \\ {}^tT & I_{2m} \end{pmatrix} \begin{pmatrix} I_n(1-u^2) & Su \\ 0 & I_{2m}-W_1u \end{pmatrix} = \begin{pmatrix} I_n - Au + Qu^2 & Su \\ 0 & I_{2m}+Ju \end{pmatrix} \begin{pmatrix} I_n & 0 \\ {}^tT - {}^tSu & I_{2m} \end{pmatrix}.$$

EXERCISE 39. Check this equality.

Take determinants to obtain

$$(1-u^2)^n \det(I - W_1 u) = \det\left(I_n - Au + Qu^2\right) \det(I_{2m} + Ju).$$

To finish the proof of formula (2-4), observe that

$$I + Ju = \begin{pmatrix} I & Iu \\ Iu & I \end{pmatrix}$$

implies

$$\begin{pmatrix} I & 0 \\ -Iu & I \end{pmatrix} (I + Ju) = \begin{pmatrix} I & Iu \\ 0 & I(1-u^2) \end{pmatrix}.$$

Thus $\det(I + Ju) = (1 - u^2)^m$. Since $r - 1 = m - n$ for a connected graph, formula (2-4) follows. $\square$

EXERCISE 40. Read about quantum graphs and consider the properties of their zeta functions. See [Horton et al. 2006a; 2006b; 2008], as well as the other papers in those volumes. Another reference is [Smilansky 2007].

4.2. The path zeta function of a graph. First we need a few definitions. A *spanning tree* T for graph X means a tree which is a subgraph of X containing all the vertices of X.

The *fundamental group* of a topological space such as our graph X has elements which are closed directed paths starting and ending at a fixed basepoint $v \in X$. Two paths are equivalent if and only if one can be continuously deformed into the other (i.e., one is homotopic to the other within X, while still starting and ending at v). The product of 2 paths a, b means the path obtained by first going around a then b.

It turns out (by the Seifert-von Kampen theorem, for example) that the fundamental group of graph X is a free group on r generators, where r is the number of edges left out of a spanning tree for X. Let us try to explain this a bit. More information can be found in [Hatcher 2002], [Massey 1967, p. 198], or [Gross and Tucker 2001].

From the graph X construct a new graph $X^\#$ by shrinking a spanning tree T of X to a point. The new graph will be a bouquet of r loops as in Figure 7. The fundamental group of X is the same as that of $X^\#$. Why? The quotient map $X \to X/T$ is what algebraic topologists call a homotopy equivalence. This

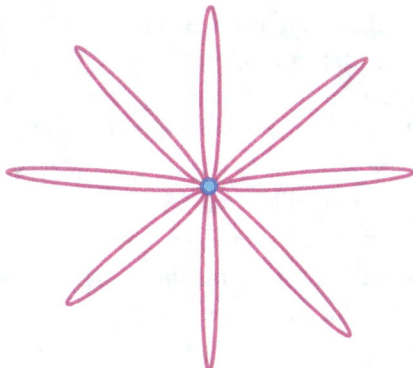

Figure 7. A bouquet of loops.

means that intuitively you can continuously deform one graph into the other without changing the topology.

The fundamental group of the bouquet of r loops in Figure 7 is the free group on r generators. The generators are the directed loops! The elements are the words in these loops.

EXERCISE 41. Show that $r - 1 = |E| - |V|$.

EXERCISE 42. The *complexity* κ_X of a graph is defined to be the number of spanning trees in X. Use the matrix-tree theorem (see [Biggs 1974]) to prove that

$$\left[\frac{d^r}{du^r}\zeta_X^{-1}(u)\right]\bigg|_{u=1} = r!(-1)^{r+1}2^r(r-1)\kappa_X.$$

This is an analog of formula (2-2) for the Dedekind zeta function of a number field at 0. The complexity is considered to be an analog of the class number of a number field.

Here we look at a zeta function invented by Stark. It has several advantages over the edge zeta. It can be used to compute the edge zeta with smaller determinants. It gives the edge zeta for a graph in which an edge has been fused; i.e., shrunk to one vertex.

Choose a spanning tree T of X. Then T has $|V| - 1 = n - 1$ edges. We denote the oriented versions of these *edges left out of the spanning tree T* (or "*deleted*" *edges* of T) and their inverses by

$$e_1, \ldots, e_r, e_1^{-1}, \ldots, e_r^{-1}.$$

Denote the remaining (oriented) *edges in the spanning tree* by T

$$t_1, \ldots, t_{n-1}, t_1^{-1}, \ldots, t_{n-1}^{-1}.$$

Any backtrackless, tailless cycle on X is uniquely (up to starting point on the tree between last and first e_k) determined by the ordered sequence of e_k's it passes through. The free group of rank r generated by the e_k's puts a group structure on backtrackless tailless cycles which is equivalent to the fundamental group of X.

There are 2 elementary reduction operations for paths written down in terms of directed edges just as there are elementary reduction operations for words in the fundamental group of X. This means that if $a_1, \ldots, a_s$ and e are taken from the e_k's and their inverses, the two *elementary reduction operations* are:

(i) $a_1 \cdots a_{i-1} e e^{-1} a_{i+2} \cdots a_s \cong a_1 \cdots a_{i-1} a_{i+2} \cdots a_s$
(ii) $a_1 \cdots a_s \cong a_2 \cdots a_s a_1$

Using the first elementary reduction operation, each equivalence class of words corresponds to a group element and a word of minimum length in an equivalence class is *reduced* word in group theory language. Since the second operation is equivalent to conjugating by a_1, an equivalence class using both elementary reductions corresponds to a conjugacy class in the fundamental group. A word of minimum length using both elementary operations corresponds to finding words of minimum length in a conjugacy class in the fundamental group. If $a_1, \ldots, a_s$ are taken from $e_1, \ldots, e_{2r}$, a word $C = a_1 \cdots a_s$ is of minimum length in its conjugacy class if and only if $a_{i+1} \neq a_i^{-1}$, for $1 \leq i \leq s-1$ and $a_1 \neq a_s^{-1}$.

This is equivalent to saying that C corresponds to a *backtrackless, tailless* cycle under the correspondence above. Equivalent cycles correspond to conjugate elements of the fundamental group. A conjugacy class $[C]$ is *primitive* if a word of minimal length in $[C]$ is not a power of another word. We will say that a word of minimal length in its conjugacy class is *reduced in its conjugacy class*. From now on, we assume a representative element of $[C]$ is chosen which is reduced in $[C]$.

DEFINITION 43. The $2r \times 2r$ *path matrix* Z has ij entry given by the complex variable z_{ij} if $e_i \neq e_j^{-1}$ and by 0 if $e_i = e_j^{-1}$.

The path matrix Z has only one zero entry in each row unlike the edge matrix W from Definition 32 which is rather sparse. Next we imitate the definition of the edge zeta function.

DEFINITION 44. Define the *path norm* for a primitive path $C = a_1 \cdots a_s$ reduced in its conjugacy class $[C]$, where $a_i \in \{e_1^{\pm 1}, \ldots, e_r^{\pm 1}\}$ as

$$N_P(C) = z_{a_1 a_2} \cdots z_{a_{s-1} a_s} z_{a_s a_1}.$$

Then the *path zeta* is defined for small $|z_{ij}|$ to be

$$\zeta_P(Z, X) = \prod_{[C]} (1 - N_P(C))^{-1},$$

where the product is over primitive reduced conjugacy classes $[C]$ other than the identity class.

We have similar results to those for the edge zeta.

THEOREM 45.
$$\zeta_P(Z, X)^{-1} = \det(I - Z).$$

PROOF. Imitate the proof of Theorem 34 for the edge zeta. □

The path zeta function is the same for all graphs with the same fundamental group. Next we define a procedure called *specializing the path matrix to the edge matrix* which will allow us to specialize the path zeta function to the edge zeta function. Use the notation above for the edges e_i left out of the spanning tree T and denote the edges of T by t_j. A prime cycle C is first written as a product of generators of the fundamental group and then as a product of actual edges e_i and t_k. Do this by inserting $t_{k_1} \cdots t_{k_s}$ which is the unique non backtracking path on T joining the terminal vertex of e_i and the starting vertex of e_j if e_i and e_j are successive deleted or non-tree edges in C. Now *specialize the path matrix Z to Z(W)* with entries

$$z_{ij} = w_{e_i t_{k_1}} w_{t_{k_1} t_{k_2}} \cdots w_{t_{k_{s-1}} t_{k_s}} w_{t_{k_s} e_j}. \quad (4\text{-}2)$$

THEOREM 46. *Using the specialization procedure just given, we have*

$$\zeta_P(Z(W), X) = \zeta_E(W, X).$$

EXAMPLE 47 (THE DUMBBELL AGAIN). Recall that the edge zeta of the dumbbell graph of Figure 6 was evaluated by a 6×6 determinant. The path zeta requires a 4×4 determinant. Take the spanning tree to be the vertical edge. One finds, using the determinant formula for the path zeta and the specialization of the path to edge zeta:

$$\zeta_E(W, X)^{-1} = \det \begin{pmatrix} w_{11}-1 & w_{12}w_{23} & 0 & w_{12}w_{26} \\ w_{35}w_{51} & w_{33}-1 & w_{35}w_{54} & 0 \\ 0 & w_{42}w_{23} & w_{44}-1 & w_{42}w_{26} \\ w_{65}w_{51} & 0 & w_{65}w_{54} & w_{66}-1 \end{pmatrix}. \quad (4\text{-}3)$$

If we shrink the vertical edge to a point (which we call *fusion* or *contraction*), the edge zeta of the new graph is obtained by replacing any $w_{x2}w_{2y}$ (for $x, y = 1, 3, 4, 6$) which appear in formula (4-3) by w_{xy} and any $w_{x5}w_{5y}$ (for $x, y = 1, 3, 4, 6$) by w_{xy}. This gives the zeta function of the new graph obtained from the dumbbell, by fusing the vertical edge.

EXERCISE 48. Compute the path zeta function for $K_4 - e$ (the tetrahedron minus one edge) and then specialize it to the edge zeta function of the graph.

EXERCISE 49. Write a Mathematica program to specialize the path matrix Z to the matrix $Z(W)$ so that $\zeta_P(Z(W), X) = \zeta_E(W, X)$.

4.3. Connections with quantum chaos. A reference with some background on random matrix theory and quantum chaos is [Terras 2007]. In Figure 5 (page 154) we saw the experimental connections between the statistics of spectra of random real symmetric matrices and the statistics of the imaginary parts of s at the poles of the Ihara zeta function $\zeta(q^{-s}, X)$ for a $(q+1)$-regular graph X. This is analogous to the connection between the statistics of the imaginary parts of zeros of the Riemann zeta function and the statistics of the spectra of random Hermitian matrices. At this point one should look at the figure produced by Bohigas and Giannoni comparing spacings of spectral lines from nuclear physics with those from number theory and billiards. Sarnak added lines from the spectrum of the Poincaré Laplacian on the fundamental domain of the modular group and I added eigenvalues of finite upper half-plane graphs. See [Terras 2007, p. 337].

Suppose you must arrange the eigenvalues E_i of a random symmetric matrix in decreasing order: $E_1 \geq E_2 \geq \cdots E_n$ and then normalize the eigenvalues so that the mean of the level spacings $E_i - E_{i+1}$ is 1. Wigner's surmise from 1957 says that the normalized level (eigenvalue) spacing histogram is approximated by the function $\frac{1}{2}\pi x \exp(-\pi x^2/4)$. In 1960 Gaudin and Mehta found the correct distribution function which is close to Wigner's. The correct distribution function is called the GOE distribution. A reference is [Mehta 1967]. The main property of this distribution is its vanishing at the origin (often called *level repulsion* in the physics literature). This differs in a big way from the spacing density of a Poisson random variable which is e^{-x}.

Many experiments have been performed with spacings of eigenvalues of the Laplace operator for a manifold such as the fundamental domain for a discrete group Γ acting on the upper half-plane H or the unit disc. The experiments of Schmit give the spacings of the eigenvalues of the Laplacian on $\Gamma \backslash H$ for an arithmetic Γ. To define "arithmetic" we must first define *commensurable subgroups* A, B of a group C. This means that $A \cap B$ has finite index both in A and B. Then suppose that Γ is an algebraic group over $\mathbb{Q}$ as in [Borel and Mostow 1966, p. 4]. One says that Γ is *arithmetic* if there is a faithful rational representation ρ into the general linear group of $n \times n$ nonsingular matrices such that ρ is defined over the rationals and $\rho(\Gamma)$ is commensurable with $\rho(\Gamma) \cap \mathrm{GL}(n, \mathbb{Z})$. Roughly we are saying that the integers are hiding somewhere in the definition of Γ. See [Borel and Mostow 1966] for more information. Arithmetic and nonarithmetic subgroups of $\mathrm{SL}(2, \mathbb{C})$ are discussed in [Elstrodt et al. 1998].

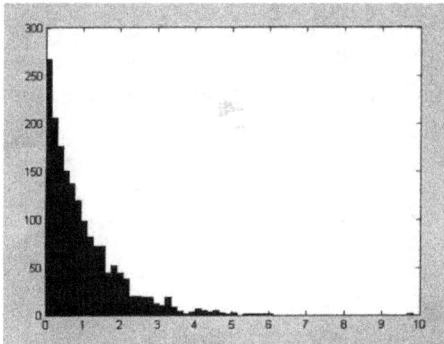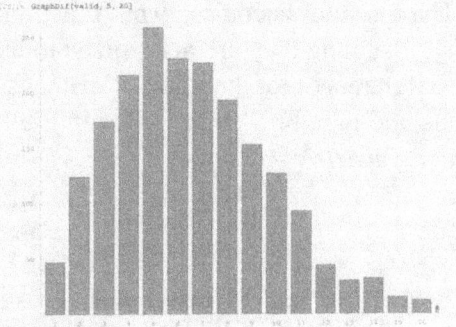

Figure 8. From [Newland 2005]. Spacings of the poles of the Ihara zeta for (left) a finite euclidean graph Euc_1999(2,1) as defined in [Terras 1999], and (right) a random regular graph from Mathematica with 2000 vertices and degree 71.

Experiments of Schmit [1991] compared spacings of eigenvalues of the Laplacian for arithmetic and nonarithmetic groups acting on the unit disc. Schmit found that the arithmetic group had spacings that were close to Poisson while the nonarithmetic group spacings looked GOE.

Newland [2005] did experiments on spacings of poles of the Ihara zeta for regular graphs. When the graph was a certain Cayley graph for an abelian group which we called a Euclidean graph in [Newland 2005], he found Poisson spacings. When the graph was random, he found GOE spacings (actually a transform of GOE coming from the relationship between the eigenvalues of the adjacency matrix of the graph and the zeta poles). Figure 8, left, shows the spacing histogram for the poles of the Ihara zeta for a finite Euclidean graph Euc1999(2,1) as in Chapter 5 of [Terras 1999]. It is a Cayley graph for a finite abelian group. The right part of the same figure shows the spacing histogram for the poles of the Ihara zeta of a random regular graph as given by Mathematica with 2000 vertices and degree 71. The moral is that the spacings for the poles of the Ihara zeta of a Cayley graph of an abelian group look Poisson while, for a random graph, the spacings look GOE. The difference between the two parts of Figures 8 is similar to that between the spacings of the Laplacian for arithmetic and nonarithmetic groups.

Our plan for the rest of this section is to investigate the spacings of the poles of the Ihara zeta function of a random graph and compare the result with spacings for covering graphs both random and with abelian Galois group. By formula (3-1), this is essentially the same as investigating the spacings of the eigenvalues of the edge adjacency matrix W_1 from Definition 16. Here, although W_1 is not symmetric, the nearest neighbor spacing can be studied. If the eigenvalues of the matrix are λ_i, $i = 1, \ldots, 2m$, we want to look at $v_i = \min\{|\lambda_i - \lambda_j| \mid j \neq i\}$.

The question becomes: what function best approximates the histogram of the v_i, assuming they are normalized to have mean 1?

References for the study of spacings of eigenvalues of nonsymmetric matrices include [Ginibre 1965; LeBoeuf 1999; Mehta 1967]. The *Wigner surmise for nonsymmetric matrices* is

$$4\Gamma\left(\frac{5}{4}\right)^4 x^3 \exp\left(-\Gamma\left(\frac{5}{4}\right)^4 x^4\right). \tag{4-4}$$

Since our matrix W_1 is real and has certain special properties, this may not be the correct Wigner surmise. In what follows some experiments are performed. The following proposition gives some of the properties of W_1.

PROPOSITION 50 (PROPERTIES OF W_1).

(i) $W_1 = \begin{pmatrix} A & B \\ C & {}^t A \end{pmatrix}$, where B and C are symmetric real, A is real with transpose ${}^t A$. The diagonal entries of B and C are 0.

(ii) *The sum of the entries of the i-th row of W_1 is the degree of the vertex which is the start of edge i.*

PROOF. See [Horton 2007] or [Terras 2010]. □

Our first experiment involves the eigenvalues of a random matrix with block form $\begin{pmatrix} A & B \\ C & {}^t A \end{pmatrix}$, where B and C are symmetric and 0 on the diagonal. We used Matlab's randn(N) command to get matrices A, B, C with normally distributed entries. There is a result known as the Girko circle law which says that the eigenvalues of a set of random $n \times n$ real matrices with independent entries with a standard normal distribution should be approximately uniformly distributed in a circle of radius $\sqrt{n}$ for large n. References are [Bai 1997; Girko 1984; Tao and Vu 2009]. A plot of the eigenvalues of a random matrix with the properties of W_1 appears in Figure 9, left. Note the symmetry with respect to the real axis, since our matrix is real. Another interesting fact is that the circle radius is not exactly that which Girko predicts. The spacing distribution for this random matrix is compared with the nonsymmetric Wigner surmise in formula (4-4) in Figure 9, right.

Our next experiments concern the spectra of actual W_1 matrices for graphs. First recall that the eigenvalues of W_1 are the reciprocals of the poles of the Ihara zeta function. You should also recall the graph theory Riemann hypothesis given in formula (2-7) as well as Theorem 22 of Kotani and Sunada. Figure 11 shows Ihara zeta poles for three graphs. The left half of Figures 12, 13, and 14 plot the eigenvalues of the W_1 matrix as well as circles of radius $\sqrt{p} \leq 1/\sqrt{R} \leq \sqrt{q}$, where $p + 1$ is the minimum degree of vertices of our graph and $q + 1$ is the maximum degree. Then R is from Definition 14 and $1/R$ is the spectral radius

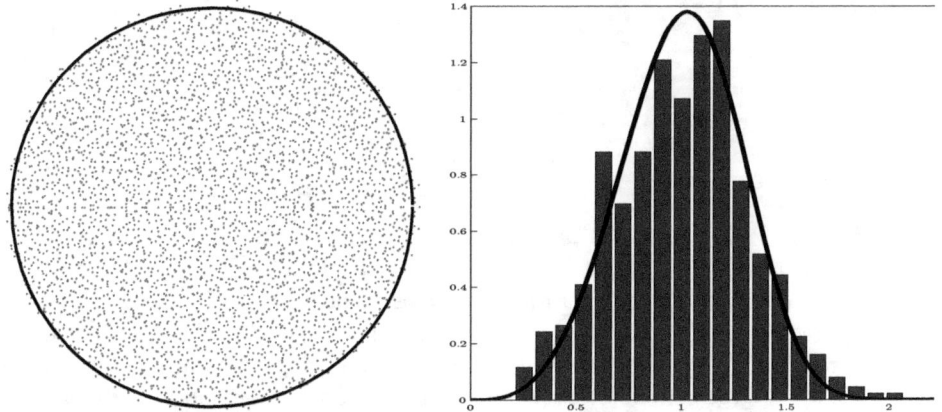

Figure 9. Left: Matlab-obtained spectrum of a random 2000×2000 matrix with the properties of W_1 except that the entries are not 0 and 1. The circle, centered at the origin, has radius $\frac{1}{2}(1+\sqrt{2})\sqrt{2000} \approx 54.0$ rather than $\sqrt{2000}$ as in Girko's circle law. Right: The normalized nearest neighbor spacing for the spectrum of the same matrix. The curve is the Wigner surmise from formula (4-4).

of W_1. Theorem 22 of Kotani and Sunada says the spectra cannot ever fill up a circle. They must lie in an annulus.

Figures 10 and 11 give the results of some Mathematica experiments on the distribution of the poles of zeta for various graphs constructed using the RealizeDegreeSequence command to create a few graphs with various degree sequences and then contracting vertices to join these graphs together. The first

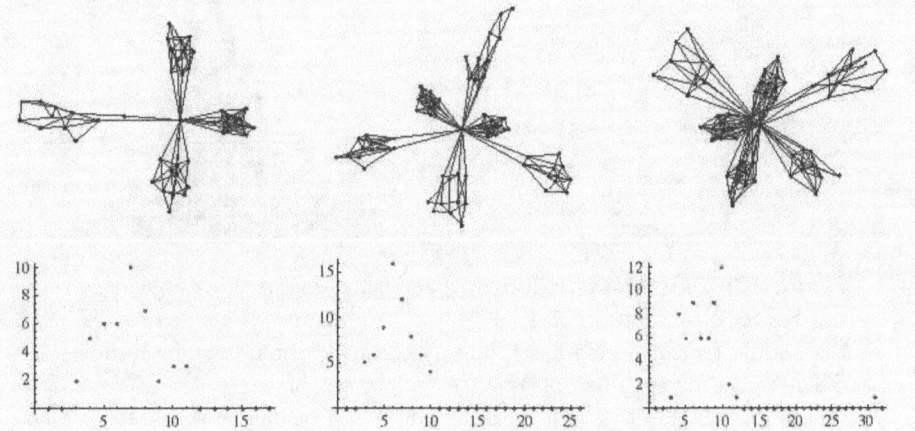

Figure 10. Mathematica-generated graphs (top row) and histograms of degrees (bottom row). The graphs were constructed using RealizeDegree-Sequence and vertex contraction.

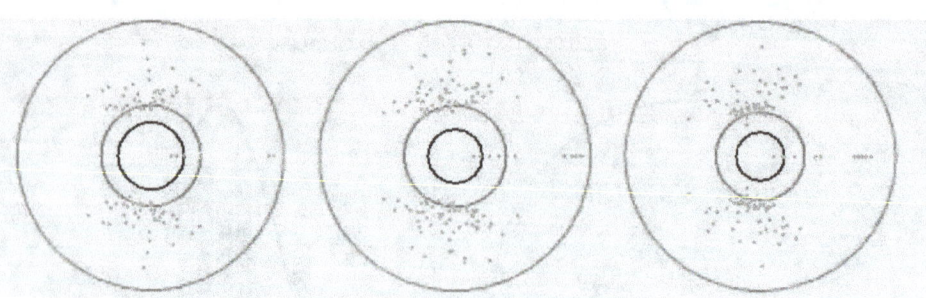

Figure 11. Mathematica-determined poles (pink points) of the Ihara zetas of the graphs in Figure 10. The middle green circle is the Riemann hypothesis circle with radius $\sqrt{R}$, for R the closest pole to 0. The inner circle has radius $1/\sqrt{q}$, where $q+1$ is the maximum degree and the outer circle has radius $1/\sqrt{p}$, where $p+1$ is the minimum degree. Many poles are inside the green (middle) circle and thus violate the Riemann hypothesis.

Figure 11 shows the poles of the Ihara zetas of the graphs in Figure 10. Many poles appear inside the green circle rather than outside as the RH would say.

Figure 12, left, shows a Matlab experiment giving the spectrum of the edge adjacency matrix W_1 for a "random graph". The inner circle has radius $\sqrt{p}$. the green circle has radius $1/\sqrt{R}$. The outer circle has radius $\sqrt{q}$. The middle green circle is the Riemann hypothesis circle. Because the eigenvalues of W_1 are reciprocals of the poles of zeta, now the RH says the spectrum should be inside the middle circle. The RH looks approximately true.

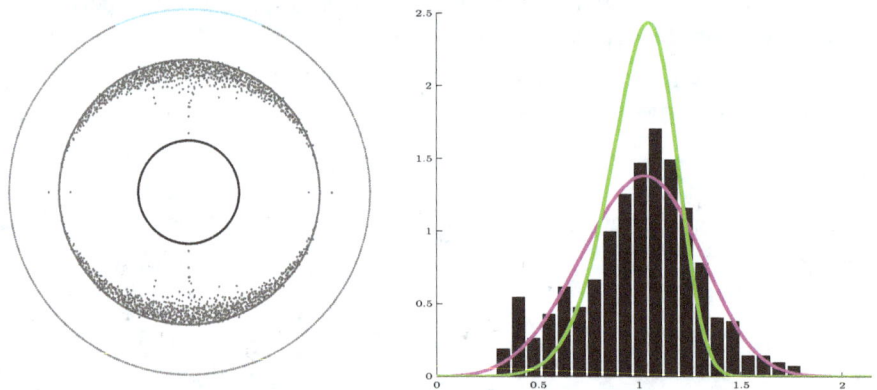

Figure 12. Left: Matlab-obtained eigenvalues of the edge adjacency matrix W_1 for a random graph (pink points). The inner circle has radius $\sqrt{p}$, the middle (green) circle has radius $1/\sqrt{R}$, and the outer one radius $\sqrt{q}$. Because the eigenvalues of W_1 are reciprocals of the poles of zeta, the Riemann hypothesis says the spectrum should be inside the green circle, which seems approximately true. The graph has 800 vertices, mean degree $\cong 13.125$, edge probability $\cong 0.0164$. Right: The histogram of the nearest neighbor spacings of the spectrum of W_1 for the same random graph, versus the modified Wigner surmise from formula (4-5) with $\omega = 3$ and 6.

Figure 12, right, shows the histogram of the nearest neighbor spacings of the spectrum of the same random graph versus a *modified Wigner surmise* (with $\omega = 3, 6$)

$$(\omega + 1)\Gamma\left(\frac{\omega+2}{\omega+1}\right)^{\omega+1} x^\omega \exp\left(-\Gamma\left(\frac{\omega+2}{\omega+1}\right)^{\omega+1} x^{\omega+1}\right). \quad (4\text{-}5)$$

When $\omega = 3$, this is the original Wigner surmise.

Next we consider some experiments involving covering graphs. An example of a graph covering is the cube covering the tetrahedron. The theory mimics the theory of extensions of algebraic number fields. In particular, there is an analog of Galois theory and Artin L-functions attached to representations of the Galois group. This helps to explain the factorizations of the zeta functions in our earlier examples. See [Stark and Terras 2000] and [Terras 2010].

DEFINITION 51. If the graph has no multiple edges and loops we can say that the graph Y is an *unramified covering* of the graph X if we have a covering map $\pi : Y \to X$ which is an onto graph mapping (i.e., taking adjacent vertices to adjacent vertices) such that for every $x \in X$ and for every $y \in \pi^{-1}(x)$, the collection of points adjacent to $y \in Y$ is mapped 1-1 onto the collection of points adjacent to $x \in X$.

The definition in the case of loops and multiple edges can be found in [Stark and Terras 2000] or [Terras 2010]. It requires directing edges and requiring the covering map to preserve local directed neighborhoods of a vertex.

DEFINITION 52. If Y/X is a d-sheeted covering with projection map $\pi : Y \longrightarrow X$, we say that it is a *normal covering* when there are d graph automorphisms $\sigma : Y \longrightarrow Y$ such that $\pi \circ \sigma = \pi$. The Galois group $G(Y/X)$ is the set of these maps σ.

For covering graphs one can say more about the expected shape of the spectrum of the edge adjacency matrix or equivalently describe the region bounding the poles of the Ihara zeta. Angel, Friedman and Hoory give in [Angel et al. 2007] a method to compute the region encompassing the spectrum of the analogous operator to the edge adjacency matrix W_1 on the universal cover of a graph X. In Section 2 we mentioned the Alon conjecture for regular graphs. Angel, Friedman and Hoory give an *analog of the Alon conjecture for irregular graphs*. Roughly their conjecture says that the new edge adjacency spectrum of a large random covering graph is near the edge adjacency spectrum of the universal covering. Here "new" means not occurring in the spectrum of W_1 for the base graph. This conjecture can be shown to imply the approximate Riemann hypothesis for the new poles of a large random cover.

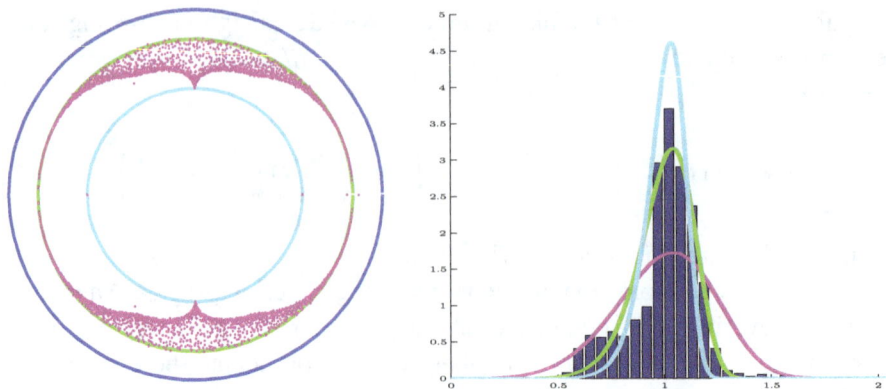

Figure 13. Left: Matlab-obtained eigenvalues (pink points) of the edge adjacency matrix of a random cover of the base graph made of two loops with an extra vertex on one loop. Thus we plot the reciprocals of the poles of zeta. The inner circle has radius 1, the middle circle has radius $1/\sqrt{R}$, and the outer one radius $\sqrt{3}$. The Riemann hypothesis is approximately true for this graph zeta. The cover has 801 sheets (copies of a spanning tree). Right: The nearest neighbor spacings for the spectrum of the edge adjacency matrix of the previous graph compared with three versions of the modified Wigner surmise from formula (4-5), with $\omega = 3, 6, 9$.

We show some examples related to this conjecture. Figure 13, left, shows the spectrum of the edge adjacency matrix of a random cover of the base graph consisting of two loops with an extra vertex on one loop. The inner circle has radius 1. The middle circle has radius $1/\sqrt{R}$. The outer circle has radius $\sqrt{3}$. The Riemann hypothesis is approximately true for this graph zeta.

Figure 13, right, shows the nearest neighbor spacings for the points in this same spectrum, compared with the modified Wigner surmise in formula (4-5), for various small values of ω.

Figure 14, left, shows the spectrum of the edge adjacency matrix for a Galois $\mathbb{Z}_{163} \times \mathbb{Z}_{45}$ covering of the base graph consisting of two loops with an extra vertex on one loop. The inner circle has radius 1. The middle circle has radius $1/\sqrt{R}$, with R as in Definition 14. The outer circle has radius $\sqrt{3}$. The Riemann hypothesis looks very false.

Figure 14, right, shows the histogram of the nearest neighbor spacings for the spectrum of the edge adjacency matrix of the graph in the preceding figure compared with spacings of a Poisson random variable (e^{-x}) and the Wigner surmise from formula (4-4).

EXERCISE 53. Compute more examples of poles of zeta functions of graphs. In particular, it would be interesting to look at graphs with degrees satisfying a power law d^{-e}, where $2 \leq e \leq 3$, say.

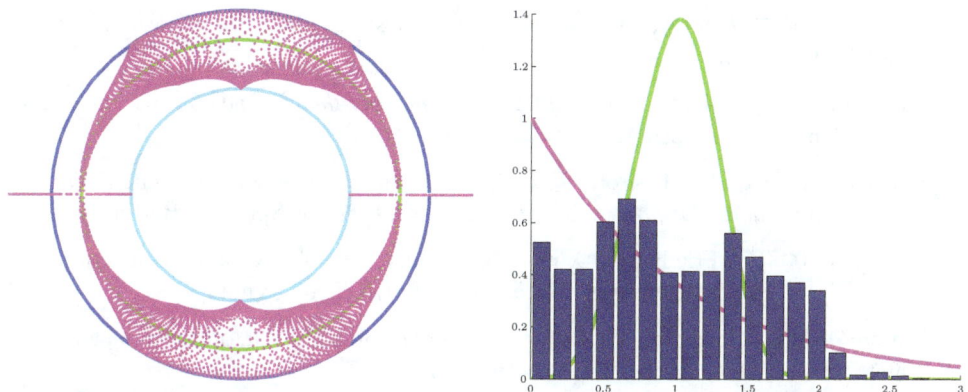

Figure 14. Left: Matlab-obtained eigenvalues (pink points) of the edge adjacency matrix W_1 for a Galois $\mathbb{Z}_{163} \times \mathbb{Z}_{45}$ covering of the base graph consisting of two loops with an extra vertex on one loop. The inner circle has radius 1. The middle circle has radius $1/\sqrt{R}$, with R as in Definition 14. The outer circle has radius $\sqrt{3}$. The Riemann hypothesis is very false. Left: The histogram of the nearest neighbor spacings for the spectrum of the same edge adjacency matrix W_1, compared with spacings of a Poisson random variable (e^{-x}) and the Wigner surmise from formula (4-4).

References

[Angel et al. 2007] O. Angel, J. Friedman, and S. Hoory, "The non-backtracking spectrum of the universal cover of a graph", preprint, 2007. Available at http://www.math.ubc.ca/~jf/pubs/web_stuff/NBRW_29.html.

[Apostol 1976] T. M. Apostol, *Introduction to analytic number theory*, Springer, New York, 1976.

[Artin and Mazur 1965] M. Artin and B. Mazur, "On periodic points", *Ann. of Math.* (2) **81** (1965), 82–99.

[Bai 1997] Z. D. Bai, "Circular law", *Ann. Probab.* **25**:1 (1997), 494–529.

[Bass 1992] H. Bass, "The Ihara–Selberg zeta function of a tree lattice", *Internat. J. Math.* **3**:6 (1992), 717–797.

[Bedford et al. 1991] T. Bedford, M. Keane, and C. Series (editors), *Ergodic theory, symbolic dynamics, and hyperbolic spaces* (Trieste, 1989), Oxford University Press, New York, 1991.

[Biggs 1974] N. Biggs, *Algebraic graph theory*, Cambridge Tracts in Mathematics **67**, Cambridge University Press, London, 1974.

[Borel and Mostow 1966] A. Borel and G. D. Mostow, *Algebraic groups and discontinuous subgroups* (Boulder, 1965), Proc. Symp. Pure Math. **9**, American Mathematical Society, Providence, 1966.

[Chung 2007] F. Chung, "Four proofs for the Cheeger inequality and graph partition algorithms", preprint II, 1-4, ICCM, 2007.

[Conrey 2003] B. Conrey, "The Riemann hypothesis", *Notices Amer. Math. Soc.* **50** (2003), 341–353.

[Davenport 1980] H. Davenport, *Multiplicative number theory*, 2nd ed., Graduate Texts in Mathematics **74**, Springer, New York, 1980.

[Elstrodt et al. 1998] J. Elstrodt, F. Grunewald, and J. Mennicke, *Groups acting on hyperbolic space: Harmonic analysis and number theory*, Springer, Berlin, 1998.

[Friedman 2008] J. Friedman, "A proof of Alon's second eigenvalue conjecture and related problems", *Mem. Amer. Math. Soc.* **195**:910 (2008), viii+100.

[Ginibre 1965] J. Ginibre, "Statistical ensembles of complex, quaternion, and real matrices", *J. Mathematical Phys.* **6** (1965), 440–449.

[Girko 1984] V. L. Girko, "The circular law", *Teor. Veroyatnost. i Primenen.* **29**:4 (1984), 669–679. In Russian; translated in *Theory Prob. Appl.* 29 (1984), 694-706.

[Gross and Tucker 2001] J. L. Gross and T. W. Tucker, *Topological graph theory*, Dover, Mineola, NY, 2001.

[Hashimoto 1989] K.-i. Hashimoto, "Zeta functions of finite graphs and representations of p-adic groups", pp. 211–280 in *Automorphic forms and geometry of arithmetic varieties*, Adv. Stud. Pure Math. **15**, Academic Press, Boston, MA, 1989.

[Hatcher 2002] A. Hatcher, *Algebraic topology*, Cambridge University Press, New York, 2002. Available at http://www.math.cornell.edu/~hatcher/AT/ATpage.html.

[Hoory et al. 2006] S. Hoory, N. Linial, and A. Wigderson, "Expander graphs and their applications", *Bull. Amer. Math. Soc. (N.S.)* **43**:4 (2006), 439–561.

[Horn and Johnson 1990] R. A. Horn and C. R. Johnson, *Matrix analysis*, Cambridge University Press, Cambridge, 1990.

[Horton 2007] M. D. Horton, "Ihara zeta functions of digraphs", *Linear Algebra Appl.* **425**:1 (2007), 130–142.

[Horton et al. 2006a] M. D. Horton, D. B. Newland, and A. A. Terras, "The contest between the kernels in the Selberg trace formula for the $(q+1)$-regular tree", pp. 265–293 in *The ubiquitous heat kernel*, Contemp. Math. **398**, Amer. Math. Soc., Providence, RI, 2006.

[Horton et al. 2006b] M. D. Horton, H. M. Stark, and A. A. Terras, "What are zeta functions of graphs and what are they good for?", pp. 173–189 in *Quantum graphs and their applications*, Contemp. Math. **415**, Amer. Math. Soc., Providence, RI, 2006.

[Horton et al. 2008] M. D. Horton, H. M. Stark, and A. A. Terras, "Zeta functions of weighted graphs and covering graphs", pp. 29–50 in *Analysis on graphs and its applications*, Proc. Sympos. Pure Math. **77**, Amer. Math. Soc., Providence, RI, 2008.

[Koetter et al. 2005] R. Koetter, W.-C. W. Li, P. O. Vontobel, and J. L. Walker, "Pseudo-codewords of cycle codes via zeta functions", preprint, 2005. Available at arxiv.org/abs/cs/0502033.

[Kotani and Sunada 2000] M. Kotani and T. Sunada, "Zeta functions of finite graphs", *J. Math. Sci. Univ. Tokyo* **7**:1 (2000), 7–25.

[Lagarias 1999] J. C. Lagarias, "Number theory zeta functions and dynamical zeta functions", pp. 45–86 in *Spectral problems in geometry and arithmetic* (Iowa City, 1997), Contemp. Math. **237**, Amer. Math. Soc., Providence, RI, 1999.

[LeBoeuf 1999] P. LeBoeuf, "Random matrices, random polynomials, and Coulomb systems", preprint, 1999. Available at arxiv.org/abs/cond-mat/9911222.

[Lorenzini 1996] D. Lorenzini, *An invitation to arithmetic geometry*, Graduate Studies in Mathematics **9**, American Mathematical Society, Providence, RI, 1996.

[Lubotzky 1995] A. Lubotzky, "Cayley graphs: eigenvalues, expanders and random walks", pp. 155–189 in *Surveys in combinatorics, 1995 (Stirling)*, London Math. Soc. Lecture Note Ser. **218**, Cambridge Univ. Press, Cambridge, 1995.

[Lubotzky et al. 1988] A. Lubotzky, R. Phillips, and P. Sarnak, "Ramanujan graphs", *Combinatorica* **8**:3 (1988), 261–277.

[Massey 1967] W. S. Massey, *Algebraic topology: An introduction*, Harcourt, Brace, New York, 1967.

[Mehta 1967] M. L. Mehta, *Random matrices and the statistical theory of energy levels*, Academic Press, New York, 1967.

[Miller and Novikoff 2008] S. J. Miller and T. Novikoff, "The distribution of the largest nontrivial eigenvalues in families of random regular graphs", *Experiment. Math.* **17**:2 (2008), 231–244. Available at http://projecteuclid.org/getRecord?id=euclid.em/1227118974.

[Montgomery 1973] H. L. Montgomery, "The pair correlation of zeros of the zeta function", pp. 181–193 in *Analytic number theory* (St. Louis, MO, 1972), Proc. Sympos. Pure Math. **24**, Amer. Math. Soc., Providence, R.I., 1973.

[Murty 2001] M. R. Murty, *Problems in analytic number theory*, Graduate Texts in Mathematics **206**, Springer, New York, 2001.

[Newland 2005] D. Newland, *Kernels in the Selberg trace formula on the k-regular tree and zeros of the Ihara zeta function*, Ph.D. thesis, University of California, San Diego, 2005.

[Rosen 2002] M. Rosen, *Number theory in function fields*, Graduate Texts in Mathematics **210**, Springer, New York, 2002.

[Ruelle 1994] D. Ruelle, *Dynamical zeta functions for piecewise monotone maps of the interval*, CRM Monograph Series **4**, American Mathematical Society, Providence, RI, 1994.

[Schmit 1991] C. Schmit, "Quantum and classical properties of some billiards on the hyperbolic plane", pp. 331–370 in *Chaos et physique quantique* (Les Houches, 1989), North-Holland, Amsterdam, 1991.

[Selberg 1989] A. Selberg, *Collected papers*, vol. 1, Springer, Berlin, 1989.

[Smilansky 2007] U. Smilansky, "Quantum chaos on discrete graphs", *J. Phys. A* **40**:27 (2007), F621–F630.

[Stark 1992] H. M. Stark, "Galois theory, algebraic number theory, and zeta functions", pp. 313–393 in *From number theory to physics* (Les Houches, 1989), Springer, Berlin, 1992.

[Stark and Terras 2000] H. M. Stark and A. A. Terras, "Zeta functions of finite graphs and coverings. II", *Adv. Math.* **154**:1 (2000), 132–195.

[Tao and Vu 2009] T. Tao and V. Vu, "From the Littlewood–Offord problem to the circular law: universality of the spectral distribution of random matrices", *Bull. Amer. Math. Soc. (N.S.)* **46**:3 (2009), 377–396.

[Terras 1985] A. Terras, *Harmonic analysis on symmetric spaces and applications. I*, Springer, New York, 1985.

[Terras 1999] A. Terras, *Fourier analysis on finite groups and applications*, London Mathematical Society Student Texts **43**, Cambridge University Press, Cambridge, 1999.

[Terras 2007] A. Terras, "Arithmetical quantum chaos", pp. 333–375 in *Automorphic forms and applications*, IAS/Park City Math. Ser. **12**, Amer. Math. Soc., Providence, RI, 2007.

[Terras 2010] A. Terras, *A stroll through the garden of graph zeta functions*, Cambridge University Press, New York, 2010. Prepublication at http://www.math.ucsd.edu/~aterras/newbook.pdf. In press.

[Terras and Stark 2007] A. A. Terras and H. M. Stark, "Zeta functions of finite graphs and coverings. III", *Adv. Math.* **208**:1 (2007), 467–489.

[Terras and Wallace 2003] A. Terras and D. Wallace, "Selberg's trace formula on the k-regular tree and applications", *Int. J. Math. Math. Sci.* no. 8 (2003), 501–526.

[Weil 1952] A. Weil, "Sur les 'formules explicites' de la théorie des nombres premiers", *Medd. Lunds Univ. Mat. Sem.*, tome supplementaire (1952), 252–265.

AUDREY TERRAS
MATHEMATICS DEPARTMENT
UNIVERSITY OF CALIFORNIA, SAN DIEGO
9500 GILMAN DRIVE LA JOLLA, CA 92093-0112
aterras@ucsd.edu

Vertex operators and modular forms

GEOFFREY MASON AND MICHAEL TUITE

CONTENTS

Correlation functions and Eisenstein series

1. The big picture	184
2. Vertex operator algebras	185
3. Modular and quasimodular forms	195
4. Characters of vertex operator algebras	200
5. Elliptic functions and 2-point functions	204

Modular-invariance and rational vertex operator algebras

6. Modules over a vertex operator algebra	212
7. Examples of regular vertex operator algebras	218
8. Vector-valued modular forms	224
9. Vertex operator algebras and modular invariance	228

Two current research areas

10. Some preliminaries	236
11. The genus-two partition function for the Heisenberg VOA	244
12. Exceptional VOAs and the Virasoro algebra	260

Appendices

13. Lie algebras and representations	271
14. The square bracket formalism	273
References	274

Mason's research is supported by the NSA and the NSF. Tuite's research is supported by Science Foundation Ireland.

Part I. Correlation functions and Eisenstein series

1. The big picture

String theory
↑ ↓
2-d conformal field theory
↑ ↓
Vertex operator algebras
↑ ↓
Modular forms and elliptic functions
↑ ↓
L-series and zeta-functions

The *leitmotif* of these notes is the idea of a *vertex operator algebra* (VOA) and the relationship between VOAs and *elliptic functions and modular forms*. This is to some extent analogous to the relationship between a finite group and its irreducible characters; the algebraic structure determines a set of numerical invariants, and arithmetic properties of the invariants provides feedback in the form of restrictions on the algebraic structure. One of the main points of these notes is to explain how this works, and to give some reasonably interesting examples.

VOAs may be construed as an axiomatization of 2-dimensional conformal field theory, and it is via this connection that vertex operators enter into physical theories. A sketch of the VOA-CFT connection via the Wightman Axioms can be found in the introduction to [K1]. Although we make occasional comments to relate our development of VOA theory to physics, no technical expertise in physics is necessary to understand these notes. As mathematical theories go, the one we are discussing here is relatively new. There are a number of basic questions which are presently unresolved, and we will get far enough in the notes to explain some of them.

To a modular form one may attach (via the Mellin transform) a Dirichlet series, or L-function, and Weil's Converse Theorem says that one can go the other way too. So there is a close connection between modular forms and certain L-functions, and this is one way in which our subject matter relates to the contents of other parts of this book. Nevertheless, as things stand at present, it is the Fourier series of a modular form, rather than its Dirichlet series, that is important in VOA theory. As a result, L-functions will not enter into our development of the subject.

The notes are divided into three parts. In Part I we give some of the foundations of VOA theory, and explain how modular forms on the full modular group (Eisenstein series in particular) and elliptic functions naturally intervene in the description of n-point correlation functions. This is a general phenomenon and the simplest VOAs, namely the free boson (Heisenberg VOA) and the Virasoro VOA, suffice to illustrate the computations. For this reason we delay the introduction of more complicated VOAs until Part II, where we describe several families of VOAs and their representations. We also cover some aspects of vector-valued modular forms, which is the appropriate language to describe the modular properties of C_2-cofinite and rational VOAs. We give some applications to holomorphic VOAs to illustrate how modularity impinges on the algebraic structure of VOAs. In Part III we describe two current areas of active research of the authors. The first concerns the development of VOA theory on a genus-two Riemann surface and the second is concerned with the relationship between exceptional VOAs and Lie algebras and the Virasoro algebra.

There are a number of exercises at the end of each subsection. They provide both practice in the ideas and also a subtext to which we often refer during the course of the notes. Some of the exercises are straightforward, others less so. Even if the reader is not intent on working out the exercises, he or she should read them over before proceeding.

These notes constitute an expansion of the lectures we gave at MSRI in the summer of 2008 during the Workshop *A window into zeta and modular physics*. We thank the organizers of the workshop, in particular Klaus Kirsten and Floyd Williams, for giving us the opportunity to participate in the program.

2. Vertex operator algebras

2.1. Notation and conventions. $\mathbb{Z}$ is the set of integers, $\mathbb{R}$ the real numbers, $\mathbb{C}$ the complex numbers, $\mathfrak{H}$ the complex upper half-plane

$$\mathfrak{H} = \{\tau \in \mathbb{C} \mid \operatorname{Im} \tau > 0\}.$$

All linear spaces V are defined over $\mathbb{C}$; linear transformations are $\mathbb{C}$-linear; $\operatorname{End}(V)$ is the space of *all* endomorphisms of V. For an indeterminate z,

$$V[\![z, z^{-1}]\!] = \left\{ \sum_{n \in \mathbb{Z}} v_n z^n \mid v_n \in V \right\}, \qquad V[\![z]\!][z^{-1}] = \left\{ \sum_{n=-M}^{\infty} v_n z^n \mid v_n \in V \right\}.$$

These are linear spaces with respect to the obvious addition and scalar multiplication. The formal residue is

$$\operatorname{Res}_z \sum_{n \in \mathbb{Z}} v_n z^n = v_{-1}.$$

For integers m, n with $n \geq 0$,
$$\binom{m}{n} = \frac{m(m-1)\dots(m-n+1)}{n!}.$$
For indeterminates x, y we adopt the convention that
$$(x+y)^m = \sum_{n \geq 0} \binom{m}{n} x^{m-n} y^n, \tag{1}$$
i.e., for $m < 0$ we formally expand in the second parameter y.

We use the following q-convention:
$$q_x = e^x, \qquad q = q_{2\pi i \tau} = e^{2\pi i \tau} \quad (\tau \in \mathfrak{H}), \tag{2}$$
where x is anything for which e^x makes sense.

2.2. Local fields. We deal with formal series
$$a(z) = \sum_{n \in \mathbb{Z}} a_n z^{-n-1} \in \mathrm{End}(V)[[z, z^{-1}]]. \tag{3}$$
$a(z)$ defines a linear map $V \to V[[z, z^{-1}]]$ by the rule
$$v \mapsto \sum_{n \in \mathbb{Z}} a_n(v) z^{-n-1}.$$
The endomorphisms a_n are called the *modes* of $a(z)$. We often refer to the elements in V as *states*, and call V the *state-space* or *Fock space*.

REMARK 2.1. The convention for powers of z in (3) is standard in mathematics. A different convention is common in the physics literature. Whenever a mathematician and physicist discuss fields, they should first agree on their conventions.

DEFINITION 2.2. $a(z) \in \mathrm{End}(V)[[z, z^{-1}]]$ is a *field* if it satisfies the following truncation condition $\forall v \in V$:
$$a(z)v \in V[[z]][z^{-1}].$$
That is, for $v \in V$ there is an integer N (depending on v) such that $a_n(v) = 0$ for all $n > N$.

Set
$$\mathfrak{F}(V) = \{a(z) \in \mathrm{End}(V)[[z, z^{-1}]] \mid a(z) \text{ is a field}\}.$$
$\mathfrak{F}(V)$ is the field-theoretic analog of $\mathrm{End}(V)$; it's a subspace of $\mathrm{End}(V)[[z, z^{-1}]]$.

The introduction of a *second* indeterminate facilitates the study of products and commutators of fields. Set

$$\left[\sum_m a_m z_1^{-m-1}, \sum_n b_n z_2^{-n-1}\right] = \sum_{m,n}[a_m, b_n] z_1^{-m-1} z_2^{-n-1},$$

which lies in $\text{End}(V)[[z_1, z_1^{-1}, z_2, z_2^{-1}]]$. The idea of *locality* is crucial.

DEFINITION 2.3. Two elements $a(z), b(z) \in \text{End}(V)[[z, z^{-1}]]$ are *mutually local* if there is a nonnegative integer k such that

$$(z_1 - z_2)^k [a(z_1), b(z_2)] = 0. \tag{4}$$

If (4) holds, we write $a(z) \sim_k b(z)$ and say that $a(z)$ and $b(z)$ are *mutually local of order k*. Write $a(z) \sim b(z)$ if k is not specified. $a(z)$ is a *local field* if $a(z) \sim a(z)$. (4) means that the coefficient of each monomial $z_1^{r-1} z_2^{s-1}$ in the expansion of the left hand side vanishes. Explicitly, this means that

$$\sum_{j=0}^{k} (-1)^j \binom{k}{j} [a_{k-j-r}, b_{j-s}] = 0. \tag{5}$$

Locality defines a *symmetric relation* which is generally neither reflexive nor transitive.

Fix a *nonzero* state $\mathbf{1} \in V$. We say that $a(z) \in \mathfrak{F}(V)$ is *creative* (with respect to $\mathbf{1}$) and *creates the state a* if

$$a(z)\mathbf{1} = a + \cdots \in V[[z]].$$

We sometimes write this in the form $a(z)\mathbf{1} = a + O(z)$. In terms of modes,

$$a_n \mathbf{1} = 0, \ n \geq 0, \qquad a_{-1}\mathbf{1} = a.$$

EXERCISE 2.4. Let $\partial a(z) = \sum_n (-n-1) a_n z^{-n-2}$ be the formal derivative of $a(z)$. Suppose that $a(z), b(z) \in \mathfrak{F}(V)$ and $a(z) \sim_k b(z)$. Prove that $\partial a(z) \in \mathfrak{F}(V)$ and $\partial a(z) \sim_{k+1} b(z)$.

EXERCISE 2.5. (Locality-truncation relation) Suppose that $a(z), b(z)$ are creative fields with $a(z) \sim_k b(z)$. By choosing $s = 1$ and $r = k - n$ for $n \geq k$ in (5), show that $a_n b = 0$ for all $n \geq k$ i.e., the order of truncation N is $k - 1$.

2.3. Axioms for a vertex algebra.
For various approaches to the contents of this subsection, see [B], [FHL], [FLM], [Go], [K1], [LL], [MN].

DEFINITION 2.6. A *vertex algebra* (VA) is a quadruple $(V, Y, \mathbf{1}, D)$, where

$Y : V \to \mathfrak{F}(V), \quad v \mapsto Y(v, z) = \sum v_n z^{-n-1}$ is a linear map,

$\mathbf{1} \in V, \; \mathbf{1} \neq 0,$

$D \in \operatorname{End}(V), \; D\mathbf{1} = 0,$

and the following hold for all $u, v \in V$:

$$\begin{aligned} \text{locality:} \quad & Y(u,z) \sim Y(v,z), \\ \text{creativity:} \quad & Y(u,z)\mathbf{1} = u + O(z), \\ \text{translation covariance:} \quad & [D, Y(u,z)] = \partial Y(u,z). \end{aligned}$$

We often refer to the Fock space V itself, rather than $(V, Y, \mathbf{1}, D)$, as a vertex algebra. The element $\mathbf{1}$ is called the *vacuum* state and Y is the *state-field correspondence*. The physical interpretation of creativity is that $Y(u, z)$ *creates* the state u from the vacuum. This set-up models the creation and annihilation of bosonic states from the vacuum. Most of the subtlety is tied to locality and its consequences.

There are a number of equivalent formulations of these axioms. We discuss some of them. Another approach, via so-called *rationality* ([FHL]) is also discussed in Section 10.1. The *Jacobi Identity* of [FLM] is equivalent to the identity

$$\sum_{i \geq 0} \binom{p}{i} (u_{r+i}v)_{p+q-i} = \sum_{i \geq 0} (-1)^i \binom{r}{i} \{ u_{p+r-i} v_{q+i} - (-1)^r v_{q+r-i} u_{p+i} \}, \tag{6}$$

which holds in a VA for all $u, v \in V$ and all $p, q, r \in \mathbb{Z}$. Conversely, if we have creative fields $Y(v, z) \in \mathfrak{F}(V)$, $v \in V$, with respect to $\mathbf{1}$ and they satisfy (6), then $(V, Y, \mathbf{1}, D)$ is a vertex algebra with $Du = u_{-2}\mathbf{1}$.

Specializing (6) in various ways leads to some particularly useful identities first written down in [B]:

$$\begin{aligned} \text{commutator:} \quad & [u_m, v_n] = \sum_{i \geq 0} \binom{m}{i} (u_i v)_{m+n-i}, \\ \text{associator:} \quad & (u_m v)_n = \sum_{i \geq 0} (-1)^i \binom{m}{i} \{ u_{m-i} v_{n+i} - (-1)^m v_{m+n-i} u_i \}, \\ \text{skew-symmetry:} \quad & u_m v = \sum_{i \geq 0} (-1)^{m+i+1} \frac{1}{i!} D^i v_{m+i} u. \end{aligned}$$

These identities may be stated more compactly using vertex operators, and it is often more efficacious to use the vertex operator format. We state one more consequence of (6), the *associativity* formula, in the operator format. For large enough k, and recalling convention (1),

$$(z_1 + z_2)^k Y(u, z_1 + z_2)Y(v, z_2)w = (z_1 + z_2)^k Y(Y(u, z_1)v, z_2)w. \quad (7)$$

THEOREM 2.7 [FKRW], [MP]. *Let V be a linear space with $0 \neq \mathbf{1} \in V$ and $D \in \mathrm{End}(V)$. Suppose $S \subseteq \mathfrak{F}(V)$ is a set of mutually local, creative, translation-covariant fields which generates V in the sense that*

$$V = \mathrm{span}\{a^1_{-n_1} \ldots a^k_{-n_k}\mathbf{1} \mid a^i(z) \in S, n_1, \ldots, n_k \geq 1, k \geq 0\}.$$

Then there is a unique vertex algebra $(V, Y, \mathbf{1}, D)$ such that $Y(a^i_{-1}\mathbf{1}, z) = a^i(z)$.

EXERCISE 2.8. Prove that the state-field correspondence is *injective*.

EXERCISE 2.9. Prove that

$$Y(u, z)\mathbf{1} = q_z^D u \quad \text{(which equals } \sum_{n \geq 0} \frac{z^n}{n!} D^n u\text{)}.$$

EXERCISE 2.10. Deduce the commutator, associator and skew-symmetry formulas from (6).

EXERCISE 2.11. Assume V is a linear space and

$$\{Y(v, z) \mid v \in V\} \subseteq \mathfrak{F}(V)$$

are mutually local fields such that $Y(v, z)$ is creative (with respect to $\mathbf{1} \neq 0$) and creates v. Prove that (6) and the associator formula are *equivalent*.

EXERCISE 2.12. A is a commutative, associative algebra with identity element 1 and derivation D. Show that there is a vertex algebra $(A, Y, 1, D)$ with

$$Y(a, z)b = \sum_{n \leq -1} \frac{(D^{-n-1}a)b}{(-n-1)!} z^{-n-1}.$$

EXERCISE 2.13. $(V, Y, \mathbf{1}, D)$ is a VA. Assume either (a) $Y(v, z) \in \mathrm{End}(V)[\![z]\!]$ for $v \in V$, (b) D is the zero map, or (c) $\dim V$ is finite. Prove in each case that V is of the type described in Exercise 2.12.

EXERCISE 2.14. Show that the commutator formula is equivalent to the identity $[u_m, Y(v, z)] = \sum_{i \geq 0} \binom{m}{i} Y(u_i v, z) z^{m-i}$.

EXERCISE 2.15. Show that the skew-symmetry formula is equivalent to the identity $Y(u, z)v = q_z^D Y(v, -z)u$.

EXERCISE 2.16. Show that $q_y^D Y(u, x) q_y^{-D} = Y(u, x+y)$.

2.4. Heisenberg algebra.

In this and the following Subsection we will use Theorem 2.7 to construct two fundamental examples of VAs. We must look for generating sets S of mutually local, creative, translation-covariant fields. In our two examples, S consists of a *single* field. The construction relies on some basic techniques from Lie theory (universal enveloping algebras, Poincaré–Birkhoff–Witt Theorem, and so on) which are reviewed in the Appendix.

Let $A = \mathbb{C}a$ be a 1-dimensional linear space. The *affine algebra*

$$\hat{A} = A[t, t^{-1}] \oplus \mathbb{C}K$$

is the Lie algebra with central element K and bracket

$$[a \otimes t^m, a \otimes t^n] = m\delta_{m,-n} K. \qquad (8)$$

REMARK 2.17. Set $p_m = \frac{1}{\sqrt{m}} a \otimes t^m$ ($m > 0$) and $q_{-m} = \frac{1}{\sqrt{-m}} a \otimes t^m$ ($m < 0$.) Then (8) reads

$$[p_m, q_n] = \delta_{m,n} K. \qquad (9)$$

These are essentially the *canonical commutator relations* of quantum mechanics.

Set $\hat{A}^{\geq} = \langle a \otimes t^n, K \mid n \geq 0 \rangle$, $\hat{A}^{-} = \langle a \otimes t^n \mid n < 0 \rangle$. These are a Lie ideal and Lie subalgebra of $\hat{A}$ respectively. Let $\mathbb{C}v_h$ be the 1-dimensional $\hat{A}^{\geq}$-module defined for a scalar h via

$$K.v_h = v_h, \quad (a \otimes t^n).v_h = h\delta_{n,0} v_h \ (n \geq 0).$$

The induced (Verma) module is

$$M_h = \mathrm{Ind}_{\mathcal{U}(\hat{A}^{\geq})}^{\mathcal{U}(\hat{A})} \mathbb{C}v_h = \mathcal{U}(\hat{A}) \otimes_{\mathcal{U}(\hat{A}^{\geq})} \mathbb{C}v_h = \mathcal{U}(\hat{A}^{-}) \otimes \mathbb{C}v_h \qquad (10)$$

where $\mathcal{U}(\ )$ denotes universal enveloping algebra and the third equality in the last display is just a linear isomorphism.

Let $a_n \in \mathrm{End}(M_h)$ be the induced action of $a \otimes t^n$ on M_h, with $a(z) = \sum_n a_n z^{-n-1}$. In what follows we identify v_h with $1 \otimes v_h$. Let

$$v = a_{-n_1} \ldots a_{-n_k}.v_h,$$

with $n_1 \geq \ldots \geq n_k \geq 1$. For $n > n_1$, a_n commutes with each a_{-n_i} by (8). Therefore, $a_n.v = a_{-n_1} \ldots a_{-n_k} a_n.v_h = 0$. This shows that $a(z) \in \mathfrak{F}(M_h)$. As for locality,

$$\sum_{j=0}^{2} (-1)^j \binom{2}{j} [a_{2-j-r}, a_{j-s}] = \sum_{j=0}^{2} (-1)^j \binom{2}{j} (2-j-r)\delta_{2-j-r, s-j} K$$

$$= \{(2-r) - 2(1-r) - r\}\delta_{r+s,2} K = 0. \qquad (11)$$

By (5) this shows that $a(z) \sim_2 a(z)$. Because
$$a(z)v_h = hv_h z^{-1} + \sum_{n \leq -1} a_n . v_h z^{-n-1},$$
we see that $a(z)$ is creative with respect to v_h if (and only if) $h = 0$. In this case, Theorem 2.7 and what we have shown imply:

THEOREM 2.18. *There is a unique vertex algebra* (M_0, Y, v_0, D) *generated by* $a(z)$ *with* $Y(a, z) = a(z)$ *with* $a = a_{-1}v_0 \in M_0$, *and* $Da_n . v_0 = -na_{n-1}v_0$.

REMARK 2.19. In terms of operators on M_0, (9) reads $[p_m, q_n] = \delta_{m,n}\mathrm{Id}$. These relations may be realized by taking $p_m = \partial/\partial x_{-m}$, $q_n = x_{-n}$ acting on the Fock space $\mathbb{C}[x_{-1}, x_{-2}, \ldots]$. This affords an alternate way to understand M_0.

M_0 is variously called the (rank 1) Heisenberg VA, Heisenberg algebra, or *free boson*. In CFT it models a single free boson. (As opposed to standard mathematical usage, *free* here means that the particle is not interacting with other particles.)

2.5. Virasoro algebra. The Virasoro algebra is the Lie algebra with underlying linear space
$$\mathrm{Vir} = \bigoplus_{n \in \mathbb{Z}} \mathbb{C}L_n \oplus \mathbb{C}K$$
and bracket relations
$$[L_m, L_n] = (m-n)L_{m+n} + \frac{m^3 - m}{12}\delta_{m,-n}K. \tag{12}$$

Set $\mathrm{Vir}^{\geq} = \langle L_n, K \mid n \geq 0 \rangle$, $\mathrm{Vir}^- = \langle L_n \mid n < 0 \rangle$, and let $\mathbb{C}v_{c,h}$ be the 1-dimensional $\mathrm{Vir}^{\geq}$-module defined via
$$K.v_{c,h} = cv_{c,h}, \quad L_n.v_{c,h} = \delta_{n,0}hv_{c,h} \ (n \geq 0).$$
with arbitrary scalars c, h. The induced (Verma) module is then
$$M_{c,h} = \mathcal{U}(\mathrm{Vir}) \otimes_{\mathcal{U}(\mathrm{Vir}^{\geq})} \mathbb{C}v_{c,h} = \mathcal{U}(\mathrm{Vir}^-) \otimes \mathbb{C}v_{c,h}. \tag{13}$$

By analogy with Theorem 2.18, Exercise 2.22 (below) suggests that there is a VA with Fock space $M_{c,0}$ and vacuum[1] $v_{c,0}$, with L_{-1} playing the rôle of D. This cannot be true as it stands because $\omega(z).v_{c,0} = L_{-1}.v_{c,0}z^{-1} + \cdots$ is *not* creative. To cure this ill requires (at the very least) that we take a quotient of $M_{c,0}$ by a Vir-submodule that contains $L_{-1}.v_{c,0}$, and indeed it suffices to quotient out the cyclic Vir-submodule generated by this state. We will abuse notation by identifying states, operators and fields associated with $M_{c,0}$ with the corresponding

---
[1] As in the case of the Heisenberg algebra, we identify $v_{c,0}$ and $1 \otimes v_{c,0}$.

states, operators and fields induced on the quotient $M_{c,0}/\mathcal{U}(\mathrm{Vir})L_{-1}.v_{c,0}$. We then arrive at

THEOREM 2.20. *Set*

$$\mathrm{Vir}_c = M_{c,0}/\mathcal{U}(\mathrm{Vir})L_{-1}.v_{c,0},$$

$\omega = L_{-2}.v_{c,0}$, *and* $Y(\omega, z) = \omega(z)$. *Then* $(\mathrm{Vir}_c, Y, v_{c,0}, L_{-1})$ *is a vertex algebra generated by* $Y(\omega, z)$.

Vir_c is called the Virasoro VA of *central charge c*.

EXERCISE 2.21. Show that L_{-1}, L_0 and L_1 span a Lie subalgebra of Vir. What Lie algebra is it?

EXERCISE 2.22. Identify elements of Vir with the endomorphisms they induce on $M_{c,h}$ and set $\omega(z) = \sum L_n z^{-n-2} \in \mathrm{End}(M_{c,h})[\![z, z^{-1}]\!]$. Prove that $\omega(z)$ is a local field of order 4, and $[L_{-1}, \omega(z)] = \partial \omega(z)$.

EXERCISE 2.23. Give the details of the proof of Theorem 2.20.

2.6. Axioms for a vertex operator algebra. There is no consensus as to nomenclature for the many variants of vertex algebra. Our definition of a *vertex operator algebra* (VOA) is the one used by many practitioners of the art, but not all.

DEFINITION 2.24. A VOA is a quadruple $(V, Y, \mathbf{1}, \omega)$, where $V = \bigoplus_{n \in \mathbb{Z}} V_n$ is a $\mathbb{Z}$-graded linear space and

$$Y : V \to \mathfrak{F}(V), \quad v \mapsto Y(v, z) = \sum v_n z^{-n-1},$$
$$\mathbf{1}, \omega \in V, \ \mathbf{1} \neq 0.$$

The fields $Y(v, z)$ are assumed to be mutually local and creative, and certain conditions must be satisfied:

- $Y(\omega, z) = \sum L_n z^{-n-2}$ with a constant c such that
$$[L_m, L_n] = (m-n)L_{m+n} + \frac{m^3 - m}{12} \delta_{m,-n} c \, \mathrm{Id}_V;$$
- $V_n = \{v \in V_n \mid L_0 v = nv\}$;
- $\dim V_n < \infty$, $V_n = 0$ for $n \ll 0$;
- $Y(L_{-1}u, z) = \partial Y(u, z)$.

In effect, a VOA is a vertex algebra with a dedicated Virasoro field. This is the field determined by the distinguished state ω, called the *conformal* or *Virasoro vector*. The modes of ω are operators L_n satisfying the Virasoro relations (12) with $K = c\mathrm{Id}_V$. As in Theorem 2.20 we call c the *central charge* of V. The mode L_0 of ω, called the *degree operator*, is required to be semisimple, to have eigenvalues lying in a subset of $\mathbb{Z}$ that is bounded below, and to have finite-dimensional eigenspaces. We often write $\text{wt}(v) = n$ if v is an eigenstate for L_0 with eigenvalue n, the conformal weight. It is not hard to see that $[L_{-1}, Y(u, z)] = \partial Y(u, z)$, so that $(V, Y, \mathbf{1}, L_{-1})$ is a vertex algebra.

It should come as no surprise that the vertex algebra Vir_c has the structure of a VOA of central charge c with vacuum vector $v_{c,0}$ and conformal vector ω. To see this, note that $\{L_{-n_1} \ldots L_{-n_k}.v_{c,0} \mid n_1 \geq \cdots \geq n_k \geq 2\}$ is a basis of the Fock space. We have

$$L_0 L_{-n_1} \ldots L_{-n_k}.v_{c,0} = n_1 L_{-n_1} \ldots L_{-n_k}.v_{c,0} + L_{-n_1} L_0 L_{-n_2} \ldots L_{-n_k}.v_{c,0}.$$

Now an easy induction shows that

$$L_0.L_{-n_1} \ldots L_{-n_k}.v_{c,0} = \left(\sum_i n_i\right).L_{-n_1} \ldots L_{-n_k}.v_{c,0},$$

so that

$$\text{wt}(L_{-n_1} \ldots L_{-n_k}.v_{c,0}) = \sum_i n_i.$$

The needed properties of L_0 required for the next result follow easily, and we obtain the following extension of Theorem 2.20:

THEOREM 2.25. *Vir_c is a VOA of central charge c.*

A pair of VOAs V, V' are called *isomorphic* if there is a linear isomorphism $\varphi : V \to V', v \mapsto v'$ such that $\varphi(\omega) = \omega'$, and $\varphi Y(v, z) \varphi^{-1} = Y'(\varphi(v), z)$.

In the following exercises, V is a VOA.

EXERCISE 2.26. Complete the proof of Theorem 2.25.

EXERCISE 2.27. Prove the following: $Y(\mathbf{1}, z) = \text{Id}, \mathbf{1} \in V_0, \omega \in V_2, L_n \mathbf{1} = 0$ for $n \geq -1, (L_{-1}v)_n = -nv_{n-1}$.

EXERCISE 2.28. Suppose that $v \in V$ satisfies $L_{-1}v = 0$. Prove that $v \in V_0$.

EXERCISE 2.29. Suppose that $V_0 = \mathbb{C}\mathbf{1}$ (cf. Exercise 2.27). Prove that $V_n = 0$ for $n < 0$.

EXERCISE 2.30. Show that $\dim V$ is *finite* if, and only if, $\omega = 0$ (cf. Exercise 2.13).

EXERCISE 2.31. Show that the Heisenberg theory M_0 (cf. Section 2.4) is a VOA with vacuum $\mathbf{1} = v_0, \omega = \frac{1}{2}a_{-1}^2 \mathbf{1} = \frac{1}{2}a_{-1}a$ and central charge $c = 1$. This is the theory of *one free boson*.

EXERCISE 2.32. Let U, V be linear spaces. Show that there is a natural injection $\mathfrak{F}(U) \otimes \mathfrak{F}(V) \to \mathfrak{F}(U \otimes V)$. Suppose in addition that U and V are Fock spaces for VOAs with vacuum vectors $\mathbf{1}, \mathbf{1}'$ and conformal vectors ω, ω' respectively. Show how to construct the *tensor product VOA* $(U \otimes V, Y, \mathbf{1} \otimes \mathbf{1}', \omega \otimes \omega')$. What is the central charge of this VOA?

EXERCISE 2.33. Let $\varphi : V \to V'$ be an isomorphism of VOAs. Prove the following: (i) V and V' have the *same* central charge; (ii) $\varphi(\mathbf{1}) = \mathbf{1}'$.

2.7. VOAs on the cylinder and the square bracket formalism.

There is a sense in which we may think of a VOA as being 'on the sphere'. This is closely related to the axiomatic approach via rationality (cf. [FHL] and Section 10.1). Here we want to describe the corresponding VOA that lives 'on the cylinder'. Roughly, this corresponds to a change of variable $z \to q_z - 1$ which we call the *square bracket formalism*. The main purpose is to construct vertex operators that are automatically periodic in z with period $2\pi i$. Let $V = (V, Y, \mathbf{1}, \omega)$ be a VOA of central charge c. For $v \in V$ introduce[2]

$$Y[v, z] = Y(q_z^{L_0} v, q_z - 1) = \sum_{n \in \mathbb{Z}} v[n] z^{-n-1}. \tag{14}$$

Here, $q_z^{L_0}$ is the operator

$$q_z^{L_0} : V \to V[\![z]\!], \quad v \mapsto q_{kz} v \ (v \in V_k), \tag{15}$$

and our q-convention (2) is in force. Similar expressions will occur frequently in what follows. The $v[n]$ are new operators on V, and for $v \in V_k$ are given by

$$v[m] = m! \sum_{i \geq m} c(k, i, m) v_i \tag{16}$$

for $m \geq 0$, with

$$\binom{k-1+x}{i} = \sum_{m=0}^{i} c(k, i, m) x^m. \tag{17}$$

From (16) and (17) we find

$$\sum_{i \geq 0} \binom{k}{i} v_i = \sum_{m \geq 0} \frac{(k+1-k)^m}{m!} v[m]. \tag{18}$$

These identities are proved in Section 13 (Appendix). We also have a new conformal vector

$$\tilde{\omega} = \omega - \frac{c}{24} \mathbf{1}, \tag{19}$$

with corresponding square bracket modes

$$Y[\tilde{\omega}, z] = \sum_n L[n] z^{-n-2}.$$

In particular, $L[0]$ provides us with an alternative $\mathbb{Z}$-grading structure on V:

$$V = \bigoplus_n V_{[n]}, \quad V_{[n]} = \{u \in V \mid L[0]u = nu\}.$$

---

[2] We write modes in the square bracket formalism as $v[n]$ rather than $v_{[n]}$.

We write $\text{wt}[v] = n$ if $v \in V_{[n]}$. The following can be proved.

THEOREM 2.34. *The quadruple $(V, Y[\,,\,], \mathbf{1}, \tilde{\omega})$ is a VOA of central charge c.*

Given a VOA V, we say that its alter ego $(V, Y[\,,\,], \mathbf{1}, \tilde{\omega})$ is 'on the cylinder'. VOAs on the cylinder play an important rôle in forging the connections with modular forms.

EXAMPLE 2.35. In the square bracket formalism, the VOA $(M_0, Y[\,,\,], \mathbf{1}, \tilde{\omega})$ is generated by a state a with $\text{wt}[a] = 1$. It has a basis of Fock vectors of the form $a[-n_1]\ldots a[-n_k]\mathbf{1}$, $n_1 \geq \ldots \geq n_k \geq 1$ satisfying

$$[a[m], a[n]] = m\delta_{m+n,0}\text{Id}.$$

EXERCISE 2.36. Show that $L[-1] = L_{-1} + L_0$.

EXERCISE 2.37. A state v in a VOA V is called *primary* of weight k with respect to the original Virasoro algebra $\{L_n\}$ if, and only if, it satisfies $L_n v = k\delta_{n,0}v$ for $n \geq 0$. Prove that v is primary of weight k with respect to $\{L_n\}$ if, and only if, it is primary of weight k with respect to $\{L[n]\}$.

EXERCISE 2.38. Prove the assertions of Example 2.35 in the more precise form that $(M_0, Y, \mathbf{1}, \omega)$ and $(M_0, Y[\,,\,], \mathbf{1}, \tilde{\omega})$ are *isomorphic* Heisenberg VOAs.

3. Modular and quasimodular forms

In this section we compile some relevant background involving *elliptic modular forms*. This is a standard part of analytic number theory, and there are many excellent texts dealing with the subject, such as [Kn], [O], [Se], [Sc]. Because it is so central to our cause, we describe what we need here, referring the reader elsewhere for more details and further development.

3.1. Modular forms on $\text{SL}_2(\mathbb{Z})$. The (homogeneous) modular group is

$$\Gamma = \text{SL}_2(\mathbb{Z}) = \left\{ \begin{pmatrix} a & b \\ c & d \end{pmatrix} \mid a, b, c, d \in \mathbb{Z}, ad - bc = 1 \right\},$$

with standard generators $S = \begin{pmatrix} 0 & -1 \\ 1 & 0 \end{pmatrix}$, $T = \begin{pmatrix} 1 & 1 \\ 0 & 1 \end{pmatrix}$. The complex upper half-plane $\mathfrak{H}$ carries a left Γ-action by Möbius transformations

$$(\gamma, \tau) \mapsto \gamma\tau = \frac{a\tau + b}{c\tau + d}, \quad \gamma = \begin{pmatrix} a & b \\ c & d \end{pmatrix} \in \Gamma. \tag{20}$$

In particular, $T : \tau \mapsto \tau + 1$ and $S : \tau \mapsto -1/\tau$. For $k \in \mathbb{Z}$, there is a right action of Γ on meromorphic functions in $\mathfrak{H}$ given by

$$f|_k \gamma(\tau) = (c\tau + d)^{-k} f(\gamma\tau). \tag{21}$$

A *weak modular form* of weight k on Γ is an *invariant* of this action. Thus $f|_k\gamma(\tau) = f(\tau)$ for $\gamma \in \Gamma$, which amounts to

$$f(\tau+1) = f(\tau), \qquad f(-1/\tau) = \tau^k f(\tau).$$

By a standard argument the first of these equalities implies that $f(\tau)$ has a *q-expansion*, or *Fourier expansion at ∞*,

$$f(\tau) = \sum_{n \in \mathbb{Z}} a_n q^n, \tag{22}$$

with constants a_n called the *Fourier coefficients* of $f(\tau)$. Here we are using our q-convention (2).

We say that $f(\tau)$ is a *meromorphic modular form of weight k* if its q-expansion has the form

$$f(\tau) = \sum_{n \geq n_0} a_n q^n \tag{23}$$

for some n_0. Assume that $f(\tau) \neq 0$ with $a_{n_0} \neq 0$. We then say that $f(\tau)$ has a pole of order n_0 at ∞ if $n_0 \leq 0$ or a zero of order n_0 if $n_0 \geq 0$. In the latter situation we also say that $f(\tau)$ is *holomorphic at ∞*. $f(\tau)$ is a *holomorphic modular form of weight k* if it is holomorphic in $\mathfrak{H} \cup \{\infty\}$. $f(\tau)$ is *almost holomorphic* if it is holomorphic in $\mathfrak{H}$ (the behaviour at ∞ being unspecified beyond being at worst a pole). Modular forms of weight 0 are often called *modular functions*, though we will not be consistent on this point. Let $\mathfrak{M}_k$ be the set of holomorphic modular forms of weight k. It is a $\mathbb{C}$-linear space, possibly equal to 0.

EXERCISE 3.1. Show that the *kernel* of the Γ-action (20) is the *center* of Γ and consists of $\pm I$, where I is the 2×2 identity matrix. (The quotient group $PSL_2(\mathbb{Z}) = \tilde{\Gamma} = \Gamma/\{\pm I\}$ is the *inhomogeneous* modular group.)

EXERCISE 3.2. (a) Show that torsion elements in $\tilde{\Gamma}$ have order *at most* 3. (b) Show that $\tilde{\Gamma}$ has a *unique* conjugacy class of subgroups of order 2 or 3.

EXERCISE 3.3. Let $z \in \mathfrak{H}$ with $\text{Stab}_{\tilde{\Gamma}}(z) = \{\gamma \in \tilde{\Gamma} \mid \gamma.z = z\}$ the *stabilizer* of z in $\tilde{\Gamma}$. Prove the following: (a) $\text{Stab}_{\tilde{\Gamma}}(z)$ is a finite cyclic subgroup, (b) each nontrivial torsion element in $\tilde{\Gamma}$ stabilizes a *unique* point in $\mathfrak{H}$.

EXERCISE 3.4. Show that $\tilde{\Gamma}$ acts *properly discontinuously* on $\mathfrak{H}$ in the following sense: every $z \in \mathfrak{H}$ has an open neighborhood N_z with the property that if $\gamma \in \tilde{\Gamma}$ then $\gamma(N_z) \cap N_z = \phi$ if $\gamma \notin \text{Stab}_{\tilde{\Gamma}}(z)$ and $\gamma(N_z) \cap N_z = N_z$ otherwise. Conclude that the *orbit space* $\Gamma \backslash \mathfrak{H}$ is a topological 2-manifold (a Hausdorff space such that each point has an open neighborhood homeomorphic to $\mathbb{R}^2$).

EXERCISE 3.5. Suppose that $f(\tau)$ is a nonzero weak modular form of weight k. Show that k is *even*.

EXERCISE 3.6. Let E be the set of meromorphic modular functions of weight zero. Show that E is a field[3] containing $\mathbb{C}$.

EXERCISE 3.7. Show that pointwise multiplication defines a bilinear product $\mathfrak{M}_k \otimes \mathfrak{M}_l \to \mathfrak{M}_{k+l}$, with respect to which $\mathfrak{M} = \bigoplus_k \mathfrak{M}_k$ is a $\mathbb{Z}$-graded commutative $\mathbb{C}$-algebra.

EXERCISE 3.8. Suppose that $f(\tau)$ is a meromorphic modular form of weight zero. Show that $f'(\tau)$ is a meromorphic modular form of weight 2.

3.2. Eisenstein series on $\mathrm{SL}_2(\mathbb{Z})$. Beyond the fact that constants in $\mathbb{C}$ are modular functions of weight 0 (cf. Exercise 3.6), it is not so easy to construct nonconstant modular functions of weight 0 or *any* nonzero modular form of nonzero weight. We content ourselves with the description of some examples chosen because of their relevance to VOA theory.

The most accessible nonconstant modular forms are the *Eisenstein series*. For an integer $k \geq 2$, set

$$E_k(\tau) = -\frac{B_k}{k!} + \frac{2}{(k-1)!} \sum_{n \geq 1} \frac{n^{k-1} q^n}{1 - q^n}$$

$$= -\frac{B_k}{k!} + \frac{2}{(k-1)!} \sum_{n \geq 1} \sigma_{k-1}(n) q^n. \qquad (24)$$

Here, $\sigma_{k-1}(n) = \sum_{d \mid n} d^{k-1}$ and B_k is the k-th Bernoulli number defined by[4]

$$\frac{z}{q_z - 1} = \sum_{k \geq 0} \frac{B_k}{k!} z^k = 1 - \tfrac{1}{2} z + \tfrac{1}{12} z^2 + \cdots \qquad (25)$$

The well-known identity of Euler

$$\zeta(k) = -\frac{(2\pi i)^k B_k}{2(k!)} \quad (k \geq 2 \text{ even}), \qquad (26)$$

permits us to reexpress the constant term of (24) in terms of zeta-values. The basic fact is this: *Let $k \geq 3$. Then $E_k(\tau)$ is a holomorphic modular form of weight k; it is identically zero if, and only if, k is odd.* We will see one way to prove this in Section 5. We emphasize that $E_2(\tau)$ is *not* a modular form.

The normalization employed in (24) is related to elliptic functions (Section 5). In fact B_{2k} never vanishes, so we can renormalize so that the q-expansion

---

[3] Of course, field here is in the algebraic sense.

[4] Several different conventions are used to define Bernoulli numbers in the literature.

begins $1 + \cdots$. We single out the first three Eisenstein series corresponding to $k = 2, 4, 6$ renormalized in this way, and rename them (following Ramanujan)

$$P = 1 - 24 \sum_{n \geq 1} \sigma_1(n) q^n, \quad Q = 1 + 240 \sum_{n \geq 1} \sigma_3(n) q^n, \quad R = 1 - 504 \sum_{n \geq 1} \sigma_5(n) q^n.$$

P, Q, R are *algebraically independent*, so that they generate a weighted polynomial algebra, denoted

$$\mathfrak{Q} = \mathbb{C}[P, Q, R], \tag{27}$$

where P, Q, R naturally have weights (degree) $2, 4, 6$ respectively. $\mathfrak{Q}$ is the algebra of *quasimodular forms*. $\mathfrak{Q}$ contains every holomorphic modular form. Indeed, we have (cf. Exercise 3.7)

THEOREM 3.9. *The graded algebra* $\mathfrak{M} = \bigoplus \mathfrak{M}_k$ *of holomorphic modular forms on* Γ *is the graded subalgebra* $\mathbb{C}[Q, R]$ *of* $\mathfrak{Q}$.

Theorem 3.9 follows from a careful study of the singularities (zeros and poles) of modular forms, but we will not discuss this here. The Theorem contains a lot of information about holomorphic modular forms. For example, there are no such nonzero forms of negative weight or weight 2, holomorphic forms of weight zero are necessarily constant, and $\dim \mathfrak{M}_k < \infty$. Indeed, inasmuch as Q and R are free generators in weights 4 and 6 respectively, the Hilbert–Poincaré series of $\mathfrak{M}$ is

$$\sum_{k \geq 0} (\dim \mathfrak{M}_k) t^k = \frac{1}{(1 - t^4)(1 - t^6)}$$

$$= 1 + t^4 + t^6 + t^8 + t^{10} + 2t^{12} + t^{14} + 2t^{16} + \cdots. \tag{28}$$

As we already mentioned, $E_2(\tau)$ is *not* a modular form. Indeed, it satisfies the transformation law

$$E_2|_2 \gamma(\tau) = E_2(\tau) - \frac{c}{2\pi i (c\tau + d)}, \quad \gamma = \begin{pmatrix} a & b \\ c & d \end{pmatrix}. \tag{29}$$

The importance of $E_2(\tau)$ for us stems from its relation to *derivatives* of modular forms. Suppose that $f(\tau)$ is a meromorphic modular form of weight k. We define the *modular derivative* of $f(\tau)$ by

$$D_k f(\tau) = D f(\tau) = (\theta + k E_2(\tau)) f_k(\tau). \tag{30}$$

where $\theta = q \, d/dq$. One can show without difficulty (cf. Exercise 3.13) that $D f_k(\tau)$ is modular of weight $k + 2$, and is holomorphic if $f_k(\tau)$ is.

EXERCISE 3.10. Prove that

$$\frac{q_z}{(1 - q_z)^2} = \frac{1}{z^2} - \sum_{k \geq 2} \frac{B_k}{k!} (k - 1) z^{k-2}.$$

Deduce that $B_k = 0$ for odd $k \geq 3$.

EXERCISE 3.11. Prove that $E_8 = \frac{3}{7}E_4^2$ and $E_{10} = \frac{5}{11}E_4 E_6$.

EXERCISE 3.12. Show that (28) is equivalent to the formula $\dim \mathfrak{M}_{2k} = [k/6]$ if $k \equiv 1 \pmod{6}$ and $1 + [k/6]$ otherwise.

EXERCISE 3.13. Prove that $(D_k f)|_{k+2}(\tau) = D_k(f|_k \gamma)(\tau)$ for any meromorphic function $f(\tau)$. Conclude that D_k induces a linear map $\mathfrak{M}_k \to \mathfrak{M}_{k+2}$.

EXERCISE 3.14. Prove that $DE_4 = 14 E_6$ and $DE_6 = \frac{60}{7} E_4^2$.

EXERCISE 3.15. Let $D : \mathfrak{M} \to \mathfrak{M}$ be the linear map whose restriction to $\mathfrak{M}_k$ is D_k. Prove that D is a *derivation* of the algebra $\mathfrak{M}$.

3.3. Cusp-forms and modular functions on $\mathrm{SL}_2(\mathbb{Z})$. Thanks to Theorem 3.9, every holomorphic modular form of weight k is equal to a unique homogeneous polynomial in Q and R. In this subsection we describe some important examples of particular relevance to VOAs. We start with the *discriminant*, defined by

$$\Delta(\tau) = \frac{Q^3 - R^2}{12^3} = q - 24q^2 + \cdots. \tag{31}$$

$\Delta(\tau)$ is evidently a holomorphic modular form of weight 12. It may alternatively be described by a q-product which goes back to Kronecker, namely

$$\Delta(\tau) = q \prod_{n \geq 1} (1 - q^n)^{24}. \tag{32}$$

This formula finds its natural place in the theory of elliptic functions. From our present vantage point, the fact that (31) and (32) coincide is miraculous. Beyond the product formula, the properties that make $\Delta(\tau)$ important for us are the following: it *does not vanish in* $\mathfrak{H}$, and (up to scalars) it is the unique nonzero holomorphic modular form of least weight that *vanishes at* ∞. The nonvanishing property has a natural explanation in the theory of elliptic functions. Concerning the second property, we introduce *cusp forms* defined by

$$\mathfrak{S}_k = \{f(\tau) \in \mathfrak{M}_k \mid f \text{ vanishes at } \infty\}, \quad \mathfrak{S} = \bigoplus_k \mathfrak{S}_k.$$

Our assertions then say that $\mathfrak{S}_k = 0$ for $k < 12$ and $\mathfrak{S}_{12} = \mathbb{C}\Delta(\tau)$. Using (31), it follows that $\Delta(\tau)^{-1}$ is an almost holomorphic modular form of weight -12 with a pole of order 1 at ∞. Applications of these facts are given in Exercise 3.16.

Closely related to $\Delta(\tau)$ is the *Dedekind η-function*, whose q-expansion is the 24th root of that for $\Delta(\tau)$:

$$\eta(\tau) = q^{1/24} \prod_{n \geq 1} (1 - q^n). \tag{33}$$

Note that $\eta(\tau)$ is not a modular form in our sense. It satisfies identity

$$\eta(\tau)^{-1} = q^{-1/24} \sum_{n\geq 0} p(n)q^n \qquad (34)$$
$$= q^{-1/24}(1 + q + 2q^2 + 3q^3 + 5q^4 + \cdots),$$

where $p(n)$ is the *unrestricted partition function*. This identity goes back to Euler.

Our next example is the famous j-function, defined by

$$j(\tau) = \frac{Q^3}{\Delta(\tau)} = q^{-1} + 744 + 196884q + \cdots \qquad (35)$$

As the quotient of two modular forms of weight 12, $j(\tau)$ has weight zero, and because Δ^{-1} is almost holomorphic, so too is $j(\tau)$. In the notation of Exercise 3.6, $j(\tau) \in E$. Let $\Gamma\backslash\mathfrak{H}$ be the orbit space for the action of Γ on $\mathfrak{H}$ (cf. Exercise 3.4). Because the weight is zero, we see from (21) that j induces a map

$$j : \Gamma\backslash\mathfrak{H} \to \mathbb{C}$$

which turns out to be a *homeomorphism*. It can be shown that $E = \mathbb{C}(j)$ is exactly the field of rational functions in j.

EXERCISE 3.16. Considered as a subspace of the algebra $\mathfrak{M}$, show that $\mathfrak{S}$ is the principal ideal generated by Δ.

EXERCISE 3.17. Show that $\mathfrak{M}_k = \mathfrak{S}_{2k} \oplus \mathbb{C}E_{2k}$ for $k \geq 2$.

EXERCISE 3.18. Prove that $\theta\eta(\tau) = -\frac{1}{2}\eta(\tau)E_2(\tau)$. Conclude that $D_{12}\Delta(\tau) = 0$. Give another proof of this by using Theorem 3.9.

EXERCISE 3.19. Regard D as a derivation of $\mathfrak{M}$ as in Exercise 3.15. Show that the *space of D-constants* (i.e., the subspace of $\mathfrak{M}$ annihilated by D) is the polynomial algebra $\mathbb{C}[\Delta]$.

EXERCISE 3.20. Prove that the *ring of almost holomorphic modular functions of weight zero* on Γ is the space $\mathbb{C}[j]$ of polynomials in $j(\tau)$.

4. Characters of vertex operator algebras

Fix a VOA $(V, Y, \mathbf{1}, \omega)$ with $\mathbb{Z}$-graded Fock space $V = \oplus V_n$ and central charge c. In this section we introduce the idea of the *character* of V as a sort of analog of the character of a group representation. This is essentially the theory of *1-point correlation functions* on V.

4.1. Zero modes.

We start with a useful calculation. Suppose that $v \in V_k, w \in V_m, n \in \mathbb{Z}$. Remembering that $L_i = \omega_{i+1}$, we have

$$L_0 v_n w = ([\omega_1, v_n] + v_n L_0)w = \left(\sum_i \binom{1}{i}(L_{i-1}v)_{n+1-i} + v_n L_0\right)w$$

$$= \left((L_{-1}v)_{n+1} + (L_0 v)_n + v_n L_0\right)w = (m+k-n-1)v_n w.$$

Here, we used the commutator formula (cf. Section 2.3) for the second equality and the last identity of Exercise 2.27 for the fourth equality. What we take from this is that *modes of homogeneous states are graded operators on V*:

$$v \in V_k \Rightarrow v_n : V_m \to V_{m+k-n-1}. \tag{36}$$

In particular, let us define[5] the *zero mode* $o(v)$ of a state $v \in V_k$ to be v_{k-1}, and extend this definition to V additively. From (36) we then have for all integers m and states v that

$$o(v) : V_m \to V_m. \tag{37}$$

The point of all this is that as an operator on V_m we can *trace* the zero mode and form a generating function (cf. (15))

$$Z(v, q) = \mathrm{Tr}_V \, o(v) q^{L_0 - c/24} = q^{-c/24} \sum_n \mathrm{Tr}_{V_n} o(v) q^n. \tag{38}$$

This is to be regarded as a *formal q-expansion* at this point. Apart from memorializing the central charge, the factor $q^{-c/24}$ may seem somewhat arbitrary. This feeling will pass. Because the homogeneous spaces V_n vanish for small enough n, we see that

$$Z(v, q) \in q^{-c/24} \mathbb{C}[\![q]\!][q^{-1}].$$

$Z = Z_V$ defines the *character* of V, i.e., the linear map

$$Z : V \to q^{-c/24} \mathbb{C}[\![q]\!][q^{-1}], \quad v \mapsto Z(v, q).$$

EXERCISE 4.1. Let $U \otimes V$ be the tensor product of VOAs U, V (Exercise 2.32). Prove that $Z_{U \otimes V} = Z_U Z_V$.

EXERCISE 4.2. Suppose that V is a VOA with $v \in V$. Prove the identity $q_x^{L_0} Y(v, z) q_x^{-L_0} = Y(q_x^{L_0} v, q_x z)$.

EXERCISE 4.3. Let a be the generating state of weight 1 for the Heisenberg VOA (Section 2.4). Prove that the zero mode $o(a)$ is zero.

EXERCISE 4.4. Suppose that $V_0 = \mathbb{C}1$ (cf. Exercise 2.29). Prove using (36) that for $a \in V_{n_1}$ and $b \in V_{n_2}$ we have $a_n b = 0$ for all $n \geq n_1 + n_2$. Using Exercise 2.5, deduce that $Y(a, z) \sim_k Y(b, z)$ with order of locality $k \leq n_1 + n_2$.

---

[5] The zero mode of v is generally not the *zeroth* mode v_0 but rather the mode which has weight zero as an operator. However, in the convention used for modes in CFT as practiced by physicists, it is the zero mode.

4.2. Graded dimension.

The most prominent Z-value is that obtained by tracing the zero mode of the vacuum. From Exercise 2.26 we have $Y(\mathbf{1}, z) = \mathrm{Id}_V$ and $\mathbf{1} \in V_0$. So the zero mode of $\mathbf{1}$ is Id_V, whence

$$Z_V(\mathbf{1}) = \mathrm{Tr}_V\, q^{L_0 - c/24} = q^{-c/24} \sum_n \dim V_n q^n. \tag{39}$$

This is variously called the *graded dimension*, *q-dimension*, *0-point function*, or *partition function* of V.

The graded dimensions of our two main examples M_0 and Vir_c are readily computed. This is because the Fock spaces are Verma modules, or closely related to them, and these are easy to handle. Let us start with the Fock space M_0 for the free boson, which has central charge $c = 1$ (Theorem 2.18 and Exercise 2.32). In the notation of (10), M_0 (considered as a $\mathbb{Z}$-graded linear space) coincides with $\mathcal{U}(\hat{A}^-)$ equipped with the natural product grading for which $a \otimes t^{-n}$ has weight n. Because of the PBW Theorem, the universal enveloping algebra is itself isomorphic as graded space to the symmetric algebra $S(\coprod_{n \geq 1} \mathbb{C} x_{-n})$ with x_{-n} having weight n. (In other words, M_0 'is' a polynomial algebra in variables x_{-n}. Compare with Remark 2.19.) As graded algebras, symmetric algebras are multiplicative over direct sums. It follows that

$$Z_{M_0}(\mathbf{1}) = q^{-1/24} \prod_{n=1}^{\infty} (q\text{-dimension of } \mathbb{C}[x_{-n}])$$

$$= q^{-1/24} \prod_{n=1}^{\infty}(1 + q^n + q^{2n} + \cdots) = q^{-1/24} \prod_{n=1}^{\infty}(1 - q^n)^{-1},$$

which is none other than the inverse eta-function (33), (34). Thus we have

$$Z_{M_0}(\mathbf{1}) = \eta(q)^{-1}. \tag{40}$$

For an integer $n \geq 1$, let $M_0^{\otimes n}$ be the n-fold tensor product of M_0 considered as a VOA as described in Exercise 2.32. This is the theory of *n free bosons*. Using Exercise 3.1 we deduce from (40) that

$$Z_{M_0^{\otimes n}}(\mathbf{1}) = \eta(q)^{-n}. \tag{41}$$

In particular, the graded dimension of the VOA $M_0^{\otimes 24}$ of 24 free bosons (the *bosonic string*) is the inverse discriminant $\Delta(\tau)^{-1}$.

The calculation of the graded dimension of Vir_c is similar. Indeed, the Fock space $M_{c,0}$ (13) is *isomorphic* as $\mathbb{Z}$-graded linear space to M_0. We must quotient out the graded submodule $\mathcal{U}(\mathrm{Vir})L_{-1}.v_{c,0}$, and this is isomorphic to $M_{c,0}[1]$, that is $M_{c,0}$ with an overall shift of $+1$ in the grading, because $L_{-1}.v_{c,0}$ has weight 1 as an element of $M_{c,0}$. We find that

$$Z_{\mathrm{Vir}_c}(\mathbf{1}) = \frac{q^{-c/24}}{\prod_{n\geq 2}(1-q^n)}, \tag{42}$$

which is *not* the q-expansion of a modular form.

Next we consider the character value $Z_V(\omega)$ for a VOA V. Because the zero mode of the conformal vector is L_0, which acts on V_n as multiplication by n, we have

$$Z_V(\omega) = q^{-c/24} \sum_n n \dim V_n q^n.$$

This is almost, but not quite, equal to $\theta Z_V(\mathbf{1})$ (θ as in (30)). If instead we use the square bracket conformal vector $\tilde{\omega} \in V_{[2]}$ (19) we find

$$Z_V(\tilde{\omega}) = q^{-c/24} \sum_n (n - \frac{c}{24}) \dim V_n q^n$$
$$= \theta(Z_V(\mathbf{1})).$$

In the case of M_0, for example, we obtain using Exercise 3.18 that

$$Z_{M_0}(\tilde{\omega}) = \theta \eta(\tau)^{-1} = \frac{E_2(\tau)}{2\eta(\tau)}.$$

This suggests that 'nicer' character values obtain by evaluating Z_V on states which are homogeneous in the square bracket formalism, i.e., lie in $V_{[k]}$ for some k.

4.3. The character of the Heisenberg algebra. It is generally a difficult problem to compute the 1-point functions $Z_V(v)$ of a VOA V for a complete basis of states. We describe the solution for the Heisenberg algebra M_0 ([MT1]). It well illustrates the principle suggested at the end of the previous Subsection.

THEOREM 4.5. *Let* $M_{[0]} = \bigoplus_{n \geq 0} (M_0)_{[n]}$ *be the Fock space for* M_0 *equipped with the square bracket grading (cf. Section 2.7). Let* $\mathfrak{Q}$ *be the graded algebra of quasimodular forms* (27). *There is a* surjection *of graded linear spaces*

$$M_{[0]} \to \mathfrak{Q}, \quad v \mapsto Q_v(\tau)$$

such that $Z_{M_0}(v) = Q_v(\tau)/\eta(\tau)$.

Up to a normalizing factor $\eta(\tau)^{-1}$ then, every 1-point function is a quasimodular form, and every quasimodular form of weight k arises in this way from a state $v \in (M_0)_{[k]}$ (cf. Exercise 4.7). There is an *explicit* description of the quasimodular form $Q_v(\tau)$ attached to a state v with wt$[v] = k$ which goes as follows. A basis of states for $(M_0)_{[k]}$ is given by

$$v_\lambda = a[-k_1]\ldots a[-k_n]\mathbf{1} \tag{43}$$

where $k = k_1 + \cdots + k_n$ and $1 \leq k_1 \leq \ldots \leq k_n$ range over the parts of a partition λ of k. The quasimodular form $Q_{v_\lambda}(\tau)$ is given by

$$Q_{v_\lambda}(\tau) = \sum_{\varphi=\ldots(rs)\ldots} \prod_{(rs)} (-1)^{r+1} \frac{(r+s-1)!}{(r-1)!(s-1)!} E_{r+s}(\tau), \tag{44}$$

where the notation is as follows. Let $\Phi = \{k_1, \ldots, k_n\}$ be the parts of the partition λ. Then φ ranges over all fixed-point-free involutions in the symmetric group $\Sigma(\Phi)$, so that φ can be represented as a product of transpositions $\ldots(rs)\ldots$ with (r, s) a pair of parts of λ. For each such φ, the product ranges over the transpositions whose product (in $\Sigma(\Phi)$) is φ. We will indicate how (44) can be proved in the next Section. A detailed proof appears later in Section 11.1.

Assume formula (44) in the following exercises.

EXERCISE 4.6. Show that $Q_{v_\lambda}(\tau)$ vanishes if λ has either an odd number of parts or an odd number of odd parts.

EXERCISE 4.7. Assume that λ has both an even number of parts and an even number of odd parts. Prove that $Q_{v_\lambda}(\tau)$ has a *nonzero* constant term, and in particular does not vanish.

EXERCISE 4.8. With the same assumptions as the previous Exercise, prove that $Q_{v_\lambda}(\tau) \in \mathfrak{M}$ if, and only if, λ has *at most one part equal to* 1.

EXERCISE 4.9. Prove the assertion that *every* quasimodular form arises as the trace of a state in M_0.

5. Elliptic functions and 2-point functions

There is an extension of the idea of 1-point functions to n-point functions for any (nonnegative) n. We mainly restrict ourselves here to the case of 2-point correlation functions, which are related to *elliptic functions*.

5.1. Elliptic functions. Throughout this section, *lattice* means an additive subgroup $\Lambda \subseteq \mathbb{C}$ of rank 2. As such it is the $\mathbb{Z}$-span of an $\mathbb{R}$-basis (ω_1, ω_2) of $\mathbb{C}$. An *elliptic function* is a function $f(z)$ which is meromorphic in $\mathbb{C}$ and satisfies $f(z + \lambda) = f(z)$ for all λ in some lattice Λ. Equivalently, $f(z + \omega_i) = f(z)$ for basis vectors ω_1, ω_2 of Λ. Λ is the *period lattice* of $f(z)$. Note that $\mathbb{C}/\Lambda$

has the structure of a *complex torus* (aka *complex elliptic curve*) and that $f(z)$ induces a map

$$f : \mathbb{C}/\Lambda \to \mathbb{CP}^1 = \mathbb{C} \cup \{\infty\}.$$

The set of all meromorphic functions with period lattice Λ is a field M_Λ (the function field of the torus). We have $\mathbb{C} \subseteq M_\Lambda$ where $\mathbb{C}$ is identified with the constants, moreover $f'(z) \in M_\Lambda$ whenever $f(z) \in M_\Lambda$.

Two lattices Λ_1, Λ_2 are *homothetic* if there is $\alpha \in \mathbb{C}$ with $\alpha \Lambda_1 = \Lambda_2$. It is usually enough to deal with some fixed lattice in a homothety class (the corresponding complex tori are *isomorphic*), and every Λ is homothetic to a lattice with basis $(2\pi i, 2\pi i \tau)$ and $\tau \in \mathfrak{H}$. We let Λ_τ denote this lattice.

The classical *Weierstrass $\wp$-function* is[6]

$$\wp(z, \tau) = \frac{1}{z^2} + \sum_{m,n \in \mathbb{Z}}{}' \left(\frac{1}{(z - \omega_{m,n})^2} - \frac{1}{\omega_{m,n}^2} \right). \tag{45}$$

Here, $\omega_{m,n} = 2\pi i (m\tau + n)$. The double sum is independent of the order of summation and absolutely convergent. It defines a function with the following properties: (a) double pole at each point of Λ_τ, (b) holomorphic in $(\mathbb{C} \times \mathfrak{H}) \setminus \Lambda_\tau$, (c) *even* in z, (d) period lattice Λ_τ. In particular, for fixed $\tau \in \mathfrak{H}$ the $\wp$-function $\wp(z, \tau)$ lies in the field M_{Λ_τ}. It turns out that

$$M_{\Lambda_\tau} = \mathbb{C}(\wp(z, \tau), \wp'(z, \tau))$$

is a *function field in one variable*. Indeed the set of *even* elliptic functions is a simple transcendental extension $\mathbb{C}(\wp)$ and $M_{\Lambda_\tau} \supseteq \mathbb{C}(\wp)$ a *quadratic extension*.

There is a natural left action of Γ on $\mathbb{C} \times \mathfrak{H}$ extending (20). It is given by

$$\gamma : (z, \tau) \mapsto \left(\frac{z}{c\tau + d}, \frac{a\tau + b}{c\tau + d} \right), \quad \gamma = \begin{pmatrix} a & b \\ c & d \end{pmatrix} \in \Gamma, \tag{46}$$

corresponding to a base change $2\pi i (\tau, 1) \mapsto 2\pi i (a\tau + b, c\tau + d)$ of Λ_τ followed by the homothety (conformal rescaling) $z \mapsto z/(c\tau + d)$. Then it follows that

$$\wp(\gamma(z, \tau)) = (c\tau + d)^2 \wp(z, \tau). \tag{47}$$

This says that $\wp(z, \tau)$ is *Jacobi form* of weight 2 [EZ], though we will neither explain nor pursue this idea here.

What we need is that $\wp(z, \tau)$ is invariant under $\tau \mapsto \tau + 1$ as well as $z \mapsto z + 2\pi i$ (from the elliptic property). It follows that $\wp(z, \tau)$ has a Fourier expansion

---

[6] Here and below, a prime appended to a summation indicates that terms rendering the sum meaningless, in this case $(m, n) = (0, 0)$, are to be omitted.

in both q and q_z. (Compare with the development in Section 3.1.) To describe this we define

$$P_1(z,\tau) = \sideset{}{'}\sum_{n\in\mathbb{Z}} \frac{q_z^n}{1-q^n} - \frac{1}{2}, \qquad (48)$$

$$P_2(z,\tau) = \frac{d}{dz} P_1(z,\tau) = \sideset{}{'}\sum_{n\in\mathbb{Z}} \frac{nq_z^n}{1-q^n}. \qquad (49)$$

The extra term $-\frac{1}{2}$ in (48) ensures that $P_1(z,\tau)$ is odd in z. For nonzero z in the fundamental parallelogram defined by the basis $(2\pi i, 2\pi i\tau)$ of Λ_τ we have $-2\pi\,\mathrm{Im}\,\tau < \mathrm{Re}\,z < 0$, so that $|q| < |q_z| < 1$. $P_1(z,\tau)$ and its z-derivatives are absolutely convergent in this domain. We can now give the Fourier expansion of the $\wp$-function, which reveals a fundamental relationship with the Eisenstein series of Section 3.2.

THEOREM 5.1. *We have*

$$\wp(z,\tau) = P_2(z,\tau) - E_2(\tau) = \frac{1}{z^2} + \sum_{k\geq 2}(2k-1)E_{2k}(\tau)z^{2k-2}.$$

We sketch the proof, which uses a key identity (cf. Exercise 5.3 below):

$$\sum_{n\in\mathbb{Z}} \frac{1}{(x-2\pi i n)^2} = \frac{q_x}{(1-q_x)^2} \qquad (50)$$

for $x \neq 0$. In the exceptional case,

$$\sideset{}{'}\sum_{n\in\mathbb{Z}} \frac{1}{(2\pi i n)^2} = \frac{2\zeta(2)}{(2\pi i)^2} = -\frac{1}{12}. \qquad (51)$$

Now

$$\wp(z,\tau) + \sum_{m\in\mathbb{Z}}\left(\sum_{n\in\mathbb{Z}} \frac{1}{\omega_{m,n}^2}\right) = \sum_{m\in\mathbb{Z}}\left(\sum_{n\in\mathbb{Z}} \frac{1}{(z-\omega_{m,n})^2}\right), \qquad (52)$$

where the convergent nested double sums depend on the order of summation. For the lhs, use (50) with $x = 2\pi i m\tau \neq 0$, (51) and $|q| < 1$ to obtain

$$\sum_{m\in\mathbb{Z}}\left(\sum_{n\in\mathbb{Z}} \frac{1}{\omega_{m,n}^2}\right) = -\frac{1}{12} + \sum_{0\neq m\in\mathbb{Z}} \frac{q^m}{(1-q^m)^2} = -\frac{1}{12} + 2\sum_{m,n>0} nq^{mn} = E_2(\tau)$$

(cf. (24)). For the rhs of (52), use (50) with $x = z - 2\pi i m\tau$ and argue similarly using $|q_z q^m|, |q_z^{-1} q^m| < 1$ for $m > 0$ to get

$$\sum_{m\in\mathbb{Z}} \frac{q_z q^m}{(1-q_z q^m)^2} = \frac{q_z}{(1-q_z)^2} + \sum_{m>0}\left(\frac{q_z q^m}{(1-q_z q^m)^2} + \frac{q_z q^{-m}}{(1-q_z q^{-m})^2}\right)$$

$$= \frac{q_z}{(1-q_z)^2} + \sum_{m>0}\sum_{n>0} n(q_z^n + q_z^{-n})q^{nm}$$

$$= \frac{q_z}{(1-q_z)^2} + \sum_{n>0} n\left(q_z^n + q_z^{-n}\right) \frac{q^n}{1-q^n}$$

$$= \sum_{n>0} n\left(\frac{q_z^n}{1-q^n} - \frac{q_z^{-n}}{1-q^{-n}}\right) = P_2(z,\tau). \tag{53}$$

This proves the first equality in the theorem. From (45) we see that

$$\wp(z,\tau) = \frac{1}{z^2} + \sum_{k\geq 3}(k-1)\tilde{E}_k(\tau) z^{k-2},$$

with

$$\tilde{E}_k(\tau) = \sideset{}{'}\sum_{m,n\in\mathbb{Z}} \frac{1}{\omega_{m,n}^k} = \frac{1}{(2\pi i)^k} \sideset{}{'}\sum_{m,n\in\mathbb{Z}} \frac{1}{(m\tau+n)^k}. \tag{54}$$

We can use (50) to identify $\tilde{E}_k$ with the corresponding Eisenstein series $E_k(\tau)$ (24), in particular $\tilde{E}_k(\tau)$ is identically zero for k odd. This completes our discussion of Theorem 5.1.

We note that P_1 is not an elliptic function (cf. Exercise 5.6). Higher z-derivatives $P_1^{(m)}(z,\tau)$ for $m \geq 1$ are elliptic functions, and are derivatives of $\wp(z,\tau)$ for $m \geq 2$. We have

$$P_1^{(m)}(z,\tau) = \sideset{}{'}\sum_{n\in\mathbb{Z}} \frac{n^m q_z^n}{1-q^n}$$

$$= m!\left(\frac{(-1)^{m+1}}{z^{m+1}} + \sum_{k\geq m+1}\binom{k-1}{m} E_k(\tau) z^{k-m-1}\right). \tag{55}$$

EXERCISE 5.2. Prove directly from the definition that an elliptic function which is *holomorphic* is necessarily constant.

EXERCISE 5.3. Verify (50) by comparing poles.

EXERCISE 5.4. Prove that for even $k \geq 4$, $\tilde{E}_k(\tau)$ coincides with $E_k(\tau)$. (Use (26).)

EXERCISE 5.5. Deduce from (47) that $E_k(\tau) \in \mathfrak{M}_k$ for even $k \geq 4$.

EXERCISE 5.6. Prove that $P_1(z + 2\pi i\tau, \tau) = P_1(z,\tau) - 1$.

5.2. 2-Point correlation functions.

Let $(V, Y, \mathbf{1}, \omega)$ be a VOA of central charge c. For an integer $n \geq 0$, the n-point correlation function for states $u^1, \ldots, u^n \in V$ is the formal expression

$$F_V((u^1, z_1), \ldots, (u^n, z_n), q)$$
$$= \text{Tr}_V Y(q_1^{L_0} u^1, q_1) \ldots Y(q_n^{L_0} u^n, q_n) q^{L_0 - c/24}, \quad (56)$$

where $q_i = q_{z_i}$ for variables $z_1, \ldots, z_n$. For $n = 0$ this reduces to the graded dimension $\text{Tr}_V q^{L_0 - c/24}$ as discussed in Section 4.2. If $n = 1$ and $u^1 \in V_k$, the expression in (56) equals

$$\text{Tr}_V Y(q_1^{L_0} u^1, q_1) q^{L_0 - c/24} = q_1^k \sum_m \text{Tr}_V u_m^1 q_1^{-m-1} q^{L_0 - c/24}$$
$$= \text{Tr}_V o(u^1) q^{L_0 - c/24} = Z(u^1, q),$$

where we used (36) to get the second equality. So for $n = 1$, (56) is the 1-point function of Section 4, as expected. There are similar modal expressions for all n-point functions, but for $n \geq 2$ they are unhelpful. Here we focus on the 2-point function

$$F_V((u, z_1), (v, z_2), \tau) = \text{Tr}_V Y(q_1^{L_0} u, q_1) Y(q_2^{L_0} v, q_2) q^{L_0 - c/24}. \quad (57)$$

We want to re-express the 2-point function as a *1-point function*, and for this we need be able to manipulate vertex operators. More precisely, we need to manipulate expressions involving vertex operators which are *traced* over V. In such a context the locality of operators (4) simplifies in the sense that

$$\text{Tr}_V Y(u, z_1) Y(v, z_2) q^{L_0} = \text{Tr}_V Y(v, z_2) Y(u, z_1) q^{L_0},$$

where the additional factor $(z_1 - z_2)^k$ (loc. cit.) has conveniently disappeared. Similar comments apply to the associativity formula (7), where we have

$$\text{Tr}_V Y(u, z_1 + z_2) Y(v, z_2) q^{L_0} = \text{Tr}_V Y(Y(u, z_1) v, z_2) q^{L_0}.$$

These and similar assertions fall under the heading of *duality* in CFT, which is discussed in [FHL]. We shall use them below without further comment. Thus with some changes of variables together with Exercise 4.2, we have

$$F_V((u, z_1), (v, z_2), \tau) = \text{Tr}_V Y(Y(q_1^{L_0} u, q_1 - q_2) q_2^{L_0} v, q_2) q^{L_0 - c/24}$$
$$= \text{Tr}_V Y(q_2^{L_0} Y((q_{z_{12}}^{L_0} u, q_{z_{12}} - 1) v, q_2) q^{L_0 - c/24}$$
$$= Z_V(Y[u, z_{12}] v, \tau), \quad (58)$$

where $z_{12} = z_1 - z_2$. This is the desired 1-point function. Similarly,

$$\begin{aligned}F_V((u,z_1),(v,z_2+2\pi i\tau),\tau) &= q^{-c/24}\operatorname{Tr}_V Y(q_1^{L_0}u,q_1)Y(q^{L_0}q_2^{L_0}v,qq_2)q^{L_0}\\ &= q^{-c/24}\operatorname{Tr}_V Y(q_1^{L_0}u,q_1)q^{L_0}Y(q_2^{L_0}v,q_2)\\ &= q^{-c/24}\operatorname{Tr}_V Y(q_2^{L_0}v,q_2)Y(q_1^{L_0}u,q_1)q^{L_0}\\ &= F_V((u,z_1),(v,z_2),\tau),\end{aligned}$$

Thus F_V is periodic in z_2 with period $2\pi i\tau$, and the same holds for z_1. It is obvious that F_V is also periodic in each z_i with period $2\pi i$. It follows that, at least formally, the 2-point function F_V (alias the 1-point function (58)) is elliptic in the variable z_{12} with period lattice Λ_τ.

5.3. First Zhu recursion formula We continue to pursue the ellipticity of the 2-point function F_V. It is the analytic of F_V which needs to be established. To this end we develop a recursion formula of Zhu (see [Z]), which finds a number of applications.

THEOREM 5.7. *We have*

$$\begin{aligned}&F_V((u,z_1),(v,z_2),\tau)\\ &= \operatorname{Tr}_V o(u)o(v)q^{L_0-c/24} - \sum_{m\geq 1}\frac{(-1)^m}{m!}P_1^{(m)}(z_{12},\tau)Z_V(u[m]v,\tau).\end{aligned} \quad (59)$$

The sum in (59) is finite since $u[m]v = 0$ for m sufficiently large, and from Section 5.1 $P_1^{(m)}(z_{12},\tau)$ is elliptic for $m \geq 1$. Thus the ellipticity of F_V is reduced to the *convergence* of $\operatorname{Tr}_V o(u)o(v)$ and the 1-point functions $Z_V(u[m]v,\tau)$. This Theorem makes clear the deep connection between elliptic functions (and therefore also modular forms) and VOAs. There is an analogous recursion for all n-point functions.

To prove Theorem 5.7 we may assume that $u \in V_k$, whence

$$F_V((u,z_1),(v,z_2),\tau) = \sum_{n\in\mathbb{Z}} q_1^{-n-1+k}\operatorname{Tr}_V\left(u_n Y(q_2^{L_0}v,q_2)q^{L_0-c/24}\right). \quad (60)$$

Using (36), Exercise 4.2 and (18) we have

$$\begin{aligned}[u_n, Y(q_2^{L_0}v,q_2)] &= \sum_{i\geq 0}\binom{n}{i}Y(u_i q_2^{L_0}v,q_2)q_2^{n-i} = q_2^r Y(q_2^{L_0}\sum_{i\geq 0}\binom{n}{i}u_i v,q_2)\\ &= q_2^r \sum_{m\geq 0}\frac{r^m}{m!}Y(q_2^{L_0}u[m]v,q_2),\end{aligned}$$

where $r = n + 1 - k$.

Hence

$$\operatorname{Tr}_V\left(u_n Y(q_2^{L_0}v, q_2) q^{L_0-c/24}\right)$$
$$= \operatorname{Tr}_V\left([u_n, Y(q_2^{L_0}v, q_2)] q^{L_0-c/24}\right) + \operatorname{Tr}_V\left(Y(q_2^{L_0}v, q_2) u_n q^{L_0-c/24}\right)$$
$$= q_2^r \sum_{m \geq 0} \frac{r^m}{m!} Z_V(u[m]v, \tau) + q^r \operatorname{Tr}_V\left(Y(q_2^{L_0}v, q_2) q^{L_0-c/24} u_n\right).$$

From this we obtain

$$q_2^r \sum_{m \geq 0} \frac{r^m}{m!} Z_V(u[m]v, \tau) = (1-q^r) \operatorname{Tr}_V\left(u_n Y(q_2^{L_0}v, q_2) q^{L_0-c/24}\right),$$

so that for $r \neq 0$ we have

$$\operatorname{Tr}_V\left(u_n Y(q_2^{L_0}v, q_2) q^{L_0-c/24}\right) = \frac{q_2^r}{1-q^r} \sum_{m \geq 1} \frac{r^m}{m!} Z_V(u[m]v, \tau).$$

Finally, the term corresponding to $r = 0$ in (60) is $\operatorname{Tr}_V o(u) o(v) q^{L_0-c/24}$. Substituting into (60), we find

$$F_V((u, z_1), (v, z_2), \tau)$$
$$= \operatorname{Tr}_V\left(o(u) o(v) q^{L_0-c/24}\right) + \sum_{m \geq 1} \frac{1}{m!} Z_V(u[m]v, \tau) \sum_{n \in \mathbb{Z}}' \frac{r^m q_{z_{12}}^{-n}}{1-q^n},$$

and the theorem follows upon comparison with (55).

EXERCISE 5.8. Let a be the generating state for the Heisenberg VOA M_0 (cf. Section 2.4). Prove that $F_{M_0}((a, z_1), (a, z_2), \tau) = P_2(z_{12}, \tau)/\eta(\tau)$.

EXERCISE 5.9. For states u, v in a VOA V, show that $Z_V(u[0]v, q) = 0$.

5.4. Second Zhu recursion formula. Theorem 5.7 allows us to obtain a related recursion formula for 1-point functions.

THEOREM 5.10. *For* $n \geq 1$,

$$Z_V(u[-n]v, \tau) = \delta_{n,1} \operatorname{Tr}_V(o(u) o(v) q^{L_0-c/24})$$
$$+ \sum_{m \geq 1} (-1)^{m+1} \binom{n+m-1}{m} E_{n+m}(\tau) Z_V(u[m]v, \tau). \quad (61)$$

To see this, note from (58) that

$$F_V((u, z_1), (v, z_2), \tau) = \sum_{n \in \mathbb{Z}} Z_V(u[-n]v, \tau) z_{12}^{n-1}.$$

Now compare this with the z_{12}-expansion of the rhs of (59) using (55). Taking $n \geq 1$ we obtain (61). (For $n \leq 0$ we get no information.)

One can apply Theorem 5.10 in a number of contexts. If we work with states $u, v, \ldots$ in V that are homogeneous with respect to the square bracket Virasoro operator $L[0]$, then the 1-point functions occurring on the rhs of (61) are those of states $u[m]v$ whose (square bracket) weight is *strictly less* than that of $u[-n]v$ for $n \geq 1$. Thus one might hope to proceed inductively (with respect to square bracket weights) to show that 1-point functions are holomorphic in $\mathfrak{H}$. To illustrate, we introduce the important class of VOAs V of *CFT-type* defined by the property that the zero weight space V_0 is *nondegenerate*, i.e., *spanned* by the vacuum vector. This implies (Exercise 2.29) that

$$V = \mathbb{C}\mathbf{1} \oplus V_1 \oplus \cdots \qquad (62)$$

Using Theorem 5.10 and the remarks following Theorem 5.7 we obtain:

LEMMA 5.11. *Suppose that V is a VOA of CFT-type, and let S be a generating set for V as in Theorem 2.7. Assume that $Tr_V o(u) o(v) q^{L_0 - c/24}$ is holomorphic in $\mathfrak{H}$ for all $u \in S$ and $v \in V$, and that the graded dimension $Z_V(\mathbf{1})$ is holomorphic in $\mathfrak{H}$. Then every 1-point function for V is holomorphic in $\mathfrak{H}$, and every 2-point function for V is elliptic.*

By way of example, consider the Heisenberg algebra M_0, which is certainly of CFT-type. It is generated by a single state a in weight 1, and $o(a) = 0$ (cf. Exercise 4.3). Furthermore $Z_{M_0}(\tau)$ is the inverse η-function (40) and hence holomorphic in $\mathfrak{H}$. So the conditions of the Lemma apply to M_0, so that all 1- and 2-point functions for M_0 have the desired analytic properties. Indeed, the vanishing of the zero mode for a means that in the recursion (61), the anomalous first term on the rhs is not present (taking $u = a$, as we may). We get a recursion for 1-point functions which may be solved with some effort, and this is how one proves Theorem 4.5 and (44). The details are described in Section 11.1.

EXERCISE 5.12. Give the details for the proof of Lemma 5.11.

EXERCISE 5.13. Show that the analysis of 1-point and 2-point functions associated to the Heisenberg algebra goes through with the same conclusions for the Virasoro algebra Vir_c.

Part II. Modular-invariance and rational vertex operator algebras

The representation theory of a VOA V, i.e., the study of V-modules and their characters (correlation functions) is fundamental. In this Section we introduce some of the ideas in this subject.

6. Modules over a vertex operator algebra

6.1. Basic definitions. Let $V = (V, Y, \mathbf{1}, \omega)$ be a VOA of central charge c. As one might expect, a V-module is (roughly speaking) linear space M admitting fields associated to states of V which satisfy axioms analogous to those satisfied by the fields $Y(v, z)$. It is useful to introduce various types of modules, the most basic of which is the following.

DEFINITION 6.1. A weak V-module is a pair (M, Y_M) where

$$Y_M : V \to \mathfrak{F}(M), \quad v \mapsto Y_M(v, z) = \sum_n v_n^M z^{-n-1} \text{ is a linear map,}$$

and the following hold for all $u, v \in V$, $w \in M$:

$$\text{vacuum} : Y_M(\mathbf{1}, z) = \text{Id}_M,$$
$$\text{locality} : Y_M(u, z) \sim Y_M(v, z),$$
$$\text{associativity} : \text{for large enough } k,$$
$$(z_1 + z_2)^k Y_M(u, z_1 + z_2) Y_M(v, z_2) w = (z_1 + z_2)^k Y_M(Y(u, z_1)v, z_2) w.$$

There is no notion of creativity or translation covariance *per se* for V-modules. It is not sufficient to assume only locality of operators here; the associativity axiom (the analog of (7)) is crucial. Locality and associativity are jointly equivalent to the analog of (6), namely

$$\sum_{i \geq 0} \binom{p}{i} (u_{r+i} v)_{p+q-i}^M = \sum_{i \geq 0} (-1)^i \binom{r}{i} \left(u_{p+r-i}^M v_{q+i}^M - (-1)^r v_{q+r-i}^M u_{p+i}^M \right).$$

As before, this is the modal version of the *Jacobi Identity*. For further details, see [FHL] and [LL]. A weak V-module is essentially a module for a vertex algebra.

DEFINITION 6.2. An *admissible* V-module is a weak V-module (M, Y_M) equipped with an $\mathbb{N}$-grading $M = \bigoplus_{n \geq 0} M_n$ such that

$$v \in V_k \Rightarrow v_n^M : M_m \to M_{m+k-n-1}. \tag{63}$$

Admissible modules are also called $\mathbb{N}$-gradable modules. Note that (63) is the analog of (36). There is *no* requirement that the homogeneous spaces M_n have finite dimension. An overall shift in the grading does not affect (63), so we may, and usually shall, assume that $M_0 \neq 0$ if $M \neq 0$. We then refer to M_0 as the *top level*.

DEFINITION 6.3. A V-module is a weak V-module (M, Y_M) equipped with a grading $M = \bigoplus_{\lambda \in \mathbb{C}} M_\lambda$ such that

$$\dim M_\lambda < \infty,$$
$$\forall \lambda, M_{\lambda+n} = 0 \text{ for } n \ll 0,$$
$$L_0 m = \lambda m, \ m \in M_\lambda.$$

We frequently call a V-module as in Definition 6.3 an *ordinary V-module* if we want to emphasize that it is not merely a weak or admissible module. There are containments

{weak V-modules} $\supseteq$ {admissible V-modules} $\supseteq$ {ordinary V-modules},

which amounts to saying that ordinary V-modules can be equipped with an $\mathbb{N}$-grading making them admissible (cf. Exercise 6.6). A (weak, admissible, or ordinary) V-module M is *irreducible* if no proper, nonzero subspace of V is invariant under all modes v_n^M. More generally, we can define *submodules* of M in the usual way, though we will not go much into this here.

A VOA V is *ipso facto* an ordinary V-module in which $Y = Y_M$. It is called the *adjoint* module. If the adjoint module is irreducible then we say that V is *simple*. This is consistent with standard algebraic usage: it can be shown using skew-symmetry that if $U \subseteq V$ is a submodule of the adjoint module V then U is a (2-sided) *ideal* in a natural sense, and that V/U has a well-defined structure of VOA. See Exercise 6.11 for further details.

We want to define the *partition function* and *character* of a V-module M along the lines of that for V itself, as discussed in Section 4. This makes no sense unless M is equipped with a suitable grading. An important case when this can be carried through is when M is an irreducible, ordinary V-module. In this case, if $\lambda \in \mathbb{C}$ satisfies $M_{\lambda+n} \neq 0$ for some integer n then $\bigoplus_{n \in \mathbb{Z}} M_{\lambda+n}$ is invariant under all modes v_n^M and hence coincides with M thanks to irreducibility. Relabeling, the grading on a (nonzero) irreducible, ordinary V-module M takes the shape

$$M = \bigoplus_{n \geq 0} M_{h+n}. \tag{64}$$

M_h is the *top level* and h a uniquely determined scalar called the *conformal weight* of M. It is an important numerical invariant of the module.

The *zero mode* $o^M(v)$ for $v \in V$ is the mode of $Y_M(v,z)$ which has weight zero as an operator on M; it is defined because of (63). We can now define the *character* Z_M of an irreducible V-module M of conformal weight h in the expected manner, namely

$$Z_M(v) = \mathrm{Tr}_M \, o^M(v) q^{L_0^M - c/24} = q^{h-c/24} \sum_{n \geq 0} \mathrm{Tr}_{M_{h+n}} o^M(v) q^n. \qquad (65)$$

The partition function of M is

$$Z_M(\mathbf{1}) = \mathrm{Tr}_M \, q^{L_0^M - c/24} = q^{h-c/24} \sum_{n \geq 0} \dim M_{h+n} q^n$$

where, naturally, L_0^M is the corresponding zero mode for the Virasoro element.

The set of *all* irreducible modules over the Heisenberg VOA M_0 is readily described. As usual, let a be the weight one state that generates M_0. In Section 2.4 we defined, for each $h \in \mathbb{C}$, the Verma module M_h and constructed a field $a(z) \in \mathfrak{F}(M_h)$. It is more precise to denote this by $a^h(z)$. Much as in the case $h = 0$, one finds that each M_h is an irreducible M_0-module of conformal weight h with $Y_{M_h}(a,z) = a^h(z)$. In particular, M_0 is a simple VOA. The Stone-von Neumann Theorem is essentially the converse: for each h, M_h is the unique (up to isomorphism[7]) irreducible module over M_0 of conformal weight h. See [FLM] for a proof. The construction of the Verma module M_h shows that

$$Z_{M_h}(\mathbf{1}) = q^h/\eta(q). \qquad (66)$$

The characters Z_{M_h} can be understood along the same lines as the special case of M_0 that we described in Sections 4 and 5. As illustrated by (66), results identical to Theorem 4.5 and (44) hold for M_h, except that an extra factor q^h must be included. Our development of the theory of 1- and 2-point functions may be carried out, with essentially no change, for general V-modules rather than just adjoint modules. It should be pointed out, however, that the extra factor *spoils* the quasimodularity of the character values.

While ordinary V-modules are perhaps natural, the reader may be wondering how and why admissible V-modules are relevant. Here we will limit ourselves here to a some general comments, and continue the discussion below. See Exercises 6.8-6.10 for some details, and [DLM3], [Z] for complete proofs. One considers certain subspaces $O_0 \subseteq O_1 \subseteq \cdots \subseteq V$, the quotient spaces $A_n(V) = V/O_n(V)$, and the inverse limit

$$A(V) = \varprojlim A_n(V).$$

---

[7] We have not defined morphisms of V-modules, but readers should be able to formulate it for themselves without diffuclty.

Each $A_n(V)$ has natural structure of *associative algebra* such that the canonical projection $A_{n+1} \to A_n(V)$ is an algebra morphism. So $A(V)$ is also an associative algebra. $A_0(V)$ is called the *Zhu algebra* of V.

The representation theory of these algebras is intimately related to that of V itself. There are functors

$$\Omega_n : \text{Adm } V\text{-Mod} \to A_n(V)\text{-Mod}$$

from the category of admissible V-modules to the category of $A_n(V)$-modules, and because of the details of the construction the quotient functor Ω_n/Ω_{n-1} makes sense (Ω_{-1} is trivial). For an admissible V-module M,

$$\Omega_n(M)/\Omega_{n-1}(M)$$

is an $A_n(V)$-module that is *not* the lift of an $A_{n-1}(V)$-module. $A_n(V)$ is designed in such a way that it acts naturally on the sum of the first n graded pieces of an admissible V-module, and this is how the functor Ω_n is defined. It turns out that there is another functor

$$L_n : A_n(V)\text{-Mod} \to \text{Adm } V\text{-Mod} \qquad (67)$$

which is a *right inverse* of the functor Ω_n/Ω_{n-1}, and which is harder to describe. This is a key point. It is the existence of L_n which motivates the introduction of admissible V-modules. L_n and Ω_n/Ω_{n-1} induce *bijections* between (isomorphism classes of) irreducible, admissible V-modules and irreducible $A_n(V)$-modules which are not lifts of $A_{n-1}(V)$-modules. For $n=0$, this is just the set of irreducible $A_0(V)$-modules. To a large extent these functors reduce the study of admissible V-modules to that of modules over the associative algebras $A_n(V)$, which are more familiar objects, and they have led to a number of theoretical advances. On the other hand, the computation of the Zhu algebra $A_0(V)$, not to mention the higher $A_n(V)$'s, is usually difficult. The complete structure has been elucidated in only a relatively few cases, and computer calculations have often been important.

Needless to say, there is much more that can be said about modules over a VOA. There is a notion of *dual* module ([B], [FHL] and Section 10.3). There is also a theory of *tensor products* of modules that is important. This is an extensive subject in its own right, and we can do no more than refer the reader to the literature (e.g., [HL], [HLZ]) for further details.

EXERCISE 6.4. Let (M, Y_M) be a weak V-module. Prove that $Y_M(L_{-1}v, z) = \partial Y_M(v, z)$.

EXERCISE 6.5. Show that

$$[L_m^M, L_n^M] = (m-n)L_{m-n}^M + (m^3 - m)/12\delta_{m,-n}c\text{Id}_M.$$

(Thus, a weak module for V is *ipso facto* a module over the Virasoro algebra with the *same* central charge as V.)

EXERCISE 6.6. Show that an ordinary V-module M is an admissible V-module as follows. Let $\Lambda \subseteq \mathbb{C}$ consist of those λ for which $M_{\lambda+k} = 0$ whenever k is a negative integer, and let $M_n = \bigoplus_{\lambda \in \Lambda} M_{\lambda+n}$. Show that $M = \bigoplus_{n \geq 0} M_n$ is an $\mathbb{N}$-grading on M satisfying (63).

EXERCISE 6.7. Give a complete proof that the Verma modules M_h are irreducible modules over the Heisenberg algebra M_0.

EXERCISE 6.8. Let M be an admissible V-module. Prove that for each $v \in V$, the zero mode $o^M(L[-1]v)$ annihilates M. (Use Exercises 6.4 and 2.36.)

EXERCISE 6.9. For $n \geq 0$, $u \in V_k$, $v \in V$ define

$$u \circ_n v = \mathrm{Res}_z Y(u,z) v \frac{(1+z)^{k+n}}{z^{2n+2}}.$$

Let $O_n(V)$ be the *span* of all states $u \circ_n v$ and $L[-1]u$.
(a) Prove that if $n = 0$, the span of the states $u \circ_0 v$ already contains $L[-1]u$.
(b) Prove that $O_0(V) \subseteq O_1(V) \subseteq \cdots$.

EXERCISE 6.10. With the notation of Exercise 6.9, introduce the product

$$u *_n v = \sum_{m=0}^{n} \binom{m+n}{n} \mathrm{Res}_z Y(u,z) v \frac{(1+z)^{k+n}}{z^{n+m+1}}.$$

(a) Show that $O_n(V)$ is a 2-sided ideal with respect to the product $*_n$.
(b) Show that $*_n$ induces a structure of associative algebra on the quotient space $A_n(V) = V/O_n(V)$.

EXERCISE 6.11. V is a VOA and $U \subseteq V$ a submodule of the adjoint module, so that $v_n u \in U$ for all $u \in U, v \in V, n \in \mathbb{Z}$. Prove that $u_n v \in V$, and deduce that if $U \neq V$ then V/U inherits the structure of VOA.

6.2. C_2-cofinite, rational and regular vertex operator algebras. We are going to focus on some important classes of VOAs V which have the property that they have only *finitely many* (inequivalent) irreducible modules. The reader might well be surprised that there are any such VOAs at all beyond those of finite dimension (cf. Exercises 2.12 and 2.13). We will also make the simplifying assumption that V is of *CFT-type* (62) throughout the rest of these notes, although for many of the results to be discussed this assumption is not necessary.

DEFINITION 6.12. (a) V is *rational* if every admissible V-module is completely reducible, i.e., a direct sum of irreducible, admissible V-modules.

(b) V is *regular* if every weak V-module is a direct sum of irreducible, ordinary V-modules.

(c) V is C_2-*cofinite* if the graded subspace $C_2(V) = \langle u_{-2}v \mid u, v \in V \rangle$ has finite codimension in V.

Based on what we said in the previous Subsection, it is easy to see that a regular VOA V is a rational VOA. Indeed, an admissible V-module is a weak module, hence a direct sum of irreducible, ordinary modules and *ipso facto* a direct sum of irreducible admissible modules. It is also known (see [ABD], [Li]) that regularity is *equivalent* to the conjunction of rationality and C_2-cofiniteness.

While (a) and (b) of Definition 6.12 both assert that certain module categories are semisimple, (c) is rather different. (a) and (b) are *external conditions* that can be difficult to verify, whereas (c) is an *internal* condition that is easier to deal with. On the other hand, regular VOAs have better modular invariance properties than those which are C_2-cofinite.

THEOREM 6.13. *Suppose that V is a C_2-cofinite VOA.*
(a) *Each $A_n(V)$ is finite-dimensional.*
(b) *Every weak V-module is an admissible module.*
(c) *V has only finitely many isomorphism classes of irreducible, admissible modules.*

Note that for a *finitely generated* VOA V, (b) is *equivalent* to C_2-cofiniteness.

For further discussion of (a), see [Z], [My1], [GN], [Bu]; (b) is proved in [My1]. The approach in [GN] produces a sort of weak analog of the PBW Theorem in Lie theory (cf. Appendix) which applies to weak modules. This idea is very useful, and is used in [ABD], [My1], [Bu] and elsewhere in the literature. (c) follows from (a) and the properties of the functors L_n and Ω_n discussed in Section 6.1.

The following omnibus result collects some of the main facts about rational VOAs.

THEOREM 6.14 [DLM1], [DLM3]. *Suppose that V is a rational VOA.*
(a) *$A_0(V)$ is semisimple.*
(b) *Each $A_n(V)$ is finite-dimensional.*
(c) *V has only finitely many isomorphism classes of irreducible, admissible V-modules,*
(d) *every irreducible, admissible V-module is an ordinary V-module.*

Note that (b) is *equivalent* to rationality (loc. cit.)

Whether a rational VOA is necessarily C_2-cofinite is presently one of the main open questions in the representation theory of VOAs. If this is so, then there would be no difference between rational and regular VOAs. In the early history of VOA theory it was possible to believe that rationality and C_2-cofiniteness were *equivalent*. That, however, has turned out to be a chimera. There are VOAs which are C_2-cofinite but have admissible (in fact ordinary) modules which are *not* completely reducible. These are *logarithmic field theories*, a name that we will justify in Section 9.

EXERCISE 6.15. Give two proofs that the Heisenberg VOA M_0 is *not* a rational VOA: (a) by using Theorem 6.14, and (b) by explicitly constructing an admissible M_0-module that is *not* completely reducible.

EXERCISE 6.16. For any VOA V, show that the quotient space $P(V) = V/C_2(V)$ carries the structure of a *Poisson algebra* in the following sense: the products $\{u, v\} = u_0 v, uv = u_{-1}v$ afflict $P(V)$ with (well-defined) structures of Lie algebra and commutative, associative algebra respectively, moreover $\{uv, w\} = u\{v, w\} + \{u, v\}w$.

EXERCISE 6.17. Calculate the Poisson algebra $P(M_0)$ associated to the Heisenberg VOA.

7. Examples of regular vertex operator algebras

It is time to describe some further examples of VOAs beyond the Heisenberg and Virasoro theories. In particular, we want to have available a selection of regular VOAs. Our examples are fairly standard, but require some effort to construct. For this reason, we will mainly limit ourselves to a description of the underlying Fock spaces and generating fields.

7.1. Vertex algebras associated to Lie algebras. The reader might want to look over Appendix 1 before reading this Subsection. We can construct a VOA from a pair $(\mathfrak{g}, (\,,\,))$ consisting of a Lie algebra $\mathfrak{g}$ equipped with a symmetric, invariant, bilinear form $(\,,\,) : \mathfrak{g} \otimes \mathfrak{g} \to \mathbb{C}$. The details amount to an elaboration of the case of the Heisenberg algebra discussed in Section 2.4, which is the 1-dimensional case. The *affine Lie algebra* or *Kac–Moody algebra* associated to $(\mathfrak{g}, (\,,\,))$ is the linear space

$$\hat{\mathfrak{g}} = \mathfrak{g} \otimes \mathbb{C}[t, t^{-1}] \oplus \mathbb{C}K = \bigoplus_n \mathfrak{g} \otimes t^n \oplus \mathbb{C}K$$

with brackets

$$[a \otimes t^m, b \otimes t^n] = [a, b] \otimes t^{m+n} + m(a, b)\delta_{m, -n} K, \quad (a, b \in \mathfrak{g}), \qquad [\hat{\mathfrak{g}}, K] = 0.$$

The element $\hat{\mathfrak{g}}$ has a triangular decomposition with $\hat{\mathfrak{g}}^{\pm} = \bigoplus_{\pm n > 0} \mathfrak{g} \otimes t^n$ and $\hat{\mathfrak{g}}^0 = \mathfrak{g} \oplus \mathbb{C}K$. Here and below, we identify $\mathfrak{g}$ with $\mathfrak{g} \otimes t^0$. Fix a $\mathfrak{g}$-module X and a scalar l. We extend X to a $\hat{\mathfrak{g}}^+ \oplus \hat{\mathfrak{g}}^0$-module by letting $\hat{\mathfrak{g}}^+$ *annihilate* X

and letting K act as multiplication by l called the *level*. We have the induced $\hat{\mathfrak{g}}$-module

$$V_{\mathfrak{g}}(l, X) = \mathrm{Ind}(X) \tag{68}$$

(notation as in (208)). Following the Heisenberg case discussed in Section 2.4, we can define fields on $V_{\mathfrak{g}}(l, X)$ for each $a \in \mathfrak{g}$ by setting

$$Y_{V_{\mathfrak{g}}(l,X)}(a, z) = \sum_n a_n z^{-n-1}$$

where a_n is the induced action of $a \otimes t^n$. As in (11) we obtain

$$\sum_{j=0}^{2} (-1)^j \binom{2}{j} [a_{2-j-r}, b_{j-s}]$$
$$= \sum_{j=0}^{2} (-1)^j \binom{2}{j} ([a,b]_{2-r-s} + (2-j-r)(a,b)l\delta_{2-j-r,s-j}\mathrm{Id})$$
$$= ((2-r) - 2(1-r) - r)(a,b)l\delta_{r+s,2}\mathrm{Id} = 0,$$

so that the fields $\{Y_{V_{\mathfrak{g}}(l,X)}(a, z) \mid a \in \mathfrak{g}\}$ are mutually local of order two. Taking $X = \mathbb{C}\mathbf{1}$ to be the *trivial* 1-dimensional $\mathfrak{g}$-module, one shows via Theorem 2.7 that the corresponding fields generate a vertex algebra with Fock space $V_{\mathfrak{g}}(l, \mathbb{C}\mathbf{1})$. Moreover, each $V_{\mathfrak{g}}(l, X)$ is an admissible module.

To describe a conformal vector in $V_{\mathfrak{g}}(l, \mathbb{C}\mathbf{1})$ and thereby obtain the structure of VOA, it is convenient at this point to specialize to the case that $\mathfrak{g}$ is a *finite-dimensional, simple Lie algebra* of dimension d, say. We will also take $(\,,\,)$ to be the *Killing form*, appropriately normalized.[8] Note that this takes us out of the regime of the Heisenberg theory, to which we return in Section 7.3. An approach that covers both cases is described in [LL]. With our assumptions, one shows that

$$\omega = \frac{1}{2} \frac{1}{l+h^\vee} \sum_{i=1}^{d} u_i(-1)u^i \tag{69}$$

is the desired conformal vector with central charge $c = ld/(l+h^\vee)$. Here, $\{u_i\}$ is a basis of $\mathfrak{g}$, $\{u^i\}$ the basis dual to $\{u_i\}$ with respect to the form $(\,,\,)$, and $h^\vee$ the *dual Coxeter number* of $\mathfrak{g}$. This is usually called the *Sugawara construction*. Needless to say, we must also assume that $l + h^\vee \neq 0$.

The L_0-grading on $V_{\mathfrak{g}}(l, \mathbb{C}\mathbf{1})$ that obtains from the Sugawara construction is the natural one in which the state $a_n\mathbf{1}$ has weight $-n$ for $a \in \mathfrak{g}$ and $n \leq 0$. In particular the zero weight space is $V_{\mathfrak{g}}(l, \mathbb{C}\mathbf{1})_0 = \mathbb{C}\mathbf{1}$, and the VOA is of CFT-type. Because an ideal in the adjoint module is a graded submodule (cf. the discussion

---

[8] The normalization is an important detail, of course, but we will not need it.

in Section 6.1), any *proper* ideal necessarily lies in $\bigoplus_{n\geq 2} V_{\mathfrak{g}}(l, \mathbb{C}\mathbf{1})_n$. It follows that there is a *unique maximal* proper ideal, call it J, and the quotient space

$$L_{\mathfrak{g}}(l, 0) = V_{\mathfrak{g}}(l, \mathbb{C}\mathbf{1})/J$$

is a simple VOA.

More generally, take X to be a finite-dimensional irreducible $\mathfrak{g}$-module. As such it is a highest-weight module $L(\lambda)$ indexed by an element λ in the weight lattice of $\mathfrak{g}$. The top level of $V_{\mathfrak{g}}(l, L(\lambda))$ is naturally identified with $L(\lambda)$, and because this is an irreducible $\mathfrak{g}$-module then there is a unique maximal proper submodule $J \subseteq V_{\mathfrak{g}}(l, L(\lambda))$ (considered as $\hat{\mathfrak{g}}$-module). The quotient spaces

$$L_{\mathfrak{g}}(l, \lambda) = V_{\mathfrak{g}}(l, L(\lambda))/J$$

are ordinary, irreducible $V_{\mathfrak{g}}(l, \mathbb{C}\mathbf{1})$-modules, and they are inequivalent for distinct choices of highest weight λ. Thus the VOA $V_{\mathfrak{g}}(l, \mathbb{C}\mathbf{1})$ has infinitely many inequivalent ordinary, irreducible modules, and in particular it cannot be rational (Theorem 6.14). Concerning the question of regularity of these VOAs, we collect the main facts ([FZ], [DL], [DM1], [DLM2], [DLM4]):

THEOREM 7.1. *Let $\mathfrak{g}$ be a finite-dimensional simple Lie algebra. The simple VOA $L_{\mathfrak{g}}(l, 0)$ is rational if, and only if, l is a positive integer. In this case it is regular, and the ordinary, irreducible modules are the spaces $L_{\mathfrak{g}}(l, \lambda)$ where λ satisfies $\lambda(\theta) \leq l$ and θ is the longest positive root.*

These theories are called WZW models in the physics literature.

7.2. Discrete series Virasoro algebras. Here we discuss some quotients of Virasoro VOAs Vir_c (cf. Theorems 2.20 and 2.25) that turn out to be regular. As in the last Subsection, it is the underlying Lie structure that makes the calculations manageable. The details are quite different, however, and depend on the *Kac determinant* (e.g. [KR]) and the structure of the Verma modules (13) $M_{c,h}$ over the Virasoro algebra (these are Vir_c-modules) ([FF]). There is no space to describe these results systematically here, although we discuss some examples of Kac determinants in Subsection 10.4. So we give less detail compared to the WZW models. The theories we are going to describe in this Subsection find important applications in the physics of phase transitions and critical phenomena. See [FMS] for further background.

The Virasoro VOA Vir_c may, or may not, be a simple VOA, but there is a *unique* maximal proper submodule J and $L_c = \text{Vir}_c/J$ is a simple vertex operator algebra of central charge c. It turns out that Vir_c is *never* rational (cf. Exercise 7.4). As for the rationality of L_c, we have the following omnibus result:

THEOREM 7.2. *The following are equivalent*:
(a) L_c *is a rational VOA*.
(b) $J \neq 0$.
(c) *c lies in the so-called discrete series, i.e., there are coprime integers* $p, q \geq 2$ *such that*
$$c = c_{pq} = 1 - \frac{6(p-q)^2}{pq}. \tag{70}$$
In this case L_c *is regular, the conformal weights of the ordinary irreducible modules are*
$$h_{r,s} = \frac{(pr-qs)^2 - (p-q)^2}{4pq}, \quad 1 \leq r \leq q-1, \ 1 \leq s \leq p-1$$
(taking only one value of h for each pair $h_{r,s}, h_{q-r,p-s}$*), and two ordinary irreducible modules are isomorphic if, and only if, they have the same conformal weight.*[9] *Thus there are just* $(p-1)(q-1)/2$ *inequivalent ordinary irreducible modules over* L_c.

See [Wa] for the proof of rationality (also [DMZ]), where the idea is to compute the Zhu algebra $A_0(L_c)$. Regularity is shown in [DLM2]. The origin of the values $c_{p,q}$ is discussed in Section 10.4

Apart from the trivial case when $p = 2$, $q = 3$, the two 'smallest' cases, i.e., those with the fewest number of ordinary irreducible modules, correspond to $(p, q) = (2, 5)$ and $(3, 4)$. In the first case (the *Yang–Lee* model in physics) we have $c = -22/5$ and conformal weights $0, -1/5$. In the second case (the *Ising model*) $c = 1/2$ with conformal weights $0, 1/2, 1/16$.

EXERCISE 7.3. Prove that Vir_c has a unique maximal proper submodule J.

EXERCISE 7.4. Suppose that $J = 0$. Prove that Vir_c is *not* a rational VOA.

EXERCISE 7.5. Opine on the statement that the case $p = 2, q = 3$ is 'trivial'.

7.3. Lattice theories. Lattice theories ([B], [FLM]) are VOAs whose connections with Lie algebras are of lesser importance compared to the examples in the last two subsections. Their basic properties are of a more combinatorial nature, and reflect features that one may expect in general rational theories. Because of this and the fact that they are amenable to computation, lattice theories occupy a central position in current VOA theory.

Let d be a positive integer and $\mathfrak{h} = \mathbb{C}^d$ a rank d linear space equipped with a nondegenerate symmetric bilinear form $(\ ,\ )$. Consideration of $\mathfrak{h}$ as an *abelian* Lie algebra leads to the affine algebra $\hat{\mathfrak{h}}$ as in Section 7.1. Let
$$M_0^d = V_\mathfrak{h}(1, \mathbb{C}\mathbf{1}) \tag{71}$$

---
[9] Generally, a VOA may have inequivalent irreducible modules with the *same* conformal weight.

be the corresponding vertex algebra of *level* 1. The conformal vector ω is defined as in (69) with $l = h^\vee = 1$. The resulting VOA has central charge $c = d$. This is nothing more than a slightly different approach to the *rank d Heisenberg VOA*, as discussed in Section 2.4 (cf. Exercise 7.7).

The irreducible $\mathfrak{h}$-modules are 1-dimensional and indexed by a weight in the dual space of $\mathfrak{h}$. Identifying $\mathfrak{h}$ with its dual via (,), we obtain M_0^d-modules (68) with underlying linear space

$$V_\mathfrak{h}(1,\alpha) = S(\hat{\mathfrak{h}}^-) \otimes e^\alpha \ (\alpha \in \mathfrak{h}).$$

Here $\mathbf{1} \otimes e^\alpha$ (or just e^α) is notation for the spanning vector of the (1-dimensional) top level of $V_\mathfrak{h}(1,\alpha)$, and

$$\beta \otimes e^\alpha = \beta_0.e^\alpha = (\beta,\alpha)\mathbf{1} \otimes e^\alpha, \ \beta \in \mathfrak{h} = \mathfrak{h} \otimes t^0. \tag{72}$$

In order to describe the Fock spaces of lattice theories we need a bit more structure. Namely, we assume that $(\mathfrak{h},(\,,\,))$ is the scalar extension of a Euclidean space. Thus, $E = \mathbb{R}^d = \mathfrak{h}_\mathbb{R}$ is a real space equipped with a *positive-definite quadratic form* $Q : E \to \mathbb{R}$, $\mathfrak{h} = \mathbb{C} \otimes_\mathbb{R} E$, and $(\,,\,)$ is the $\mathbb{C}$-linear extension of the bilinear form on E defined by Q, also denoted by $(\,,\,)$. In particular, $Q(\alpha) = (\alpha,\alpha)/2$ for $\alpha \in E$. A *lattice* $L \subseteq E$ is the additive subgroup spanned by a *basis* of E. L is an *even* lattice if $(\alpha,\alpha) \in 2\mathbb{Z}$ for all $\alpha \in L$, i.e., the restriction of Q to L is integral.

For an even lattice $L \subseteq E$ we introduce the linear space

$$V_L = \bigoplus_{\alpha \in L} V_\mathfrak{h}(1,\alpha). \tag{73}$$

Identifying $\bigoplus_\alpha \mathbb{C}e^\alpha$ with the *group algebra*[10] $\mathbb{C}[L]$ of the lattice, we can write (73) more compactly as

$$V_L = S(\hat{\mathfrak{h}}^-) \otimes \mathbb{C}[L]. \tag{74}$$

There is a natural grading on V_L that turns out to be the one defined by the L_0 operator. We take the tensor product grading on (74) in which $S(\hat{\mathfrak{h}}^-)$ has the grading of the Fock space of the rank d Heisenberg algebra that it is, and where e^α has weight $Q(\alpha)$. Using (41), the partition function of V_L is

$$Z_{V_L}(\mathbf{1}) = \frac{\sum_{\alpha \in L} q^{Q(\alpha)}}{\eta(q)^d}. \tag{75}$$

The numerator here is the *theta function* of L, a topic to which we shall return in Section 8.

So far then, we have described the Fock space V_L as a sum of Heisenberg modules. We define $Y(v,z)$ for $v \in M_0^d$ to be the operator whose restriction to

---

[10] We only explicitly use the linear structure of $\mathbb{C}[L]$, although the algebra structure also plays a rôle.

$V_{\mathfrak{h}}(1,\alpha)$ is just $Y_{V_{\mathfrak{h}}(1,\alpha)}(v,z)$. In order to impose the structure of VOA on V_L, we must construct fields for all of the states in the Fock space (74). Because of Theorem 2.7 it suffices to define $Y(e^\alpha, z)$ for $\alpha \in L$ and establish locality, but nothing that has come so far has prepared us for this. The generating fields we have considered in detail for the Heisenberg, WZW and Virasoro theories have modes a_n that are closely related to some Lie algebra, but in theories such as V_L this will generally not be the case. We content ourselves with the prescription for $Y(e^\alpha, z)$, referring the reader to [FLM], [K1] for further background and motivation:

$$Y(e^\alpha, z) = \exp\left(\sum_{n>0} \frac{\alpha_{-n}}{n} z^n\right) \exp\left(\sum_{n<0} \frac{\alpha_{-n}}{n} z^n\right) e^\alpha z^\alpha. \qquad (76)$$

Beyond the modes α_n of $Y(\alpha, z)$, z^α is a shift operator $z^\alpha : v \otimes e^\beta \mapsto z^{(\alpha,\beta)} v \otimes e^\beta$ ($v \in \hat{\mathfrak{h}}^-$), and $e^\alpha : v \otimes e^\beta \mapsto \varepsilon(\alpha, \beta) v \otimes e^{\alpha+\beta}$ for a certain bilinear 2-cocycle $\varepsilon : L \otimes L \to \{\pm 1\}$ (loc. cit.)

The ordinary, irreducible modules over V_L are constructed in [Do]. The underlying Fock spaces are very similar to (74), and are indexed by the *cosets* of L in its $\mathbb{Z}$-dual L^0 (cf. Exercise 7.13). Precisely, they are

$$V_{L+\lambda} = \bigoplus_{\alpha \in L} V_{\mathfrak{h}}(1, \alpha + \lambda) = \hat{\mathfrak{h}}^- \otimes \mathbb{C}[L + \lambda]$$

for $\lambda \in L^0$, with partition functions

$$Z_{V_{L+\lambda}}(\mathbf{1}) = \frac{\sum_{\alpha \in L} q^{Q(\alpha+\lambda)}}{\eta(q)^d}. \qquad (77)$$

The fields $Y_{V_{L+\lambda}}(v,z)$ are similarly analogous to (76) (loc. cit.) Indeed, one can usefully combine *all* of these fields and Fock spaces into a bigger and better edifice. For this, see [DL]. For rationality and C_2-cofiniteness, see [Do] and [DLM4] respectively. Summarizing,

THEOREM 7.6. *Let L be an even lattice. Then V_L is a regular VOA, and its ordinary, irreducible modules are the Fock spaces $V_{L+\lambda}$. It thus has just $|L^0 : L|$ distinct ordinary, irreducible modules.*

In the following exercises, $L \subseteq E$ is an even lattice in Euclidean space as above.

EXERCISE 7.7. Show that the VOA (71) is *isomorphic* to the tensor product $M_0^{\otimes d}$ of d copies of the Heisenberg VOA M_0 (cf. Exercise 2.32).

EXERCISE 7.8. Show that M_0^d is a *simple* VOA.

EXERCISE 7.9. Verify that if $\alpha \in L$ then $\mathbf{1} \otimes e^\alpha$ has L_0-weight $Q(\alpha)$.

EXERCISE 7.10. In the definition of V_L, what is the purpose of requiring L to be an *even* lattice? What about positive-definiteness?

EXERCISE 7.11. Let $\alpha \in L$.
(a) Prove that $Y(e^\alpha, z)$ is a creative field in $\mathfrak{F}(V_L)$.
(b) Prove that $Y(e^\alpha, z)$ and $Y(v, z)$ are mutually local $(v \in \hat{\mathfrak{h}}^-)$.

EXERCISE 7.12. Let L be an even lattice with $L_0 = \{\alpha \in L \mid Q(\alpha) = 1\}$. Prove that L_0 is a *semisimple root system* with components of type ADE.

EXERCISE 7.13. The *dual lattice* of L is defined via
$$L^0 = \{\beta \in E \mid (\alpha, \beta) \in \mathbb{Z} \text{ for all } \alpha \in L\}.$$
Prove that $L \subseteq L^0$ is a subgroup of *finite index*.

EXERCISE 7.14. Let $\mathfrak{g}$ be a finite-dimensional simple Lie algebra of type ADE.
(a) Show that the WZW model $L_\mathfrak{g}(1, 0)$ of level 1 is (isomorphic to) the lattice theory V_L where L is the root lattice associated to $\mathfrak{g}$.
(b) Compute the number of inequivalent ordinary, irreducible modules over $L_\mathfrak{g}(1, 0)$ both by using Theorem 7.1, and by using Theorem 7.6.

EXERCISE 7.15. Let L_1, L_2 be a pair of even lattices.
(a) Show that the *orthogonal direct sum* $L_1 \perp L_2$ is an even lattice.
(b) Prove that $V_{L_1 \perp L_2} \cong V_{L_1} \otimes V_{L_2}$ (cf. Exercise 2.32).

8. Vector-valued modular forms

In order to formulate *modular invariance* for C_2-cofinite and regular VOAs, the idea of a *vector-valued modular form* is useful. This generalizes the theory of modular forms that we discussed in Section 3, and includes as a special case the theory of modular forms on a finite-index subgroup of $\mathrm{SL}_2(\mathbb{Z})$. We use the notation of Section 3.

8.1. Basic definitions. Fix an integer k and let $\mathfrak{F}_k$ be the space of holomorphic functions[11] in $\mathfrak{H}$ regarded as a right Γ-module with respect to the action defined in (20), (21). A weak vector-valued modular form of weight k may be taken to be a finite-dimensional Γ-submodule $V \subseteq \mathfrak{F}_k$. Let[12] $F(\tau) = (f_1(\tau), \ldots, f_p(\tau))^t$ where the component functions $f_i(\tau)$ are a set of (not necessarily linearly independent) generators for V. There is then a representation $\rho : \Gamma \to GL_p(\mathbb{C})$ such that
$$\rho(\gamma) F(\tau) = F|_k \gamma(\tau), \quad \gamma \in \Gamma, \tag{78}$$

---

[11] We could equally well deal with *meromorphic functions*.

[12] Superscript t denotes *transpose*.

where $|_k$ is the obvious extension of the stroke operator to vector-valued functions. We also call the pair (F, ρ) a weak vector-valued modular form of weight k. Given a pair (F, ρ) satisfying (78), we recover V as the span of the component functions of $F(\tau)$. The classical modular forms of Section 3 correspond to the case when ρ is the trivial 1-dimensional representation of Γ.

To describe the extension of (22) to the vector-valued case, decompose V into a direct sum of T-invariant indecomposable subspaces

$$V = V_1 \oplus \cdots \oplus V_r$$

corresponding to the Jordan decomposition of the action $T : f(\tau) \mapsto f(\tau + 1)$. The characteristic polynomial on V_i is $(x - e^{2\pi i \mu_i})^{\dim V_i}$. The basic fact is

THEOREM 8.1. *There are q-expansions $g_j(\tau) = q^{\mu_i} \sum_{n \in \mathbb{Z}} a_{ijn} q^n$, $(0 \leq j \leq n_i - 1)$ such that the functions*

$$g_0(\tau) + g_1(\tau) \log q + \cdots + g_m(\tau)(\log q)^m, \quad 0 \leq m \leq n_i - 1, \qquad (79)$$

are a basis of V_i. In particular, V has a basis of functions of this form.

We call (79) a *logarithmic*, or *polynomial*,[13] q-expansion.

Suppose that (F, ρ) is a weak vector-valued modular form. Then the component functions of $F(\tau)$ are linear combinations of polynomial q-expansions (79). We say that (F, ρ), or simply $F(\tau)$, is *almost holomorphic* if the component functions are holomorphic in $\mathfrak{H}$ and if the q-expansions $g_j(\tau)$ are *left-finite* or *meromorphic at ∞*, i.e., for all i, j the Fourier coefficients a_{ijn} vanish for $n \ll 0$. Similarly, $F(\tau)$ is *holomorphic* if it is almost holomorphic and if $a_{ijn} = 0$ whenever $\mathrm{Re}(\mu_i) + n < 0$. These definitions are independent of the choice of $g_j(\tau)$.

Fix an integer $N \geq 1$. We set

$$\Delta(N) = \langle \gamma T^N \gamma^{-1} \mid \gamma \in \Gamma \rangle.$$

This is the smallest normal subgroup of Γ that contains T^N. We say that a subgroup $G \subseteq \Gamma$ has level N if $\Delta(N) \subseteq G$. A representation $\rho : \Gamma \to GL_p(\mathbb{C})$ has *level N* if $\ker \rho$ has level N (equivalently, $\rho(T)$ has finite order dividing N). A vector-valued modular form (F, ρ) has level N if ρ has level N. Now recall that finite-order operators are *diagonalizable*. It follows from Theorem 8.1 that if (F, ρ) has level N then the component functions of $F(\tau)$ have q-expansions that are free of logarithmic terms. Indeed, the eigenvalues of $\rho(T)$ are N-th. roots of unity, so that the q-expansions (79) reduce to a single q-expansion of the form

$$g_j(\tau) = q^{r/N} \sum_{n \geq 0} a_{jn} q^n \qquad (80)$$

---

[13] We may rewrite (79) using powers of τ, or other polynomials in τ, instead of powers of $\log q$.

for some integer r.

The *principal congruence subgroup of level N* is the subgroup of Γ given by

$$\Gamma(N) = \{\gamma \in \Gamma \mid \gamma \equiv I_2 \pmod{N}\}.$$

We have $\Delta(N) \trianglelefteq \Gamma(N) \trianglelefteq \Gamma$. While $\Gamma(N)$ always has finite index in Γ, $\Delta(N)$ has finite index if, and only if $N \leq 5$ ([KLN], [Wa]). A subgroup $G \subseteq \Gamma$ is a *congruence subgroup* if $\Gamma(N) \subseteq G$ for some N; ρ and (F, ρ) are called *modular* if kerρ is a congruence subgroup. It follows that (F, ρ) is modular if, and only if, the component functions $g_j(\tau)$ of $F(\tau)$ are such that $g_j|_k\gamma(\tau)$ has a q-expansion of shape (80) for every $\gamma \in \Gamma$. This is precisely the definition of a *classical modular form* of weight k and level N (we are assuming holomorphy in $\mathfrak{H}$ for convenience). The case of level 1 again reduces to the theory discussed in Section 3.

Because $\Gamma(N)$ has finite index in Γ it follows that the *image* $\rho(\Gamma)$ is *finite* whenever ρ is modular. However, the *converse* is *false*: it may be that the image $\rho(\Gamma)$ is *finite*, so that kerρ has finite index in Γ and therefore has some finite level, yet it is not a congruence subgroup. The existence of such subgroups goes back to Klein and Fricke. In this case, a vector-valued modular form (F, ρ) will have some finite level N and its component functions have q-expansions (80), however not all of them will be classical modular forms in the previous sense. This is essentially the theory of modular forms on *noncongruence subgroups*. Modular forms on noncongruence subgroups, and more generally component functions of vector valued modular forms, share many properties in common with classical modular forms and the differences between them can be subtle. It can be difficult to determine whether a given vector-valued modular form (F, ρ) is modular. A fundamental problem in this direction is the following:

Conjecture: Let (F, ρ) be a vector-valued modular form of level N and weight k, and suppose that the component functions of $F(\tau)$ are linearly independent[14] *and have rational integers Fourier coefficients. Then (F, ρ) is modular.*

We shall see how this fits into VOA theory in the next Section.

EXERCISE 8.2. Prove the following: (a) $\Delta(N) \subseteq \Gamma(N) \trianglelefteq \Gamma$, (b) if $G \subseteq \Gamma$ is a subgroup of finite index then $\Delta(N) \subseteq G$ for some N.

EXERCISE 8.3. Let $\rho \colon \Gamma \to GL_p(\mathbb{C})$ be a representation of level N. Show that ρ is modular if, and only if, $\Gamma(N) \subseteq$ kerρ.

EXERCISE 8.4. : Let $\tilde{\Gamma}$ be the inhomogeneous modular group (Exercise 3.1) and let $\tilde{\Gamma}(N)$ be the image of $\Gamma(N)$ under the natural projection $\Gamma \to \tilde{\Gamma}$. Prove that $\tilde{\Gamma}(N)$ is *torsion-free* if, and only if, $N \geq 2$.

---

[14]This condition is harmless in practice, but is necessary to avoid trivial counterexamples, e.g. when $F = 0$.

EXERCISE 8.5. It is known that Γ can be abstractly defined by generators and relations $\langle x, y \mid x^4 = y^6 = x^2 y^{-3} = 1 \rangle$. Use this to prove the following: (a) $\Gamma/\Gamma' \cong \mathbb{Z}_{12}$, (b) Γ' is a congruence subgroup of level 12. (Γ' is the *commutator subgroup* of Γ.)

EXERCISE 8.6. Let $V \subseteq \mathfrak{F}_k$ be a finite-dimensional Γ-submodule and let $(f_1, \ldots, f_p)$ be a sequence of functions in V that contains a basis. Prove the existence of a representation ρ satisfying (78). (Hint: first do the case that $(f_1, \ldots, f_p)$ is a linearly independent set.)

8.2. Examples of vector-valued modular forms. One can construct a slew of almost holomorphic vector-valued modular forms using *modular linear differential equations* (MLDE) [M]. We briefly explain this. Let k, n be integers with n positive. The n-th iterate D_k^n of the differential operator (30) is the intertwining map

$$D_k^n = D_{k+2n-2} \circ \cdots \circ D_{k+2} \circ D_k : \mathfrak{F}_k \to \mathfrak{F}_{k+2n}.$$

For justification of the notation, see Exercise 3.13. A modular linear differential equation is a differential equation of the form

$$(D_k^n + g_2(\tau) D_k^{n-2} + \cdots + g_{2n}(\tau)) f = 0, \quad g_i(\tau) \in \mathfrak{M}_{2i}. \tag{81}$$

Using (30) one can write (81) as an ordinary differential equation with coefficients in the algebra of quasimodular forms $\mathfrak{Q}$. We can also write everything in terms of the variable q (in the interior of the unit disk in the q-plane)

$$(\theta^n + h_1(q) \theta^{n-1} + \cdots + h_{2n}(q)) f = 0, \quad h_i(q) \in \mathfrak{Q}_{2i}, \tag{82}$$

where we recall that $\theta = q\, d/dq$. Then one sees that $q = 0$ is a regular singular point ([H], [I]). By the theory of ODE, the space of solutions is an n-dimensional linear space, and because the coefficients are holomorphic in $\mathfrak{H}$, so too are the solutions. One sees that the space of solutions is a Γ-submodule of $\mathfrak{F}_{k+2n}$, and the theory of Frobenius–Fuchs (loc. cit.) shows that the solutions have q-expansions which are meromorphic at ∞ in the sense of Section 8.1. A disadvantage of this approach is that it is hard to get information about the representation of Γ furnished by the space of solutions.

We have seen that vector-valued modular forms naturally incorporate the classical theory of level N modular forms. We complete this Subsection with a discussion of an important class of such forms, namely *theta functions*. Let L be an even lattice of rank d with associated positive-definite quadratic form Q (Section 7.3). The theta function of L is defined by

$$\theta_L(\tau) = \sum_{\alpha \in L} q^{Q(\alpha)} = \sum_{n \geq 0} |L_n| q^n$$

where $L_n = \{\alpha \in L \mid Q(\alpha) = n\}$ (cf. (75)). Hecke and Schoeneberg proved ([O], [Se], [Sc]) that if d is *even* then $\theta_L(\tau)$ is a holomorphic modular form of weight $d/2$ and a certain level N. A precise description of the level would take us too far afield, but it divides twice the exponent of the finite abelian group L^0/L (cf. Exercise 7.13). In particular, suppose that L is *self-dual* in the sense that $L = L^0$. Then the level is 1, and as we explained this means that $\theta_L(\tau)$ is a holomorphic modular form of weight $d/2$ on the full group Γ.

There are various ways to prove the modularity of $\theta_L(\tau)$. One method that is useful in many other contexts is that of *Poisson summation* ([O], [Se]). The approach in ([Sc]) shows that the space spanned by the theta functions corresponding to the *cosets* of L in L^0, i.e., the numerators of the expressions on the rhs of (77), is a Γ-submodule of $\mathfrak{F}_{d/2}$. Note that the theta functions of such cosets arise as the numerator in the expression (77) of the character of an ordinary, irreducible module over a lattice VOA.

The reader may be wondering about the case when the rank d of L is odd. One still has holomorphic theta functions as above, however they are of *half-integral weight* and do not qualify as modular forms as we have defined them. Odd powers of the eta function also have half-integral weight. These and other examples demonstrate the significance of half-integer weight (vector-valued) modular forms to our subject, but there is no time to develop the subject here.

EXERCISE 8.7. Use your knowledge of the theory of ODEs to verify the details of the assertions following (81) leading to the result that the solution space is a Γ-submodule of $\mathfrak{F}_{k+2n}$.

EXERCISE 8.8. Why is there no term $g_1(\tau)D_k^{n-1}$ in (81)?

EXERCISE 8.9. For a positive-definite, even lattice L of rank d, prove the estimate $|L_n| = O(n^{d/2})$, and deduce that $\theta_L(\tau)$ is *holomorphic* in $\mathfrak{H}$.

EXERCISE 8.10. Show that E_8 is the *only* finite dimensional simple Lie algebra whose root lattice is even and self-dual.

EXERCISE 8.11. Show that the theta function $\theta_{E_8}(\tau)$ of the E_8 root lattice coincides with the Eisenstein series Q of Section 3.

EXERCISE 8.12. Show that the partition functions of a lattice theory V_L and its ordinary, irreducible modules are *classical, almost holomorphic, modular functions of weight zero* of some level N.

9. Vertex operator algebras and modular invariance

In this section we describe some of the main results concerning the connections between (vector-valued) modular forms and VOAs. We are concerned here exclusively with regular and C_2-cofinite VOAs as discussed in Sections 6 and 7. We recall that V is always assumed to be of CFT-type.

9.1. The regular case. It is convenient to assume at the outset that V is a C_2-cofinite (but not necessarily rational) VOA of central charge c. By Theorem 6.13 there are only finitely many inequivalent, ordinary, irreducible V-modules, and we denote them $V = M^1, M^2, \ldots, M^r$. Let the conformal weight of M^i be h^i (cf. (64) and attendant discussion), and let Z_i be the character of M^i (65).

The first basic fact is that 1-*point functions are holomorphic in* $\mathfrak{H}$. For example, it follows from this and Theorems 5.7 and 5.10 that the 2-point functions $F_V((u_1, z_1), (u_2, z_2))$ are elliptic functions. There are two approaches to the holomorphy of 1-point functions. The first ([Z]) is to find a modular linear differential equation (82) satisfied by $f = Z_i(v, q)$. In this case, because the coefficients of the MLDE are holomorphic in $\mathfrak{H}$, then so are the solutions (cf. Exercise 8.7). The second approach ([GN]) uses the PBW-type bases that we already mentioned in Section 6.2.

We now take V to be regular. The main properties *vis-à-vis* modular invariance are as follows:

THEOREM 9.1. *Let the notation be as above, and assume that V is regular.*
(a) *The central charge c and conformal weights h^i are rational numbers.*
(b) *There is a representation $\rho : \tilde{\Gamma} \to GL_r(\mathbb{C})$ of the inhomogeneous modular group (cf. Exercise* 3.1*) with the following property: if $v \in V$ has $L[0]$-weight k and we set $F_v = (Z_1(v), \ldots, Z_r(v))$, then (F_v, ρ) is an* almost holomorphic *vector-valued modular form of weight k and finite level N.*

We have already discussed the holomorphy of $Z_i(v)$. The heart of the matter - that there is ρ such that (F_v, ρ) is a vector-valued modular form of weight k - is more difficult. It ultimately depends on the complete reducibility of admissible V-modules into ordinary irreducible V-modules. See [Z], [DLM4] for details. The argument shows that the representation ρ is *independent of the state v.* Once the vector-valued modular form is available, one uses the theory of ODEs with regular singular points [MA] to show that (a) holds. The argument, which is arithmetic in nature, makes use of the fact that if v is taken to be the vacuum vector then the component functions $Z_i(\mathbf{1})$ of $F_\mathbf{1}$ are just the partition functions of the ordinary irreducible modules over V, and as such have *integral* Fourier coefficients. Also, because $F_\mathbf{1}$ has weight zero (because $\mathbf{1} \in V_{[0]}$), $\ker \rho$ contains $\pm I_2$ and so ρ descends to a representation of $\tilde{\Gamma}$. The rationality of conformal weights and central charge implies that (F_v, ρ) has finite level N (e.g., one can take N to be the gcd of the denominators of the rational numbers $h_i - c/24$). There is a basic open problem here:

Modularity conjecture. In the context of Theorem 9.1*, (F_v, ρ) is modular.*

This is an article of faith in the physics literature. There are compelling arguments (e.g., [Ba1], [Ba2], [FMS]) which, however, are not (yet) mathematically rigorous. Note that this Conjecture follows from the conjectured modularity of vector-valued modular forms of level N with integral Fourier coefficients stated at the end of Section 8.1. There are other avenues via which the modularity of (F_v, ρ) might be established, in particular using the theory of tensor products of modules over a VOA and tensor categories (cf. [HL]).

It hardly needs to be said that all known regular VOAs satisfy the Modularity Conjecture. The case of lattice theories follows from Exercise 8.12. The case of WZW models was studied prior to the advent of VOA theory using Lie theory (cf. [KP], [K2]). A discussion of this case as well as that of the simple Virasoro VOAs L_c in the discrete series may be found in [FMS].

9.2. The C_2-cofinite case. One desires an analog of Theorem 9.1 for the more general case of C_2-cofinite VOAs, but any generalization must deal with the fact that the span of the partition functions $Z_i(1)$ of the ordinary irreducible modules is generally *not* a Γ-module unless V is a regular VOA. Miyamoto's solution [My1] (see also [Fl]) involves generalized or *pseudo trace functions*. The idea is to utilize the admissible V-modules $L_n(X)$ constructed from a finite-dimensional module X over the algebra $A_n(V)$ (67). C_2-cofiniteness implies that $A_n(V)$ is finite-dimensional (Theorem 6.13), and this leads to the fact that each of the homogeneous pieces $L_n(X)_m$ are also finite-dimensional. Because $L_n(X)$ is admissible then the zero mode $o(\omega) = L_0$ of the conformal vector operates on these homogeneous pieces (63). However, in the present context L_0 may not be the degree operator, indeed L_0 *may not be a semisimple operator*.

We decompose $L_n(X)_m$ into a direct sum of Jordan blocks for the action of L_0. On such a block B there is an L_0-eigenvector with eigenvalue $m + \lambda$, $\lambda \in \mathbb{C}$, $L_0 - (m + \lambda)I$ is *nilpotent*, and the exponential operator

$$q^{L_0} = q^{m+\lambda} \sum_{t \geq 0} \frac{(2\pi i \tau (L_0 - m - \lambda))^t}{t!} \qquad (83)$$

on B reduces to a *finite* sum. If X is indecomposable, λ is determined by the action of ω, which (when regarded as an element of $A_n(V)$) turns out to be a central element and thus acts on X as a scalar. One can piece together the exponentials (83) and incorporate zero modes $o(v)$ of other states as before. However, the details are subtle, as one needs pseudotraces [My1], which is a type of symmetric function on $A_n(V)$ which replaces the usual trace.

The upshot of the analysis sketched above is this: we can define[15] (pseudo) trace functions $\text{Tr}^{\phi}_{L_n(X)} o(v) q^{L_0 - c/24}$. Once these gadgets are introduced, one

---

[15] ϕ denotes 'pseudo'.

can use the arguments in the regular case described in the previous Subsection together with additional arguments (to account for the failure of $A_n(V)$ to be semisimple) to show that for each n and for $v \in V_{[k]}$, the pseudo trace functions define a (finite-dimensional) almost holomorphic vector-valued modular form of weight k. Alternatively, they span a finite-dimensional Γ-submodule of $\mathfrak{F}_k$ (notation as in Section 8.1). In particular, the pseudo characters $\text{Tr}^{\phi}_{L_n(X)} q^{L_0-c/24}$ are seen to be linear combinations of characters of ordinary, irreducible V-modules with coefficients in $\mathbb{C}[\tau]$. That is, they are polynomial q-expansions in the sense of Section 8.1. This is, of course, fully consistent with Theorem 8.1. Furthermore, one finds as in the regular case that the central charge and conformal weights of the ordinary, irreducible V-modules again lie in $\mathbb{Q}$.

It would take as too far afield to try to describe any VOAs for which the pseudo trace functions actually involve log terms. Such theories are, naturally, called *logarithmic field theories* in the physics literature. For some examples, see e.g., [GK], [A] and references therein.

EXERCISE 9.2. Prove that the (image of) the conformal vector ω is a central element of $A_n(V)$ (cf. Exercises 6.9, 6.10).

9.3. The holomorphic case. We call a simple, regular VOA V *holomorphic* if it has a *unique* irreducible module, namely the adjoint module V. It seems likely that a simple VOA with a unique ordinary irreducible module is necessarily regular, and therefore holomorphic, but this appears to be unknown. Be that as it may, in the case of holomorphic VOAs Theorem 9.1 can be refined, and in particular the Modularity Conjecture of Section 9.1 holds in this case. This is because if a vector-valued modular form of weight k has a single component $f(\tau)$ then it affords a 1-dimensional representation of Γ and so there is a character $\alpha : \Gamma \to \mathbb{C}^*$ such that

$$f|_k \gamma(\tau) = \alpha(\gamma) f(\tau), \quad \gamma \in \Gamma. \tag{84}$$

Since Γ' is a congruence subgroup of level 12 (Exercise 8.5) it follows that $f(\tau)$ is a classical modular form of level dividing 12. Thanks to Theorem 9.1 all of this applies with $f = Z_V(v,q)$, indeed a bit more is true in this case: the group of characters of Γ is cyclic of order 12 (Exercise 8.5) hence that of $\tilde{\Gamma}$ is cyclic of order 6; and one can argue (cf. Exercise 9.4) that $S \in \ker \alpha$, so that in fact α has order dividing 3 and each $\alpha(\gamma)$ in (84) is a cube root of unity. We thus arrive at

THEOREM 9.3. *Suppose that V is a holomorphic VOA of central charge c.*
(a) *If $v \in V_{[k]}$ then $Z_V(v,\tau)$ is an almost holomorphic modular form of weight k and level 1 or 3.*

(b) c is an integer divisible by 8. It is divisible by 24 if, and only if, $Z_V(v,q)$ has level 1.

Lattice theories provide a large number of holomorphic VOAs. From Theorem 7.6 it is immediate that V_L is holomorphic if, and only if, $L = L^0$ is self-dual. The partition function is $\theta_L(\tau)/\eta^c(\tau)$ where c is the rank of L (75), and in this case the modularity of the partition function follows directly from comments in Section 8.2.

We also mention that the modules over a tensor product $U \otimes V$ of VOAs (Exercise 2.32) are just the tensor products $M \otimes N$ of modules M over U and N over V ([FHL]). In particular, if U, V are holomorphic then so too is $U \otimes V$.

EXERCISE 9.4. Let V be a holomorphic VOA, and let α be the character of Γ satisfying (*) $Z_V(\mathbf{1})|_0\gamma(\tau) = \alpha(\gamma)Z_V(\mathbf{1})$. Prove that $\alpha(S) = 1$. (Hint: take $\gamma = S$ and evaluate (*) at $\tau = i$.) Using this, give the details of the proofs of (a) and (b) in Theorem 9.3.

EXERCISE 9.5. Let V be a holomorphic VOA of central charge c, and let $v \in V_{[k]}$. Prove that $Z_V(v,\tau) = g(\tau)/\eta^c(\tau)$ where $g(\tau)$ is an almost holomorphic modular form on Γ of weight $k + c/2$.

9.4. Applications of modular invariance.
Theorem 9.1 places strong conditions on the 1-point trace functions of a regular VOA, and in particular on the partition function. If V is a holomorphic VOA the conditions are even stronger. In this subsection we give a few illustrations of how modular invariance can be used to study the structure of holomorphic VOAs.

By Exercise 9.5, $Z_V(\mathbf{1}) = g(\tau)/\eta^c(\tau)$ where $g(\tau) = 1 + \cdots \in \mathfrak{M}_{c/2}$ is a holomorphic modular form on Γ of weight $c/2$. There are no (nonzero) such forms of negative weight, so we have $c \geq 0$. If $c = 0$ then $g(\tau) = 1 = Z_V(\tau)$, corresponding to the 1-dimensional VOA $\mathbb{C}\mathbf{1}$ (cf. Exercise 2.30) which is indeed holomorphic.

Since $8|c$, the next two cases are $c = 8, 16$, when $g(\tau)$ has weight 4 and 8 respectively. Because of the structure of the algebra $\mathfrak{M}$ of modular forms on Γ (Theorem 3.9 and (28)) there is only one choice for $g(\tau)$ in these cases, namely $g(\tau) = Q$ or Q^2, so the partition function is *uniquely determined* as $Z_V(\mathbf{1}) = Q/\eta^8(\tau)$ or $Q^2/\eta^{16}(\tau) = (Q/\eta^8(\tau))^2$ (Exercise 8.11 is relevant here). We have already seen holomorphic VOAs with these partition functions in Section 9.3, namely the lattice theories V_{E_8} and $V_{E_8 \perp E_8} \equiv V_{E_8}^{\otimes 2}$ (E_8 refers to the root lattice of type E_8). In fact, there is a second even, self-dual lattice L_2 of rank 16 not isometric to $E_8 \perp E_8$ and we obtain in this way a second holomorphic VOA V_{L_2}. It turns out that these are the *only* holomorphic VOAs (up to isomorphism) with $c = 8$ or 16. This result requires additional techniques based on applications of the recursion in Theorem 5.10 and analytic properties of vector-valued modular forms ([DM2], [DM3]). To summarize:

THEOREM 9.6. *Suppose that V is a holomorphic VOA of central charge $c \leq 16$. Then one of the following holds*:
(a) $c = 0$ and $V = \mathbb{C}\mathbf{1}$.
(b) $c = 8$ and $V = V_{E_8}$ is the E_8-lattice theory.
(c) $c = 16$ and $V = V_{E_8 \perp E_8}$ or V_{L_2} is a lattice theory.

We now consider holomorphic VOAs V of central charge $c = 24$. In some ways, this is the most interesting case. If $c \geq 32$ the number of isometry classes of even, self-dual lattices of rank c is very large (see [Se] for further comments), so there are a correspondingly large number of isomorphism classes of holomorphic VOAs. For rank 24 there are just 24 isometry classes of even, self-dual lattices (cf. [CS], [Se]), so one might hope that there are not too many holomorphic VOAs with $c = 24$. In fact, Schellekens has conjectured that there are just 71 such theories [Sch]. Now $Z_V(1) = q^{-1} + \cdots$ is an almost holomorphic modular function of weight zero and level 1 by Theorem 9.3. As such it is a polynomial in the modular function $j(\tau) = q^{-1} + 744 + \cdots$ (cf. (35) and Exercise 3.20). So there is an integer d such that

$$Z_V(1) = j(\tau) + (d - 744) = q^{-1} + d + 196884q + \cdots$$

and the partition function is *determined uniquely* by d. Obviously $d = \dim V_1$, so it is a nonnegative integer, but one cannot say more about d on the basis of modular invariance alone because $j(\tau) + c'$ is a modular function for *any* constant c'. It can in fact be proved that there are only finitely many choices of d that correspond to possible holomorphic VOAs[16]. The arguments use Lie algebra theory, starting with the Lie algebra structure on V_1 (Exercise 9.7) as well as modular forms (see [DM1], [DM2], [DM3], [Sch]). Of the 71 conjectured holomorphic $c = 24$ VOAs, it seems that only 39 are known to exist. Beyond the 24 lattice theories, the other 15 are constructed as so-called $\mathbb{Z}_2$-orbifold models of lattice theories [DGM]. The first construction of this type [FLM] leads to the famous *Moonshine Module*, about which we will shortly say a bit more. It is a major problem to decide whether the others also exist, and to develop construction techniques when they do.

As a final example, we mention some recent work of E. Witten [Wi] where certain holomorphic vertex operator algebras $V^{(k)}$ are posited to exist which are related, via the *AdS-CFT correspondence*, to phenomena concerning gravity with a negative cosmological constant. $V^{(k)}$ has central charge $c_k = 24k, k = 1, 2, \ldots$ and a *minimal structure* compatible with the requirements of modular

---

[16]No more than a few hundred.

invariance imposed by Theorem 9.3. To explain what this is supposed to mean, recall (cf. Theorem 7.2) that $\text{Vir}_{c_k} = L_{c_k}$ is simple, and the L_{c_k}-submodule of $V^{(k)}$ generated by **1** is a graded subspace U naturally identified as the Fock space for L_{c_k}. By (42), the graded dimension of U is

$$q^{-k} \prod_{n \geq 2} (1-q^n)^{-1} = q^{-k} \sum_{n=0}^{k} d_n q^n + O(q)$$

for integers $d_0, \ldots d_k$. The posited minimal structure of $V^{(k)}$ means that the partition function of $V^{(k)}$ also satisfies

$$Z_{V^{(k)}}(\mathbf{1}) = q^{-k} \sum_{n=0}^{k} d_n q^n + O(q).$$

In other words, the first $k+1$ graded subspaces $V_n^{(k)}$ ($0 \leq n \leq k$) of $V^{(k)}$ coincide with the corresponding graded pieces of U, so that they are as small as they can be. We know that $Z_{V^{(k)}}(\mathbf{1})$ is a monic polynomial $\Phi_k(j)$ of degree k in $j(\tau)$, and it is clear that Φ_k is *uniquely determined* by $d_0, \ldots, d_k$, and hence by k.

As in the case of the 'missing' holomorphic $c = 24$ theories, the main question here for the VOA theorist is whether $V^{(k)}$ exists or not. The answer is unknown for any k with the notable exception of the *Moonshine module* $V^\natural$ ([B], [FLM], [DGM], [My2]) corresponding to $k = 1$. In this case the graded dimension of U is $q^{-1} + O(q)$, the partition function of $V^\natural$ is

$$Z_{V^\natural}(q) = j(q) - 744 = q^{-1} + 0 + 196884q + \cdots,$$

and the minimal structure is reflected in the vanishing of the constant term. In this case the Lie algebra structure on the weight 1 subspace is absent, and one must exploit instead the *Griess algebra*, i.e., the commutative algebra structure on $V_2^\natural$ (cf. Exercise 9.9).

One of the main features of the Moonshine Module is its automorphism group, which is the Monster sporadic simple group ([FLM], [G1], [G2]). In order to develop this aspect of $V^\natural$ as well as the $\mathbb{Z}_2$-orbifold construction that we mentioned above and other features of VOAs, it would be necessary to develop the theory of *automorphism groups* of VOAs. This will have to wait for another day. A brief description of some of the connections between automorphism groups and generalized modular forms can be found in [KM].

EXERCISE 9.7. Let V be a VOA of CFT-type. Prove the following:
(a) The product $[a, b] = a_0 b$ equips V_1 with the structure of a *Lie algebra*.
(b) $\langle a, b \rangle = a_1 b$ defines a symmetric, invariant, bilinear form on V_1.

EXERCISE 9.8. Prove that the Fourier coefficients of the q-expansion of $\Phi_k(j)$ are nonnegative integers (a necessary condition for the existence of $V^{(k)}$).

EXERCISE 9.9. Show that the product $a_1 b$ ($a, b \in V_2^\natural$) equips the weight 2 subspace of $V^\natural$ with the structure of a *commutative, nonassociative* algebra.

Part III. Two current research areas

10. Some preliminaries

10.1. VOAs and rational matrix elements. As noted in Section 2.6 there are a number of equivalent sets of axioms for VOA theory. Here we discuss one of these equivalent approaches wherein the properties of a VOA are expressed in terms of the properties of matrix elements which turn out to be rational functions of the formal vertex operator parameters. In many ways, this is the closest approach to CFT (see [FMS], for example) in that the formal parameters can be taken to be complex numbers with the matrix elements considered as rational functions on the Riemann sphere.

We begin by defining matrix elements. In order to simplify the discussion, we always assume that the VOA is of CFT-type (62). This condition is satisfied in all examples we consider. We define the *restricted dual space* of V by [FHL]

$$V' = \bigoplus_{n \geq 0} V_n^*, \qquad (85)$$

where V_n^* is the dual space of linear functionals on the finite dimensional space V_n. Let $\langle \, , \, \rangle_d$ denote the canonical pairing between V' and V. Define *matrix elements* for $a' \in V'$, $b \in V$ and vertex operators $Y(u^1, z_1), \ldots Y(u^n, z_n)$ by

$$\langle a', Y(u^1, z_1) \ldots Y(u^n, z_n) b \rangle_d. \qquad (86)$$

In particular, choosing $b = \mathbf{1}$ and $a' = \mathbf{1}'$ we obtain the (genus zero) n-point correlation function

$$F_V^{(0)}((u^1, z_1), \ldots, (u^n, z_n)) = \langle \mathbf{1}', Y(u^1, z_1) \ldots Y(u^n, z_n) \mathbf{1} \rangle_d. \qquad (87)$$

One can show in general that every matrix element is a homogeneous rational function of $z_1, \ldots, z_n$ [FHL], [DGM]. Thus the formal parameters of VOA theory can be replaced by complex parameters on (appropriate subdomains of) the genus zero Riemann sphere $\mathbb{CP}^1$. We illustrate this by considering matrix elements containing one or two vertex operators. Recall from (36) that, for $u \in V_n$,

$$u_k : V_m \to V_{m+n-k-1}. \qquad (88)$$

Hence it follows that for $a' \in V'_{m'}$, $b \in V_m$ and $u \in V_n$ we obtain a monomial

$$\langle a', Y(u, z) b \rangle_d = C_{a'b}^u z^{m'-m-n}, \qquad (89)$$

where $C_{a'b}^u = \langle a', u_{m+n-m'-1} b \rangle_d$.

We next consider the matrix element of two vertex operators to find (recalling convention (1)):

THEOREM 10.1. *Let* $a' \in V'_{m'}, b \in V_m, u^1 \in V_{n_1}$ *and* $u^2 \in V_{n_2}$. *Then*

$$\langle a', Y(u^1, z_1)Y(u^2, z_2)b \rangle_d = \frac{f(z_1, z_2)}{z_1^{m+n_1} z_2^{m+n_2} (z_1 - z_2)^{n_1+n_2}}, \quad (90)$$

$$\langle a', Y(u^2, z_2)Y(u^1, z_1)b \rangle_d = \frac{f(z_1, z_2)}{z_1^{m+n_1} z_2^{m+n_2} (-z_2 + z_1)^{n_1+n_2}}, \quad (91)$$

where $f(z_1, z_2)$ *is a homogeneous polynomial of degree* $m + m' + n_1 + n_2$.

REMARK 10.2. The matrix elements (90), (90) are thus determined by a unique homogeneous rational function which can be evaluated on $\mathbb{CP}^1$ in the domains $|z_1| > |z_2|$ and $|z_2| > |z_1|$ respectively.

PROOF. Consider

$$\langle a', Y(u^1, z_1)Y(u^2, z_2)b \rangle_d = \sum_{k \geq 0} \sum_{c \in V_k} \langle a', Y(u^1, z_1)c \rangle_d \langle c', Y(u^2, z_2)b \rangle_d,$$

where c ranges over any basis of V_k and $c' \in V_k^*$ is dual to c. From (89) it follows that

$$\langle a', Y(u^1, z_1)Y(u^2, z_2)b \rangle_d = \frac{z_1^{m'-n_1}}{z_2^{m+n_2}} G\left(\frac{z_2}{z_1}\right),$$

for infinite series

$$G(x) = \sum_{k \geq 0} \sum_{c \in V_k} C_{a'c}^{u^1} C_{c'b}^{u^2} x^k.$$

Hence the matrix element is homogeneous of degree $m' - m - n_1 - n_2$. Similarly

$$\langle a', Y(u^2, z_2)Y(u^1, z_1)b \rangle_d = \frac{z_2^{m'-n_2}}{z_1^{m+n_1}} H\left(\frac{z_1}{z_2}\right),$$

for the infinite series

$$H(y) = \sum_{k \geq 0} \sum_{c \in V_k} C_{a'c}^{u^2} C_{c'b}^{u^1} y^k.$$

But $Y(u^2, z_2)$ and $Y(u^1, z_1)$ are local of order at most $n_1 + n_2$ (cf. Exercise 4.4) and hence

$$\frac{(z_1 - z_2)^{n_1+n_2}}{z_1^{m+n_1} z_2^{m+n_2}} z_1^{m'+m} G\left(\frac{z_2}{z_1}\right) = \frac{(z_1 - z_2)^{n_1+n_2}}{z_1^{m+n_1} z_2^{m+n_2}} z_2^{m'+m} H\left(\frac{z_1}{z_2}\right). \quad (92)$$

It follows that

$$f(z_1, z_2) = z_1^{m'+m}(z_1 - z_2)^{n_1+n_2} G\left(\frac{z_2}{z_1}\right) = z_2^{m'+m}(z_1 - z_2)^{n_1+n_2} H\left(\frac{z_1}{z_2}\right)$$

is a homogeneous polynomial of degree $m + m' + n_1 + n_2$. $\square$

Properties (90) and (90) are equivalent to locality of $Y(u^1, z_1)$ and $Y(u^2, z_2)$ so that the axioms of a VOA can be alternatively formulated in terms of rational matrix elements [DGM], [FHL]. Theorem 10.1 can also be generalized for all matrix elements. Furthermore, using the vertex commutator property (Exercise 2.14) one can also derive a recursive relationship in terms of rational functions between matrix elements for n vertex operators and $n-1$ vertex operators that is the genus zero version of Zhu's first recursion formula (Theorem 5.7).

EXERCISE 10.3. Prove (89).

EXERCISE 10.4. Show (92) implies that $f(z_1, z_2)$ is a polynomial.

10.2. Genus-zero Heisenberg correlation functions. We illustrate these structures by considering the example of the rank one Heisenberg VOA M_0 generated by a weight one vector a. Let

$$G_n^{(0)}(z_1, \ldots, z_n) \equiv F_{M_0}^{(0)}((a, z_1), \ldots, (a, z_n)). \tag{93}$$

denote the n-point correlation function for n Heisenberg vectors. This must be a symmetric rational function in z_i with poles of order two at $z_i = z_j$ for all $i \neq j$ from locality. We now determine its exact form. Since $a_0 \mathbf{1} = 0$ it follows that $G_1^{(0)}(z_1) = 0$. The 2-point function is

$$G_2^{(0)}(z_1, z_2) = \sum_{m \geq 0} z_1^{-m-1} \langle \mathbf{1}', a_m Y(a, z_2) \mathbf{1} \rangle_d,$$

where (88) implies that there is no contribution for $m < 0$. Commuting a_m we find

$$G_2^{(0)}(z_1, z_2) = \sum_{m \geq 0} z_1^{-m-1} \langle \mathbf{1}', [a_m, Y(a, z_2)] \mathbf{1} \rangle_d,$$

using $a_m \mathbf{1} = 0$ for $m \geq 0$. But the Heisenberg commutation relations imply

$$[a_m, Y(a, z_2)] = m z_2^{m-1},$$

so that

$$G_2^{(0)}(z_1, z_2) = \sum_{m \geq 0} m z_1^{-m-1} z_2^{m-1} = \frac{1}{(z_1 - z_2)^2}. \tag{94}$$

The general n-point function is similarly given by

$$G_n^{(0)}(z_1, \ldots, z_n) = \sum_{m \geq 0} z_1^{-m-1} \sum_{i=2}^{n} \langle \mathbf{1}', Y(a, z_2) \ldots [a_m, Y(a, z_i)] \ldots Y(a, z_n) \mathbf{1} \rangle_d,$$

leading to a recursive identity

$$G_n^{(0)}(z_1, \ldots, z_n) = \sum_{i=2}^{n} \frac{1}{(z_1 - z_i)^2} G_{n-2}^{(0)}(z_2, \ldots, \hat{z}_i, \ldots, z_n), \tag{95}$$

where $\hat{z}_i$ is deleted. Thus we may recursively solve to find $G_n^{(0)} = 0$ for n odd whereas for n even, $G_n^{(0)}$ is expressed as multiples of rational terms of the form $1/(z_i - z_j)^2$ for all possible pairings z_i, z_j. This can be equivalently described in terms of the subset, denoted by $F(\Phi)$, of the permutations of the label set $\Phi = \{1, \ldots n\}$ consisting of *fixed-point-free involutions*. Thus a typical element $\varphi \in F(\Phi)$ is given by $\varphi = \ldots (ij) \ldots$, a product of $n/2$ disjoint cycles. We then find (95) implies

THEOREM 10.5. $G_n^{(0)}$ *vanishes for n odd, whereas for n even*

$$G_n^{(0)}(z_1, \ldots, z_n) = \sum_{\varphi \in F(\Phi)} \prod \frac{1}{(z_i - z_j)^2}, \qquad (96)$$

where the product ranges over all the cycles of $\varphi = \ldots (ij) \ldots$.

REMARK 10.6. Using associativity one can show that $G_n^{(0)}(z_1, \ldots, z_n)$ is in fact a generating function for all matrix elements of the Heisenberg VOA.

EXERCISE 10.7. Show that $|F(\Phi)| = (n-1)!! = (n-1).(n-3).(n-5)\ldots$.

EXERCISE 10.8. For $n = 4$ show that $F(\Phi) = \{(12)(34), (13)(24), (14)(23)\}$ and $G_4^{(0)}(z_1, z_2, z_3, z_4)$ is given by

$$\frac{1}{(z_1 - z_2)^2(z_3 - z_4)^2} + \frac{1}{(z_1 - z_3)^2(z_2 - z_4)^2} + \frac{1}{(z_1 - z_4)^2(z_2 - z_3)^2}.$$

10.3. Adjoint vertex operators. The Virasoro subalgebra $\{L_{-1}, L_0, L_1\}$ generates a natural action on vertex operators associated with $SL(2, \mathbb{C})$ Möbius transformations on z (cf. [B], [DGM], [FHL], [K1] and Exercise 2.21). Thus under the translation $z \mapsto z + \lambda$ generated by L_{-1} we have (cf. Exercise 2.16)

$$q_\lambda^{L_{-1}} Y(u, z) q_\lambda^{-L_{-1}} = Y(u, z + \lambda). \qquad (97)$$

Under $z \mapsto q_\lambda z$ generated by L_0 we have (cf. Exercise 4.2)

$$q_\lambda^{L_0} Y(u, z) q_\lambda^{-L_0} = Y(q_\lambda^{L_0} u, q_\lambda z). \qquad (98)$$

Finally, under the transformation $z \mapsto z/(1 - \lambda z)$ generated by L_1 we find

$$q_\lambda^{L_1} Y(u, z) q_\lambda^{-L_1} = Y(q_{\lambda(1-\lambda z)}^{L_1}(1 - \lambda z)^{-2L_0} u, \frac{z}{1 - \lambda z}). \qquad (99)$$

Combining these it follows that the transformation $z \mapsto -\lambda^2 z^{-1}$ is described by $T_\lambda \equiv q_\lambda^{L_{-1}} q_{\lambda^{-1}}^{L_1} q_\lambda^{L_{-1}}$ with

$$T_\lambda Y(u, z) T_\lambda^{-1} = Y(q_{-z\lambda^{-2}}^{L_1}(\lambda^{-2} z^2)^{-L_0} u, -\lambda^2 z^{-1}). \qquad (100)$$

Taking $\lambda = \sqrt{-1}$ in (100) corresponding to the inversion $z \mapsto z^{-1}$ we find

$$Y^\dagger(u,z) \equiv T_{\sqrt{-1}} Y(u,z) T_{\sqrt{-1}}^{-1} = Y(q_z^{L_1}(-z^2)^{-L_0}u, z^{-1}). \tag{101}$$

We call $Y^\dagger(u,z)$ the *adjoint* vertex operator[17]. For u of weight $\text{wt}(u)$ it follows that $Y^\dagger(u,z) = \sum_n u_n^\dagger z^{-n-1}$ has modes

$$u_n^\dagger = (-1)^{\text{wt}(u)} \sum_{k=0}^{\text{wt}(u)} \frac{1}{k!}(L_1^k u)_{2\text{wt}(u)-n-k-2}. \tag{102}$$

For a quasiprimary state u (102) simplifies to

$$u_n^\dagger = (-1)^{\text{wt}(u)} u_{2\text{wt}(u)-n-2}. \tag{103}$$

Thus for a weight one Heisenberg vector a we find

$$a_n^\dagger = -a_{-n}. \tag{104}$$

and for the weight two Virasoro vector ω we find that for $L_n^\dagger \equiv \omega_{n+1}^\dagger$

$$L_n^\dagger = L_{-n}. \tag{105}$$

We also note that the adjoint vertex operators can be used to construct a canonical V-module as follows. Define vertex operators $Y_{V'} : V \to \mathcal{F}(V')$ by

$$\langle Y'(u,z)a', b\rangle_d = \langle a', Y^\dagger(u,z)b\rangle_d, \tag{106}$$

for $a' \in V'$ and $b, u \in V$. Then $(V', Y_{V'})$ can be shown to be a V-module called the *dual or contragradient module* [FHL].

EXERCISE 10.9. Prove (100).

EXERCISE 10.10. Show for a quasiprimary state u (i.e., $L_1 u = 0$) of weight $\text{wt}(u)$ that under a Möbius transformation $z \to \phi(z) = (az+b)/(cz+d)$

$$Y(u,z) \to \left(\frac{d\phi}{dz}\right)^{\text{wt}(u)} Y(u, \phi(z)). \tag{107}$$

EXERCISE 10.11. Hence show for n quasiprimary vectors u^i of weight $\text{wt}(u^i)$ that the rational n-point function (87) is associated with a (formal) Möbius-invariant differential form on $\mathbb{CP}^1$

$$\mathcal{F}_V^{(0)}(u^1, \ldots, u^n) = F_V^{(0)}((u^1, z_1), \ldots, (u^n, z_n)) \prod_{1 \leq i \leq n} dz_i^{\text{wt}(u^i)}. \tag{108}$$

---

[17] This terminology differs from that of [FHL].

REMARK 10.12. $\mathcal{F}_V^{(0)}(u^1,\ldots,u^n)$ is a conformally invariant global meromorphic differential form on $\mathbb{CP}^1$ if $u^1,\ldots,u^n$ are primary vectors i.e., $L_n u^i = 0$ for all $n > 0$.

EXERCISE 10.13. Prove (102).

EXERCISE 10.14. Show that $(Y^\dagger(u,z))^\dagger = Y(u,z)$.

10.4. Invariant bilinear forms. In this subsection we consider the construction of a canonical bilinear form on V motivated by (106). We say a bilinear form $\langle\,,\,\rangle: V \times V \longrightarrow \mathbb{C}$ is *invariant* if for all $a,b,u \in V$

$$\langle Y(u,z)a, b \rangle = \langle a, Y^\dagger(u,z)b \rangle, \tag{109}$$

with $Y^\dagger(a,z)$ the adjoint operator of (101). In terms of modes, (109) reads

$$\langle u_n a, b \rangle = \langle a, u_n^\dagger b \rangle, \tag{110}$$

using (102). Applying (105) it follows that

$$\langle L_0 a, b \rangle = \langle a, L_0 b \rangle. \tag{111}$$

Thus for homogeneous a and b then $\langle a,b \rangle = 0$ for $\mathrm{wt}(a) \neq \mathrm{wt}(b)$.

Next consider a, b with $\mathrm{wt}(a) = \mathrm{wt}(b)$. Invariance and skew-symmetry (see Exercise 2.15) give

$$\langle \mathbf{1}, Y^\dagger(a,z)b \rangle = (-z^2)^{-\mathrm{wt}(a)} \langle \mathbf{1}, Y(q_z^{L_1}a, z^{-1})b \rangle$$
$$= (-z^2)^{-\mathrm{wt}(b)} \langle \mathbf{1}, q_{z^{-1}}^{L_{-1}} Y(b, -z^{-1}) q_z^{L_1} a \rangle$$
$$= \langle \mathbf{1}, q_{z^{-1}}^{L_{-1}} Y^\dagger(q_z^{L_1}b, -z) q_z^{L_1} a \rangle.$$

But (105) implies this is

$$\langle q_{z^{-1}}^{L_1}\mathbf{1}, Y^\dagger(q_z^{L_1}b, -z) q_z^{L_1} a \rangle = \langle \mathbf{1}, Y^\dagger(q_z^{L_1}b, -z) q_z^{L_1} a \rangle.$$

Using invariance this becomes

$$\langle Y(q_z^{L_1}b, -z)\mathbf{1}, q_z^{L_1} a \rangle.$$

Finally, using Exercise 2.9 and (105) this is

$$\langle q_{-z}^{L_{-1}} q_z^{L_1} b, q_z^{L_1} a \rangle = \langle b, q_z^{L_{-1}} a \rangle = \langle b, Y(a,z)\mathbf{1} \rangle.$$

Thus we have shown

$$\langle Y(a,z)\mathbf{1}, b \rangle = \langle b, Y(a,z)\mathbf{1} \rangle.$$

In particular, considering the z^0 term, this implies that the bilinear form is symmetric:

$$\langle a, b \rangle = \langle b, a \rangle. \tag{112}$$

Consider again a, b with $\text{wt}(a) = \text{wt}(b)$. Using the creation axiom $a_{-1}\mathbf{1} = a$ we obtain

$$\langle a, b \rangle = \langle \mathbf{1}, a^\dagger_{-1} b \rangle. \tag{113}$$

with $a^\dagger_{-1} b \in V_0$. Thanks to the assumption that V is of CFT-type[18] we have $a^\dagger_{-1} b = \alpha \mathbf{1}$ for some $\alpha \in \mathbb{C}$ with $\langle a, b \rangle = \alpha \langle \mathbf{1}, \mathbf{1} \rangle$. Hence either $\langle \mathbf{1}, \mathbf{1} \rangle = 0$ so that $\langle a, b \rangle = 0$ for all a, b or else $\langle a, b \rangle$ is non-trivial and is uniquely determined up to the value of $\langle \mathbf{1}, \mathbf{1} \rangle \neq 0$ in which case we choose the normalization $\langle \mathbf{1}, \mathbf{1} \rangle = 1$.

It is straightforward to show that if $\langle \mathbf{1}, \mathbf{1} \rangle \neq 0$ then $L_1 V_1 = 0$ (cf. Exercise 10.16). Li has shown [Li] that the converse is also true: for a VOA of CFT-type, then $\langle \mathbf{1}, \mathbf{1} \rangle \neq 0$ if and only if $L_1 V_1 = 0$. We say that a VOA is of *Strong CFT-type* if it is of CFT-type and $L_1 V_1 = 0$. Such a VOA therefore has a *unique* normalized invariant bilinear form.

The pairing $\langle \, , \, \rangle$ determines a standard map from V to the restricted dual space V' defined by

$$a \mapsto \langle a, \cdot \rangle. \tag{114}$$

Let $\mathcal{K}$ denote the kernel of this map. $\langle \, , \, \rangle$ is nondegenerate with $\mathcal{K}$ trivial if, and only if, V is isomorphic to V' (in other words, V is self-dual). In this case, we may identify $\langle \, , \, \rangle$ with the canonical pairing $\langle \, , \, \rangle_d$ and the dual module (106) is isomorphic to the original VOA.

The nondegeneracy of $\langle \, , \, \rangle$ is also related to the simplicity of the VOA V in much that same way that nondegeneracy of the Killing form determines semi-simplicity in Lie theory [Li]. Let $\mathcal{J} \subset V$ denote the maximal proper ideal of V (cf. Exercise 6.11), so that

$$u_n b \in \mathcal{J}, \tag{115}$$

for all $b \in \mathcal{J}, u \in V$. V is simple if $\mathcal{J}$ is trivial (cf. Section 6). We now show that assuming V is of strong CFT-type then $\mathcal{J} = \mathcal{K}$ and hence V is simple if, and only if, $\langle \, , \, \rangle$ is nondegenerate.

We firstly note that $\mathbf{1} \notin \mathcal{J}$ (otherwise $u = u_{-1}\mathbf{1} \in \mathcal{J}$ for all $u \in V$). Because V is of CFT-type, then for all $b \in \mathcal{J}$ it follows $b \notin V_0$ and so

$$\langle \mathbf{1}, b \rangle = 0. \tag{116}$$

Consider $u \in V$ and $b \in \mathcal{J}$. Then $u^\dagger_{-1} b \in \mathcal{J}$ from (102) and so

$$\langle u, b \rangle = \langle \mathbf{1}, u^\dagger_{-1} b \rangle = 0,$$

for all u from (116). Hence we find $\mathcal{J} \subseteq \mathcal{K}$. Conversely, suppose that $c \in \mathcal{K}$. Then

$$\langle Y^\dagger(u, z)v, c \rangle = 0,$$

---
[18] The general situation is discussed in [Li]).

for all $u, v \in V$. Invariance implies $\langle v, Y(u, z)c \rangle = 0$ and hence $u_n c \in \mathcal{K}$ for all u_n. But given V is of strong CFT-type then $\langle \, , \, \rangle$ is nontrivial so that $\mathcal{K} \neq V$ and hence $\mathcal{K} \subseteq \mathcal{I}$. Thus we conclude $\mathcal{I} = \mathcal{K}$.

Altogether we may summarize these results as follows:

THEOREM 10.15. *Let V be a VOA. An invariant bilinear form $\langle \, , \, \rangle$ on V is symmetric and diagonal with respect to the canonical L_0-grading. Furthermore, if V is of strong CFT-type, $\langle \, , \, \rangle$ is unique up to normalization and is nondegenerate if and only if V is simple.*

The invariant bilinear form is equivalent to the chiral part of the Zamolodchikov metric in CFT ([BPZ; FMS; P]) where (abusing notation)

$$\begin{aligned}\langle a, b \rangle &= \lim_{z_1 \to 0} \lim_{z_2 \to 0} \langle Y(a, z_1)\mathbf{1}, Y(b, z_2)\mathbf{1} \rangle \\ &= \lim_{z_1 \to 0} \lim_{z_2 \to 0} \langle \mathbf{1}, Y^\dagger(a, z_1) Y(b, z_2)\mathbf{1} \rangle \\ &= \text{``}\langle \mathbf{1}, Y(a, w_1 = \infty) Y(b, z_2 = 0)\mathbf{1}\rangle\text{''},\end{aligned} \quad (117)$$

for $w_1 = 1/z_1$ following (101). We thus refer to the nondegenerate bilinear form as the *Li–Zamolodchikov metric* on V or LiZ-metric for short[19].

Consider the rank one Heisenberg VOA M_0 generated by a weight one state a with V spanned by Fock vectors

$$v = a_{-1}^{e_1} a_{-2}^{e_2} \ldots a_{-p}^{e_p} \mathbf{1}, \quad (118)$$

for nonnegative integers e_i. Using (104), we find that the Fock basis consisting of vectors of the form (118) is orthogonal with respect to the LiZ-metric with

$$\langle v, v \rangle = \prod_{1 \leq i \leq p} (-i)^{e_i} e_i!. \quad (119)$$

Clearly $\langle \, , \, \rangle$ is nondegenerate, so by Theorem 10.15 it follows that M_0 is a simple VOA (as already discussed in Section 6).

Consider the Virasoro VOA Vir_c generated by the Virasoro vector ω of central charge c. Using (111) it is sufficient to consider the nondegeneracy of $\langle \, , \, \rangle$ on each homogeneous space V_n. In particular, let $M_n(c) = (\langle a, b \rangle)$ be the Gram matrix of $(\text{Vir}_c)_n$ with respect to some basis. The *Kac determinant* (see [KR]) is $\det M_n(c)$, which is conveniently considered as a polynomial in c. By Theorem 10.15, Vir_c is simple if, and only if, $\det M_n(c) \neq 0$ for all n. For $n = 2$ we have $V_2 = \mathbb{C}\omega$ with Kac determinant

$$\det M_2(c) = \langle \omega, \omega \rangle = \langle \mathbf{1}, L_2 L_{-2} \mathbf{1} \rangle = \frac{c}{2}, \quad (120)$$

---

[19] Although we use the term metric here, the bilinear form is not necessarily positive definite.

with a zero at $c = 0$. For $n = 4$ we have $V_4 = \mathbb{C}\langle L_{-2}^2 \mathbf{1}, L_{-4}\mathbf{1}\rangle$ with

$$M_4(c) = \begin{bmatrix} c(4 + \tfrac{1}{2}c) & 3c \\ 3c & 5c \end{bmatrix}, \tag{121}$$

and Kac determinant

$$\det M_4(c) = \tfrac{1}{2}c^2(5c + 22) \tag{122}$$

with zeros at $c = 0, -\tfrac{22}{5}$.

There is a general formula for the Kac determinant $\det M_n(c)$ which turns out to have zeros for central charge

$$c = c_{p,q} = 1 - \frac{6(p-q)^2}{pq}, \tag{123}$$

where $(p-1)(q-1) = n$ for coprime $p, q \geq 2$. Thus Vir_c is a simple VOA iff $c \neq c_{p,q}$ for some coprime $p, q \geq 2$ (cf. Theorem 7.2).

EXERCISE 10.16. Show that if $\langle \mathbf{1}, \mathbf{1}\rangle \neq 0$ then $L_1 V_1 = 0$.

EXERCISE 10.17. Suppose that $a \in V_m, b \in V_n$ and at least one of a or b is quasiprimary. Prove that the 2-point correlation function is given by

$$\langle \mathbf{1}, Y(a, z_1)Y(b, z_2)\mathbf{1}\rangle = \frac{\langle a, b\rangle}{(z_1 - z_2)^{2m}}\delta_{m,n}.$$

(The Zamolodchikov metric in CFT is often introduced in this way.)

EXERCISE 10.18. Verify (119).

11. The genus-two partition function for the Heisenberg VOA

In this section we will discuss some recent research by the authors wherein we develop a theory of partition and n-point correlation functions on a Riemann surface of genus-two [T1; MT2; MT3; MT4]. The basic idea is to construct a genus-two Riemann surface by specific sewing schemes where we either sew together two once punctured tori or self-sew a twice punctured torus (i.e., attach a handle). The partition and n-point functions on the genus-two surface are then defined in terms of correlation functions on the lower genus surfaces combined together in an appropriate way. We will not explore the full details entailed in this programme. Instead we will consider the example of the Heisenberg VOA M_0 and compute the partition function on the genus-two surface formed from two tori.

11.1. Genus-one Heisenberg 1-point functions. We first discuss the genus-one 1-point correlation function for all elements of the Heisenberg VOA M_0 generated by the weight one Heisenberg vector a [MT1]. We make heavy use of the Zhu recursion formulas (Theorems 5.7 and 5.10). In particular, we prove Theorem 4.5 by considering the 1-point function $Z_{M_0}(v, \tau)$ for a Fock vector in the square bracket formulation

$$v = a[-k_1] \ldots a[-k_n]\mathbf{1}, \tag{124}$$

for $k_i \geq 1$. The Fock vector v is of square bracket weight $\text{wt}[v] = \sum_i k_i$. We want to show that

$$Z_{M_0}^{(1)}(v, \tau) = \frac{Q_v(\tau)}{\eta(q)}, \tag{125}$$

for $Q_v(\tau) \in \mathfrak{Q}$, the algebra of quasimodular forms. $Q_v(\tau)$ is of weight $\text{wt}[v]$ and is expressed in terms of

$$C(k, l) = C(k, l, \tau) = (-1)^{l+1} \frac{(k+l-1)!}{(k-1)!(l-1)!} E_{k+l}(\tau), \tag{126}$$

for $k, l \geq 1$. Here $E_n(\tau)$ is the Eisenstein series of (24). We recall that $E_n = 0$ for n odd, $E_2(\tau)$ is a quasimodular form of weight 2 and E_n is a modular form of weight n for even $n \geq 4$. Thus $C(k, l, \tau)$ is a quasimodular form of weight $k + l$. We also note that $C(k, l) = C(l, k)$.

Each Fock vector v is described by a label set $\Phi_\lambda = \{k_1, \ldots, k_n\}$ which corresponds in a natural 1-1 manner with unrestricted partitions $\lambda = \{1^{e_1}, 2^{e_2}, \ldots\}$ of $\text{wt}[v]$ (where $e_i \geq 0$). We write $v = v(\lambda)$ to indicate this correspondence, which will play a significant rôle later on. Define $F(\Phi_\lambda)$ to be the subset of all permutations on Φ_λ consisting only of *fixed-point-free involutions*. Let $\varphi = \ldots (k_i k_j) \ldots$, a product of disjoint cycles, denote a typical element of $F(\Phi_\lambda)$.

We can now describe the 1-point function $Z_{M_0}^{(1)}(v(\lambda), \tau)$ of (125) [MT1]:

THEOREM 11.1. *For even n*

$$Q_v(\tau) = \sum_{\phi \in F(\Phi_\lambda)} \Gamma(\phi, \tau), \tag{127}$$

$$\Gamma(\phi, \tau) = \prod_{(k_i k_j)} C(k_i, k_j, \tau). \tag{128}$$

for C of (126), where the product ranges over all the cycles of $\varphi = \ldots (k_i k_j) \ldots$ in $F(\Phi_\lambda)$. Moreover $Q_v(\tau)$ lies in $\mathfrak{Q}$ and is of weight $\text{wt}[v]$. For n odd $Q_v(\tau)$ vanishes.

PROOF. Let $v(\lambda) = a[-k_1]w$ for $w = a[-k_2]\ldots a[-k_n]\mathbf{1}$ and use the second Zhu recursion formula (Theorem 5.10) to find

$$Z_{M_0}^{(1)}(a[-k_1]w, \tau) = \delta_{k_1,1} \operatorname{Tr}_{M_0}(o(a)o(w)q^{L_0-1/24})$$
$$+ \sum_{m \geq 1} (-1)^{m+1} \binom{k_1+m-1}{m} E_{k_1+m}(\tau) Z_{M_0}^{(1)}(a[m]w, \tau).$$

But $o(a)u = 0$ for all $u \in M$ and the Heisenberg commutation relations imply

$$Z_{M_0}^{(1)}(a[-k_1]w, \tau) = 0 + \sum_{j=2}^{n} (-1)^{k_j+1} \binom{k_1+k_j-1}{k_j} E_{k_1+k_j}(\tau) k_j Z_{M_0}^{(1)}(\hat{w}, \tau)$$
$$= \sum_{j=2}^{n} C(k_1, k_j, \tau) Z_{M_0}^{(1)}(\hat{w}, \tau),$$

where $\hat{w}$ denotes the Fock vector with label set $\{k_2, \ldots, \hat{k}_j \ldots, k_n\}$ with the index k_j deleted. The result follows by repeated application of this recursive formula until we obtain $\hat{w} = \mathbf{1}$ for which $Z_{M_0}^{(1)}(\mathbf{1}, \tau) = 1/\eta(q)$. The resulting expression for $Q_v(\tau)$ is clearly a quasimodular form of weight $\operatorname{wt}[v] = \sum_i k_i$. Thus Theorem 11.1 follows. □

Some further insight into the combinatorial structure of $Q_v(\tau)$ can be garnered by a consideration of the n-point function for n Heisenberg vectors which we denote by

$$G_n^{(1)}(z_1, \ldots, z_n, \tau) \equiv F_{M_0}^{(1)}((a, z_1), \ldots (a, z), \tau). \qquad (129)$$

This is a symmetric function in z_i with a pole of order two at $z_i = z_j$ for all $i \neq j$ (from locality). For $n = 1$ we immediately find

$$G_1^{(1)}(z_1, \tau) = \operatorname{Tr}_{M_0} o(a) q^{L_0-1/24} = 0.$$

The 2-point function is easily computed via the first Zhu recursion formula (Theorem 5.7):

$$G_2^{(1)}(z_1, z_2, \tau)$$
$$= \operatorname{Tr}_{M_0} o(a)o(a) q^{L_0-1/24} - \sum_{m \geq 1} \frac{(-1)^m}{m!} P_1^{(m)}(z_{12}, \tau) Z_{M_0}^{(1)}(a[m]a, \tau)$$
$$= 0 + P_2(z_{12}, \tau) \frac{1}{\eta(q)}, \qquad (130)$$

since $a[m]a = \mathbf{1}\delta_{m,1}$ and where, from Theorem 5.1, we recall

$$P_2(z, \tau) = \frac{d}{dz} P_1(z, \tau) = \frac{1}{z^2} + \sum_{n=2}^{\infty} (n-1) E_n(\tau) z^{n-2}.$$

(130) is the elliptic analogue of the genus zero formula (94) and reflects a deeper geometrical structure underlying the Heisenberg VOA e.g. [MT4].

Using the n-point correlation function version of the first Zhu recursion we can similarly obtain the genus-one analogue of Theorem 10.5 to find [MT1]:

THEOREM 11.2. *For n even*

$$G_n^{(1)}(z_1,\ldots,z_n,\tau) = \frac{1}{\eta(q)} \sum_{\varphi \in F(\Phi)} \prod_{(ij)} P_2(z_{ij},\tau), \tag{131}$$

where the product ranges over all the cycles of $\varphi = \ldots(ij)\ldots$ *for* $\Phi = \{1,2,\ldots,n\}$ *whereas for n odd $G_n^{(1)}$ vanishes.*

We may use this result to compute *any* genus-one n-point correlation function for M_0 by considering an appropriate analytic expansion of $G_n^{(1)}(z_1,\ldots,z_n,\tau)$ [MT1]. In particular, we can rederive (127) by making use of the identity

$$\begin{aligned}G_n^{(1)}(z_1,\ldots,z_n,\tau) &= Z_{M_0}^{(1)}(Y[a,z_1]\ldots Y[a,z_n]\mathbf{1},\tau)\\ &= \sum_{k_1,\ldots k_n} Z_{M_0}^{(1)}(v,\tau) z_1^{k_1-1}\ldots z_n^{k_n-1},\end{aligned} \tag{132}$$

for Fock vector $v = a[-k_1]\ldots a[-k_n]\mathbf{1}$ for all k_i. We may extract the nonnegative values of $k_1,\ldots,k_n$ from the expansion

$$P_2(z_{ij},\tau) = \frac{1}{(z_i-z_j)^2} + \sum_{k_i,k_j \geq 1}^{\infty} C(k_i,k_j,\tau) z_i^{k_i-1} z_j^{k_j-1}, \tag{133}$$

for C of (126). Thus (131) implies the formula (127) of Theorem 11.1 found for $Q_v(\tau)$.

It is very useful to recast Theorem 11.1 in terms of graph theory as follows. Consider a Fock vector $v(\lambda)$ with label set $\Phi_\lambda = \{k_1,\ldots,k_n\}$ and let $\phi \in F(\Phi_\lambda)$ be a fixed-point-free involution of Φ_λ leading to a contribution $\Gamma(\phi,\tau)$ to $Q_v(\tau)$ in (127). We may then associate to each $\phi \in F(\Phi_\lambda)$ a ϕ-graph γ_ϕ consisting of n vertices labelled by Φ_λ of unit valence with $n/2$ unoriented edges connecting the pairs of vertices (k_i,k_j) determined by $\varphi = \ldots(k_ik_j)\ldots$. Following Exercise 10.7 there are $(n-1)!!$ such graphs for a given label set Φ_λ. Thus, in Exercise 11.4 with $v = a[-1]^3 a[-2]^2 a[-5]\mathbf{1}$ there are 15 independent ϕ-graphs (cf. Exercise 11.5). A ϕ-graph for a fixed point involution $\phi = (11)(22)(15)$ is shown in Figure 1.[20]

Given a ϕ-graph γ_ϕ we define a *weight function*

$$\kappa : \{\gamma_\phi\} \longrightarrow \mathfrak{Q},$$

---

[20] Note that there are 3 distinct fixed point involutions notated by $(11)(22)(15)$.

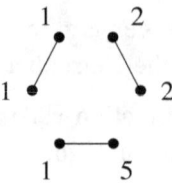

Figure 1. A ϕ-graph for $(11)(22)(15)$.

as follows: for every edge E labeled as $\overset{k}{\bullet} - \overset{l}{\bullet}$ define

$$\kappa(E, \tau) = C(k, l, \tau), \tag{134}$$

with

$$\kappa(\gamma_\phi, \tau) = \prod \kappa(E, \tau), \tag{135}$$

where the product is taken over all edges of γ_ϕ. Thus the ϕ-graph of Figure 1 has weight $C(1, 1)C(2, 2)C(1, 5) = -30 E_2(\tau) E_4(\tau) E_6(\tau)$.

Clearly Theorem 11.1 can now be restated in terms of graphs:

THEOREM 11.3. *For a Fock vector $v(\lambda)$ with label set $\Phi_\lambda = \{k_1, \ldots, k_n\}$ and even n*

$$Q_v(\tau) = \sum_{\gamma_\phi} \kappa(\gamma_\phi, \tau), \tag{136}$$

where the sum is taken over all independent ϕ-graphs for Φ_λ.

EXERCISE 11.4. For $v = a[-1]^3 a[-2]^2 a[-5]\mathbf{1}$ of weight $\text{wt}[v] = 12$ with $\Phi_\lambda = \{1, 1, 1, 2, 2, 5\}$ and $|F(\Phi_\lambda)| = 5!! = 15$ (cf. Exercise 10.7) show that

$$Q_v(\tau) = 6C(1, 1)C(1, 2)C(2, 5) + 3C(1, 1)C(2, 2)C(1, 5) + 6C(1, 2)^2 C(1, 5)$$
$$= 0 - 90 E_2(\tau) E_4(\tau) E_6(\tau) + 0.$$

Thus only 3 elements of $F(\Phi_\lambda)$ make a nonzero contribution to $Q_v(\tau)$.

EXERCISE 11.5. Find all the ϕ-graphs for $v = a[-1]^3 a[-2]^2 a[-5]\mathbf{1}$.

11.2. Sewing two tori. In this section we digress from VOA theory to briefly review some aspects of Riemann surface theory and the construction of a genus-two surface by sewing together two punctured tori. A genus-two Riemann surface can also be constructed by sewing a handle to a torus but we do not consider that situation here. For more details see [MT2], [MT4].

Let $\mathcal{S}^{(2)}$ denote a compact Riemann surface of genus-two and let a_1, a_2, b_1, b_2 be the canonical homology basis (see [FK], for example). There exists two holomorphic 1-forms v_i, $i = 1, 2$, which we may normalize by

$$\oint_{a_i} v_j = 2\pi i \delta_{ij}.$$

The genus-two period matrix Ω is defined by

$$\Omega_{ij} = \frac{1}{2\pi i}\oint_{b_i} v_j, \tag{137}$$

for $i, j = 1, 2$. Using the Riemann bilinear relations, one finds that Ω is a complex symmetric matrix with positive-definite imaginary part, i.e., $\Omega \in \mathfrak{H}_2$, the genus-two Siegel complex upper half-space.

The intersection form Ξ is a natural nondegenerate symplectic bilinear form on the first homology group $H_1(\mathcal{S}^{(2)}, \mathbb{Z}) \cong \mathbb{Z}^4$, satisfying

$$\Xi(a_i, a_j) = \Xi(b_i, b_j) = 0, \quad \Xi(a_i, b_j) = \delta_{ij}, \quad i, j = 1, 2.$$

The mapping class group is given by the symplectic group

$$\mathrm{Sp}(4, \mathbb{Z}) = \{\gamma = \begin{pmatrix} A & B \\ C & D \end{pmatrix} \in \mathrm{SL}(4, \mathbb{Z}) \mid$$
$$AB^T = BA^T, \, CD^T = D^T C = AD^T - BC^T = I_2\},$$

where A^T denotes the transpose of A. The group $\mathrm{Sp}(4, \mathbb{Z})$ acts on $\mathfrak{H}_2$ via

$$\gamma.\Omega = (A\Omega + B)(C\Omega + D)^{-1}, \tag{138}$$

and naturally on $H_1(\mathcal{S}, \mathbb{Z})$, where it preserves Ξ.

We now briefly review a general method originally due to Yamada [Y] and discussed at length in [MT2] for calculating the period matrix (and other structures) on a Riemann surface formed by sewing together two other Riemann surfaces. In particular, we wish to describe Ω_{ij} on a genus-two Riemann surface formed by sewing together two tori $\mathcal{S}_a$ for $a = 1, 2$ (Figure 11.2). Consider an oriented torus $\mathcal{S}_a = \mathbb{C}/\Lambda_{\tau_a}$ with lattice Λ_{τ_a} with basis $(2\pi i, 2\pi i \tau_a)$ for $\tau_a \in \mathfrak{H}$, the complex upper half plane. For local coordinate $z_a \in \mathcal{S}_a$ consider the closed disk $|z_a| \leq r_a$. This is contained in $\mathcal{S}_a$ provided $r_a < \frac{1}{2}D(q_a)$ where

$$D(q_a) = \min_{\lambda \in \Lambda_{\tau_a}, \lambda \neq 0} |\lambda|, \tag{139}$$

is the minimal lattice distance.

Introduce a sewing parameter $\varepsilon \in \mathbb{C}$ where $|\varepsilon| \leq r_1 r_2 < \frac{1}{4}D(q_1)D(q_2)$ and excise the disk $\{z_a, |z_a| \leq |\varepsilon|/r_{\bar{a}}\}$ centered at $z_a = 0$ to form a punctured torus

$$\hat{\mathcal{S}}_a = \mathcal{S}_a \setminus \{z_a, |z_a| \leq |\varepsilon|/r_{\bar{a}}\},$$

where we use the convention

$$\bar{1} = 2, \quad \bar{2} = 1. \tag{140}$$

Define the annulus

$$\mathcal{A}_a = \{z_a, |\varepsilon|/r_{\bar{a}} \leq |z_a| \leq r_a\} \subset \hat{\mathcal{S}}_a, \tag{141}$$

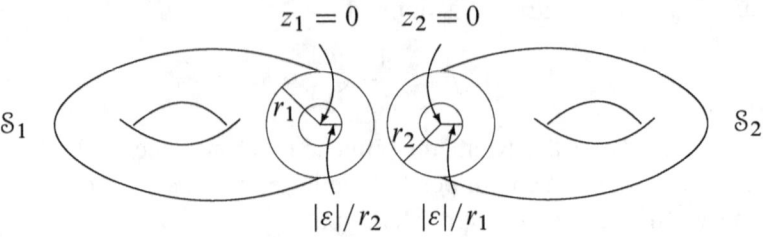

Figure 2. Sewing two tori.

We then identify $\mathcal{A}_1$ with $\mathcal{A}_2$ via the sewing relation

$$z_1 z_2 = \varepsilon, \tag{142}$$

to obtain an explicit construction of a genus-two Riemann surface

$$\mathcal{S}^{(2)} = \hat{\mathcal{S}}_1 \cup \hat{\mathcal{S}}_2 \cup (\mathcal{A}_1 \simeq \mathcal{A}_2),$$

which is parameterized by the domain

$$\mathcal{D}^\varepsilon = \{(\tau_1, \tau_2, \varepsilon) \in \mathfrak{H} \times \mathbb{H}_1 \times \mathbb{C} \mid |\varepsilon| < \frac{1}{4} D(q_1) D(q_2)\}. \tag{143}$$

In [Y], Yamada describes a general method for computing the period matrix on the sewn Riemann surface $\mathcal{S}^{(2)}$ in terms of data obtained from the two tori. This is described in detail in [MT2] where we obtain the explicit form for Ω in terms of the infinite matrix $A_a(\tau_a, \varepsilon) = (A_a(k, l, \tau_a, \varepsilon))$ for $k, l \geq 1$ where

$$A_a(k, l, \tau_a, \varepsilon) = \frac{\varepsilon^{(k+l)/2}}{\sqrt{kl}} C(k, l, \tau_a), \tag{144}$$

and where $C(k, l, \tau_a)$ is given in (126). Thus, dropping the subscript,

$$A(\tau, \varepsilon) = \begin{pmatrix} \varepsilon E_2(\tau) & 0 & \sqrt{3}\varepsilon^2 E_4(\tau) & 0 & \cdots \\ 0 & -3\varepsilon^2 E_4(\tau) & 0 & -5\sqrt{2}\varepsilon^3 E_6(\tau) & \cdots \\ \sqrt{3}\varepsilon^2 E_4(\tau) & 0 & 10\varepsilon^3 E_6(\tau) & 0 & \cdots \\ 0 & -5\sqrt{2}\varepsilon^3 E_6(\tau) & 0 & -35\varepsilon^4 E_8(\tau) & \cdots \\ \vdots & \vdots & \vdots & \vdots & \ddots \end{pmatrix}.$$

The matrices A_1, A_2 not only play a central rôle here but also later on in our discussion of the genus-two partition for the Heisenberg VOA M_0. In particular, the matrix $I - A_1 A_2$ and $\det(I - A_1 A_2)$ (where I is the infinite identity matrix here) are important, where $\det(I - A_1 A_2)$ is defined by

$$\log \det(I - A_1 A_2) = \operatorname{Tr} \log(I - A_1 A_2) = -\sum_{n \geq 1} \frac{1}{n} \operatorname{Tr}((A_1 A_2)^n). \tag{145}$$

These expressions are power series in $\mathfrak{Q}[\![\varepsilon]\!]$. One finds:

THEOREM 11.6 [MT2].
(a) *The infinite matrix*

$$(I - A_1 A_2)^{-1} = \sum_{n \geq 0} (A_1 A_2)^n, \qquad (146)$$

is convergent for $(\tau_1, \tau_2, \varepsilon) \in \mathcal{D}^\varepsilon$.
(b) $\det(I - A_1 A_2)$ *is nonvanishing and holomorphic on* $\mathcal{D}^\varepsilon$.

Furthermore we may obtain an explicit formula for the genus-two period matrix on $\mathcal{S}^{(2)}$:

THEOREM 11.7. *The sewing procedure determines a holomorphic map*

$$F^\varepsilon : \mathcal{D}^\varepsilon \to \mathfrak{H}_2, \quad (\tau_1, \tau_2, \varepsilon) \mapsto \Omega(\tau_1, \tau_2, \varepsilon), \qquad (147)$$

where $\Omega = \Omega(\tau_1, \tau_2, \varepsilon)$ *is given by*

$$2\pi i \Omega_{11} = 2\pi i \tau_1 + \varepsilon (A_2 (I - A_1 A_2)^{-1})(1, 1),$$
$$2\pi i \Omega_{22} = 2\pi i \tau_2 + \varepsilon (A_1 (I - A_2 A_1)^{-1})(1, 1),$$
$$2\pi i \Omega_{12} = -\varepsilon (I - A_1 A_2)^{-1}(1, 1).$$

Here $(1, 1)$ *refers to the* $(1, 1)$-*entry of a matrix.*

$\mathcal{D}^\varepsilon$ is preserved under the action of

$$G \simeq (\mathrm{SL}(2, \mathbb{Z}) \times \mathrm{SL}(2, \mathbb{Z})) \rtimes \mathbb{Z}_2,$$

the direct product of the left and right torus modular groups, which are interchanged upon conjugation by an involution β as follows:

$$\gamma_1.(\tau_1, \tau_2, \varepsilon) = \left(\frac{a_1 \tau_1 + b_1}{c_1 \tau_1 + d_1}, \tau_2, \frac{\varepsilon}{c_1 \tau_1 + d_1} \right),$$
$$\gamma_2.(\tau_1, \tau_2, \varepsilon) = \left(\tau_1, \frac{a_2 \tau_2 + b_2}{c_2 \tau_2 + d_2}, \frac{\varepsilon}{c_2 \tau_2 + d_2} \right),$$
$$\beta.(\tau_1, \tau_2, \varepsilon) = (\tau_2, \tau_1, \varepsilon), \qquad (148)$$

for $(\gamma_1, \gamma_2) \in \mathrm{SL}(2, \mathbb{Z}) \times \mathrm{SL}(2, \mathbb{Z})$ with $\gamma_i = \begin{pmatrix} a_i & b_i \\ c_i & d_i \end{pmatrix}$.

There is a natural injection $G \to \mathrm{Sp}(4, \mathbb{Z})$ in which the two $\mathrm{SL}(2, \mathbb{Z})$ subgroups are mapped to

$$\Gamma_1 = \left\{ \begin{bmatrix} a_1 & 0 & b_1 & 0 \\ 0 & 1 & 0 & 0 \\ c_1 & 0 & d_1 & 0 \\ 0 & 0 & 0 & 1 \end{bmatrix} \right\}, \quad \Gamma_2 = \left\{ \begin{bmatrix} 1 & 0 & 0 & 0 \\ 0 & a_2 & 0 & b_2 \\ 0 & 0 & 1 & 0 \\ 0 & c_2 & 0 & d_2 \end{bmatrix} \right\}, \qquad (149)$$

and the involution is mapped to

$$\beta = \begin{bmatrix} 0 & 1 & 0 & 0 \\ 1 & 0 & 0 & 0 \\ 0 & 0 & 0 & 1 \\ 0 & 0 & 1 & 0 \end{bmatrix}. \qquad (150)$$

Thus as a subgroup of $\mathrm{Sp}(4,\mathbb{Z})$, G also has a natural action on the Siegel upper half plane $\mathfrak{H}_2$ as given in (138). This action is compatible with respect to the map (147) which is directly related to the observation that $A_a(k, l, \tau_a, \varepsilon)$ of (144) is a modular form of weight $k+l$ for $k+l > 2$, whereas $A_a(1, 1, \tau_a, \varepsilon) = \varepsilon E_2(\tau_a)$ is a quasimodular form. The exceptional modular transformation property of the latter term (29) leads via Theorem 11.7 to the following result:

THEOREM 11.8. *F^ε is equivariant with respect to the action of G; i.e., there is a commutative diagram for $\gamma \in G$,*

$$\begin{array}{ccc} \mathcal{D}^\varepsilon & \xrightarrow{F^\varepsilon} & \mathfrak{H}_2 \\ \gamma \downarrow & & \downarrow \gamma \\ \mathcal{D}^\varepsilon & \xrightarrow{F^\varepsilon} & \mathfrak{H}_2 \end{array}$$

EXERCISE 11.9. Show that to $O(\varepsilon^4)$

$$2\pi i \Omega_{11} = 2\pi i \tau_1 + E_2(\tau_2)\varepsilon^2 + E_2(\tau_1)E_2(\tau_2)^2\varepsilon^4,$$
$$2\pi i \Omega_{22} = 2\pi i \tau_2 + E_2(\tau_1)\varepsilon^2 + E_2(\tau_1)^2 E_2(\tau_1)^2 \varepsilon^4,$$
$$2\pi i \Omega_{12} = -\varepsilon + E_2(\tau_1)E_2(\tau_2)\varepsilon^3.$$

11.3. The genus-two partition function for the Heisenberg VOA. In this section we define and compute the genus-two partition function for the Heisenberg VOA M_0 on the genus-two Riemann surface $\mathcal{S}^{(2)}$ described in the last section. The partition function is defined in terms of the genus-one 1-point functions $Z^{(1)}_{M_0}(v, \tau_a)$ on $\mathcal{S}_a = \mathbb{C}/\Lambda_{\tau_a}$ for all $v \in M$. The rationale behind this definition, which is strongly influenced by ideas in CFT, can be motivated by considering the following trivial sewing of a torus $\mathcal{S}_1 = \mathbb{C}/\Lambda_{\tau_1}$ to a Riemann sphere $\mathbb{CP}^1$. Let $z_1 \in \mathcal{S}_1$ and $z_2 \in \mathbb{CP}^1$ be local coordinates and define the sewing by identifying the annuli $r_a \geq |z_a| \geq |\varepsilon|r_{\bar{a}}^{-1}$ via the sewing relation $z_1 z_2 = \varepsilon$ (adopting the same notation as above). The resulting surface is a torus described by the same modular parameter τ_1.

Let V be a VOA with LiZ metric $\langle\,,\,\rangle$ and consider an n-point function[21] $F^{(1)}_V((v^1, x_1), \ldots (v^n, x_n), \tau_1)$ for $x_i \in \mathcal{A}_1$, the torus annulus (141). This can

---

[21] Here and below we include a superscript (1) to indicate the genus of the Riemann torus.

be expressed in terms of a 1-point function ([MT1], Lemma 3.1) by

$$F_V^{(1)}((v^1,x_1),\ldots(v^n,x_n),\tau_1) = Z_V^{(1)}(Y[v^1,x_1]\ldots Y[v^n,x_n]\mathbf{1},\tau_1)$$
$$= Z_V^{(1)}(Y[v^1,x_{1n}]\ldots Y[v^{n-1},x_{n-1n}]v^n,\tau_1), \quad (151)$$

for $x_{in} = x_i - x_n$ (see (132)). Denote the square bracket LiZ metric by $\langle\,,\,\rangle_{sq}$, and choose a basis $\{u\}$ of $V_{[r]}$ with dual basis $\{\bar{u}\}$ with respect to $\langle\,,\,\rangle_{sq}$. Expanding in this basis we find that for any $0 \leq k \leq n-1$

$$Y[v^{k+1},x_{k+1}]\ldots Y[v^n,x_n]\mathbf{1} = \sum_{r\geq 0}\sum_{u\in V_{[r]}} \langle \bar{u},Y[v^{k+1},x_{k+1}]\ldots Y[v^n,x_n]\mathbf{1}\rangle_{sq} u,$$

so that

$$F_V^{(1)}((v^1,x_1),\ldots(v^n,x_n),\tau_1) = \sum_{r\geq 0}\sum_{u\in V_{[r]}} Z_V^{(1)}(Y[v^1,x_1]\ldots Y[v^k,x_k]u,\tau_1)$$
$$\cdot \langle \bar{u},Y[v^{k+1},x_{k+1}]\ldots Y[v^n,x_n]\mathbf{1}\rangle_{sq}.$$

Using (151) we have

$$Z_V^{(1)}(Y[v^1,x_1]\ldots Y[v^k,x_k]u,\tau_1)$$
$$= \operatorname{Res}_{z_1} z_1^{-1} F_V^{(1)}((v^1,x_1),\ldots(v^k,x_k),(u,z_1),\tau_1). \quad (152)$$

Let us now assume that each v^i is quasiprimary of $L[0]$ weight $\operatorname{wt}[v^i]$ and let $y_i = \varepsilon/x_i \in \mathbb{CP}^1$. Then (109), (112), (98) and (103) respectively imply

$$\langle \bar{u},Y[v^{k+1},x_{k+1}]\ldots Y[v^n,x_n]\mathbf{1}\rangle_{sq}$$
$$= \langle \mathbf{1},Y^\dagger[v^n,x_n]\ldots Y^\dagger[v^{k+1},x_{k+1}]\bar{u}\rangle_{sq}$$
$$= \langle \mathbf{1},\varepsilon^{L[0]}Y^\dagger[v^n,x_n]\varepsilon^{-L[0]}\ldots \varepsilon^{L[0]}Y^\dagger[v^{k+1},x_{k+1}]\varepsilon^{-L[0]}\varepsilon^{L[0]}\bar{u}\rangle_{sq}$$
$$= \varepsilon^r \langle \mathbf{1},Y[v^n,y_n]\ldots Y[v^{k+1},y_{k+1}]\bar{u}\rangle_{sq} \prod_{k+1\leq j\leq n}\left(-\frac{\varepsilon}{x_j^2}\right)^{\operatorname{wt}[v^j]}$$
$$= \varepsilon^r \operatorname{Res}_{z_2} z_2^{-1} Z_V^{(0)}((v^n,y_n),\ldots(v^{k+1},y_{k+1}),(\bar{u},z_2)) \prod_{k+1\leq j\leq n}\left(\frac{dy_j}{dx_j}\right)^{\operatorname{wt}[v^j]}.$$

We are also making use here of the isomorphism between the round and square bracket formalisms in the identification of the genus zero correlation function. The result of these calculations is that, for any $0 \leq k \leq n-1$,

$$\mathcal{F}_V^{(1)}(v^1,\ldots,v^n;\tau_1) \equiv F_V^{(1)}((v^1,x_1),\ldots,(v^n,x_n),\tau_1) \prod_{1\leq i\leq n} dx_i^{\operatorname{wt}[v^i]} =$$

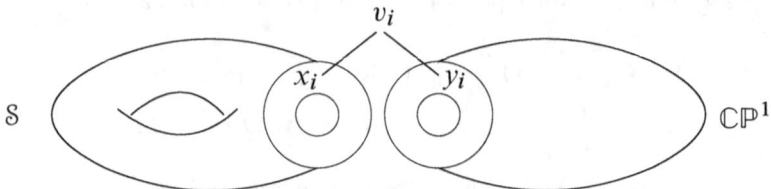

Figure 3. Equivalent insertion of v^i at x_i or $y_i = \varepsilon/x_i$.

$$= \sum_{r \geq 0} \varepsilon^r \sum_{u \in V_{[r]}} \left(\text{Res}_{z_1} z_1^{-1} F_V^{(1)}((v^1, x_1), \ldots (v^k, x_k), (u, z_1), \tau_1) \right.$$
$$\times \text{Res}_{z_2} z_2^{-1} F_V^{(0)}((v^{k+1}, y_{k+1}), \ldots (v^1, y_1), (\bar{u}, z_2))$$
$$\left. \times \prod_{1 \leq i \leq k} dx_i^{\text{wt}[v^i]} \prod_{k+1 \leq j \leq n} dy_j^{\text{wt}[v^j]} \right). \quad (153)$$

Following Exercises 10.10 and 10.11 the (formal) form $\mathcal{F}_V^{(1)}(v^1, \ldots, v^n; \tau_1)$ is invariant with respect to Möbius transformations. (Similarly to Remark 10.12, we note that $\mathcal{F}_V^{(1)}(v^1, \ldots v^n; \tau_1)$ is a conformally invariant global form on $\mathcal{S}_1$ for primary $v^1, \ldots, v^n$). Geometrically, (153) is telling us that we express $\mathcal{F}_V^{(1)}(v^1, \ldots, v^n; \tau_1)$ via the sewing procedure in terms of data arising from $\mathcal{F}_V^{(1)}(v^1, \ldots, v^k, u; \tau_1)$ and $\mathcal{F}_V^{(0)}(v^{k+1}, \ldots, v^1, \bar{u})$ (cf. (108)). Furthermore, we may choose to consider the contribution from a quasiprimary vector v^i as arising from either an "insertion" at $x_i \in \mathcal{S}_1$ or at the identified point $y_i = \varepsilon/x_i \in \mathbb{CP}^1$.

A special case of (153) is the partition (0-point) function for which we find the trivial identity

$$Z_V^{(1)}(\tau_1) = \sum_{r \geq 0} \varepsilon^r \sum_{u \in V_{[r]}} Z_V^{(1)}(u, \tau_1) \text{Res}_{z_2} z_2^{-1} F_V^{(0)}(\bar{u}, z_2) = Z_V^{(1)}(\tau_1) + 0, \quad (154)$$

since $F_V^{(0)}(\bar{u}, z_2) = 0$ for $\bar{u} \notin V_{[0]}$.

Motivated by this example, we define the genus-two partition function where we effectively replace the Riemann sphere in Figure 3, right, by a second torus $\mathcal{S}_2 = \mathbb{C}/\Lambda_{\tau_2}$ as described in the Section 11.2. Thus replacing the genus-zero 1-point function $F_V^{(0)}(\bar{u}, 0)$ of (154) by $Z_V^{(1)}(\bar{u}, \tau_2)$ we define the genus-two partition function for a VOA V with a LiZ metric by

$$Z_V^{(2)}(\tau_1, \tau_2, \varepsilon) = \sum_{r \geq 0} \varepsilon^r \sum_{u \in V_{[r]}} Z_V^{(1)}(u, \tau_1) Z_V^{(1)}(\bar{u}, \tau_2). \quad (155)$$

The inner sum is taken over any basis $\{u\}$ for $V_{[r]}$ with dual basis $\{\bar{u}\}$ with respect to the square bracket LiZ metric. Although the definition is associated with the specific genus-two sewing scheme, it is regarded at this stage as a purely formal expression which can be computed to any given order in ε. One can also define genus-two correlation functions by inserting appropriate genus-one correlation functions in (155). We do not consider these here.

Let us now compute the genus-two partition function for the rank one Heisenberg VOA M_0 generated by a of weight 1. We employ the square bracket Fock basis of (124) which we alternatively notate here (cf. (118)) by

$$v = v(\lambda) = a[-1]^{e_1} \ldots a[-p]^{e_p} \mathbf{1}, \quad (156)$$

for nonnegative integers e_i. We recall that $v(\lambda)$ is of square bracket weight $\mathrm{wt}[v] = \sum_i i e_i$ and is described by a label set $\Phi_\lambda = \{1, \ldots, p\}$ with $n = \sum e_i$ elements corresponding to an unrestricted partition $\lambda = \{1^{e_1} \ldots p^{e_p}\}$ of $\mathrm{wt}[v]$. The Fock vectors (156) form a diagonal basis for the LiZ metric $\langle \, , \, \rangle_{\mathrm{sq}}$ with

$$\bar{v} = \frac{1}{\prod_{1 \leq i \leq p} (-i)^{e_i} e_i!} v, \quad (157)$$

from (119). Following (155), we find

$$Z_{M_0}^{(2)}(\tau_1, \tau_2, \varepsilon) = \sum_{v \in V} \frac{\varepsilon^{\mathrm{wt}[v]}}{\prod_i (-i)^{e_i} e_i!} Z_{M_0}^{(1)}(v, \tau_1) Z_{M_0}^{(1)}(v, \tau_2), \quad (158)$$

where the sum is taken over the basis (156). $Z_{M_0}^{(2)}(\tau_1, \tau_2, \varepsilon)$ is given by the following closed formula [MT4]:

THEOREM 11.10. *The genus-two partition function for the rank one Heisenberg VOA is*

$$Z_{M_0}^{(2)}(\tau_1, \tau_2, \varepsilon) = \frac{1}{\eta(\tau_1)\eta(\tau_2)} (\det(I - A_1 A_2))^{-1/2}, \quad (159)$$

with A_a of (144).

PROOF. The proof relies on an interesting graph-theoretic interpretation of (158). This follows the technique introduced in Theorem 11.3 for graphically interpreting the genus-one 1-point function $Z_{M_0}^{(1)}(v(\lambda), \tau_1)$ in terms the sum of weights for the ϕ-graphs. We sketch the main features of the proof leaving the interested reader to explore the details in [MT4].

Since $v(\lambda)$ is indexed by unrestricted partitions $\lambda = \{1^{e_1}, 2^{e_2}, \ldots\}$ we may write (158) as

$$Z_{M_0}^{(2)}(\tau_1, \tau_2, \varepsilon) = \sum_{\lambda = \{i^{e_i}\}} \frac{1}{\prod_i e_i!} \prod_i \left(\frac{\varepsilon^i}{-i}\right)^{e_i} Z_{M_0}^{(1)}(v(\lambda), \tau_1) Z_{M_0}^{(1)}(v(\lambda), \tau_2). \quad (160)$$

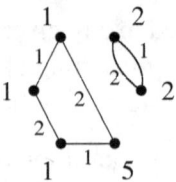

Figure 4. A chequered diagram.

Theorem 11.3 implies $Z^{(1)}_{M_0}(v(\lambda), \tau_1) = 0$ for odd $n = \sum e_i$ whereas for n even

$$Z^{(1)}_{M_0}(v(\lambda), \tau_1) Z^{(1)}_{M_0}(v(\lambda), \tau_2) = \frac{1}{\eta(\tau_1)\eta(\tau_2)} \sum_{\gamma_{\phi_1}} \sum_{\gamma_{\phi_2}} \kappa(\gamma_{\phi_1}, \tau_1) \kappa(\gamma_{\phi_2}, \tau_2),$$

where $\gamma_{\phi_1}, \gamma_{\phi_2}$ independently range over the ϕ–graphs for Φ_λ. Any pair $\gamma_{\phi_1}, \gamma_{\phi_2}$ can be naturally combined to form a *chequered diagram* D consisting of n vertices labelled by Φ_λ of valence 2 with n unoriented edges $\overset{k\ a\ l}{\bullet\!-\!\bullet}$ consecutively labelled by $a = 1, 2$ as specified by $\phi_a = \ldots (kl) \ldots$. Following Exercise 10.7 there are $(n!!)^2$ chequered diagrams for a given $v(\lambda)$. We illustrate an example of such a diagram in Figure 4 for $v = a[-1]^3 a[-2]^2 a[-5]\mathbf{1}$ with ϕ_1 of Figure 1 and a separate choice for ϕ_2 with cycle shape $(11)(22)(15)$

For $\lambda = \{1^{e_1} \ldots p^{e_p}\}$ the symmetric group $\Sigma(\Phi_\lambda)$ acts on the chequered diagrams which have Φ_λ as underlying set of labeled nodes. We define $\mathrm{Aut}(D)$, the *automorphism group of* D, to be the subgroup of $\Sigma(\Phi_\lambda)$ which preserves node labels. $\mathrm{Aut}(D)$ is isomorphic to $\Sigma_{e_1} \times \cdots \times \Sigma_{e_p}$ of order $|\mathrm{Aut}(D)| = \prod_i e_i!$. We may thus express (160) as a sum over the isomorphism classes of chequered diagrams D with

$$Z^{(2)}_{M_0}(\tau_1, \tau_2, \varepsilon) = \frac{1}{\eta(\tau_1)\eta(\tau_2)} \sum_D \frac{\zeta(D)}{|\mathrm{Aut}(D)|},$$

and

$$\zeta(D) = \prod_i \left(\frac{\varepsilon^i}{i}\right)^{e_i} \kappa(\gamma_{\phi_1}, \tau_1) \kappa(\gamma_{\phi_2}, \tau_2), \tag{161}$$

where D is determined by $\gamma_{\phi_1}, \gamma_{\phi_2}$ and noting that $\prod_i (-1)^{e_i} = 1$ for n even. From (135) we recall that $\kappa(\gamma_{\phi_a}, \tau_1)$ is a product of the weights of the a labelled edges. Then $\zeta(D)$ can be more naturally expressed in terms of a weight function on chequered diagrams defined by

$$\zeta(D) = \Pi_E \zeta(E), \tag{162}$$

where the product is taken over the edges E of D and where for an edge E

labeled $\overset{k\ a\ l}{\bullet-\bullet}$ we define

$$\zeta(E) = \frac{\varepsilon^{\frac{k+l}{2}}}{\sqrt{kl}} C(k,l,\tau_a) = A_a(k,l,\tau_a,\varepsilon),$$

for A_a of (144).

Every chequered diagram can be formally represented as a product

$$D = \prod_i L_i^{m_i},$$

with D a disjoint union of unoriented chequered cycles (connected diagrams) L_i with multiplicity m_i (e.g. the chequered diagram of Figure 4 is the product of two disjoint cycles). Then $\mathrm{Aut}(D)$ is isomorphic to the direct product of the groups $\mathrm{Aut}(L_i^{m_i})$ of order $|\mathrm{Aut}(L_i^{m_i})| = |\mathrm{Aut}(L_i)|^{m_i}$ so that

$$|\mathrm{Aut}(D)| = \prod_i |\mathrm{Aut}(L_i)|^{m_i} m_i!.$$

But from (162) it is clear that $\zeta(D)$ is multiplicative over disjoint unions of diagrams, and we find

$$\sum_D \frac{\zeta(D)}{|\mathrm{Aut}(D)|} = \prod_L \sum_{m \geq 0} \frac{\zeta(L)^m}{|\mathrm{Aut}(L)|^m m!} = \exp \sum_L \frac{\zeta(L)}{|\mathrm{Aut}(L)|},$$

where L ranges over isomorphism classes of unoriented chequered cycles. Further analysis shows that [MT4]

$$\sum_L \frac{\zeta(L)}{|\mathrm{Aut}(L)|} = \tfrac{1}{2} \mathrm{Tr} \sum_{n \geq 1} \frac{1}{n}(A_1 A_2)^n = -\tfrac{1}{2} \mathrm{Tr} \log(1 - A_1 A_2),$$

so that we find

$$\sum_D \frac{\zeta(D)}{|\mathrm{Aut}(D)|} = (\det(1 - A_1 A_2))^{-1/2},$$

following (145). Thus Theorem 11.10 holds. □

The convergence and holomorphy of the determinant is the subject of Theorem 11.6 (b) so that having computed the closed formula (159) we may conclude that $Z_{M_0}^{(2)}(\tau_1, \tau_2, \varepsilon)$ is not just a formal function but can be evaluated on $\mathcal{D}^\varepsilon$ to find

THEOREM 11.11. *$Z_{M_0}^{(2)}(\tau_1, \tau_2, \varepsilon)$ is holomorphic on the domain $\mathcal{D}^\varepsilon$.*

We next consider the automorphic properties of $Z_{M_0}^{(2)}(\tau_1, \tau_2, \varepsilon)$ with respect to the modular group $G \subset \mathrm{Sp}(4, \mathbb{Z})$ of (148) which acts on $\mathcal{D}^\varepsilon$. We first recall a little from the classical theory of modular forms (cf. Section 3). For a meromorphic function $f(\tau)$ on $\mathfrak{H}$, $k \in \mathbb{Z}$ and $\gamma = \begin{pmatrix} a & b \\ c & d \end{pmatrix} \in \mathrm{SL}(2,\mathbb{Z})$, we define the right action

$$f(\tau)|_k \gamma = f(\gamma\tau)(c\tau + d)^{-k}, \tag{163}$$

where, as usual

$$\gamma\tau = \frac{a\tau + b}{c\tau + d}.$$

$f(\tau)$ is called a weak modular form for a subgroup $\Gamma \subseteq \mathrm{SL}(2,\mathbb{Z})$ of weight k if $f(\tau)|_k \gamma = f(\tau)$ for all $\gamma \in \Gamma$.

We have already discussed the (genus-one) partition function for the rank n Heisenberg VOA $V = M_0^{\otimes n}$ in Section 4.2 (cf. (41)). In particular, for $n = 2$ we have

$$Z_{M_0^2}^{(1)}(\tau) = Z_{M_0}^{(1)}(\tau)^2 = \frac{1}{\eta(\tau)^2}.$$

Then we find

$$Z_{M_0^2}^{(1)}(\tau)|_{-1}\gamma = \chi(\gamma) Z_{M_0^2}^{(1)}(\tau), \tag{164}$$

where χ is a character of $\mathrm{SL}(2,\mathbb{Z})$ of order 12 (cf. Exercise 8.5 and [Se]), and

$$Z_{M_0^{24}}^{(1)}(\tau)^{-1} = \Delta(\tau). \tag{165}$$

Similarly, we consider the genus-two partition function for the rank two Heisenberg VOA given by

$$Z_{M_0^2}^{(2)}(\tau_1, \tau_2, \varepsilon) = Z_{M_0}^{(2)}(\tau_1, \tau_2, \varepsilon)^2 = \frac{1}{\eta(\tau_1)^2 \eta(\tau_2)^2 \det(I - A_1 A_2)}. \tag{166}$$

Analogously to (163), we define for all $\gamma \in G$

$$f(\tau_1, \tau_2, \varepsilon)|_k \gamma = f(\gamma(\tau_1, \tau_2, \varepsilon)) \det(C\Omega + D)^{-k}, \tag{167}$$

where the action of γ on the right-hand-side is as in (148) and $\Omega(\tau_1, \tau_2, \varepsilon)$ is determined by Theorem 11.7. Then (167) defines a right action of G on functions $f(\tau_1, \tau_2, \varepsilon)$. We next obtain a natural genus-two extension of (164). Define the a character $\chi^{(2)}$ of G by

$$\chi^{(2)}(\gamma_1 \gamma_2 \beta^m) = (-1)^m \chi(\gamma_1) \chi(\gamma_2), \quad \gamma_i \in \Gamma_i, \ i = 1, 2.$$

with Γ_i, β of (149) and (150). $\chi^{(2)}$ takes values which are twelfth roots of unity. Then, much as for Theorem 11.7, the exceptional transformation law of $A_a(1, 1, \tau_a, \varepsilon) = E_2(\tau_a)$ implies that

THEOREM 11.12. *If $\gamma \in G$ then*

$$Z_{M_0^2}^{(2)}(\tau_1, \tau_2, \varepsilon)|_{-1}\gamma = \chi^{(2)}(\gamma) Z_{M_0^2}^{(2)}(\tau_1, \tau_2, \varepsilon).$$

The definition (167) is analogous to that for a *Siegel modular form* for the symplectic group $\mathrm{Sp}(4, \mathbb{Z})$ defined as follows (e.g. [Fr]). For a meromorphic function $F(\Omega)$ on $\mathfrak{H}_2$, $k \in \mathbb{Z}$ and $\gamma \in \mathrm{Sp}(4, \mathbb{Z})$, we define the right action

$$F(\Omega)|_k \gamma = F(\gamma.\Omega) \det(C\Omega + D)^{-k}, \tag{168}$$

with $\gamma.\Omega$ of (138). $F(\Omega)$ is called a modular form for $\Gamma \subseteq \mathrm{Sp}(4, \mathbb{Z})$ of weight k if $F(\Omega)|_k \gamma = F(\Omega)$ for all $\gamma \in \Gamma$.

Theorem 11.12 implies that for the rank 24 Heisenberg VOA M_0^{24}

$$Z^{(2)}_{M_0^{24}}(\tau_1, \tau_2, \varepsilon)|_{-12}\gamma = Z^{(2)}_{M_0^{24}}(\tau_1, \tau_2, \varepsilon), \tag{169}$$

for all $\gamma \in G$. This might lead one to speculate that, analogously to the genus-one case in (165), $Z^{(2)}_{M_0^{24}}(\tau_1, \tau_2, \varepsilon)^{-1}$ is a holomorphic Siegel modular form of weight 12. Indeed, there does exist a unique holomorphic Siegel 12 form, $\Delta^{(2)}_{12}(\Omega)$, such that

$$\Delta^{(2)}_{12}(\Omega) \to \Delta(\tau_1)\Delta(\tau_2),$$

as $\varepsilon \to 0$, but explicit calculations show that $Z^{(2)}_{M_0^{24}}(\tau_1, \tau_2, \varepsilon)^{-1} \neq \Delta^{(2)}_{12}(\Omega)$. In any case, we cannot naturally extend the action of G on $\mathcal{D}^\varepsilon$ to $\mathrm{Sp}(4, \mathbb{Z})$. These observations are strongly expected to be related to the conformal anomaly [BK] in string theory and to the non-existence of a global section for the Hodge line bundle in algebraic geometry [Mu2].

Siegel modular forms do arise in the determination of the genus-two partition function for a lattice VOA V_L for even lattice L of rank l (and conjecturally for all rational theories) as follows. We recall the genus-one partition function for V_L is (cf. Section 7.3)

$$Z^{(1)}_{V_L}(\tau) = Z^{(1)}_{M_0^l}(\tau)\theta^{(1)}_L(\tau), \tag{170}$$

for $\theta^{(1)}_L(\tau) = \sum_\alpha q^{(\alpha,\alpha)/2}$. In the genus-two case, we may define the Siegel lattice theta function by [Fr]

$$\theta^{(2)}_L(\Omega) = \sum_{\alpha,\beta \in L} \exp(\pi i ((\alpha,\alpha)\Omega_{11} + 2(\alpha,\beta)\Omega_{12} + (\beta,\beta)\Omega_{22})).$$

$\theta^{(2)}_L(\Omega)$ is a Siegel modular form of weight $l/2$ for a subgroup of $\mathrm{Sp}(4, \mathbb{Z})$. The genus-one result (170) is naturally generalized to find [MT4]:

THEOREM 11.13. *For a lattice VOA V_L we have*

$$Z^{(2)}_{V_L}(\tau_1, \tau_2, \varepsilon) = Z^{(2)}_{M_0^l}(\tau_1, \tau_2, \varepsilon)\theta^{(2)}_L(\Omega).$$

EXERCISE 11.14. Show that $Z^{(2)}_{M_0}(\tau_1, \tau_2, \varepsilon)$ to $O(\varepsilon^4)$ is given by

$$\frac{1}{\eta(\tau_1)\eta(\tau_2)}\Big(1 + \tfrac{1}{2}E_2(\tau_1)E_2(\tau_2)\varepsilon^2 \\ + \big(\tfrac{3}{8} E_2(\tau_1)^2 E_2(\tau_2)^2 + \tfrac{15}{2} E_4(\tau_1)E_4(\tau_2)\big)\varepsilon^4\Big).$$

EXERCISE 11.15. Verify (159) to $O(\varepsilon^4)$ by showing that

$$\det(I - A_1 A_2) = 1 - E_2(\tau_1)E_2(\tau_2)\varepsilon^2 - 15\, E_4(\tau_1)E_4(\tau_2)\varepsilon^4 + O(\varepsilon^6).$$

12. Exceptional VOAs and the Virasoro algebra

In this section we review some recent research concerning a rôle played by the Virasoro algebra in certain exceptional VOAs [T2], [T3]. We will mainly concern ourselves here with simple VOAs V of strong CFT-type for which $\dim V_1 > 0$. We construct certain quadratic Casimir vectors from the elements of V_1 and examine the constraints on V arising from the assumption that the Casimir vectors of low weight are Virasoro descendants of the vacuum. This sort of assumption is similar to that of 'minimality' in the holomorphic VOAs $V^{(k)}$ that we discussed in Section 9.4. In particular we discuss how a special set of simple Lie algebras: $A_1, A_2, G_2, D_4, F_4, E_6, E_7, E_8$, known as Deligne's exceptional series [De], arises in this context. We also show that the genus-one partition function is determined by the same Virasoro condition. These constraints follow from an analysis of appropriate genus zero matrix elements and genus-one 2-point functions. In particular, we will make a relatively elementary use of rational matrix elements, the LiZ metric, the Zhu reduction formula and modular differential equations. As such, this example offers a useful and explicit application of many of the concepts reviewed in these notes.

12.1. Quadratic Casimirs and genus-zero constraints.

Consider a simple VOA V of strong CFT-type of central charge c with $d = \dim V_1 > 0$. From Theorem 10.15, V possesses an LiZ metric $\langle\,,\,\rangle$, i.e., a unique (nondegenerate) normalized symmetric bilinear form. For $a, b \in V_1$ define $[a,b] \equiv a_0 b$. From Exercise 9.7 this defines a Lie algebra on V_1 with invariant bilinear form $\langle\,,\,\rangle$. We denote this Lie algebra by $\mathfrak{g}$. The modes of elements of V_1 satisfy the Kac–Moody algebra (cf. Exercise 12.7)

$$[a_m, b_n] = [a,b]_{m+n} - m\langle a,b\rangle \delta_{m+n,0}. \tag{171}$$

which we denote by $\hat{\mathfrak{g}}$.

Let $\{u^\alpha \mid \alpha = 1\ldots d\}$ and $\{\bar{u}^\beta \mid \beta = 1\ldots d\}$ denote a $\mathfrak{g}$-basis and LiZ dual basis respectively. Define the *quadratic Casimir vectors* by

$$\lambda^{(n)} = u^\alpha_{1-n}\bar{u}^\alpha \in V_n, \tag{172}$$

where α is summed. Since $u^\alpha \in V_1$ is a primary vector it follows that $[L_m, u_n^\alpha] = -nu_{m+n}^\alpha$ and hence

$$L_m \lambda^{(n)} = (n-1)\lambda^{(n-m)} \text{ for } m > 0. \tag{173}$$

Let Vir_c denote the subVOA of V generated by the Virasoro vector ω. We then find:

LEMMA 12.1. *The LiZ metric is nondegenerate on* Vir_c.

PROOF. Let $v = L_{-n_1} L_{-n_2} \ldots L_{-n_k} \mathbf{1} \in \text{Vir}_c$. Then (105) gives

$$\langle v, a \rangle = \langle \mathbf{1}, L_{n_k} \ldots L_{n_2} L_{n_1} a \rangle = 0,$$

for $a \in V \setminus \text{Vir}_c$. Since $\langle \, , \, \rangle$ is nondegenerate on V it must be nondegenerate on Vir_c. □

REMARK 12.2. This implies from Theorem 10.15 that Vir_c is simple with $c \neq c_{p,q}$ of (123).

We now consider the constraints on $\mathfrak{g}$ that follow from assuming that $\lambda^{(n)} \in \text{Vir}_c$ for small n.[22] Firstly let us note [Mat]

LEMMA 12.3. *If* $\lambda^{(n)} \in \text{Vir}_c$ *then* $\lambda^{(m)} \in \text{Vir}_c$ *and is uniquely determined for all* $m \leq n$.

PROOF. If $\lambda^{(n)} \in \text{Vir}_c$ then $\lambda^{(n)} = \sum_{v \in (\text{Vir}_c)_n} \langle \bar{v}, \lambda^{(n)} \rangle v$ summing over a basis for $(\text{Vir}_c)_n$. But $\langle \bar{v}, \lambda^{(n)} \rangle$ is uniquely determined by repeated use of (173) and Exercise 12.8. Furthermore, for $m \leq n$ we have $\lambda^{(m)} = \frac{1}{n-1} L_{n-m} \lambda^{(n)} \in (\text{Vir}_c)_m$. □

It follows that $\lambda^{(2)} \in \text{Vir}_c$ implies

$$\lambda^{(2)} = -\frac{2d}{c}\omega, \tag{174}$$

where $c \neq 0$ following Remark 12.2. Note that for $\mathfrak{g}$ simple, this is the standard Sugawara construction for ω of (69). Similarly $\lambda^{(4)} \in \text{Vir}_c$ implies

$$\lambda^{(4)} = -\frac{3d}{c(5c+22)}\left(4L_{-2}^2 \mathbf{1} + (2+c) L_{-4} \mathbf{1}\right), \tag{175}$$

with $c \neq 0, -\frac{22}{5}$ following Remark 12.2. □

---

[22] The original motivation, due to Matsuo [Mat], for considering quadratic Casimirs is that both they and Vir_c are invariant under the automorphism group of V. Matsuo considered VOAs for which the automorphism invariants of V_n consist *only* of Virasoro descendents for small n. Hence for these VOAs it necessarily follows that $\lambda^{(n)} \in \text{Vir}_c$.

We next consider the constraints on $\mathfrak{g}$ if either (174) or (175) hold. We do this by analysing the following genus zero matrix element

$$F(a, b; x, y) = \langle a, Y(u^\alpha, x)Y(\bar{u}^\alpha, y)b \rangle, \tag{176}$$

where α is summed and $a, b \in V_1$. Using associativity and (172) we find

$$F(a, b; x, y) = \langle a, Y(Y(u^\alpha, x-y)\bar{u}^\alpha, y)b \rangle$$
$$= \frac{1}{(x-y)^2} \sum_{n \geq 0} \langle a, o(\lambda^{(n)})b \rangle \left(\frac{x-y}{y}\right)^n, \tag{177}$$

where $o(\lambda^{(n)}) = \lambda_{n-1}$ from (37). Thus Exercise 12.8 implies

$$F(a, b; x, y) = \frac{1}{(x-y)^2}\left(-d\langle a, b \rangle + 0 + \langle a, o(\lambda^{(2)})b \rangle \left(\frac{x-y}{y}\right)^2 + \cdots\right). \tag{178}$$

Alternatively, we also have

$$F(a, b; x, y) = \langle a, Y(u^\alpha, x)e^{yL_{-1}}Y(b, -y)\bar{u}^\alpha \rangle$$
$$= \langle a, e^{yL_{-1}}Y(u^\alpha, x-y)Y(b, -y)\bar{u}^\alpha \rangle$$
$$= \langle e^{yL_1}a, Y(u^\alpha, x-y)Y(b, -y)\bar{u}^\alpha \rangle$$
$$= \langle a, Y(u^\alpha, x-y)Y(b, -y)\bar{u}^\alpha \rangle$$
$$= \frac{1}{y^2} \sum_{m \geq 0} \langle a, u^\alpha_{m-1}b_{1-m}\bar{u}^\alpha \rangle \left(\frac{-y}{x-y}\right)^m$$
$$= \frac{1}{y^2}\left(\langle a, u^\alpha_{-1}b_1\bar{u}^\alpha \rangle - \langle a, u^\alpha_0 b_0 \bar{u}^\alpha \rangle \frac{y}{x-y} + \cdots\right),$$

using skew-symmetry and translation (cf. Exercises 2.15 and 2.16), invariance of the LiZ metric and that a is primary. The leading term is

$$\langle a, u^\alpha_{-1}b_1\bar{u}^\alpha \rangle = -\langle a, u^\alpha \rangle\langle b, \bar{u}^\alpha \rangle = -\langle a, b \rangle.$$

The next to leading term is

$$-\langle a, u^\alpha_0 b_0 \bar{u}^\alpha \rangle = \langle u^\alpha, a_0 b_0 \bar{u}^\alpha \rangle = K(a, b),$$

the Lie algebra Killing form

$$K(a, b) = Tr_\mathfrak{g}(a_0 b_0). \tag{179}$$

Thus we have

$$F(a, b; x, y) = \frac{1}{y^2}\left(-\langle a, b \rangle + K(a, b)\frac{y}{x-y} + \cdots\right). \tag{180}$$

From Theorem 10.1 we know that $F(a, b; x, y)$ is given by a rational function

$$F(a, b; x, y) = \frac{f(a, b; x, y)}{x^2 y^2 (x-y)^2}, \tag{181}$$

where $f(a, b; x, y)$ is a homogeneous polynomial of degree 4. Furthermore $f(a, b; x, y)$ is clearly symmetric in x, y so that it may parameterized

$$f(a, b; x, y) = p(a, b) x^2 y^2 + q(a, b) xy(x-y)^2 + r(a, b)(x-y)^4, \tag{182}$$

for some bilinears $p(a, b)$, $q(a, b)$ and $r(a, b)$. We find:

PROPOSITION 12.4. $p(a, b), q(a, b), r(a, b)$ are given by

$$p(a, b) = -d\langle a, b\rangle, \tag{183}$$
$$q(a, b) = K(a, b) - 2\langle a, b\rangle, \tag{184}$$
$$r(a, b) = -\langle a, b\rangle. \tag{185}$$

PROOF. Expanding (181) in $(x-y)/y$ we have

$$F(a, b; x, y) = \frac{1}{(x-y)^2} \left[p(a, b) + q(a, b) \left(\frac{x-y}{y} \right)^2 + \cdots \right], \tag{186}$$

whereas expanding (181) in $y/(x-y)$ gives

$$F(a, b; x, y) = \frac{1}{y^2} \left[r(a, b) + (-2r(a, b) + q(a, b)) \frac{y}{x-y} + \cdots \right]. \tag{187}$$

Comparing to (178) and (180) gives the result. □

We next show that if $\lambda^{(2)} \in \mathrm{Vir}_c$ then the Killing form is proportional to the LiZ metric:

PROPOSITION 12.5. If $\lambda^{(2)} \in \mathrm{Vir}_c$ then

$$K(a, b) = -2\langle a, b\rangle \left(\frac{d}{c} - 1 \right), \tag{188}$$

so that

$$f(a, b; x, y) = -\langle a, b\rangle \left(dx^2 y^2 + \frac{2d}{c} xy(x-y)^2 + (x-y)^4 \right). \tag{189}$$

PROOF. Equation (174) implies $o(\lambda^{(2)}) = -\frac{2d}{c} L_0$. Comparing the next to leading terms in (178) and (186) we find

$$q(a, b) = \langle a, o(\lambda^{(2)}) b\rangle = -\frac{2d}{c} \langle a, b\rangle,$$

which implies the result. □

Since the LiZ metric is nondegenerate, it follows from Cartan's criterion in Lie theory that $\mathfrak{g}$ is solvable for $d = c$ and is semisimple for $d \neq c$, i.e.,

$$\mathfrak{g} = \mathfrak{g}^1 \oplus \mathfrak{g}^2 \oplus \cdots \oplus \mathfrak{g}^r,$$

for simple components $\mathfrak{g}^i$ of dimension d^i. The corresponding Kac–Moody algebra $\hat{\mathfrak{g}}^i$ has level $l^i = -\frac{1}{2}\langle \alpha^i, \alpha^i \rangle$ where α^i is a long root[23] so that the dual Coxeter number is

$$h_i^\vee = l^i \left(\frac{d}{c} - 1 \right). \tag{190}$$

Furthermore, (174) implies that $\omega = \sum_{1 \le i \le r} \omega^i$ with ω^i the Sugawara Virasoro vector for central charge $c^i = l^i d^i / (l^i + h_i^\vee)$ for the simple component $\hat{\mathfrak{g}}^i$. It follows that for each component

$$\frac{d^i}{c^i} = \frac{d}{c}, \tag{191}$$

so that

$$\lambda^{i(2)} = -\frac{2d}{c} \omega^i, \tag{192}$$

for the quadratic Casimir on $\mathfrak{g}^i$.

We next show that if $\lambda^{(4)} \in \mathrm{Vir}_c$ then $\mathfrak{g}$ must be simple. Let L_n^i denote the modes of ω^i and $L_n = \sum_i L_n^i$ denote the modes of ω with $[L_m^i, L_n^j] = 0$ for $i \neq j$. Using $\lambda^{(4)} = \sum_i \lambda^{i(4)}$ (for quadratic Casimirs on $\mathfrak{g}^i$) it follows from (173) that

$$L_2^i \lambda^{(4)} = 3 \lambda^{i(2)}. \tag{193}$$

Since L_n^i satisfies the Virasoro algebra of central charge c^i we find

$$L_2^i L_{-2}^2 \mathbf{1} = 8\omega^i + c^i \omega, \qquad L_2^i L_{-4} \mathbf{1} = 6\omega^i.$$

If $\lambda^{(4)} \in \mathrm{Vir}_c$ then (175) holds and hence

$$L_2^i \lambda^{(4)} = -\frac{3d}{c(5c+22)} \left((44 + 6c)\omega^i + 4c^i \omega \right).$$

Equating to (193) and using (192) implies that

$$\omega^i = \frac{c^i}{c} \omega.$$

But since the Virasoro vectors $\omega^1, \ldots \omega^r$ are independent it follows that $r = 1$; i.e., $\mathfrak{g}$ is a simple Lie algebra.

If (175) holds one also finds that

$$\langle a, o(\lambda^{(4)}) b \rangle = -\frac{9d(6+c)}{c(5c+22)} \langle a, b \rangle.$$

---

[23] Then $(a,b)_i \equiv -\langle a,b \rangle / l_i$ is the unique nondegenerate form on $\hat{\mathfrak{g}}_i^{(l_i)}$ with normalization $(\alpha_i, \alpha_i)_i = 2$.

Comparing to the corresponding term in (177) this results in a further constraint on the parameters d, c in (189) given by

$$d = \frac{c(5c+22)}{10-c}. \tag{194}$$

Notice that the numerator vanishes for $c = 0, -22/5$, the zeros of the Kac determinant $\det M_4(c)$ (122).

For integral $d > 0$ there are only 42 rational values of c satisfying (194). This list is further restricted by the possible values of d for $\mathfrak{g}$ simple. The level l is necessarily rational from (190). Restricting l to be integral (for example, if V is assumed to be C_2-cofinite [DM1]) we find that $l = 1$ and $\mathfrak{g}$ must be one of Deligne's exceptional Lie algebras:

THEOREM 12.6. *Suppose $\lambda^{(4)} \in \mathrm{Vir}_c$.*
(a) *$\mathfrak{g}$ is a simple Lie algebra.*
(b) *If c is rational and the level l of $\hat{\mathfrak{g}}$ is integral then*

$$\mathfrak{g} = A_1, A_2, G_2, D_4, F_4, E_6, E_7 \text{ or } E_8,$$

with dual Coxeter number

$$h^\vee = \frac{d}{c} - 1 = \frac{12 + 6c}{10 - c},$$

for central charge $c = 1, 2, \frac{14}{5}, 4, \frac{26}{5}, 6, 7, 8$ respectively and level $l = 1$.

The simple Lie algebras appearing in Theorem 12.6 are known as Deligne's exceptional Lie algebras [De]. These algebras are of particular interest because not only is the dimension d of the adjoint representation $\mathfrak{g}$ described by a rational function of c in (194) but also the dimensions of the irreducible representations that arise in decomposition of up to four tensor products of $\mathfrak{g}$. In Deligne's original calculations, these dimensions were expressed as rational functions of a convenient parameter λ. In this VOA setting we instead employ the canonical parameter c, where

$$\lambda = \frac{c-10}{2+c}.$$

EXERCISE 12.7. Verify (171).

EXERCISE 12.8. Show that $\lambda^{(0)} = -d\mathbf{1}$ and $\lambda^{(1)} = 0$.

EXERCISE 12.9. Verify (174).

EXERCISE 12.10. Verify (175) using (121).

12.2. Genus-one constraints from quadratic Casimirs.

We next consider the constraints on the genus-one partition function $Z_V(\tau)$ that follow if $\lambda^{(4)} \in \mathrm{Vir}_c$. We will show that in this case, $Z_V(\tau)$ is the unique solution to a second order Modular Linear Differential Equation (MLDE) (cf. Section 8.2). As a consequence, we prove that $V = L_{\mathfrak{g}}(1,0)$, the level 1 WZW VOA where $\mathfrak{g}$ is an Deligne exceptional series. To prove this we apply both versions of Zhu's recursion formulas (Theorems 5.7 and 5.10). In particular, we evaluate the 1-point correlation function for a Virasoro descendent of the vacuum from where an MLDE naturally arises. This is similar in spirit to Zhu's [Z] analysis of correlation functions for the modules of C_2-cofinite VOAs but has the advantage of being considerably less technical.

We recall the genus-one partition function

$$Z_V(\tau) = \mathrm{Tr}_V(q^{L_0-c/24}),$$

the 1-point correlation function for $a \in V$

$$Z_V(a,\tau) = \mathrm{Tr}_V o(a) q^{L_0-c/24}, \quad (195)$$

and the 2-point correlation function which can be expressed in terms of 1-point functions by

$$F_V((a,x),(b,y),\tau) = Z_V(Y[a,x]Y[b,y]\mathbf{1},\tau) \quad (196)$$
$$= Z_V(Y[a,x-y]b,\tau), \quad (197)$$

for square bracket vertex operators $Y[a,z] = Y(q_z^{L_0}a, q_z - 1)$.

We define quadratic Casimir vectors in the square bracket VOA formalism

$$\lambda^{[n]} = u^\alpha[1-n]\bar{u}^\alpha \in V_{[n]},$$

(for α summed) for basis $\{u^\alpha\}$ and square bracket LiZ dual basis $\{\bar{u}^\alpha\}$. Consider the genus-one analogue of (176) given by the 2-point function

$$F_V((u^\alpha,x),(\bar{u}^\alpha,y),\tau) = Z_V(Y[u^\alpha,x]Y[\bar{u}^\alpha,y]\mathbf{1},\tau),$$

(α summed). Associativity (196) implies the genus-one analogue of (177) so that

$$F_V((u^\alpha,x),(\bar{u}^\alpha,y),\tau) = \sum_{n \geq 0} Z_V(\lambda^{[n]},\tau)(x-y)^{n-2}. \quad (198)$$

From Zhu's first recursion formula (Theorem 5.7) we may alternatively expand $F((u^\alpha,x),(\bar{u}^\alpha,y),\tau)$ in terms of Weierstrass functions as follows:

$$F_V((u^\alpha, x), (\bar{u}^\alpha, y), \tau) = \text{Tr}_V\left(o(u^\alpha)o(\bar{u}^\alpha)q^{L_0-c/24}\right)$$
$$+ \sum_{m \geq 1} \frac{(-1)^{m+1}}{m!} P_1^{(m)}(x-y, \tau) Z_V(u^\alpha[m]\bar{u}^\alpha, \tau)$$
$$= \text{Tr}_V\left(o(u^\alpha)o(\bar{u}^\alpha)q^{L_0-c/24}\right) - dP_2(x-y, \tau) Z_V(\tau).$$

Recalling Theorem 5.1 we may compare the $(x-y)^2$ terms in this expression and (198) to obtain

$$Z_V(\lambda^{[4]}, \tau) = -3dE_4(\tau)Z_V(\tau). \tag{199}$$

Since $(V, Y(\ ), \mathbf{1}, \omega)$ is isomorphic to $(V, Y[\ ], \mathbf{1}, \tilde{\omega})$ it follows that $\lambda^{(n)} \in \text{Vir}_c$ iff $\lambda^{[n]} \in \text{Vir}_c$. Thus assuming $\lambda^{(4)} \in \text{Vir}_c$ we have

$$Z_V(\lambda^{[4]}, \tau) = \frac{-3d}{c(5c+22)}\left(4Z_V(L[-2]^2\mathbf{1}, \tau) + (2+c)Z_V(L[-4]\mathbf{1}, \tau)\right), \tag{200}$$

by (175). The Virasoro 1-point functions $Z_V(L[-2]^2\mathbf{1}, \tau)$, $Z_V(L[-4]\mathbf{1}, \tau)$ can be evaluated via Zhu's second recursion formula (Theorem 5.10). In particular taking $u = \tilde{\omega}$ and v of $L[0]$ weight k in (61) we obtain the general Virasoro recursion formula

$$Z_V(L[-n]v, \tau) = \delta_{n,2}\text{Tr}_V(o(\tilde{\omega})o(v)q^{L_0-c/24})$$
$$+ \sum_{0 \leq m \leq k} (-1)^m \binom{m+n-1}{m+1} E_{m+n}(\tau) Z_V(L[m]v, \tau). \tag{201}$$

But $o(\tilde{\omega}) = L_0 - c/24$ and hence

$$\text{Tr}_V(o(\tilde{\omega})o(v)q^{L_0-c/24}) = \theta Z_V(v, \tau),$$

where $\theta = q\,d/dq$. It follows that

$$Z_V(L[-2]v, \tau) = D_k Z_V(v, \tau) + \sum_{2 \leq m \leq k} E_{2+m}(\tau) Z_V(L[m]v, \tau), \tag{202}$$

where $D_k = \theta + kE_2(\tau)$ is the modular derivative (30). (Zhu makes extensive use of the identities (201) and (202) in his analysis of correlation functions for C_2-cofinite VOAs [Z]. This is the origin of MLDEs as discussed in Section 9).

We immediately find from (201) that $Z_V(L[-4]\mathbf{1}, \tau) = 0$ and

$$Z_V(L[-2]^2\mathbf{1}, \tau) = D_2 Z_V(L[-2]\mathbf{1}, \tau) + E_4(\tau)Z_V(L[2]L[-2]\mathbf{1}, \tau)$$
$$= \left(D^2 + \tfrac{1}{2}cE_4(\tau)\right)Z_V(\tau),$$

where $D^2 = D_2 D_0 = (q\,d/dq)^2 + 2E_2(\tau)q\,d/dq$. Substituting into (200) we find $Z_V(\tau)$ satisfies the following second order MLDE:

$$\left(D^2 - \tfrac{5}{4}c(c+4)E_4(\tau)\right)Z_V(\tau) = 0. \tag{203}$$

(203) has a regular singular point at $q = 0$ with indicial roots $-c/24$ and $(c+4)/24$. Applying (194) it follows that there exists a unique solution with leading q expansion $Z_V(\tau) = q^{-c/24}(1 + O(q))$. Furthermore, since $E_4(\tau)$ is holomorphic then $Z_V(\tau)$ is also holomorphic for $0 < |q| < 1$. In summary, we find:

THEOREM 12.11. *If $\lambda^{(4)} \in \mathrm{Vir}_c$ then $Z_V(\tau)$ is a uniquely determined holomorphic function in $\mathfrak{H}$.*

An immediate consequence of Theorems 12.6 and 12.11 is:

THEOREM 12.12. *$V = L_\mathfrak{g}(1,0)$ the level one WZW model generated by $\mathfrak{g}$.*

PROOF. Clearly $L_\mathfrak{g}(1,0) \subseteq V$ with $\omega, \lambda^{(2)}, \lambda^{(4)} \in L_\mathfrak{g}(1,0)$. Thus $L_\mathfrak{g}(1,0)$ satisfies the conditions of Theorem 12.11 for the same central charge c. Hence $Z_{L_\mathfrak{g}(1,0)}(\tau) = Z_V(\tau)$ and so $L_\mathfrak{g}(1,0) = V$. □

It is straightforward to substitute $Z(\tau) = q^{-c/24} \sum_n \dim V_n q^n$ into (203) and solve recursively for $\dim V_n$ as a rational function in c. In this way we recover $\dim V_1 = d$ of (194). The next two terms are

$$\dim V_2 = \frac{c(804 + 508c + 175c^2 + 25c^3)}{2(22-c)(10-c)},$$

$$\dim V_3 = \frac{c(33344 + 148872c + 68308c^2 + 10330c^3 + 975c^4 + 125c^5)}{6(34-c)(22-c)(10-c)}.$$

These dimension formulas can be further refined as follows. Consider the Virasoro decomposition of V_2:

$$V_2 = \mathbb{C}\omega \oplus L_{-1}\mathfrak{g} \oplus P_2, \tag{204}$$

where P_2 is the space of weight two primary vectors. Let $p_2 = \dim P_2$. Then $\dim V_2 = 1 + d + p_2$ with

$$p_2 = \frac{5(5c+22)(c+2)^2(c-1)}{2(22-c)(10-c)}. \tag{205}$$

Comparing with Deligne's analysis of the irreducible decomposition of tensor products of $\mathfrak{g}$ we find that

$$p_2 = \dim Y_2^*,$$

where Y_2^* denotes an irreducible representation of $\mathfrak{g}$ in Deligne's notation [De]. This is explored further in [T3].

Similarly for V_3 we find

$$V_3 = \mathbb{C}[L_{-1}\omega] \oplus L_{-1}^2 \mathfrak{g} \oplus L_{-2}\mathfrak{g} \oplus L_{-1} P_2 \oplus P_3,$$

where P_3 is the space of weight three primary vectors. Let $p_3 = \dim P_3$. Then $\dim V_3 = 1 + 2d + p_2 + p_3$ with

$$p_3 = \frac{5c(5c+22)(c-1)(c+5)(5c^2+268)}{6(34-c)(22-c)(10-c)} = \dim X_2 + \dim Y_3^*,$$

where X_2, Y_3^* denote two other irreducible representations of $\mathfrak{g}$ in Deligne's notation of dimension

$$\dim X_2 = \frac{5c(5c+22)(c+6)(c-1)}{2(10-c)^2},$$

$$\dim Y_3^* = \frac{5c(5c+22)(c+2)^2(c-8)(5c-2)(c-1)}{6(10-c)^2(22-c)(34-c)}.$$

12.3. Higher-weight constructions. We can generalize the arguments given above to consider a VOA V with $\dim V_1 = 0$. Here we construct Casimir vectors from the weight two primary space P_2 (provided $\dim P_2 > 0$) and obtain constraints on V that follow from such Casimirs being Virasoro vacuum descendents. If $\dim P_2 = 0$ we consider primaries of weight 3 and so on. In general, let V be a VOA with primary vector space P_K of lowest weight K; i.e., $V_n = (\text{Vir}_c)_n$ for all $n < K$, so that

$$Z_V(\tau) = q^{-c/24} \left(\sum_{n<K} \dim(\text{Vir}_c)_n q^n + O(q^K) \right). \tag{206}$$

(Recall from (42) that $\sum_{n \geq 0} \dim(\text{Vir}_c)_n q^n = \prod_{m \geq 2}(1-q^m)^{-1} = 1 + q^2 + q^3 + 2q^3 + \cdots$.) We construct Casimir vectors, as in (172), from a P_k basis $\{u^\alpha\}$ and LiZ dual basis $\{\bar{u}^\alpha\}$

$$\lambda^{(n)} = u^\alpha_{2K-1-n} \bar{u}^\alpha \in V_n.$$

We find the following natural generalization of Theorems 12.11 and 12.12:

THEOREM 12.13. *Let V be a VOA with primary vectors of lowest weight $K = 2$ or 3. If $\lambda^{(2K+2)} \in \text{Vir}_c$, then*
(a) *$Z_V(\tau)$ of (206) is a holomorphic function in $\mathfrak{H}$ and is the unique solution to a MLDE of order $K+1$; and*
(b) *V is generated by P_K.*

REMARK 12.14. We conjecture that Theorem 12.13 holds for all K.

For $K = 2$ the elements of P_2 satisfy a commutative nonassociative algebra with invariant (LiZ) form known as a Griess algebra (cf. Exercise 9.9). Theorem 12.13 implies the dimension of the Griess algebra is

$$\dim P_2 = \frac{1}{2} \frac{(5c+22)(2c-1)(7c+68)}{c^2 - 55c + 748}. \tag{207}$$

This result originally appeared in [Mat] subject to stronger assumptions. Following Remark 12.2 we note that the zeros of the numerator are the zeros $c_{p,q}$ of the Kac determinant $\det M_n(c)$ for $n \leq 6$. There are 37 rational values of c for which $\dim P_2$ is a positive integer. Furthermore, we may solve $Z_V(\tau)$ iteratively for $\dim V_n$ as rational functions in c. There are 9 values of c for which $\dim V_n$ is a positive integer for $n \leq 400$ given by [T3]:

c:	$-\frac{44}{5}$	8	16	$\frac{47}{2}$	24	32	$\frac{164}{5}$	$\frac{236}{7}$	40
$\dim P_2$:	1	155	2295	96255	196883	139503	90117	63365	20619

The first five cases can all be realized by explicit constructions. Of particular interest is the case $c = 24$ realized by the FLM Moonshine Module $V^\natural$ with $Z_{V^\natural}(\tau) = j(\tau) - 744$ for which P_2 is the original Griess algebra of dimension 196883 and whose automorphism group is the Monster group (cf. Section 9.4). There are constructions for $c = 32$ and 40 with the appropriate partition function but it is not known if $\lambda^{(6)} \in \text{Vir}_c$. There are no known constructions for $c = \frac{164}{5}$ and $\frac{236}{7}$.

For $K = 3$ we find

$$\dim P_3 = \frac{(5c+22)(2c-1)(7c+68)(5c+3)(3c+46)}{-5c^4 + 703c^3 - 32992c^2 + 517172c - 3984},$$

where the zeros of the numerator are Kac determinant zeros $c_{p,q}$ for $(p-1) \times (q-1) = n \leq 8$. Iteratively solving the appropriate MLDE for $Z_V(\tau)$ we find $\dim P_3$ and $\dim V_n$ are positive integral for only 3 rational values of c:

c:	$-\frac{114}{7}$	$\frac{4}{5}$	48
$\dim P_3$:	1	1	42987519

The first two examples can be realized by known VOAs. For $c = 48$ we find $Z_V(\tau) = J(\tau)^2 - 393767$ which, intruigingly, is the partition function of the minimal holomorphic VOA $V^{(2)}$ briefly discussed in Section 9.4.

Part IV. Appendices
13. Lie algebras and representations

An *associative algebra* is a linear space A equipped with a bilinear, associative product $A \otimes A \to A$, denoted by juxtaposition. Thus $a \otimes b \mapsto ab$ and
$$(ab)c = a(bc).$$

A *Lie algebra* is a linear space L equipped with a bilinear product (usually called bracket) $[\ ] : L \otimes L \to L$ such that

$$[ab] = -[ba] \qquad \text{(skew-commutativity)}$$
$$[a[bc]] + [b[ca]] + [c[ab]] = 0 \quad \text{(Jacobi identity)}$$

An associative algebra A gives rise to a Lie algebra A^- on the *same* linear space by defining $[ab] = ab - ba$. A basic example is $\text{End}(V)$ for a linear space V, where the associative product is composition of endomorphisms. This situation can be exploited using another basic associative algebra, the *tensor algebra*

$$T(V) = \bigoplus_{n \geq 0} V^{\otimes n} = \mathbb{C} \oplus V \oplus V \otimes V \oplus \cdots$$

over V. Let $\iota : V \to T(V)$ be canonical identification of V with the degree 1 piece of $T(V)$. The *universal mapping property* (UMP) for tensor algebras says that any linear map $f : V \to A$ into an associative algebra A has a *unique* extension to a morphism of associative algebras $\alpha : T(V) \to A$:

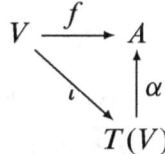

with $f = \alpha \circ \iota$.

A *representation* of a Lie algebra L is a linear map $\pi : L \to \text{End}(V)$ for some V such that
$$\pi([ab]) = \pi(a)\pi(b) - \pi(b)\pi(a).$$
That is, $\pi : L \to \text{End}(V)^-$ is a morphism of Lie algebras. We call V an L-module.

UMP provides an extension of π to a morphism of associative algebras $\alpha : T(L) \to \text{End}(V)$. Identifying $a \in L$ with its image in $T(L)$, we see that for $a, b \in L$

$$\alpha(a \otimes b - b \otimes a - [ab]) = \alpha(a)\alpha(b) - \alpha(b)\alpha(a) - \alpha([ab])$$
$$= \pi(a)\pi(b) - \pi(b)\pi(a) - \pi([ab]) = 0.$$

Let $J \subseteq T(L)$ be the 2-sided ideal generated by $a \otimes b - b \otimes a - [ab]$, $a, b \in L$, and set

$$\mathcal{U}(L) = T(L)/J.$$

This is the *universal enveloping algebra* of L. Thus every representation π of L extends canonically to a representation of the universal enveloping algebra:

$$L \xrightarrow{\pi} \text{End}(V)$$
$$\searrow_{\iota'} \quad \uparrow \alpha$$
$$\mathcal{U}(L)$$

where ι' is the composition $L \xrightarrow{\iota} T(L) \to \mathcal{U}(L)$.

THEOREM 13.1 (POINCARÉ–BIRKHOFF–WITT, OR PBW). *Fix an ordered basis $x_1, x_2, \ldots$ of L, with $\bar{x}_i$ the image of x_i in $\mathcal{U}(L)$. Then*

$$\{\bar{x}_{i_1} \bar{x}_{i_2} \ldots \bar{x}_{i_k} \mid i_1 \geq i_2 \geq \cdots \geq i_k \geq 1\}$$

is a basis for $\mathcal{U}(L)$.

From PBW we see that ι' is *injective*. Then for a representation of $\mathcal{U}(L)$, restriction to the subspace $L = \iota(L)$ furnishes a representation of L. In this way, representations of L and $\mathcal{U}(L)$ determine each other in a canonical fashion - a statement that can be better stated using categories of modules.

The Lie algebra L has a *triangular decomposition* if it decomposes as

$$L = L^+ \oplus L^0 \oplus L^-$$

such that $L^\pm, L^0$ are Lie subalgebras, and the bracket satisfies

$$[L^+ L^-] \subseteq L^0, \quad [L^\pm L^0] \subseteq L^\pm.$$

Use of PBW and an appropriate choice of (ordered) basis leads to an identification

$$\mathcal{U}(L) = \mathcal{U}(L^-) \otimes \mathcal{U}(L^0) \otimes \mathcal{U}(L^+).$$

Noting that $L^0 \oplus L^+ \subseteq L$ is a Lie subalgebra, let $\pi : L^0 \oplus L^+ \to \text{End}(V)$ be a representation. The *induced module* is

$$\text{Ind}(V) = \text{Ind}_{\mathcal{U}(L^0 \oplus L^+)}^{\mathcal{U}(L)} V := \mathcal{U}(L) \otimes_{\mathcal{U}(L^0 \oplus L^+)} V = \mathcal{U}(L^-) \otimes V. \quad (208)$$

It is a $\mathcal{U}(L)$-module, hence also an L-module upon restriction. A ubiquitous special case occurs when V is an L^0-module, which then becomes an $L^0 \oplus L^+$-module by letting L^+ annihilate V.

EXERCISE 13.2. Show that the following Lie algebras have natural triangular decompositions:

(a) Heisenberg algebra $\hat{A}$ with

$$\hat{A}^+ = \bigoplus_{n>0} \mathbb{C}a \otimes t^n, \ \hat{A}^- = \bigoplus_{n<0} \mathbb{C}a \otimes t^n, \ \hat{A}^0 = \mathbb{C}a \otimes t^0 \oplus \mathbb{C}K.$$

(b) Virasoro algebra Vir with

$$\text{Vir}^+ = \bigoplus_{n>0} \mathbb{C}L_n, \ \text{Vir}^- = \bigoplus_{n<0} \mathbb{C}L_n, \ \text{Vir}^0 = \mathbb{C}L_0 \oplus \mathbb{C}K.$$

(c) Finite-dimensional simple Lie algebra (equipped with a choice of Cartan subalgebra and root system) with $L^+ = \{\text{positive root spaces}\}$, $L^- = \{\text{negative root spaces}\}$, $L^0 = \{\text{Cartan subalgebra}\}$.

14. The square bracket formalism

We prove Equations (16)–(18) of Section 2.7. The square bracket vertex operator (14), (15) is

$$Y[v, z] = q_z^{\text{wt}(v)} Y(v, q_z - 1).$$

Thus the square bracket modes of $Y[v, z] = \sum_{m \in \mathbb{Z}} v[m] z^{-m-1}$ are given by

$$v[m] = \text{Res}_z Y(v, q_z - 1) z^m q_z^{\text{wt}(v)}$$
$$= \text{Res}_z Y(v, q_z - 1) \frac{d}{dz}(q_z - 1) z^m q_z^{\text{wt}(v)-1}.$$

We may rewrite this in terms of $w = q_z - 1 = z + O(z^2)$ by means of a (formal) chain rule [FHL], [Z] so that

$$v[m] = \text{Res}_w Y(v, w) z(w)^m q_{z(w)}^{\text{wt}(v)-1}$$
$$= \text{Res}_w Y(v, w) \ln(1+w)^m (1+w)^{\text{wt}(v)-1}.$$

Defining $c(\text{wt}(v), i, m)$ for $i \geq m \geq 0$ by

$$\sum_{i \geq m} c(\text{wt}(v), i, m) w^i = \frac{1}{m!} \ln(1+w)^m (1+w)^{\text{wt}(v)-1},$$

we obtain (16).

Next note that $\sum_{m \geq 0} \frac{1}{m!} \ln(1+w)^m x^m = (1+w)^x$. Hence we find

$$\sum_{i \geq 0} \sum_{m=0}^{i} c(\text{wt}(v), i, m) w^i x^m = (1+w)^{\text{wt}(v)-1+x},$$

from which (17) follows. Finally,

$$\sum_{m \geq 0} \frac{(k+1-\mathrm{wt}(v))^m}{m!} v[m] = \sum_{m \geq 0} \sum_{i \geq m} c(\mathrm{wt}(v), i, m)(k+1-\mathrm{wt}(v))^m v_i$$

$$= \sum_{i \geq 0} v_i \sum_{m=0}^{i} c(\mathrm{wt}(v), i, m)(k+1-\mathrm{wt}(v))^m,$$

$$= \sum_{i \geq 0} \binom{k}{i} v_i.$$

giving (18).

References

[A] T. Abe, A $\mathbb{Z}_2$-orbifold model of the symplectic fermionic vertex operator superalgebra, Math. Z. **255**:4 (2007), 755–792.

[ABD] T. Abe, G. Buhl and C. Dong, Rationality, regularity and C_2-cofiniteness, TAMS. **356**:8 (2004), 3391-3402.

[B] R. Borcherds, Vertex algebras, Kac–Moody algebras, and the Monster, Proc. Nat. Acad. Sc. **83** (1986), 3068-3071.

[Ba1] P. Bantay, Permutation orbifolds and their applications, in *Vertex operator algebras in mathematics and physics* (Toronto. 2000), Fields Inst. Comm. **39**, AMS, 2003.

[Ba2] P. Bantay, The kernel of the modular representation and the Galois action in RCFT, CMP. **233**:3 (2003), 423-438.

[BK] A. A. Belavin and V. G. Knizhnik, Algebraic geometry and the geometry of quantum strings, Phys. Lett **168B** (1986) 202–206.

[BPZ] A. Belavin, A. Polyakov and A. Zamolodchikov, Infinite conformal symmetry in two-dimensional quantum field theory, Nucl. Phys. **B241** (1984), 333–380.

[Bu] G. Buhl, A spanning set for VOA modules, J. Alg. **254** (2002), 125–151.

[CS] J. Conway and N. Sloan, *Sphere-packings, lattices and groups*, Springer, New York, 1988.

[Do] C. Dong, Vertex algebras associated with even lattices, J. Alg. **161** (1993), 245–265.

[De] P. Deligne, La série exceptionnelle de groupes de Lie, C. R. Acad. Sci. Paris Sér. I Math. **322** (1996), 321–326.

[DGM] L. Dolan, P. Goddard and P. Montague, Conformal field theories, representations and lattice constructions, Commun. Math. Phys. **179** (1996), 61–129.

[DL] C. Dong and J. Lepowsky, *Generalized vertex algebras and relative vertex operators*, Progress in Math. **112**, Birkhäuser, Boston, 1993.

[DLM1] C. Dong, H. Li and G. Mason, Twisted representations of vertex operator algebras, Math. Ann. **310** (1998), 571–600.

[DLM2] C. Dong, H. Li and G. Mason, Regularity of rational vertex operator algebras, Adv. in Math. **132** (1997), 148–166.

[DLM3] C. Dong, H. Li and G. Mason, Vertex operator algebras and associative algebras, J. Alg. **206** (1998), 67–96.

[DLM4] C. Dong, H. Li and G. Mason, Modular-invariance of trace functions in orbifold theory and generalized Moonshine, Comm. Math. Phys. **214** (2000), 1–56.

[DM1] C. Dong and G. Mason, Integrability of C_2-cofinite vertex operator algebras, Int. Math. Res. Notices **2006**, Article ID 80468, 1–15.

[DM2] C. Dong and G. Mason, Holomorphic vertex operator algebras of small central charge, PJM. **213**:2 (2004), 253–266.

[DM3] C. Dong and G. Mason, Rational vertex operator algebras and the effective central charge, Int. Math. Res. Notices. (2004) No. 56, 2989–3008.

[DMZ] C. Dong, G. Mason and Y. Zhu, Discrete series of the Virasoro algebra and the moonshine module, in *Algebraic groups and their generalizations*, Proc. Symp. Pure Math. **56**, Part 2 (1994), 295–316.

[EZ] M. Eichler and D. Zagier, Jacobi forms, Birkhäuser, Boston.

[Fl] M. Flohr, On modular invariant partition functions of conformal field theories with logarithmic operators, Int. J. Mod. Phys. A **11** (1996), 4147–4172.

[Fr] E. Freitag, *Siegelische Modulfunktionen*, Springer, Berlin, 1983).

[FF] B. Feigin and D. Fuchs, Verma modules over the Virasoro algebra, in *Topology*, Lect. Notes in Math. **1060**, 230–245, Springer, Berlin, 1984.

[FHL] I. Frenkel, J. Lepowsky and Y.-Z. Huang, *On axiomatic approaches to vertex operator algebras and modules*, Mem. AMS. **494**, 1993.

[FK] H. M. Farkas and I. Kra, *Riemann surfaces*, Springe, New York, 1980.

[FKRW] E. Frenkel, V. Kac, A. Radul and W. Wang, $W_{1+\infty}$ and $W(gl_\infty)$ with central charge N, Comm. Math. Phys. **170** (1995), 337–357.

[FLM] I. Frenkel, J. Lepowsky and A. Meurman, *Vertex operator algebras and the Monster*, Pure and Appl. Math. **134**, Academic Press, Boston, 1988.

[FMS] P. Di Francesco, P. Mathieu, and D. Senechal, *Conformal field theory*, Springer Graduate Texts in Contemporary Physics.

[FZ] I. Frenkel and Y. Zhu, Vertex operator algebras associated to representations of affine and Virasoro algebras, Duke Math. J. **66**:1 (1992), 123–168.

[G1] R. Griess, The Friendly Giant, Invent. Math. **69** (1982), 1–102.

[G2] R. Griess, The Monster and its nonassociative algebra, in *Finite groups: coming of age* (Montreal, 1982), Contemp. Math., AMS, Providence, 1985.

[Go] P. Goddard, Meromorphic conformal field theory, in *Infinite-dimensional Lie algebras and groups* (Luminy), World Sci., Singapore, 1989.

[GK] M. Gaberdiel and H. Kausch, A rational logarithmic conformal field theory, Phys. Lett. B **386** (1996), 305–331.

[GN] M. Gaberdiel and A. Neitzke, Rationality, quasirationality and finite W-algebras, Commun. Math. Phys. **238** (2003), 305–331.

[H] E. Hille, *Ordinary differential equations in the complex domain*, Dover, New York, 1956.

[HL] Y.-Z. Huang and J. Lepowsky, A theory of tensor products for module categories for a vertex operator algebra I, II, Sel. Math. (N. S.) **1**:4 (1995), 699-756, 757-786.

[HLZ] Y.-Z. Huang, J. Lepowsky and L. Zhang, A logarithmic generalization of tensor product theory for modules for a vertex operator algebra, IJM. **17**:8 (2006), 975–1012.

[I] E. Ince, *Ordinary differential equations*, Dover, New York, 1976.

[K1] V. Kac, *Vertex algebras for beginners*, 2nd ed., Univ. Lect. Ser. **10**, AMS, 1998.

[K2] V. Kac, *Infinite dimensional Lie algebras* 2nd ed., Cambridge University Press. 1985.

[KLN] M. Knopp, J. Lehner and M. Newman, A bounded automorphic form of dimension zero is constant, Duke Math. J. **32** (1965), 457-460.

[Kn] M. Knopp, Modular functions in analytic number theory, AMS. Chelsea Publ., Providence, 1993.

[KM] W. Kohnen and G. Mason, On generalized modular forms and their applications, Nagoya M. J. **192** (2008), 119-136.

[KP] V. Kac and D. Peterson, Infinite dimensional Lie algebras, theta functions and modular forms, Adv. in Math. **53** (1984), 125–264.

[KR] V. Kac and A. Raina, *Highest weight representations of infinite dimensional lie algebras*, Adv. Ser. in Math. Phys. **2**, World Scientific, Singapore, 1987.

[L] S. Lang, *Introduction to modular forms*, Springer, Berlin, 1976.

[Li] H. Li, Some finiteness properties of regular vertex operator algebras, J. Alg. **212** (1999), 495–514.

[LL] J. Lepowsky and H. Li, *Introduction to vertex algebras*, Progress in Math. **227**, Birkhäuser, Boston, 2004.

[M] G. Mason, Vector-valued modular forms and linear differential operators, IJNT. **3**:3, (2007), 377–390.

[Mat] A. Matsuo, Norton's trace formula for the Griess algebra of a vertex operator algebra with large symmetry, Commun. Math. Phys. **224** (2001), 565–591.

[My1] M. Miyamoto, Modular Invariance of Vertex Operator Algebras Satisfying C_2-Cofiniteness, Duke Math. J. **122**:1 (2004), 51–91.

[My2] M. Miyamoto, A new construction of the Moonshine vertex operator algebra over the real number field, Ann. Math. **159**:2 (2004), 535–596.

[MA] G. Anderson and G. Moore, Rationality in conformal field theory, Commun. Math. Phys. **117** (1988), 441–450.

[MN] A. Matsuo and K. Nagatomo, *Axioms for a vertex algebra and the locality of quantum fields*, Math. Soc. of Japan Memoirs **4**, Tokyo, 1999.

[MP] A. Meurman and M. Primc, Vertex operator algebras and representations of affine Lie algebras, Acta Applic. Math. **44** (1996), 207–215.

[MT1] G. Mason and M. P. Tuite, Torus chiral n-point functions for free boson and lattice vertex operator algebras, Commun. Math. Phys. **235** (2003), 47–68.

[MT2] G. Mason, and M. P. Tuite, On genus-two Riemann surfaces formed from sewn tori, Commun. Math. Phys. **270** (2007), 587–634.

[MT3] G. Mason, and M. P. Tuite, Partition functions and chiral algebras, in *Lie algebras, vertex operator algebras and their applications* (in honor of Jim Lepowsky and Robert Robert L. Wilson), Contemporary Mathematics **442**, AMS (2007).

[MT4] G. Mason and M. P. Tuite, The genus two partition function for free bosonic and lattice vertex operator algebras, arXiv:0712.0628.

[Mu2] D. Mumford, Stability of projective varieties, L. Ens. Math. **23** (1977) 39–110.

[O] A. Ogg, *Modular forms and Dirichlet series*, Benjamin, 1969.

[P] J. Polchinski, *String theory*, Vol. I, Cambridge University Press, (Cambridge, 1998).

[Se] J.-P. Serre, *A course in arithmetic*, Springer, New York, 1973.

[Sc] B. Schoeneberg, *Elliptic modular forms*, Springer, Berlin, 1973.

[Sch] A. Schellekens, Meromorphic $c = 24$ conformal field theories, Commun. Math. Phys. **153**:1 (1993), 159–185.

[T1] M. P. Tuite, Genus two meromorphic conformal field theory, CRM Proceedings and Lecture Notes **30** (2001), 231–251.

[T2] M. P. Tuite, *The Virasoro algebra and some exceptional Lie and finite groups*, SIGMA **3** (2007) 008.

[T3] M. P. Tuite, *Exceptional vertex operator algebras and the virasoro algebra*, to appear in *Proceedings of the Conference on Vertex Operator Algebras*, edited by M. Bergvelt, G. Yamskulna, and W. Zhao, arXiv:0811.4523.

[Wa] W.-Q. Wang, Rationality of Virasoro vertex operator algebras, Int. Math. Res. Notices. **7** (1993), 197–211.

[Wi] E. Witten: Three-dimensional gravity revisited, arXiv:0706.3359.

[Wo] K. Wohlfahrt, An extension of F. Klein's level concept, IJM. **13** (1964), 529–535.

[Y] A. Yamada, Precise variational formulas for abelian differentials. Kodai Math J. **3** (1980), 114–143.

[Z] Y. Zhu, Modular-invariance of characters of vertex operator algebras, J. Amer. Math. Soc. **9**:1 (1996), 237–302.

GEOFFREY MASON
DEPARTMENT OF MATHEMATICS
UNIVERSITY OF CALIFORNIA
SANTA CRUZ, CA 95064
UNITED STATES
  gem@cats.ucsc.edu

MICHAEL TUITE
DEPARTMENT OF APPLIED MATHEMATICS
NATIONAL UNIVERSITY OF IRELAND, GALWAY
UNIVERSITY ROAD
GALWAY
IRELAND
  michael.tuite@nuigalway.ie

Applications of elliptic and theta functions to Friedmann–Robertson–Lemaître–Walker cosmology with cosmological constant

JENNIE D'AMBROISE

1. Introduction

Elliptic functions are known to appear in many problems, applied and theoretical. A less known application is in the study of exact solutions to Einstein's gravitational field equations in a Friedmann–Robertson–Lemaître–Walker, or FRLW, cosmology [Abdalla and Correa-Borbonet 2002; Aurich and Steiner 2001; Aurich et al. 2004; Basarab-Horwath et al. 2004; Kharbediya 1976; Kraniotis and Whitehouse 2002]. We will show explicitly how Jacobi and Weierstrass elliptic functions arise in this context, and will additionally show connections with theta functions. In Section 2, we review the definitions of various elliptic functions. In Section 3, we record relations between elliptic functions and theta functions. In Section 4 we introduce the FRLW cosmological model and we then proceed to show how elliptic functions appear as solutions to Einstein's gravitational equations in sections 5 and 6. The author thanks Floyd Williams for helpful discussions.

2. Elliptic functions

An *elliptic integral* is one of the form $\int R(x, \sqrt{P(x)})\, dx$, where $P(x)$ is a polynomial in x of degree three or four and R is a rational function of its arguments. Such integrals are called elliptic since an integral of this kind arises in the computation of the arclength of an ellipse. Legendre showed that any elliptic integral can be written in terms of the three fundamental or *normal elliptic integrals*

$$F(x,k) \stackrel{\text{def.}}{=} \int_0^x \frac{dt}{\sqrt{(1-t^2)(1-k^2 t^2)}},$$

$$E(x,k) \stackrel{\text{def.}}{=} \int_0^x \sqrt{\frac{1-k^2t^2}{1-t^2}}\,dt, \qquad (2.1)$$

$$\Pi(x,\alpha^2,k) \stackrel{\text{def.}}{=} \int_0^x \frac{dt}{(1-\alpha^2t^2)\sqrt{(1-t^2)(1-k^2t^2)}},$$

which are referred to as normal elliptic integrals of the first, second and third kind, respectively. The parameter α is any real number and k is referred to as the *modulus*. For many problems in which real quantities are desired, $0 < k, x < 1$, although this is not required in the above definitions (for $k = 0$ and $k = 1$, the integral can be expressed in terms of elementary functions and is therefore *pseudo-elliptic*).

Elliptic functions are inverse functions of elliptic integrals. They are known to be the simplest of nonelementary functions and have applications in the study of classical equations of motion of various systems in physics including the pendulum. One can easily show that if $f(u)$ denotes the inverse function of an elliptic integral $y(x) = \int R(x, \sqrt{P(x)})\,dx$, then since $y(f(u)) = u$, which implies $y'(f(u)) = 1/f'(u)$, we have

$$f'(u)^2 = \frac{1}{R(f(u), \sqrt{P(f(u))})^2}. \qquad (2.2)$$

The *Jacobi elliptic function* $\operatorname{sn}(u,k)$ is the inverse of $F(x,k)$ defined above, and eleven other Jacobi elliptic functions can be written in terms of $\operatorname{sn}(u,k)$: $\operatorname{cn}(u,k)$ and $\operatorname{dn}(u,k)$ satisfy $\operatorname{sn}^2 u + \operatorname{cn}^2 u = 1$ and $k^2 \operatorname{sn}^2 u + \operatorname{dn}^2 u = 1$, respectively, and

$$\operatorname{ns} \stackrel{\text{def.}}{=} \frac{1}{\operatorname{sn}}, \quad \operatorname{nc} \stackrel{\text{def.}}{=} \frac{1}{\operatorname{cn}}, \quad \operatorname{nd} \stackrel{\text{def.}}{=} \frac{1}{\operatorname{dn}},$$

$$\operatorname{sc} \stackrel{\text{def.}}{=} \frac{\operatorname{sn}}{\operatorname{cn}}, \quad \operatorname{sd} \stackrel{\text{def.}}{=} \frac{\operatorname{sn}}{\operatorname{dn}}, \quad \operatorname{cd} \stackrel{\text{def.}}{=} \frac{\operatorname{cn}}{\operatorname{dn}}, \qquad (2.3)$$

$$\operatorname{cs} \stackrel{\text{def.}}{=} \frac{1}{\operatorname{sc}}, \quad \operatorname{ds} \stackrel{\text{def.}}{=} \frac{1}{\operatorname{sd}}, \quad \operatorname{dc} \stackrel{\text{def.}}{=} \frac{1}{\operatorname{cd}}.$$

By (2.2) and (2.3), one can see that the Jacobi elliptic functions satisfy the differential equation

$$f'(u)^2 + af(u)^2 + bf(u)^4 = c \qquad (2.4)$$

for a, b, c in terms of the modulus k according to the following table.

$f(u)$	a	b	c
$\operatorname{sn}(u,k)$	$1+k^2$	$-k^2$	1
$\operatorname{cn}(u,k)$	$1-2k^2$	k^2	$1-k^2$
$\operatorname{dn}(u,k)$	k^2-2	1	k^2-1
$\operatorname{ns}(u,k)$	$1+k^2$	-1	k^2
$\operatorname{nc}(u,k)$	$1-2k^2$	k^2-1	$-k^2$
$\operatorname{nd}(u,k)$	k^2-2	$1-k^2$	-1
$\operatorname{sc}(u,k)$	k^2-2	k^2-1	1
$\operatorname{sd}(u,k)$	$1-2k^2$	$k^2(1-k^2)$	1
$\operatorname{cd}(u,k)$	$1+k^2$	$-k^2$	1
$\operatorname{cs}(u,k)$	k^2-2	-1	$1-k^2$
$\operatorname{ds}(u,k)$	$1-2k^2$	-1	$k^2(k^2-1)$
$\operatorname{dc}(u,k)$	$1+k^2$	-1	k^2

(The a-value for $f(u) = \operatorname{dn}(u,k)$ seen above corrects an error in [Basarab-Horwath et al. 2004].)

The *Weierstrass elliptic function*

$$\wp(z;\omega_1,\omega_2) = \frac{1}{z^2} + \sum_{(m,n)\in\mathbb{Z}\times\mathbb{Z}-\{(0,0)\}} \frac{1}{(z-m\omega_1-n\omega_2)^2} - \frac{1}{(m\omega_1+n\omega_2)^2} \quad (2.5)$$

is a doubly periodic elliptic function of $z \in \mathbb{C}$ with periods $\omega_1, \omega_2 \in \mathbb{C}$ such that $Im(\omega_1/\omega_2) > 0$. The function $\wp$ is the inverse of the elliptic integral

$$\wp^{-1}(x;g_2,g_3) = \int_x^\infty \frac{1}{\sqrt{4t^3 - g_2 t - g_3}}\, dt \quad (2.6)$$

where $g_2, g_3 \in \mathbb{C}$ are known as *Weierstrass invariants*. Given periods ω_1, ω_2, the invariants are

$$\begin{aligned} g_2 &= 60 \sum_{(m,n)\in\mathbb{Z}\times\mathbb{Z}-\{(0,0)\}} \frac{1}{(m\omega_1+n\omega_2)^4}, \\ g_3 &= 140 \sum_{(m,n)\in\mathbb{Z}\times\mathbb{Z}-\{(0,0)\}} \frac{1}{(m\omega_1+n\omega_2)^6}. \end{aligned} \quad (2.7)$$

Alternately given invariants g_2, g_3, periods ω_1, ω_2 can be constructed if the *discriminant* $\Delta \stackrel{\text{def.}}{=} g_2^3 - 27g_3^2$ is nonzero — that is, when the *Weierstrass cubic* $4t^3 - g_2 t - g_3$ does not have repeated roots (see 21·73 of [Whittaker and Watson 1927]). Therefore we refer to $\wp(z;\omega_1,\omega_2)$ as either $\wp(z)$ or $\wp(z;g_2,g_3)$ and consider only cases where the invariants are such that $g_2^3 \neq 27g_3^2$. By (2.2) and

(2.6) one can see that the Weierstrass elliptic function satisfies

$$\wp'(z)^2 = 4\wp(z)^3 - g_2\wp(z) - g_3. \tag{2.8}$$

Note that in Michael Tuite's lecture in this volume, $\omega_{m,n}$ there is equal to $m\omega_1 + n\omega_2$ here with our ω_1, ω_2 specialized to $2\pi i \tau$ and $2\pi i$ in his lecture.

In the special case that the discriminant $\Delta > 0$, the roots of $4t^3 - g_2 t - g_3$ are real and distinct, and are conventionally notated by $e_1 > e_2 > e_3$ for $e_1 + e_2 + e_3 = 0$. In this case $4t^3 - g_2 t - g_3 = 4(t-e_1)(t-e_2)(t-e_3)$, the Weierstrass invariants are given in terms of the roots by

$$g_2 = -4(e_2 e_3 + e_1 e_3 + e_1 e_2), \qquad g_3 = 4 e_1 e_2 e_3, \tag{2.9}$$

and $\wp$ can be written in terms of the Jacobi elliptic functions by

$$\begin{aligned}\wp(z) &= e_3 + \gamma^2 \operatorname{ns}^2(\gamma z, k), \\ \wp(z) &= e_2 + \gamma^2 \operatorname{ds}^2(\gamma z, k), \\ \wp(z) &= e_1 + \gamma^2 \operatorname{cs}^2(\gamma z, k)\end{aligned} \tag{2.10}$$

for $e_1 < z \in \mathbb{R}$, $\gamma^2 \stackrel{\text{def.}}{=} e_1 - e_3$ and modulus k such that $k^2 = \frac{e_2 - e_3}{e_1 - e_3}$ (similar equations hold if $z \in \mathbb{R}$ and z is in a different range in relation to the real roots e_1, e_2, e_3, and alternate relations hold for nonreal roots when $\Delta < 0$; see chapter II of [Greenhill 1959]).

Note that the Jacobi elliptic functions solve a differential equation which contains only even powers of $f(u)$, and $\wp$ solves an equation with no squared or quartic powers of $\wp$. Weierstrass elliptic functions have the advantage of being easily implemented in the case that the cubic $4t^3 - g_2 t - g_3$ is not factored in terms of its roots. Elliptic integrals of the type $\int R(x)/\sqrt{P(x)}dx$, where $P(x)$ is a cubic polynomial and R is a rational function of x, can be written in terms of three fundamental *Weierstrassian normal elliptic integrals* although we will not record the details here (see Appendix of [Byrd and Friedman 1954]). In Section 6 we will see a method which allows one to write the elliptic integral $\int_{x_0}^{x} 1/\sqrt{F(t)}dt$, for $F(t)$ a quartic polynomial, in terms of the Weierstrassian normal elliptic integral *of the first kind* (2.6) by reducing the quartic to a cubic.

3. Jacobi theta functions

Jacobi theta functions are functions of two arguments, $z \in \mathbb{C}$ a complex number and $\tau \in \mathbb{H}$ in the upper-half plane. Every elliptic function can be written as the ratio of two theta functions. Doing so elucidates the meromorphic nature of elliptic functions and is useful in the numerical evaluation of elliptic functions. One must be cautious with the notation of theta functions, since many different conventions are used. We will use the notation of [Whittaker and Watson 1927] to define

$$\theta_1(z,\tau) \stackrel{\text{def.}}{=} 2q^{1/4} \sum_{n=0}^{\infty} (-1)^n q^{n(n+1)} \sin((2n+1)z),$$

$$\theta_2(z,\tau) \stackrel{\text{def.}}{=} 2q^{1/4} \sum_{n=0}^{\infty} q^{n(n+1)} \cos((2n+1)z), \quad (3.1)$$

$$\theta_3(z,\tau) \stackrel{\text{def.}}{=} 1 + 2 \sum_{n=1}^{\infty} q^{n^2} \cos(2nz),$$

$$\theta_4(z,\tau) \stackrel{\text{def.}}{=} 1 + 2 \sum_{n=1}^{\infty} (-1)^n q^{n^2} \cos(2nz),$$

where $q \stackrel{\text{def.}}{=} e^{\pi i \tau}$ is called the *nome*. We also define the special values $\theta_i \stackrel{\text{def.}}{=} \theta_i(0, \tau)$.

In terms of theta functions, the Jacobi elliptic functions are

$$\text{sn}(u,k) = \frac{\theta_3 \, \theta_1(u/\theta_3^2, \tau)}{\theta_2 \, \theta_4(u/\theta_3^2, \tau)},$$

$$\text{cn}(u,k) = \frac{\theta_4 \, \theta_2(u/\theta_3^2, \tau)}{\theta_2 \, \theta_4(u/\theta_3^2, \tau)}, \quad (3.2)$$

$$\text{dn}(u,k) = \frac{\theta_4 \, \theta_3(u/\theta_3^2, \tau)}{\theta_3 \, \theta_4(u/\theta_3^2, \tau)},$$

where τ is chosen such that $k^2 = \theta_2^4/\theta_3^4$. By [Whittaker and Watson 1927, 22·11], if $0 < k^2 < 1$ there exists a value of τ for which the quotient θ_2^4/θ_3^4 equals k^2.

4. The FRLW cosmological model

The Friedmann–Robertson–Lemaître–Walker cosmological model assumes that our current expanding universe is on large scales homogeneous and isotropic. On a d-dimensional spacetime this assumption translates into a metric of the form

$$ds^2 = -dt^2 + \tilde{a}(t)^2 \left(\frac{dr^2}{1 - k'r^2} + r^2 d\Omega_{d-2}^2 \right), \quad (4.1)$$

where $\tilde{a}(t)$ is the *cosmic scale factor* and $k' \in \{-1, 0, 1\}$ is the *curvature parameter*.

The Einstein field equations

$$G_{ij} = -\kappa_d T_{ij} + \Lambda g_{ij}$$

then govern the evolution of the universe over time. In these equations, the Einstein tensor

$$G_{ij} \stackrel{\text{def.}}{=} R_{ij} - \tfrac{1}{2} R g_{ij}$$

is computed directly from the metric g_{ij} by calculating the Ricci tensor R_{ij} and the scalar curvature R. Also $\kappa_d = 8\pi G_d$, where G_d is a generalization of Newton's constant to d-dimensional spacetime and $\Lambda > 0$ is the cosmological constant. The form of the energy-momentum tensor T_{ij} depends on what sort of matter content one is assuming, and in this lecture will be that of a perfect-fluid — that is, $T_{ij} = (p+\rho)g_{i0}g_{j0} + p g_{ij}$, where $\rho(t)$ and $p(t)$ are the density and pressure of the fluid, respectively.

For the metric (4.1), Einstein's equations are

$$\frac{(d-1)(d-2)}{2}\left(H^2 + \frac{k'}{\tilde{a}^2}\right) = \kappa_d \rho(t) + \Lambda, \qquad \text{(i)}$$

$$(d-2)\dot{H} + \frac{(d-1)(d-2)}{2}H^2 + \frac{(d-2)(d-3)}{2}\frac{k'}{\tilde{a}^2} = -\kappa_d p(t) + \Lambda, \qquad \text{(ii)}$$

for $H(t) \stackrel{\text{def.}}{=} \dot{\tilde{a}}(t)/\tilde{a}(t)$ and where dot denotes differentiation with respect to t. In this lecture, only equation (i) will be required to relate $\tilde{a}(t)$ to elliptic and theta functions. We rewrite equation (i) in terms of *conformal time* η by defining the *conformal scale factor* $a(\eta) \stackrel{\text{def.}}{=} \tilde{a}(f(\eta))$, where $f(\eta)$ is the inverse function of $\eta(t)$, which satisfies $\dot{\eta}(t) = 1/\tilde{a}(t)$. In terms of $a(\eta)$, (i) becomes

$$a'(\eta)^2 = \tilde{\Lambda} a(\eta)^4 + \tilde{\kappa}_d \rho(f(\eta)) a(\eta)^4 - k' a(\eta)^2, \qquad (4.2)$$

where we use notation $\tilde{\Lambda} \stackrel{\text{def.}}{=} 2\Lambda/(d-1)(d-2)$, $\tilde{\kappa}_d \stackrel{\text{def.}}{=} 2\kappa_d/(d-1)(d-2)$ and we take spacetime dimension $d > 2$.

5. FRLW and Jacobi elliptic and theta functions

In general, if $f(u)$ is a solution to $f'(u)^2 + af(u)^2 + bf(u)^4 = c$, then $g(u) = \beta f(\alpha u)$ is a solution to

$$g'(u)^2 + Ag(u)^2 + Bg(u)^4 = \frac{A^2 bc}{a^2 B} \qquad (5.1)$$

for $\alpha = \sqrt{A/a}$ and $\beta = \sqrt{Ab/(aB)}$, where we may choose either the positive or negative square root for each of α and β. We will construct solutions to (4.2), given that Jacobi elliptic functions solve (2.4), and also proceed to write these solutions in terms of theta functions by the relations in (3.2).

For the special case of density $\rho(t) = D/\tilde{a}(t)^4$ with $D > 0$, (4.2) becomes
$$a'(\eta)^2 + k'a(\eta)^2 - \tilde{\Lambda}a(\eta)^4 = \tilde{\kappa}_d D. \tag{5.2}$$

Therefore in (5.1) we take $A = k'$, $B = -\tilde{\Lambda}$, and a, b, c as in the table in Section 2. To restrict to real solutions we only consider entries in the table for which the ratios a/A and b/B are positive, and we also restrict D to be such that the right side of (5.2) agrees with the right side of (5.1).

For positive curvature $k' = 1$, and for $D = \dfrac{k^2}{\tilde{\kappa}_d \tilde{\Lambda}(1+k^2)^2}$ with $0 < k < 1$, we solve
$$a'(\eta)^2 + a(\eta)^2 - \tilde{\Lambda}a(\eta)^4 = \frac{k^2}{\tilde{\Lambda}(1+k^2)^2} \tag{5.3}$$

in conformal time in terms of Jacobi elliptic functions to obtain

$$a_{\text{sn}}(\eta) = \frac{k}{\sqrt{\tilde{\Lambda}(1+k^2)}} \, \text{sn}\left(\frac{\eta}{\sqrt{1+k^2}}, k\right) = \frac{\theta_2 \theta_3}{\sqrt{\tilde{\Lambda}(1+\theta_2^4 \theta_3^4)}} \frac{\theta_1\left(\eta/\sqrt{\theta_2^4+\theta_3^4}, \tau\right)}{\theta_4\left(\eta/\sqrt{\theta_2^4+\theta_3^4}, \tau\right)},$$

$$a_{\text{ns}}(\eta) = \frac{1}{\sqrt{\tilde{\Lambda}(1+k^2)}} \, \text{ns}\left(\frac{\eta}{\sqrt{1+k^2}}, k\right) = \frac{\theta_2 \theta_3}{\sqrt{\tilde{\Lambda}(1+\theta_2^4 \theta_3^4)}} \frac{\theta_4\left(\eta/\sqrt{\theta_2^4+\theta_3^4}, \tau\right)}{\theta_1\left(\eta/\sqrt{\theta_2^4+\theta_3^4}, \tau\right)},$$

$$a_{\text{cd}}(\eta) = \frac{k}{\sqrt{\tilde{\Lambda}(1+k^2)}} \, \text{cd}\left(\frac{\eta}{\sqrt{1+k^2}}, k\right) = \frac{\theta_2 \theta_3}{\sqrt{\tilde{\Lambda}(1+\theta_2^4 \theta_3^4)}} \frac{\theta_2\left(\eta/\sqrt{\theta_2^4+\theta_3^4}, \tau\right)}{\theta_3\left(\eta/\sqrt{\theta_2^4+\theta_3^4}, \tau\right)},$$

$$a_{\text{dc}}(\eta) = \frac{1}{\sqrt{\tilde{\Lambda}(1+k^2)}} \, \text{dc}\left(\frac{\eta}{\sqrt{1+k^2}}, k\right) = \frac{\theta_2 \theta_3}{\sqrt{\tilde{\Lambda}(1+\theta_2^4 \theta_3^4)}} \frac{\theta_3\left(\eta/\sqrt{\theta_2^4+\theta_3^4}, \tau\right)}{\theta_2\left(\eta/\sqrt{\theta_2^4+\theta_3^4}, \tau\right)}, \tag{5.4}$$

where τ is chosen such that $k^2 = \theta_2^4/\theta_3^4$.

The first two solutions, $a_{\text{sn}}(\eta)$ and $a_{\text{ns}}(\eta)$, reduce to hyperbolic trigonometric functions in the case of modulus $k = 1$, since $\text{sn}(u, 1) = \tanh(u)$ and $\text{ns}(u, 1) = \coth(u)$. That is, two additional solutions in terms of elementary functions are

$$a_1(\eta) = \frac{1}{\sqrt{2\tilde{\Lambda}}} \tanh(\eta/\sqrt{2}), \qquad a_2(\eta) = \frac{1}{\sqrt{2\tilde{\Lambda}}} \coth(\eta/\sqrt{2}). \tag{5.5}$$

For these two solutions, one may solve the differential equation $\dot{\eta}(t) = 1/\tilde{a}(t) = 1/a(\eta(t))$ for $\eta(t)$, and therefore obtain the cosmic scale factor $\tilde{a}(t) = a(\eta(t))$ which solves the Einstein field equation (i) in Section 4 for $\rho(t) = D/\tilde{a}(t)^4$ with special value $D = k^2/\tilde{\Lambda}\tilde{\kappa}_d(1+k^2)^2$. Doing so, we obtain

$$\tilde{a}_1(t) = a_1(\eta(t)) = \frac{1}{\sqrt{2\tilde{\Lambda}}} \tanh \ln\left(e^{\sqrt{\tilde{\Lambda}}t} + \sqrt{e^{2\sqrt{\tilde{\Lambda}}t} - 1}\right),$$

$$\tilde{a}_2(t) = a_2(\eta(t)), = \frac{1}{\sqrt{2\tilde{\Lambda}}} \coth \ln\left(e^{\sqrt{\tilde{\Lambda}}t} + \sqrt{e^{2\sqrt{\tilde{\Lambda}}t} + 1}\right), \tag{5.6}$$

for $t > 0$.

For negative curvature $k' = -1$, and for

$$D = \frac{1-k^2}{\tilde{\Lambda}\,\tilde{\kappa}_d(k^2-2)^2}$$

with $0 < k < 1$, equation (5.2) becomes

$$a'(\eta)^2 - a(\eta)^2 - \tilde{\Lambda}a(\eta)^4 = \frac{1-k^2}{(k^2-2)^2\tilde{\Lambda}}. \tag{5.7}$$

In terms of Jacobi elliptic functions and theta functions, we obtain the two solutions for the scale factor in conformal time,

$$a_{\text{sc}}(\eta) = \frac{\sqrt{1-k^2}}{\sqrt{\tilde{\Lambda}(2-k^2)}}\,\text{sc}\left(\frac{\eta}{\sqrt{2-k^2}},k\right) = \frac{\sqrt{\theta_3^4-\theta_2^4}}{\sqrt{\tilde{\Lambda}(2\theta_3^4-\theta_2^4)}}\,\frac{\theta_3\,\theta_1\left(\eta/\theta_3^2\sqrt{2\theta_3^4-\theta_2^4},\tau\right)}{\theta_4\,\theta_2\left(\eta/\theta_3^2\sqrt{2\theta_3^4-\theta_2^4},\tau\right)},$$

$$a_{\text{cs}}(\eta) = \frac{1}{\sqrt{\tilde{\Lambda}(2-k^2)}}\,\text{cs}\left(\frac{\eta}{\sqrt{2-k^2}},k\right) = \frac{\theta_3\,\theta_4}{\sqrt{\tilde{\Lambda}(2\theta_3^4-\theta_2^4)}}\,\frac{\theta_2\left(\eta/\theta_3^2\sqrt{2\theta_3^4-\theta_2^4},\tau\right)}{\theta_1\left(\eta/\theta_3^2\sqrt{2\theta_3^4-\theta_2^4},\tau\right)}, \tag{5.8}$$

where τ is such that $k^2 = \theta_2^4/\theta_3^4$.

Note that by the comments following equation (2.10), it is possible to express the solutions obtained in this section in terms of Weierstrass functions. E. Abdalla and L. Correa-Borbonet [2002] have also considered the Einstein equation (i) with $\rho(t) = D/\tilde{a}(t)^4$, and have found connections with Weierstrass functions in cosmic time (as opposed to the conformal time argument given here). In Section 6 of this lecture we will find more general solutions to the conformal time equation (4.2) in terms of $\wp$, for arbitrary curvature k' and D-value. We will also consider the density functions $\rho(t) = D/\tilde{a}(t)^3$ and $\rho(t) = D_1/\tilde{a}(t)^3 + D_2/\tilde{a}(t)^4$ in Section 6.

6. FRLW and Weierstrass elliptic functions

In general, if $g(0) = x_0$ and $g(u)$ satisfies

$$g'(u)^2 = F(g(u)) \tag{6.1}$$

for $F(x) = A_4 x^4 + A_3 x^3 + A_2 x^2 + A_1 x + A_0$ any quartic polynomial with no repeated roots, then the inverse function $y(x)$ of $g(u)$ is the elliptic integral

$$y(x) = \int_{x_0}^{x} \frac{dt}{\sqrt{F(t)}}. \tag{6.2}$$

For the initial condition $g'(0) = 0$, x_0 is a root of the polynomial $F(x)$ by (6.1). In this case the integral (6.2) can be rewritten as $\int_{\xi}^{\infty} dz/\sqrt{P(z)}$, where $\xi = 1/(x-x_0)$ and $P(z)$ is a cubic polynomial. To do this, first expand $F(t)$

into its Taylor series about x_0 and then perform the change of variables $z = 1/(t - x_0)$. Furthermore one may obtain the form $\int_\tau^\infty d\sigma/\sqrt{Q(\sigma)}$ for $\tau = \frac{1}{4}(F'(x_0)\xi + \frac{1}{6}F''(x_0))$, where $Q(\sigma) = 4\sigma^3 - g_2\sigma - g_3$. This is done by setting $z = (4\sigma - B_2/3)/B_3$ where $B_2 = F''(x_0)/2$ and $B_3 = F'(x_0)$ are the quadratic and cubic coefficients of $P(z)$ respectively. Note that since $F(x)$ has no repeated roots, x_0 is not a double root, $F'(x_0) \neq 0$ and the variable z is well-defined. Therefore we have obtained

$$y\left(\frac{6F'(x_0)}{24\tau - F''(x_0)} + x_0\right) = \wp^{-1}(\tau).$$

Writing this in terms of x, setting $x = g(u)$ and solving for $g(u)$, one obtains the solution to (6.1):

$$g(u) = x_0 + \frac{F'(x_0)}{4\wp(u; g_2, g_3) - \frac{1}{6}F''(x_0)}, \quad (6.3)$$

where

$$g_2 = A_0 A_4 - \frac{1}{4}A_1 A_3 + \frac{1}{12}A_2^2 \quad \text{and}$$
$$g_3 = \frac{1}{6}A_0 A_2 A_4 + \frac{1}{48}A_1 A_2 A_3 - \frac{1}{16}A_1^2 A_4 - \frac{1}{16}A_0 A_3^2 - \frac{1}{216}A_2^3 \quad (6.4)$$

are referred to as the *invariants of the quartic* $F(x)$. Since $F(x)$ has no repeated roots, the discriminant $\Delta = g_2^3 - 27g_3^2 \neq 0$. Here if $x_0 = 0$, (6.3) becomes

$$g(u) = \frac{A_1}{4\wp(u; g_2, g_3) - \frac{1}{3}A_2}. \quad (6.5)$$

If the initial condition on the first derivative is such that $g'(0) \neq 0$ then x_0 is not a root of $F(x)$ and a more general solution to (6.1) is due to Weierstrass. The proof (which we will not include here) was published by Biermann in 1865 (see [Biermann 1865; Reynolds 1989]). The solution is

$$g(u) = x_0 + \frac{\sqrt{F(x_0)}\wp'(u) + \frac{1}{2}F'(x_0)\left(\wp(u) - \frac{1}{24}F''(x_0)\right) + \frac{1}{24}F(x_0)F'''(x_0)}{2\left(\wp(u) - \frac{1}{24}F''(x_0)\right)^2 - \frac{1}{48}F(x_0)F''''(x_0)} \quad (6.6)$$

where $\wp$ is formed with the invariants of the quartic seen in (6.4) such that $\Delta \neq 0$. Here if $x_0 = 0$, (6.6) becomes

$$g(u) = \frac{\sqrt{A_0}\wp'(u) + \frac{1}{2}A_1\left(\wp(u) - \frac{1}{12}A_2\right) + \frac{1}{4}A_0 A_3}{2\left(\wp(u) - \frac{1}{12}A_2\right)^2 - \frac{1}{2}A_0 A_4}. \quad (6.7)$$

To generate a number of examples, we consider the conformal time Einstein equation (4.2) for density $\rho(t) = D_1/\tilde{a}(t)^3 + D_2/\tilde{a}(t)^4$ with $D_1, D_2 > 0$. In this case (4.2) becomes

$$a'(\eta)^2 = \tilde{\Lambda}a(\eta)^4 - k'a(\eta)^2 + \tilde{\kappa}_d D_1 a(\eta) + \tilde{\kappa}_d D_2 \qquad (6.8)$$

and we take $A_4 = \tilde{\Lambda}$, $A_3 = 0$, $A_2 = -k'$, $A_1 = \tilde{\kappa}_d D_1$ and $A_0 = \tilde{\kappa}_d D_2$. The most general solution seen here is with initial conditions $a'(0) \neq 0$ so that $a(0) \stackrel{\text{def.}}{=} a_0$ is not a root of the polynomial

$$F(t) = \tilde{\Lambda}t^4 - k't^2 + \tilde{\kappa}_d D_1 t + \tilde{\kappa}_d D_2. \qquad (6.9)$$

The solution is given by (6.6) as

$$a(\eta) = a_0 + \frac{\sqrt{F(a_0)}\wp'(\eta) + \frac{1}{2}F'(a_0)\left(\wp(\eta) - \frac{1}{24}F''(a_0)\right) + \frac{1}{24}F(a_0)F'''(a_0)}{2\left(\wp(\eta) - \frac{1}{24}F''(a_0)\right)^2 - \frac{1}{48}F(a_0)F''''(a_0)}, \qquad (6.10)$$

with Weierstrass invariants

$$\begin{aligned} g_2 &= \tilde{\Lambda}\tilde{\kappa}_d D_2 + \tfrac{1}{12}(k')^2 \quad \text{and} \\ g_3 &= -\tfrac{1}{6}\tilde{\Lambda}\tilde{\kappa}_d D_2 k' - \tfrac{1}{16}\tilde{\Lambda}\tilde{\kappa}_d^2 D_1^2 + \tfrac{1}{216}(k')^3 \end{aligned} \qquad (6.11)$$

restricted to be such that $\Delta = g_2^3 - 27g_3^2 \neq 0$ so that $F(t)$ does not have repeated roots.

Since $D_2, \tilde{\kappa}_d > 0$, by (6.9) zero is not a root of $F(t)$ and therefore for initial condition $a'(0) \neq 0$ and $a_0 = 0$, the solution to (6.8) is given by (6.7) as

$$a(\eta) = \frac{\sqrt{\tilde{\kappa}_d D_2}\wp'(\eta) + \frac{1}{2}\tilde{\kappa}_d D_1\left(\wp(\eta) + \frac{k'}{12}\right)}{2\left(\wp(\eta) + \frac{k'}{12}\right)^2 - \frac{1}{2}\tilde{\Lambda}\tilde{\kappa}_d D_2} \qquad (6.12)$$

for invariants g_2, g_3 as in (6.11) with $\Delta \neq 0$. One can compare this with the results in papers by Aurich, Steiner and Then, where curvature is taken to be $k' = -1$ [Aurich and Steiner 2001; Aurich et al. 2004].

For $a'(0) = 0$ and a_0 a root of $F(t)$ in (6.9), the solution to (6.8) is given by (6.3),

$$a(\eta) = a_0 + \frac{F'(a_0)}{4\wp(\eta) - \frac{1}{6}F''(a_0)} \qquad (6.13)$$

again with invariants (6.11) such that $\Delta \neq 0$.

For a more concrete example, consider the density function $\rho(t) = D/\tilde{a}(t)^3$ for $D > 0$ so that conformal time equation (4.2) becomes

$$a'(\eta)^2 = \tilde{\Lambda}a(\eta)^4 - k'a(\eta)^2 + \tilde{\kappa}_d D a(\eta). \qquad (6.14)$$

Here zero is a root of the polynomial $F(t)$ with $A_4 = \tilde{\Lambda}$, $A_3 = A_0 = 0$, $A_2 = -k'$ and $A_1 = \tilde{\kappa}_d D$. Therefore with initial conditions $a'(0) = a(0) = 0$, the solution to (6.14) is given by (6.5) as

$$a(\eta) = \frac{3\tilde{\kappa}_d D}{12\wp(\eta) + k'} \tag{6.15}$$

with invariants

$$g_2 = \tfrac{1}{12}(k')^2 \quad \text{and} \quad g_3 = -\tfrac{1}{16}\tilde{\Lambda}\tilde{\kappa}_d^2 D^2 + \tfrac{1}{216}(k')^3 \tag{6.16}$$

restricted to be such that $\Delta \neq 0$. As noted in the comments following equations (2.10), one can write this solution in terms of Jacobi elliptic functions (by using equations (2.10) if the roots of the reduced cubic $4t^3 - g_2 t - g_3$ are real). To demonstrate this, we choose

$$D = \frac{1}{\tilde{\kappa}_d}\sqrt{\frac{2}{27\tilde{\Lambda}}}$$

and $k' = 1$, so that $g_3 = 0$ and $g_2 = \tfrac{1}{12}$. For this positive curvature case, (6.15) becomes

$$a(\eta) = \frac{\sqrt{2}}{\sqrt{3\tilde{\Lambda}}(12\wp(\eta) + 1)}, \tag{6.17}$$

and the reduced cubic is $4t^3 - (1/12)t = 4t(t - 1/4\sqrt{3})(t + 1/4\sqrt{3})$. Applying (2.10) with $e_3 = -1/4\sqrt{3}$, $e_2 = 0$, $e_1 = 1/4\sqrt{3}$, (6.17) can be equivalently written in terms of Jacobi elliptic functions as

$$a(\eta) = \frac{\sqrt{2/\tilde{\Lambda}}}{\sqrt{3} - 3 + 6\,\mathrm{ns}^2\left(\eta/\sqrt{2\sqrt{3}}, 1/\sqrt{2}\right)} = \frac{\sqrt{2/\tilde{\Lambda}}}{\sqrt{3} + 6\,\mathrm{ds}^2\left(\eta/\sqrt{2\sqrt{3}}, 1/\sqrt{2}\right)}$$

$$= \frac{\sqrt{2/\tilde{\Lambda}}}{\sqrt{3} + 3 + 6\,\mathrm{cs}^2\left(\eta/\sqrt{2\sqrt{3}}, 1/\sqrt{2}\right)}, \tag{6.18}$$

since $k^2 = \frac{e_2 - e_3}{e_1 - e_3} = \frac{1}{2}$ and $\gamma^2 = e_1 - e_3 = \frac{1}{2\sqrt{3}}$. In terms of theta functions, this is

$$a(\eta) = \frac{\sqrt{2/\tilde{\Lambda}}\,\theta_3^2\,\theta_1^2(\eta/\sqrt{2\sqrt{3}}\,\theta_3^2, \tau)}{(\sqrt{3} - 3)\,\theta_3^2\,\theta_1^2(\eta/\sqrt{2\sqrt{3}}\,\theta_3^2, \tau) + 6\,\theta_2^2\,\theta_4^2(\eta/\sqrt{2\sqrt{3}}\,\theta_3^2, \tau)}$$

$$= \frac{\sqrt{2/\tilde{\Lambda}}\,\theta_3^4\,\theta_1^2(\eta/\sqrt{2\sqrt{3}}\,\theta_3^2, \tau)}{\sqrt{3}\,\theta_3^4\,\theta_1^2(\eta/\sqrt{2\sqrt{3}}\,\theta_3^2, \tau) + 6\,\theta_2^2\,\theta_4^2\,\theta_3^2(\eta/\sqrt{2\sqrt{3}}\,\theta_3^2, \tau)}$$

$$= \frac{\sqrt{2/\tilde{\Lambda}}\,\theta_3^2\,\theta_1^2(\eta/\sqrt{2\sqrt{3}}\,\theta_3^2, \tau)}{(\sqrt{3} + 3)\,\theta_3^2\,\theta_1^2(\eta/\sqrt{2\sqrt{3}}\,\theta_3^2, \tau) + 6\,\theta_4^2\,\theta_2^2(\eta/\sqrt{2\sqrt{3}}\,\theta_3^2, \tau)} \tag{6.19}$$

where τ is taken such that $\tfrac{1}{2} = \theta_2^4/\theta_3^4$.

As a final example, we return to $\rho(t) = D/\tilde{a}(t)^4$ for $D > 0$ considered in Section 5. That is, we will obtain alternate solutions to equation (5.2). Since $D > 0$, zero is not a root of the polynomial $F(t)$ with $A_4 = \tilde{\Lambda}$, $A_3 = A_1 = 0$, $A_2 = -k'$ and $A_0 = \tilde{\kappa}_d D$. Therefore for initial conditions $a'(0) \neq 0$ and $a(0) = 0$, (6.7) gives the solution

$$a(\eta) = \frac{\sqrt{\tilde{\kappa}_d D} \wp'(\eta)}{2(\wp(\eta) + \frac{1}{12}k')^2 - \frac{1}{2}\tilde{\Lambda}\tilde{\kappa}_d D} \tag{6.20}$$

for invariants

$$g_2 = \tilde{\Lambda}\tilde{\kappa}_d D + \tfrac{1}{12}(k')^2 \quad \text{and} \quad g_3 = -\tfrac{1}{6}\tilde{\Lambda}\tilde{\kappa}_d D k' + \tfrac{1}{216}(k')^3 \tag{6.21}$$

restricted to be such that $\Delta \neq 0$. (6.20) is more general than the solutions to (5.2) in Section 5, since here the curvature k' and the constant D are unspecified. To see this solution expressed in terms of Jacobi elliptic functions, take curvature $k' = 1$ and $D = 1/(36\tilde{\Lambda}\tilde{\kappa}_d)$ for $0 < k < 1$ so that $g_3 = 0$ and $g_2 = \tfrac{1}{9}$. Then the reduced cubic is $4t^3 - \tfrac{1}{9}t = 4(t - \tfrac{1}{6})(t + \tfrac{1}{6})$ so that $e_3 = -\tfrac{1}{6}$, $e_2 = 0$, $e_1 = \tfrac{1}{6}$ and (6.20) becomes

$$a(\eta) = \frac{(12/\sqrt{\tilde{\Lambda}})\, \wp'(\eta)}{144(\wp(\eta) + \tfrac{1}{12})^2 - 1}. \tag{6.22}$$

By (2.8) and (2.10), we write this solution in terms of Jacobi elliptic functions and obtain

$$a(\eta) = \frac{\sqrt{2 - 3\,\mathrm{sn}^2\left(\tfrac{\eta}{\sqrt{3}}, \tfrac{1}{\sqrt{2}}\right) + \mathrm{sn}^4\left(\tfrac{\eta}{\sqrt{3}}, \tfrac{1}{\sqrt{2}}\right)}}{\sqrt{6\tilde{\Lambda}}\left(2\,\mathrm{ns}\left(\tfrac{\eta}{\sqrt{3}}, \tfrac{1}{\sqrt{2}}\right) - \mathrm{sn}\left(\tfrac{\eta}{\sqrt{3}}, \tfrac{1}{\sqrt{2}}\right)\right)}$$

$$= \frac{1}{2\sqrt{3\tilde{\Lambda}}\,\mathrm{ds}\left(\tfrac{\eta}{\sqrt{3}}, \tfrac{1}{\sqrt{2}}\right)}\sqrt{\frac{2\,\mathrm{ds}^2\left(\tfrac{\eta}{\sqrt{3}}, \tfrac{1}{\sqrt{2}}\right) - 1}{2\,\mathrm{ds}^2\left(\tfrac{\eta}{\sqrt{3}}, \tfrac{1}{\sqrt{2}}\right) + 1}}$$

$$= \frac{\mathrm{cs}\left(\tfrac{\eta}{\sqrt{3}}, \tfrac{1}{\sqrt{2}}\right)}{\sqrt{6\tilde{\Lambda}}\sqrt{\left(1 + \mathrm{cs}^2\left(\tfrac{\eta}{\sqrt{3}}, \tfrac{1}{\sqrt{2}}\right)\right)\left(1 + 2\,\mathrm{cs}^2\left(\tfrac{\eta}{\sqrt{3}}, \tfrac{1}{\sqrt{2}}\right)\right)}} \tag{6.23}$$

where each of the positive and negative square roots solve (5.2) for $k' = 1$, $D = 1/(36\tilde{\Lambda}\tilde{\kappa}_d)$ and where $\gamma^2 = e_1 - e_3 = \tfrac{1}{3}$ and $k^2 = (e_2 - e_3)/(e_1 - e_3) = \tfrac{1}{2}$. Writing (6.23) in terms of theta functions, and defining $\mu = \eta/\sqrt{3\theta_3^3}$, we find that

$$a(\eta) = \frac{\theta_3\theta_1(\mu)\sqrt{2\theta_2^4\theta_4^4(\mu) - 3\theta_2^2\theta_3^2\theta_1^2(\mu)\theta_4^2(\mu) + \theta_3^4\theta_1^4(\mu)}}{\sqrt{6\tilde{\Lambda}}\,\theta_2\theta_4(\mu)\left(2\theta_2^2\theta_4^2(\mu) - \theta_3^2\theta_1^2(\mu)\right)}$$

$$= \frac{\theta_3^2\theta_1(\mu)}{2\sqrt{3\tilde{\Lambda}}\,\theta_2\theta_4\theta_3(\mu)}\sqrt{\frac{2\theta_2^2\theta_4^2\theta_3^2(\mu) - \theta_3^4\theta_1^2(\mu)}{2\theta_2^2\theta_4^2\theta_3^2(\mu) + \theta_3^4\theta_1^2(\mu)}}$$

$$= \frac{\theta_4\theta_3\theta_2(\mu)\theta_1(\mu)}{\sqrt{6\tilde{\Lambda}\left(\theta_3^2\theta_1^2(\mu) + \theta_4^2\theta_2^2(\mu)\right)\left(\theta_3^2\theta_1^2(\mu) + 2\theta_4^2\theta_2^2(\mu)\right)}} \quad (6.24)$$

by the theta function representations for sn, ds and cs respectively. Here the τ that forms the theta functions is suppressed and is taken to satisfy $\frac{1}{2} = \theta_2^4/\theta_3^4$.

7. Summary

There are a number of ways to see that elliptic and theta functions solve the d-dimensional Einstein gravitational field equations in a FRLW cosmology with a cosmological constant. Here we considered a scenario with no scalar field and with density functions $\rho(t) = D_1/\tilde{a}(t)^3 + D_2/\tilde{a}(t)^4$, $\rho(t) = D/\tilde{a}(t)^3$ and $\rho(t) = D/\tilde{a}(t)^4$ scaling in inverse proportion to the scale factor $\tilde{a}(t)$. In these cases the first Einstein equation (i) takes the form $\dot{\tilde{a}}(t)^2 =$ an expression containing negative powers of the cosmic scale factor $\tilde{a}(t)$. At this point, one could have introduced the inverse function $y(x)$ of $\tilde{a}(t)$ to obtain an expression for $y(x)$ as the integral of a power of x divided by the square root of a polynomial in x. That is, $y(x)$ would be an elliptic integral that is not normal; other authors have taken this approach [Abdalla and Correa-Borbonet 2002; Kraniotis and Whitehouse 2002]. Here, we switched to conformal time by a change of variables $a(\eta) \stackrel{\text{def.}}{=} \tilde{a}(f(\eta))$. This produced an equation of the form $a'(\eta)^2 =$ an expression containing nonnegative powers of the conformal scale factor $a(\eta)$.

After reviewing the definitions and properties of elliptic and theta functions in sections 2 and 3, we introduced the FRLW cosmological model in Section 4. In Section 5 for $\rho(t) = D/\tilde{a}(t)^3$, we obtained a differential equation for $a(\eta)$ containing only even powers of $a(\eta)$ and constructed solutions in terms of Jacobi elliptic functions, restricted to particular values of the constant D, parameterized by modulus $0 < k < 1$. The equivalent theta function representations for these solutions were recorded, and we noted the special cases for which the elliptic solutions reduce to elementary functions and the corresponding solution in cosmic time was also computed. In Section 6, we considered each of $\rho(t) = D_1/\tilde{a}(t)^3 + D_2/\tilde{a}(t)^4$, $\rho(t) = D/\tilde{a}(t)^3$ and $\rho(t) = D/\tilde{a}(t)^4$

with various initial conditions and obtained solutions in terms of Weierstrass functions for general curvature k' and constants $D_1, D_2, D > 0$. By considering these solutions restricted to certain D-values (again, parameterized by modulus $0 < k < 1$), we wrote $a(\eta)$ equivalently in terms of Jacobi elliptic and theta functions.

In current joint work with Floyd Williams [D'Ambroise and Williams $\geq$ 2010], we have seen that elliptic functions also appear in the presence of a scalar field $\phi(t)$, for both the FRLW and Bianchi I d-dimensional cosmological models with a $\Lambda \neq 0$ and with a similar density function scaling in inverse proportion to $\tilde{a}(t)$. There we note that the equations of each of these cosmological models can be rewritten in terms of a generalized Ermakov–Milne–Pinney differential equation [Lidsey 2004; D'Ambroise and Williams 2007], a type which the square root of the second moment of the wave function of the Bose–Einstein condensate (BEC) also satisfies. On the cosmological side of the FRLW-BEC correspondence, imposing an equation of state $\rho_\phi(t) = w p_\phi(t)$ (w constant) on the density $\rho_\phi(t)$ and pressure $p_\phi(t)$ of the scalar field $\phi(t)$ allows one to obtain the differential equation for an elliptic function on the side of the BECs.

References

[Abdalla and Correa-Borbonet 2002] E. Abdalla and L. Correa-Borbonet, "The elliptic solutions to the Friedmann equation and the Verlinde's maps", preprint, 2002. arXiv:hep-th/0212205.

[Aurich and Steiner 2001] R. Aurich and F. Steiner, "The cosmic microwave background for a nearly flat compact hyperbolic universe", *Monthly Not. Royal Astron. Soc.* **323**:4 (2001), 1016–1024. Also see arXiv:astro-ph/0007264.

[Aurich et al. 2004] R. Aurich, F. Steiner, and H. Then, "Numerical computation of Maass waveforms and an application to cosmology", preprint, 2004. arXiv:gr-qc/0404020.

[Basarab-Horwath et al. 2004] P. Basarab-Horwath, W. I. Fushchych, and L. F. Barannyk, "Solutions of the relativistic nonlinear wave equation by solutions of the nonlinear Schrödinger equation", pp. 81–99 in *Scientific works of W. I. Fushchych*, vol. 6, 2004.

[Biermann 1865] G. G. A. Biermann, *Problemata quaedam mechanica functionum ellipticarum ope soluta*, Dissertatio Inauguralis, Friedrich-Wilhelm-Universität, 1865. Available at http://edoc.hu-berlin.de/ebind/hdiss/BIERPROB_PPN313151385/XML.

[Byrd and Friedman 1954] P. F. Byrd and M. D. Friedman, *Handbook of elliptic integrals for engineers and physicists*, Grundlehren der math. Wissenschaften **67**, Springer, Berlin, 1954.

[D'Ambroise and Williams 2007] J. D'Ambroise and F. L. Williams, "A non-linear Schrödinger type formulation of FLRW scalar field cosmology", *Int. J. Pure Appl. Math.* **34**:1 (2007), 117–127.

[D'Ambroise and Williams ≥ 2010] J. D'Ambroise and F. Williams, "A dynamic correspondence between FRLW cosmology with cosmological constant and Bose–Einstein condensates", to appear.

[Greenhill 1959] A. G. Greenhill, *The applications of elliptic functions*, Dover, New York, 1959.

[Kharbediya 1976] L. I. Kharbediya, "Some exact solutions of the Friedmann equations with the cosmological term", *Astronom. Ž.* **53** (1976), 1145–1152.

[Kraniotis and Whitehouse 2002] G. V. Kraniotis and S. B. Whitehouse, "General relativity, the cosmological constant and modular forms", *Classical Quantum Gravity* **19**:20 (2002), 5073–5100. arXiv:gr-qc/0105022.

[Lidsey 2004] J. Lidsey, "Cosmic dynamics of Bose–Einstein condensates", *Classical and Quantum Gravity* **21** (2004), 777–785. arXiv:gr-qc/0307037.

[Reynolds 1989] M. J. Reynolds, "An exact solution in nonlinear oscillations", *J. Phys. A* **22**:15 (1989), L723–L726. Available at http://stacks.iop.org/0305-4470/22/L723.

[Whittaker and Watson 1927] E. T. Whittaker and G. N. Watson, *A course of modern analysis*, Cambridge Mathematical Library, Cambridge University Press, Cambridge, 1927.

JENNIE D'AMBROISE
DEPARTMENT OF MATHEMATICS AND STATISTICS
LEDERLE GRADUATE RESEARCH TOWER
UNIVERSITY OF MASSACHUSETTS
AMHERST, MA 01003-9305
UNITED STATES
dambroise@math.umass.edu

Integrable systems and 2D gravitation: How a soliton illuminates a black hole

SHABNAM BEHESHTI

1. Introduction

The interlacing of number theory with modern physics has a long and fruitful history. Indeed, the inital seeds were sown by Riemann himself, from paving the way to the Einstein equations with his introduction of the curvature tensor to setting the stage for quantum correction to black hole entropy with his careful study of the zeta function. This note indicates several elegant connections between two-dimensional gravitation, an eigenvalue problem of interest, extended objects (1D bosonic strings), and zeta regularization in 2D quantum gravity; we also note the presence of modular forms when possible and connect our results to the classical three-dimensional theory. It is our aim to find points of tangency with themes from the 2008 MSRI Summer School on Zeta and Modular Physics and motivate the reader for further exploration.

2. JT Gravitation: A simple 2D metric-scalar field theory

Consider the vacuum Einstein equations with vanishing cosmological constant

$$R_{ij} - \frac{R}{2} g_{ij} = 0, \qquad 1 \le i, j \le n \qquad (1)$$

where R_{ij} and R denote the Ricci tensor and scalar curvature, respectively, and the solution (M^n, g) is an n-dimensional Riemannian manifold with metric tensor $g = g_{ij}$. A simple calculation shows that for dim $M = n = 2$, Equation (1) is trivially satisfied. Thus, to make a meaningful interpretation

---

The title of this communication is both in reference and in honour of Geoffrey Mason's lecture *Vertex operators and arithmetic: how a single photon illuminates number theory*, at the 2004 ICMS Conference on Moonshine Conjectures and Vertex Algebras.

of these field equations, one is compelled to modify the Einstein-Hilbert action $\int_{M^n} R(g)\sqrt{|\det g|}d^n x$ from which they arise [40]. Introducing a dilaton $\tau = \tau(x) = \tau(x_1, x_2)$, a scalar field in the two variables of the manifold, a potential function $V = V(y)$, and a nonzero constant m, the modified action becomes

$$\int_{M^2} (R(g)\tau - m^2 V \circ \tau)\sqrt{|\det g|}\, d^2 x. \quad (2)$$

Standard variational principles yield a corresponding set of field equations

$$R(g) - m^2(V' \circ \tau) = 0,$$
$$\nabla_i \nabla_j \tau - \frac{m^2}{2} g_{ij}(V \circ \tau) = 0. \quad 1 \leq i, j \leq 2 \quad (3)$$

Here, $\nabla_i \nabla_j \tau$ denotes the Hessian of the field τ computed with respect to the metric g_{ij} [9]. The first equation in (3) is referred to as the Einstein equation of the system, with the remainder being called equations of motion for the dilaton τ. In contrast to (1), solutions to the modified two-dimensional model consist of a metric-dilaton pair (g, τ). This toy model has proved useful in understanding several key problems of interest, including

- relating exact solutions of system (3) to nonlinear equations having known special function solutions [5; 6; 41; 44],
- understanding the statistical origin for black hole entropy [20; 30],
- studying the endpoint of gravitational collapse [14; 21],
- examining the thermodynamics of black hole solutions in two and three dimensions [11; 12; 37],
- comparing 2D string gravity with higher dimensional counterparts [28; 45],
- providing a stepping stone for finding a consistent theory of quantum gravity (e.g. computing a one-loop effective action in a 2D model [16; 18]).

To better understand (3), we will assume the potential takes the form $V(y) = 2y$ until otherwise indicated. Independently studied by Jackiw [25] in the context of Liouville theory and Teitelboim [36] in the context of Hamiltonian dynamics, this specific case of the action is known as the *JT action* having *JT field equations*. Notice the Einstein equation of this system is a constant curvature condition on the manifold, namely $R(g) - 2m^2 = 0$. Having reduced the problem considerably with this choice of potential, we shall state a few solutions without proof.

EXAMPLE 1. Let $(x_1, x_2) = (T, r)$. If one makes the simplification $\tau(T, r) = \tau(r)$, then the remaining field equations may be solved to find a static solution to the JT field equations:

$$ds_{\text{bh}}^2 = (M - m^2 r^2)\, dT^2 - \frac{1}{M - m^2 r^2} dr^2, \quad \tau_{\text{bh}}(T, r) = mr, \quad (4)$$

for M a constant. Excluding the surface $\tau = 0$, one may show that the Penrose diagram of this metric is identical to a two-dimensional section of the Schwarzschild black hole [1; 11; 27]. We use this to justify using the subscript "bh" in (4) and to refer to ds^2_{bh} as a *black hole metric*.

On the other hand, setting $(x_1, x_2) = (x, t)$ and making the metric ansatz

$$ds^2 \stackrel{g}{=} \cos^2 \frac{u(x,t)}{2} dx^2 - \sin^2 \frac{u(x,t)}{2} dt^2 \tag{5}$$

for an arbitrary function $u = u(x,t)$, one finds the Einstein equation in (3) is satisfied if and only if

$$\Delta u = m^2 \sin u, \tag{6}$$

i.e., u solves the *elliptic* sine-Gordon equation. This well-studied nonlinear partial differential equation is completely integrable in the sense that it has infinitely many conservation laws, a Lax formulation, a Bäcklund Transformation and can be successfully treated with the inverse scattering technique [26; 32; 35; 48]. Consequently, equation (6) falls into a special class of nonlinear equations possessing *soliton* solutions, or localised wave solutions which maintain their shape and velocity upon collisions. Solving system (3) thus reduces to fixing a soliton solution u of (6) in the metric ansatz above and solving the equations of motion for the dilaton $\tau = \tau(x,t)$. We thus use the subscript "sol" and refer to ds^2_{sol} as a *soliton metric*.

EXAMPLE 2. Set u to be the simplest nontrivial solution to (6), namely the *kink soliton* $u(x,t) = 4 \arctan e^{m(x-vt)/a}$, with a, v constants such that $a^2 = 1 + v^2$. Then a solution of (3) is given by

$$ds^2_{sol} = \cos^2 \frac{u}{2} dx^2 - \sin^2 \frac{u}{2} dt^2 \qquad \tau_{sol}(x,t) = a \operatorname{sech} \frac{m}{a}(x-vt). \tag{7}$$

EXAMPLE 3. Choosing a slightly more complicated solution to (6), the oscillating kink-antikink soliton

$$u = u(x,t) = 4 \arctan \frac{v \sinh amx}{a \cos vmt},$$

with a and v as before, one may verify the pair

$$ds^2_{sol} = \cos^2 \frac{u}{2} dx^2 - \sin^2 \frac{u}{2} dt^2, \qquad \tau_{sol}(x,t) = \frac{4v^2 am \sin vmt \sinh amx}{a^2 \cos^2 vmt + v^2 \sinh^2 amx} \tag{8}$$

solves system (3). Further details as to the derivation of these two examples may be found in [5; 44].

Since one expects the two dimensional metrics ds_{bh}^2 and ds_{sol}^2 to be locally equivalent, it is reasonable to pose whether it is possible to find an explicit correspondence between the solution spaces $(ds_{\text{bh}}^2, \tau_{\text{bh}})$ and $(ds_{\text{sol}}^2, \tau_{\text{sol}})$. When $M = v^2$ in (4), an explicit map is known between the black hole metric and the kink soliton metric described in Example 2 [44]. PDE conditions for a general mapping $\Theta(x,t)$ were later found, establishing a correspondence between $(ds_{\text{sol}}^2, \tau(x,t))$ for an arbitrary solution $u(x,t)$ of (6) and a generalised black hole solution $(ds_{\text{bh}}^2, \tau(T,r) = mr)$ of the form

$$ds_{\text{bh}}^2 = -\left(|\nabla\tau|_{\text{sol}}^2 \circ \Psi\right)/m^2 dT^2 + m^2/\left(|\nabla\tau|_{\text{sol}}^2 \circ \Psi\right) dr^2.$$

The notation $|\nabla\tau|_{\text{sol}}^2$ denotes the length of the gradient of τ with respect to the soliton metric ds_{sol}^2 and $\Psi(T,r) = \Theta(x,t)^{-1}$. To further elevate the status of the dilaton, it is also worth noting that τ plays a crucial role in determining the geometry of the two-dimensional black hole, as the Killing vectors are known once τ is given [5; 20]. Remarkably, the mappings Θ, Ψ constructed in [5; 6] turn out to be isometries. Specifically, they are transformations of the solution spaces of the field equations defined by the Laplace Beltrami operators of the soliton and black hole metrics.

3. Application of special functions to JT theory

3.1. Illuminating an eigenvalue problem. We rephrase the final statment of the last section in a particularly useful way. If $f = f(T,r)$, then a mapping Θ satisfying the PDE system found in [5] satisfies

$$\Delta_{\text{sol}}(f \circ \Theta) = (\Delta_{\text{bh}} f) \circ \Theta. \tag{9}$$

Therefore, $\Delta_{\text{bh}} f = \nu f$ if and only if $\Delta_{\text{sol}} F = \nu F$ with $F = f \circ \Theta$. This gives us a mechanism by which to solve eigenvalue problems in soliton coordinates by examining the vastly simpler equation $\Delta_{\text{bh}} f = \nu f$. Using the separation of variables $f(T,r) = e^{\omega T} h(r)$, one obtains a *differential equation of hypergeometric type* in r

$$\sigma(r)h''(r) + \sigma(r)\tilde{\tau}(r)h'(r) + \tilde{\sigma}(r)h(r) = 0, \tag{10}$$

where σ, and $\tilde{\tau}$ and $\tilde{\sigma}$ are polynomials in r satisfying $\deg \sigma, \deg \tilde{\sigma} \le 2$, $\deg \tilde{\tau} \le 1$. Using the methods in [31; 43], Equation (10) is expressed in canonical form and quantization conditions are derived, from which infinite families of solutions may be written down. In special cases, the final solutions will involve functions such as Jacobi elliptic functions, or Gauss' hypergeometric functions, among others [5; 6; 43].

EXAMPLE 4. Under the mentioned separation of variables, the eigenvalue problem $\Delta_{\text{bh}} f = \nu f$ reduces to

$$B(r)h''(r) - 2m^2 r B(r)h'(r) + (2m^2(B(r) - \nu))h(r) = 0,$$

where $B(r) = M - m^2 r^2$. Setting $\nu = \omega^2$, it is possible to reduce the equation to one of the form $r(1-r)v''(r) + [\gamma + r(\alpha + \beta + 1)]v'(r) - \alpha\beta v(r) = 0$, with $m_0 = 1/(m\sqrt{M})$, $\alpha = 2 + |\omega|m_0$, $\beta = -1 + |\omega|m_0$, $\gamma = 1 + |\omega|m_0$ and $v(r) = \frac{1}{2}h(m^2 m_0 r + 1)$. The equation is now in the standard form of Gauss' hypergeometric equation, having as solutions generalised hypergeometric functions $F(\alpha, \beta, \gamma; r)$:

$$F(\alpha, \beta, \gamma; z) = \sum_{n=0}^{\infty} \frac{(\alpha)_n (\beta)_n}{(\gamma)_n} z^n \qquad \text{for} |z| < 1, \gamma \neq \mathbb{Z}_{\leq 0}, \qquad (11)$$

and where $(a)_n := \Gamma(a+n)/\Gamma(a) = a(a+1)(a+2)\cdots(a+n-1)$ is the Pochhammer symbol [22; 43]. Thus, one solves the eigenvalue problem as $f(T, r) = e^{\omega T} h(r)$, with $h(r) = F(\alpha, \beta, \gamma; mr/(2\sqrt{M}) + 1)$, and α, β, γ defined above. It is important to note that properties of this F (e.g., Saalschütz Theorem, Dougall-Ramanujan identity, etc.), and consequently properties of the solution f are intimately dependent on number theoretic and complex analytic results of the Gamma function [15; 34; 39].

Interestingly, hypergeometric differential equations also lend themselves to the study zero-weight modular forms and vertex operator algebras [38; 47].

3.2. Solitons and black hole entropy. A second way in which one may examine the interplay between the two-dimensional black hole and the soliton gauge, is by computing quantities of physical interest, such as entropy. In explicating a correspondence between ds^2_{bh} and ds^2_{sol}, one finds a relationship between the black hole mass M and several soliton parameters (the constants a, v in (7), for instance). In particular, we find the M is nonnegative in all the cases studied. More, [20; 27] compute the ADM energy ($mM/2G$, with G=Newton's coupling constant), Hawking temperature ($m\sqrt{M}/2\pi$) and associated Bekenstein-Hawking entropy ($2\pi\sqrt{M}/G$) for the black hole solutions ds^2_{bh} given in (4), so each of these quantities may be expressed in terms of the soliton parameters as well. The physical interpretation of these correspondences is still under investigation. Attempts have also been made to recover the asymptotic behaviour of the entropy using N-solitons and partition functions [19]. We shall outline a *fairly speculative* argument by Gegenberg and Kunstatter here with the hope that further discussion can shed light on the matter. Takhtadjan and Faddeev [35] compute the total energy for an N-soliton to be $E = \sum_{j=1}^{N} (m^2/\beta^2 + p_j^2)^2$, where p_j is the canonical momentum of the jth wave packet and β is a constant.

The rest energy of the state is thus $E_0 = Nm/\beta$. The claim is that "degeneracy of the state comes from the fact that the wave packets of an N-soliton are indistinguishable"; that is, degeneracy is the number of different ways to write N as a sum of non-negative integers. The Hardy–Ramanujan partition function p(N) counts precisely this value and is given asymptotically by

$$p(N) \sim \frac{1}{4N\sqrt{3}} e^{K\sqrt{N}},$$

where $K = \pi\sqrt{2/3}$; see [23]. Thus, for large N, the entropy grows as $S \sim \log p(N) \sim \sqrt{N} \sim \sqrt{E_0}$, up to order one multiplicative constant factors; this coincides with the Bekenstein-Hawking entropy stated above, found in [3; 20]. We remark that in order to make any of the above discussion rigourous, it is first necessary to make an argument which will include solutions of the sine-Gordon equation which do not fit the form of an N-soliton (e.g., breathers, non-soliton solutions). Furthermore, black hole energy has not been proven to be given by the rest energy of the N-soliton solution. We remark that the partition function can be cast as a special case of the Rademacher-Zuckerman formula for the coefficients of a modular form of negative weight -$\frac{1}{2}$ [33]. It is possible that a modular forms perspective will clarify these points.

4. Other two-dimensional considerations: Strings and quantum gravity

Two-dimensional theory is not restricted to the study of the JT field equations, of course. We touch upon two possible directions of exploration by altering the potential function V appearing in the action (2) and consequently, the resulting field equations (3).

4.1. 1D bosonic strings. If we now assume $V(y) = \gamma y^\alpha$, the original model not only encompasses the JT Theory ($\gamma = 2$, $\alpha = 1$), but several other gravitational theories of interest, including string-inspired gravity and spherically symmetric gravity as well. We shall only discuss the first of these two. Let $\alpha = 0$ so that $V(y) = \gamma \geq 0$. Correspondingly, the action appearing in (2) is modified to

$$I[g, \tau] = \int_{M^2} \left(R(g)\tau - m^2\gamma \right) \sqrt{|\det(g)|} \, d^2x, \tag{12}$$

and first field equation becomes $R(g) = 0$. Thus, the metric ansatz in (5) gives rise to the harmonicity condition $\triangle u = 0$, rather than the sine-Gordon equation. In this sense, the integrable systems content of the field equations changes qualitatively. However, under the conformal change of coordinates $\hat{g} = ge^\varphi$, one may, in fact, recover the classical Polyakov (bosonic) string action [7; 28; 29]

$$I[\hat{g}, \varphi, \beta] = \int_{M^2} (R(\hat{g}) - 4|\nabla\varphi|_{\hat{g}}^2 + \beta)e^{-2\varphi}\sqrt{|\det \hat{g}|}\, d^2x, \qquad (13)$$

for $\beta = m^2\gamma$ and $\tau = e^{-2\varphi}$. The physical and geometric role of the dilaton appears in a new context, as the square of the conformal factor. Although the two actions are related, black hole solutions exist in the string model having vastly different geometry than the static Schwarzschild-type case we have discussed. An example of this follows.

EXAMPLE 5. Consider the target space action given in [28]

$$S(g, \varphi, T) = \int_{M^2} (R(g) - 4|\nabla\varphi|^2 + |\nabla T|^2 + V(T))e^{-2\varphi}\sqrt{|\det g|}\, d^2x, \qquad (14)$$

where g is a two-dimensional metric, φ and T are scalar fields, known as the dilaton field and the tachyon field, respectively, and V is a polynomial potential satisfying $V(0) = 0$. Supposing the tachyon field vanishes and $\varphi(T, r) = \kappa r$ for some constant κ, the field equations reduce to an inhomogeneous second order ODE, which yield the metric-dilaton solution

$$ds^2 = -(1 - ae^{Qr})\, dT^2 + \frac{1}{1 - ae^{Qr}}\, dr^2 \qquad \varphi(T, r) = \frac{Q}{2}r, \qquad (15)$$

where $Q^2 = -C$ is related to the central charge. The asymptotic and topological behaviour of this solution is clearly not of Schwarzschild type. Investigations of this example and string theories in general are detailed in [2; 28; 45].

In connection to the previous section, we comment that the parition function $p(N)$ has been used to count the microstates of a bosonic string; for further details, see [14; 46] and references therein.

4.2. From classical to quantum gravity: Zeta regularization. In a dimensionally reduced model, it is often possible to exactly compute various quantities of interest, both classically and quantum mechanically. We mention the value of two-dimensional models in the context of quantum gravity and zeta functions. Elizalde and Odintsov consider the action

$$\int_{M^2} \left(R\frac{1}{\Delta}R + \Lambda\right)\sqrt{|\det g|}\, d^2x \qquad (16)$$

of induced two-dimensional gravity on the background $M^2 = \mathbb{R}^1 \times \mathbb{S}^1$; $R = R(g)$ is the scalar curvature, $\frac{1}{\Delta}$ is the resolvent operator, and Λ is a constant [16; 18]. Upon consideration of the one-loop gauge-independent effective action, the effective potential V is computed via regularization. Derived in terms of the differential operator Δ (see [17]) and a particular constant β, the zeta function is given by

$$\zeta_{-\Delta+m^2}\left(\frac{s}{2}\right) = -\frac{S}{\pi}\int_0^\infty \sum_{n=-\infty}^{+\infty}\left(k^2 + \frac{2\pi n}{\beta} + m^2\right)^{-s/2} dk. \qquad (17)$$

Defining the variables $x = \Lambda/4(2-a)$ and $y = R\sqrt{x}$, with $a =$ constant, the effective potential is computed as

$$V = \sqrt{x}\left(8\pi(2-a)y + \frac{y}{8}(1 - \ln x) - \frac{1}{4} + \frac{1}{24y} - F(y)\right), \qquad (18)$$

for

$$F(y) = \frac{1}{4\pi}\sum_{k=0}^\infty \left(\frac{(16\pi)^{-k}}{k!} y^{-k-\frac{1}{2}} \prod_{j=1}^k (4-(2j-1)^2)\sum_{n=1}^\infty n^{-k-\frac{3}{2}} e^{-2\pi n y}\right).$$

From this, a minimum for V is found and the authors conclude the compactification is stable [18]. The result is in marked contast to multidimensional quantum gravity on $\mathbb{R}^d \times \mathbb{S}^1$, which is known to be one-loop unstable [10; 24].

5. Relation to the 3D BTZ black hole

Two-dimensional models are studied with the ultimate goal of understanding higher dimensional theories. Naturally, we would like to connect the dilaton theory to higher dimensions in an explicit fashion. The Einstein Equations, arising from the classical Einstein-Hilbert action from the first section have also been examinined for *three dimensions* via

$$\int_{M^3} (R(g) - 2\Lambda)\sqrt{|\det(g)|} d^3 x, \qquad (19)$$

with $x = (x_1, x_2, x_3)$ and Λ a constant. One solution of particular interest is the black hole metric discovered by Bañados, Teitelboim and Zanelli

$$ds^2_{BTZ} = -N(r)^2 dT^2 + \frac{1}{N(r)^2} dr^2 + r^2(N^\phi(r)\, dT + d\phi)^2, \qquad (20)$$

where $x = (T, r, \phi)$, $N(r) = \Lambda r^2 - M + J^2/(4r^2)$ and $N^\phi(r) = -J/(2r)$; see [4]. The constants M and J correspond to the mass and angular momentum of the black hole, respectively. The field equations afforded by the three-dimensional case have been carefully studied, with the geometry and physics of the BTZ black hole outlined in [1; 3; 4; 11]. We notice an immediate relationship between the BTZ black hole and the JT black hole from Section 1. Keeping the ϕ-coordinate constant and setting $J = 0$, $\Lambda = m^2$, the two-dimensional metric in (4) is recovered as a static slice of (20). This motivates the following discussion: let $x = (x_1, x_2, \phi)$ and impose axial symmetry on the 3D metric g as $ds^2 = h_{ij}(\tilde{x}) \, dx_i \, dx_j + \tau(\tilde{x}) \, d\phi^2$, for $\tilde{x} = (x_1, x_2)$, $1 \leq i, j \leq 2$. Then (19) reduces to a two-dimensional action from which the JT field equations arise

$$I[g, \tau] = \int (R(h) - 2m^2) \tau(\tilde{x}) \sqrt{|\det(h)|} \, d^2 \tilde{x}, \qquad (21)$$

where $R(h)$ is the scalar curvature of the two dimensional metric $h = h_{ij}(x_1, x_2)$ and $\tau(\tilde{x}) = \tau(x_1, x_2)$ is the dilaton, as before; compare with (2) for $V(y) = 2y$. In this way, the scalar field τ can be viewed as a radius along the direction of symmetry (the $\phi\phi$ direction) of the the surface defined by the metric h. Clearly then, the soliton content of the three-dimensional case can be considered, as well as pertinent questions on the presence of exact solutions involving special functions and physical quantities of interest. In this context, modular forms of negative weight also appear. The Rademacher-Zuckerman formula asymptotically yields the Cardy entropy formula of conformal field theory and as a special case, the statistical derivation of the Bekenstein-Hawking entropy of the BTZ black hole [8; 13; 14]. Further, quantum correction to entropy can be realised as a deformation of zeta and thus close connections between zeta functions and BTZ black hole thermodynamics have been suggested [42]. It is an interesting question whether the integrability structure in two dimensions sheds any light on the three dimensional case. Such avenues are currently under exploration and will be discussed in a future communication.

6. Conclusion

In the context of classical two-dimensional gravitation, we have only touched upon the possible mergers of pure mathematics with black hole physics and cosmology. For further exploration, consult the references.

References

[1] Ana Achúcarro and Miguel E. Ortiz. Relating black holes in two and three dimensions. *Phys. Rev. D* (3), 48(8):3600–3605, 1993.

[2] Sergio Albeverio, Jürgen Jost, Sylvie Paycha, and Sergio Scarlatti. *A mathematical introduction to string theory*, volume 225 of *London Mathematical Society Lecture Note Series*. Cambridge University Press, Cambridge, 1997. Variational problems, geometric and probabilistic methods.

[3] Máximo Bañados, Marc Henneaux, Claudio Teitelboim, and Jorge Zanelli. Geometry of the $2 + 1$ black hole. *Phys. Rev. D* (3), 48(4):1506–1525, 1993.

[4] Máximo Bañados, Claudio Teitelboim, and Jorge Zanelli. Black hole in three-dimensional spacetime. *Phys. Rev. Lett.*, 69(13):1849–1851, 1992.

[5] Shabnam Beheshti. Solutions to the dilaton field equations with applications to the soliton-black hole correspondence in generalised JT gravity. Ph.D. Thesis, University of Massachusetts, Amherst, MA (2008). Department of Mathematics.

[6] Shabnam Beheshti and Floyd L. Williams. Explicit soliton–black hole correspondence for static configurations. *J. Phys. A*, 40(14):4017–4024, 2007.

[7] A. Belavin and A. Polyakov. Metastable states of two-dimensional isotropic ferromagnets. *JETP Letters*, 22:245–247, (1975).

[8] Danny Birmingham and Siddhartha Sen. Exact black hole entropy bound in conformal field theory. *Phys. Rev. D* (3), 63(4):047501, 2001.

[9] William M. Boothby. *An introduction to differentiable manifolds and Riemannian geometry*, volume 120 of *Pure and Applied Mathematics*. Academic Press Inc., Orlando, FL, second edition, 1986.

[10] I. L. Buchbinder, P. M. Lavrov, and S. D. Odintsov. Unique effective action in Kaluza-Klein quantum theories and spontaneous compactification. *Nuclear Phys. B*, 308(1):191–202, 1988.

[11] Mariano Cadoni. 2D extremal black holes as solitons. *Phys. Rev. D* (3), 58(10): 104001, 1998.

[12] Mariano Cadoni and Salvatore Mignemi. Nonsingular four-dimensional black holes and the Jackiw-Teitelboim theory. *Phys. Rev. D* (3), 51(8):4319–4329, 1995.

[13] John L. Cardy. Operator content of two-dimensional conformally invariant theories. *Nuclear Phys. B*, 270(2):186–204, 1986.

[14] S. Carlip. Logarithmic corrections to black hole entropy, from the Cardy formula. *Classical Quantum Gravity*, 17(20):4175–4186, 2000.

[15] J. Dougall. On Vandermonde's theorem and some more general expansions. *Proc. Edinburgh Math. Soc.*, 25:114–132, 1907.

[16] E. Elizalde and S. D. Odintsov. Spontaneous compactification in 2D induced quantum gravity. *Modern Phys. Lett. A*, 7(26):2369–2376, 1992.

[17] E. Elizalde and A. Romeo. Regularization of general multidimensional Epstein zeta-functions. *Rev. Math. Phys.*, 1(1):113–128, 1989.

[18] Emilio Elizalde. *Ten physical applications of spectral zeta functions*, volume 35 of *Lecture Notes in Physics. New Series m: Monographs*. Springer-Verlag, Berlin, 1995.

[19] J. Gegenberg and G. Kunstatter. From two-dimensional black holes to sine-Gordon solitons. In *Solitons (Kingston, ON, 1997)*, CRM Ser. Math. Phys., pages 99–106. Springer, New York, 2000.

[20] J. Gegenberg, G. Kunstatter, and D. Louis-Martinez. Classical and quantum mechanics of black holes in generic 2-D dilaton gravity. In *Heat kernel techniques and quantum gravity (Winnipeg, MB, 1994)*, volume 4 of *Discourses Math. Appl.*, pages 333–346. Texas A & M Univ., College Station, TX, 1995.

[21] J. Gegenberg, G. Kunstatter, and D. Louis-Martinez. Observables for two-dimensional black holes. *Phys. Rev. D (3)*, 51(4):1781–1786, 1995.

[22] I. S. Gradshteyn and I. M. Ryzhik. *Table of integrals, series, and products.* Elsevier/Academic Press, Amsterdam, seventh edition, 2007. Translated from the Russian, Translation edited and with a preface by Alan Jeffrey and Daniel Zwillinger, With one CD-ROM (Windows, Macintosh and UNIX).

[23] G. H. Hardy. *Ramanujan: twelve lectures on subjects suggested by his life and work.* Chelsea Publishing Company, New York, 1959.

[24] S. R. Huggins, G. Kunstatter, H. P. Leivo, and D. J. Toms. The Vilkovisky-DeWitt effective action for quantum gravity. *Nuclear Phys. B*, 301(4):627–660, 1988.

[25] R. Jackiw. Liouville field theory: a two-dimensional model for gravity? In *Quantum theory of gravity*, pages 403–420. Hilger, Bristol, 1984.

[26] Peter D. Lax. Integrals of nonlinear equations of evolution and solitary waves. *Comm. Pure Appl. Math.*, 21:467–490, 1968.

[27] José P. S. Lemos and Paulo M. Sá. Black holes of a general two-dimensional dilaton gravity theory. *Phys. Rev. D (3)*, 49(6):2897–2908, 1994.

[28] Gautam Mandal, Anirvan M. Sengupta, and Spentra R. Wadia. Classical solutions of 2-dimensional string theory. *Modern Phys. Lett. A*, 6(18):1685–1692, 1991.

[29] Robert Marnelius. Canonical quantization of Polyakov's string in arbitrary dimensions. *Nuclear Phys. B*, 211(1):14–28, 1983.

[30] A. J. M. Medved. Quantum-corrected entropy for $(1+1)$-dimensional gravity revisited. *Classical Quantum Gravity*, 20(11):2147–2156, 2003.

[31] Arnold F. Nikiforov and Vasilii B. Uvarov. *Special functions of mathematical physics.* Birkhäuser Verlag, Basel, 1988. A unified introduction with applications, Translated from the Russian and with a preface by Ralph P. Boas, With a foreword by A. A. Samarskiĭ.

[32] S. Novikov, S. V. Manakov, L. P. Pitaevskiĭ, and V. E. Zakharov. *Theory of solitons.* Contemporary Soviet Mathematics. Consultants Bureau [Plenum], New York, 1984. The inverse scattering method, Translated from the Russian.

[33] Hans Rademacher and Herbert S. Zuckerman. On the Fourier coefficients of certain modular forms of positive dimension. *Ann. of Math. (2)*, 39(2):433–462, 1938.

[34] L. Saalschütz. Eine summationsformel. *Z. fÃ¼r Math. u. Phys.*, 35:186–188, 1890.

[35] L. A. Takhtadzhyan and L. D. Faddeev. Essentially nonlinear one-dimensional model of classical field theory. *Theoretical and Mathematical Physics*, 22:1046–1057, February 1975.

[36] Claudio Teitelboim. The Hamiltonian structure of two-dimensional space-time and its relation with the conformal anomaly. In *Quantum theory of gravity*, pages 327–344. Hilger, Bristol, 1984.

[37] Jennie Traschen. An introduction to black hole evaporation. In *Mathematical methods in physics (Londrina, 1999)*, pages 180–208. World Sci. Publ., River Edge, NJ, 2000.

[38] Hiroyuki Tsutsumi. Modular differential equations of second order with regular singularities at elliptic points for $SL_2(\mathbb{Z})$. *Proc. Amer. Math. Soc.*, 134(4):931–941, 2006.

[39] N. Ja. Vilenkin. *Special functions and the theory of group representations*. Translated from the Russian by V. N. Singh. Translations of Mathematical Monographs, Vol. 22. American Mathematical Society, Providence, R. I., 1968.

[40] Robert M. Wald. *General relativity*. University of Chicago Press, Chicago, IL, 1984.

[41] Floyd Williams. On solitons, non-linear sigma-models and two-dimensional gravity. Fourth International Winter Conference on Mathematical Physics (Rio de Janeiro, 2005). PoS(WC2004)003. http://pos.sissa.it,

[42] Floyd Williams. Remarks on the btz instanton with conical singularity. Fifth International Conference on Mathematical Methods in Physics (Rio de Janeiro, 2006). PoS(IC2006)006. http://pos.sissa.it,

[43] Floyd Williams. *Topics in quantum mechanics*, volume 27 of *Progress in Mathematical Physics*. Birkhäuser Boston Inc., Boston, MA, 2003.

[44] Floyd Williams. Further thoughts on first generation solitons and J-T gravity. In *Trends in Soliton Research, L. Chen, Ed.*, pages 1–14. Nova Sci. Publ., New York, 2006.

[45] Edward Witten. String theory and black holes. *Phys. Rev. D* (3), 44(2):314–324, 1991.

[46] Donam Youm. Black holes and solitons in string theory. *Phys. Rep.*, 316(1-3):232, 1999.

[47] Don Zagier. Modular forms and differential operators. *Proc. Indian Acad. Sci. Math. Sci.*, 104(1):57–75, 1994. K. G. Ramanathan memorial issue.

[48] V. E. Zakharov, L. A. Tahtadžjan, and L. D. Faddeev. A complete description of the solutions of the "sine-Gordon" equation. *Dokl. Akad. Nauk SSSR*, 219:1334–1337, 1974.

SHABNAM BEHESHTI
DEPARTMENT OF MATHEMATICS
110 FRELINGHUYSEN ROAD
RUTGERS UNIVERSITY
PISCATAWAY, NJ 08854-8019
UNITED STATES
beheshti@math.rutgers.edu

Functional determinants in higher dimensions using contour integrals

KLAUS KIRSTEN

ABSTRACT. In this contribution we first summarize how contour integration methods can be used to derive closed formulae for functional determinants of ordinary differential operators. We then generalize our considerations to partial differential operators. Examples are used to show that also in higher dimensions closed answers can be obtained as long as the eigenvalues of the differential operators are determined by transcendental equations. Examples considered comprise of the finite temperature Casimir effect on a ball and the functional determinant of the Laplacian on a two-dimensional torus.

1. Introduction

Functional determinants of second-order differential operators are of great importance in many different fields. In physics, functional determinants provide the one-loop approximation to quantum field theories in the path integral formulation [21; 48]. In mathematics they describe the analytical torsion of a manifold [47].

Although there are various ways to evaluate functional determinants, the zeta function scheme seems to be the most elegant technique to use [9; 16; 17; 31]. This is the method introduced by Ray and Singer to define analytical torsion [47]. In physics its origin goes back to ambiguities in dimensional regularization when applied to quantum field theory in curved spacetime [11; 29].

For many second-order ordinary differential operators surprisingly simple answers can be given. The determinants for these situations have been related to boundary values of solutions of the operators, see, e.g., [8; 10; 12; 22; 23; 26; 36; 39; 40]. Recently, these results have been rederived with a simple and accessible method which uses contour integration techniques [33; 34; 35]. The main advantage of this approach is that it can be easily applied to general kinds of boundary conditions [35] and also to cases where the operator has zero modes [34; 35]; see also [37; 38; 42]. Equally important, for some higher dimensional situations the task of finding functional determinants remains feasible. Once again closed answers can be found but compared to one dimension technicalities are significantly more involved [13; 14]. It is the aim of this article to choose specific higher dimensional examples where technical problems remain somewhat confined. The intention is to illustrate that also for higher dimensional situations closed answers can be obtained which are easily evaluated numerically.

The outline of this paper is as follows. In Section 2 the essential ideas are presented for ordinary differential operators. In Section 3 and 4 examples of functional determinants for partial differential operators are considered. The determinant in Section 3 describes the finite temperature Casimir effect of a massive scalar field in the presence of a spherical shell [24; 25]. The calculation in Section 4 describes determinants for strings on world-sheets that are tori [46; 50] and it gives an alternative derivation of known answers. Section 5 summarizes the main results.

2. Contour integral formulation of zeta functions

In this section we review the basic ideas that lead to a suitable contour integral representation of zeta functions associated with ordinary differential operators. This will form the basis of the considerations for partial differential operators to follow later.

We consider the simple class of differential operators

$$P := -\frac{d^2}{dx^2} + V(x)$$

on the interval $I = [0, 1]$, where $V(x)$ is a smooth potential. For simplicity we consider Dirichlet boundary conditions. From spectral theory [41] it is known that there is a spectral resolution $\{\phi_n, \lambda_n\}_{n=1}^\infty$ satisfying

$$P\phi_n(x) = \lambda_n \phi_n(x), \qquad \phi_n(0) = \phi_n(1) = 0.$$

The spectral zeta function associated with this problem is then defined by

$$\zeta_P(s) = \sum_{n=1}^\infty \lambda_n^{-s}, \qquad (2\text{-}1)$$

where by Weyl's theorem about the asymptotic behavior of eigenvalues [49] this series is known to converge for $\operatorname{Re} s > \frac{1}{2}$.

If the potential is not a very simple one, eigenfunctions and eigenvalues will not be known explicitly. So how can the zeta function in equation (2-1), and in particular the determinant of P defined via

$$\det P = e^{-\zeta'_P(0)},$$

be analyzed? From complex analysis it is known that series can often be evaluated with the help of the argument principle or Cauchy's residue theorem by rewriting them as contour integrals. In the given context this can be achieved as follows. Let $\lambda \in \mathbb{C}$ be an arbitrary complex number. From the theory of ordinary differential equations it is known that the initial value problem

$$(P - \lambda)u_\lambda(x) = 0, \quad u_\lambda(0) = 0, \quad u'_\lambda(0) = 1, \qquad (2\text{-}2)$$

has a unique solution. The connection with the boundary value problem is made by observing that the eigenvalues λ_n follow as solutions to the equation

$$u_\lambda(1) = 0; \qquad (2\text{-}3)$$

note that $u_\lambda(1)$ is an analytic function of λ.

With the help of the argument principle, equation (2-3) can be used to write the zeta function, equation (2-1), as

$$\zeta_P(s) = \frac{1}{2\pi i}\int_\gamma d\lambda\, \lambda^{-s}\frac{d}{d\lambda}\ln u_\lambda(1). \qquad (2\text{-}4)$$

Here, γ is a counterclockwise contour and encloses all eigenvalues which we assume to be positive; see Figure 1. The pertinent remarks when finitely many eigenvalues are nonpositive are given in [35].

The asymptotic behavior of $u_\lambda(1)$ as $|\lambda| \to \infty$, namely

$$u_\lambda(1) \sim \frac{\sin\sqrt{\lambda}}{\sqrt{\lambda}},$$

implies that this representation is valid for $\operatorname{Re} s > \frac{1}{2}$. To find the determinant of P we need to construct the analytical continuation of equation (2-4) to a neighborhood about $s = 0$. This is best done by deforming the contour to enclose

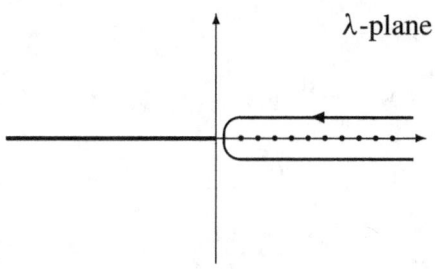

Figure 1. Contour γ used in equation (2-4).

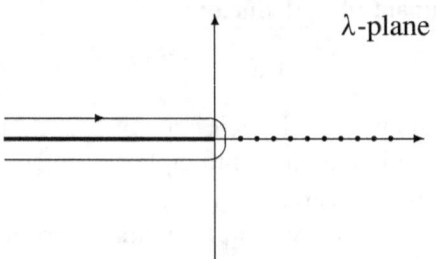

Figure 2. Contour γ used in equation (2-4) after deformation.

the branch cut along the negative real axis and then shrinking it to the negative real axis; see Figure 2.

The outcome is

$$\zeta_P(s) = \frac{\sin \pi s}{\pi} \int_0^\infty d\lambda \, \lambda^{-s} \frac{d}{d\lambda} \ln u_{-\lambda}(1). \tag{2-5}$$

To see where this representation is well defined notice that for $\lambda \to \infty$ the behavior follows from [41]

$$u_{-\lambda}(1) \sim \frac{\sin(i\sqrt{\lambda})}{i\sqrt{\lambda}} = \frac{e^{\sqrt{\lambda}}}{2\sqrt{\lambda}}\left(1 - e^{-2\lambda}\right).$$

The integrand, to leading order in λ, therefore behaves like $\lambda^{-s-1/2}$ and convergence at infinity is established for $\operatorname{Re} s > \frac{1}{2}$. As $\lambda \to 0$ the behavior λ^{-s} follows. Therefore, in summary, (2-5) is well defined for $\frac{1}{2} < \operatorname{Re} s < 1$. To shift the range of convergence to the left we add and subtract the leading $\lambda \to \infty$ asymptotic behavior of $u_{-\lambda}(1)$. The whole point of this procedure will be to obtain one piece that at $s = 0$ is finite, and another piece for which the analytical continuation can be easily constructed.

Given we want to improve the $\lambda \to \infty$ behavior without worsening the $\lambda \to 0$ behavior, we split the integration range. In detail we write

$$\zeta_P(s) = \zeta_{P,f}(s) + \zeta_{P,\mathrm{as}}(s), \tag{2-6}$$

where

$$\zeta_{P,f}(s) = \frac{\sin \pi s}{\pi} \int_0^1 d\lambda \, \lambda^{-s} \frac{d}{d\lambda} \ln u_{-\lambda}(1)$$
$$+ \frac{\sin \pi s}{\pi} \int_1^\infty d\lambda \, \lambda^{-s} \frac{d}{d\lambda} \ln\!\left(u_{-\lambda}(1) 2\sqrt{\lambda} e^{-\sqrt{\lambda}}\right), \tag{2-7}$$

$$\zeta_{P,\mathrm{as}}(s) = \frac{\sin \pi s}{\pi} \int_1^\infty d\lambda \, \lambda^{-s} \frac{d}{d\lambda} \ln \frac{e^{\sqrt{\lambda}}}{2\sqrt{\lambda}}. \tag{2-8}$$

By construction, $\zeta_{P,f}(s)$ is analytic about $s = 0$ and its derivative at $s = 0$ is trivially obtained,

$$\zeta'_{P,f}(0) = \ln u_{-1}(1) - \ln u_0(1) - \ln\left(u_{-1}(1)2e^{-1}\right) = -\ln\frac{2u_0(1)}{e}. \quad (2\text{-}9)$$

Although the representation (2-8) is only defined for $\operatorname{Re} s > \frac{1}{2}$, the analytic continuation to a meromorphic function on the complex plane is found using

$$\int_1^\infty d\lambda\, \lambda^{-\alpha} = \frac{1}{\alpha - 1} \quad \text{for} \quad \operatorname{Re}\alpha > 1.$$

This shows that

$$\zeta_{P,\text{as}}(s) = \frac{\sin \pi s}{2\pi}\left(\frac{1}{s - 1/2} - \frac{1}{s}\right),$$

and furthermore

$$\zeta'_{P,\text{as}}(0) = -1.$$

Adding up, the final answer reads

$$\zeta'_P(0) = -\ln(2u_0(1)). \quad (2\text{-}10)$$

For the numerical evaluation of the determinant, not even one eigenvalue is needed. The only relevant information is the boundary value of the unique solution to the initial value problem

$$\left(-\frac{d^2}{dx^2} + V(x)\right)u_0(x) = 0, \qquad u_0(0) = 0, \qquad u'_0(0) = 1.$$

General boundary conditions can be dealt with as easily. The best formulation results by rewriting the second-order differential equation as a first-order system in the usual way. Namely, we define $v_\lambda(x) = du_\lambda(x)/dx$ such that the differential equation (2-2) turns into

$$\frac{d}{dx}\begin{pmatrix} u_\lambda(x) \\ v_\lambda(x) \end{pmatrix} = \begin{pmatrix} 0 & 1 \\ V(x) - \lambda & 0 \end{pmatrix}\begin{pmatrix} u_\lambda(x) \\ v_\lambda(x) \end{pmatrix}. \quad (2\text{-}11)$$

Linear boundary conditions are given in the form

$$M\begin{pmatrix} u_\lambda(0) \\ v_\lambda(0) \end{pmatrix} + N\begin{pmatrix} u_\lambda(1) \\ v_\lambda(1) \end{pmatrix} = \begin{pmatrix} 0 \\ 0 \end{pmatrix}, \quad (2\text{-}12)$$

where M and N are 2×2 matrices whose entries characterize the nature of the boundary conditions. For example, the previously described Dirichlet boundary conditions are obtained by choosing

$$M = \begin{pmatrix} 1 & 0 \\ 0 & 0 \end{pmatrix}, \qquad N = \begin{pmatrix} 0 & 0 \\ 1 & 0 \end{pmatrix}.$$

In order to find an implicit equation for the eigenvalues like equation (2-3) we use the fundamental matrix of (2-11). Let $u_\lambda^{(1)}(x)$ and $u_\lambda^{(2)}(x)$ be linearly independent solutions of (2-11). Suitably normalized, these define the fundamental matrix

$$H_\lambda(x) = \begin{pmatrix} u_\lambda^{(1)}(x) & u_\lambda^{(2)}(x) \\ v_\lambda^{(1)}(x) & v_\lambda^{(2)}(x) \end{pmatrix}, \qquad H_\lambda(0) = \text{Id}_{2\times 2}.$$

The solution of (2-11) with initial value $(u_\lambda(0), v_\lambda(0))$ is then obtained as

$$\begin{pmatrix} u_\lambda(x) \\ v_\lambda(x) \end{pmatrix} = H_\lambda(x) \begin{pmatrix} u_\lambda(0) \\ v_\lambda(0) \end{pmatrix}.$$

The boundary conditions (2-12) can therefore be rewritten as

$$(M + NH_\lambda(1)) \begin{pmatrix} u_\lambda(0) \\ v_\lambda(0) \end{pmatrix} = \begin{pmatrix} 0 \\ 0 \end{pmatrix}. \qquad (2\text{-}13)$$

This shows that the condition for eigenvalues to exist is

$$\det(M + NH_\lambda(1)) = 0,$$

which replaces (2-3) in case of general boundary conditions. The zeta function associated with the boundary condition (2-12) therefore takes the form

$$\zeta_P(s) = \frac{1}{2\pi i} \int_\gamma d\lambda\, \lambda^{-s} \frac{d}{d\lambda} \ln \det(M + NH_\lambda(1))$$

and the analysis proceeds from here depending on M and N. If P represents a system of operators one can proceed along the same lines. Note that we have replaced the task of evaluating the determinant of a differential operator by one of computing the determinant of a finite matrix.

The procedure just outlined is by no means confined to be applied to ordinary differential operators only. In fact, the zeta function associated with many boundary value problems allowing for a separation of variables can be analyzed using this contour integral technique. In more detail, starting off with some coordinate system (see [43], for example), eigenvalues are often determined by

$$F_j(\lambda_{j,n}) = 0,$$

where j is a suitable quantum number depending on the coordinate system considered and F_j is a given special function depending on the coordinate system; e.g. for ellipsoidal coordinate systems the relevant special function is the Mathieu function. Continuing along the lines described above, denoting by d_j an appropriate degeneracy that might be present, we write somewhat symbolically

$$\zeta_P(s) = \sum_j d_j \frac{1}{2\pi i} \int_\gamma d\lambda\, \lambda^{-s} \frac{d}{d\lambda} \ln F_j(\lambda), \qquad (2\text{-}14)$$

the task being to construct the analytical continuation of this object to $s = 0$. The details of the procedure will depend very much on the properties of the special function F_j that enters. For example, on balls Bessel functions are relevant [4; 6; 7], the spherical suspension [3], or sphere-disc configurations [27; 32] involve Legendre functions, ellipsoidal boundaries involve Mathieu functions etc. For many examples relevant properties of $F_j(\lambda)$ are not available in the literature and need to be derived using techniques of asymptotic analysis [41; 44; 45]. For quite common coordinate systems like the polar coordinates this is not necessary. When the asymptotics is known, the relevant integrals resulting in (2-14) need to be evaluated and closed expressions representing the determinant of partial differential operators are found. Although the remaining sums in general cannot be explicitly performed, the results obtained are very suitable for numerical evaluation.

3. Finite temperature Casimir energy on the ball

Let us now apply the above remarks about higher dimensions using the general formalism described in [14]. As a concrete example we consider the finite temperature theory of a massive scalar field on the three dimensional ball. Using the zeta function scheme we have to consider the eigenvalue problem

$$P\phi_\lambda(\tau, \vec{x}) := \left(-\frac{d^2}{d\tau^2} - \Delta + m^2\right)\phi_\lambda(\tau, \vec{x}) = \lambda^2 \phi_\lambda(\tau, \vec{x}), \qquad (3\text{-}1)$$

where τ is the imaginary time and $\vec{x} \in B^3 := \{\vec{x} \in \mathbb{R}^3 \mid |\vec{x}| \leq 1\}$. We have written λ^2 for the eigenvalues to avoid the occurrence of square roots in arguments of Bessel functions later on.

For finite temperature theory we impose periodic boundary conditions in the imaginary time,

$$\phi_\lambda(\tau, \vec{x}) = \phi_\lambda(\tau + \beta, \vec{x}),$$

where β is the inverse temperature, and for simplicity we choose Dirichlet boundary conditions on the boundary of the ball,

$$\phi_\lambda(\tau, \vec{x})|_{|\vec{x}|=1} = 0.$$

The zeta function associated with this boundary value problem is then

$$\zeta_P(s) = \sum_\lambda \lambda^{-2s}, \qquad (3\text{-}2)$$

and the energy of the system is defined by

$$E := -\frac{1}{2}\frac{\partial}{\partial \beta}\zeta'_{P/\mu^2}(0), \qquad (3\text{-}3)$$

where μ is an arbitrary parameter with dimension of a mass introduced in order to get the correct dimension for the energy. For a full discussion of its relevance in the renormalization process in this model at zero temperature see [5]. That discussion remains completely unchanged at finite temperature and we will put $\mu = 1$ henceforth.

Given the radial symmetry of the problem we separate variables in polar coordinates according to

$$\phi_\lambda(\tau, r, \theta, \varphi) = \frac{1}{\sqrt{r}} e^{i(2\pi n\tau/\beta)} J_{\ell+1/2}(\omega_{\ell j} r) Y_{\ell m}(\theta, \varphi),$$

with the spherical surface harmonics $Y_{\ell m}(\theta, \varphi)$ [20] solving

$$-\frac{1}{\sin^2\theta} \frac{\partial^2}{\partial\varphi^2} - \frac{1}{\sin\theta} \frac{\partial}{\partial\theta} \sin\theta \frac{\partial}{\partial\theta} Y_{\ell m}(\theta, \varphi) = \ell(\ell+1) Y_{\ell m}(\theta, \varphi),$$

and with the Bessel function $J_\nu(z)$, which is the regular solution of the differential equation [28]

$$\frac{d^2 J_\nu(z)}{dz^2} + \frac{1}{z} \frac{d J_\nu(z)}{dz} + \left(1 - \frac{\nu^2}{z^2}\right) J_\nu(z) = 0.$$

Imposing the boundary condition on the unit sphere,

$$J_{\ell+1/2}(\omega_{\ell j}) = 0, \qquad (3\text{-}4)$$

determines the eigenvalues. Namely,

$$\lambda_{n\ell j}^2 = \left(\frac{2\pi n}{\beta}\right)^2 + \omega_{\ell j}^2 + m^2, \qquad n \in \mathbb{Z},\ \ell \in \mathbb{N}_0,\ j \in \mathbb{N}. \qquad (3\text{-}5)$$

This leads to the analysis of the zeta function

$$\zeta_P(s) = \sum_{n=-\infty}^{\infty} \sum_{\ell=0}^{\infty} \sum_{j=1}^{\infty} (2\ell+1)(p_n^2 + \omega_{\ell j}^2 + m^2)^{-s}, \qquad (3\text{-}6)$$

where we have used the standard abbreviation $p_n = 2\pi n/\beta$. The factor $2\ell+1$ represents the multiplicity of eigenvalues for angular momentum ℓ.

The zeroes $\omega_{\ell j}$ of the Bessel functions $J_{\ell+1/2}(\omega_{\ell j})$ are not known in closed form and thus we represent the j-summation using contour integrals. Starting with equation (3-4) and following the argumentation of the previous section, this gives the identity

$$\zeta_P(s) = \sum_{n=-\infty}^{\infty} \sum_{\ell=0}^{\infty} (2\ell+1) \int_\gamma \frac{d\lambda}{2\pi i} (p_n^2 + \lambda^2 + m^2)^{-s} \frac{d}{d\lambda} \ln J_{\ell+1/2}(\lambda), \qquad (3\text{-}7)$$

valid for $\operatorname{Re} s > 2$. The contour γ runs counterclockwise and must enclose all the solutions of (3-4) on the positive real axis. The next step is to shift the contour and place it along the imaginary axis. As $\lambda \to 0$ we observe that to leading order $J_\nu(\lambda) \sim \lambda^\nu/(2^\nu \Gamma(\nu+1))$ such that the integrand diverges in this limit. Therefore, we include an additional factor $\lambda^{-\ell-1/2}$ in the logarithm in order to avoid contributions coming from the origin. Because there is no additional pole enclosed, this does not change the result. Furthermore we should note that the integrand has branch cuts starting at $\lambda = \pm i(p_n^2 + m^2)$. Leaving out the n, ℓ summations for the moment and considering the λ-integration alone, we then obtain, with $\nu = \ell + \frac{1}{2}$,

$$\zeta_{P,n\ell}(s) := \int_\gamma \frac{d\lambda}{2\pi i}(p_n^2 + \lambda^2 + m^2)^{-s} \frac{d}{d\lambda}\ln\left(\lambda^{-\nu} J_\nu(\lambda)\right)$$

$$= \frac{\sin \pi s}{\pi} \int_{\sqrt{p_n^2+m^2}}^\infty dk\, (k^2 - p_n^2 - m^2)^{-s} \frac{d}{dk}\ln\left(k^{-\nu} I_\nu(k)\right), \quad (3\text{-}8)$$

where $J_\nu(ik) = e^{i\pi\nu} J_\nu(-ik)$ and $I_\nu(k) = e^{-i\nu\pi/2} J_\nu(ik)$ has been used [28].

The next step is to add and subtract the asymptotic behavior of the integrand in (3-8). The relevant uniform asymptotics, after substituting $k = \nu z$ in the integral, is the Debye expansion of the Bessel functions [1]. We have

$$I_\nu(\nu z) \sim \frac{1}{\sqrt{2\pi\nu}} \frac{e^{\nu\eta}}{(1+z^2)^{1/4}} \left(1 + \sum_{k=1}^\infty \frac{u_k(t)}{\nu^k}\right), \quad (3\text{-}9)$$

with $t = 1/\sqrt{1+z^2}$ and $\eta = \sqrt{1+z^2} + \ln(z/(1+\sqrt{1+z^2}))$. The first few coefficients are listed in [1], higher coefficients are immediately obtained by using the recursion [1]

$$u_{k+1}(t) = \frac{1}{2}t^2(1-t^2)u_k'(t) + \frac{1}{8}\int_0^t d\tau\, (1-5\tau^2)u_k(\tau), \quad (3\text{-}10)$$

starting with $u_0(t) = 1$. As is clear, all the $u_k(t)$ are polynomials in t. The same holds for the coefficients $D_n(t)$ defined by

$$\ln\left(1 + \sum_{k=1}^\infty \frac{u_k(t)}{\nu^k}\right) \sim \sum_{n=1}^\infty \frac{D_n(t)}{\nu^n}. \quad (3\text{-}11)$$

The polynomials $u_k(t)$ as well as $D_n(t)$ are easily found with the help of a simple computer program. As we will see below, we need the first three terms in the expansion (3-11). Explicitly,

$$D_1(t) = \tfrac{1}{8}t - \tfrac{5}{24}t^3,$$
$$D_2(t) = \tfrac{1}{16}t^2 - \tfrac{3}{8}t^4 + \tfrac{5}{16}t^6, \qquad (3\text{-}12)$$
$$D_3(t) = \tfrac{25}{384}t^3 - \tfrac{531}{640}t^5 + \tfrac{221}{128}t^7 - \tfrac{1105}{1152}t^9.$$

Adding and subtracting these terms in (3-8) allows us to rewrite the zeta function as

$$\zeta_P(s) = \zeta_{P,f}(s) + \zeta_{P,\mathrm{as}}(s),$$

where

$$\zeta_{P,f}(s) = \frac{\sin \pi s}{\pi} \sum_{n=-\infty}^{\infty} \sum_{\ell=0}^{\infty} (2\ell+1) \int_{\sqrt{p_n^2+m^2}/\nu}^{\infty} dz \, (z^2 \nu^2 - p_n^2 - m^2)^{-s}$$
$$\times \frac{d}{dz} \left(\ln(z^{-\nu} I_\nu(\nu z)) - \ln \frac{z^{-\nu} e^{\nu \eta}}{\sqrt{2\pi \nu}(1+z^2)^{1/4}} - \frac{D_1(t)}{\nu} - \frac{D_2(t)}{\nu^2} - \frac{D_3(t)}{\nu^3} \right), \quad (3\text{-}13)$$

$$\zeta_{P,\mathrm{as}}(s) = \frac{\sin \pi s}{\pi} \sum_{n=-\infty}^{\infty} \sum_{\ell=0}^{\infty} (2\ell+1) \int_{\sqrt{p_n^2+m^2}/\nu}^{\infty} dz \, (z^2 \nu^2 - p_n^2 - m^2)^{-s}$$
$$\times \frac{d}{dz} \left(\ln \frac{z^{-\nu} e^{\nu \eta}}{\sqrt{2\pi \nu}(1+z^2)^{1/4}} + \frac{D_1(t)}{\nu} + \frac{D_2(t)}{\nu^2} + \frac{D_3(t)}{\nu^3} \right). \quad (3\text{-}14)$$

The number of terms subtracted in (3-13) is chosen so that $\zeta_{P,f}(s)$ is analytic about $s = 0$. The contributions from the asymptotics collected in (3-14) are simple enough for an analytical continuation to be found. Although it would be possible to proceed just with the contribution from inside the ball, in order to make the calculation as transparent and unambiguous as possible (as far as the interpretation of results goes) let us add the contribution from outside the ball.

The exterior of the ball, once the free Minkowski space contribution is subtracted, yields the starting point (3-8) with the replacement $k^{-\nu} I_\nu \to k^\nu K_\nu$ [5]. In this case the relevant uniform asymptotics is [1]

$$K_\nu(\nu z) \sim \sqrt{\frac{\pi}{2\nu}} \frac{e^{-\nu \eta}}{(1+z^2)^{1/4}} \left(1 + \sum_{k=1}^{\infty} (-1)^k \frac{u_k(t)}{\nu^k} \right), \qquad (3\text{-}15)$$

where the notation is as in (3-9). This produces the analogous splitting of the zeta function for the exterior space. Due to the characteristic sign changes in the asymptotics of I_ν and K_ν, adding up the interior and exterior contributions several cancellations take place. As a result, the zeta function for the total space has the form

$$\zeta_{\mathrm{tot}}(s) = \zeta_{\mathrm{tot},f}(s) + \zeta_{\mathrm{tot},\mathrm{as}}(s)$$

with

$$\zeta_{\text{tot},f}(s) = \frac{\sin \pi s}{\pi} \sum_{n=-\infty}^{\infty} \sum_{\ell=0}^{\infty} (2\ell+1) \int_{\sqrt{p_n^2+m^2}/\nu}^{\infty} dz \, (z^2\nu^2 - p_n^2 - m^2)^{-s}$$
$$\times \frac{d}{dz} \left(\ln(I_\nu(\nu z) K_\nu(\nu z)) + \ln(2\nu) + \tfrac{1}{2}\ln(1+z^2) - 2\nu^{-2} D_2(t) \right), \quad (3\text{-}16)$$

$$\zeta_{\text{tot,as}}(s) = \frac{\sin \pi s}{\pi} \sum_{n=-\infty}^{\infty} \sum_{\ell=0}^{\infty} (2\ell+1) \int_{\sqrt{p_n^2+m^2}/\nu}^{\infty} dz \, (z^2\nu^2 - p_n^2 - m^2)^{-s}$$
$$\times \frac{d}{dz} \left(-\ln(2\nu) - \tfrac{1}{2}\ln(1+z^2) + 2\nu^{-2} D_2(t) \right). \quad (3\text{-}17)$$

By construction, $\zeta_{\text{tot},f}(s)$ is analytic about $s = 0$ and one immediately finds

$$\zeta'_{\text{tot},f}(0) = -\sum_{n=-\infty}^{\infty} \sum_{\ell=0}^{\infty} (2\ell+1) \Big(\ln(I_\nu(\nu z) K_\nu(\nu z)) + \ln(2\nu) \quad (3\text{-}18)$$
$$+ \tfrac{1}{2}\ln(1+z^2) - 2\nu^{-2} D_2(t) \Big) \Big|_{z=\sqrt{p_n^2+m^2}/\nu},$$

with $t = 1/\sqrt{1+z^2}$ as defined earlier. Although one could use (3-18) for numerical evaluation, further simplifications are possible. Following [14] we rewrite this expression according to

$$1 + z^2 = 1 + \frac{p_n^2 + m^2}{\nu^2} = \left(1 + \frac{p_n^2}{\nu^2}\right)\left(1 + \frac{m^2}{\nu^2 + p_n^2}\right). \quad (3\text{-}19)$$

The advantage of the right-hand side is that it can be expanded further for $\nu^2 \to \infty$ or $p_n^2 \to \infty$ or both. This will allow us to subtract exactly the behavior that makes the double series convergent; the oversubtraction immanent in (3-18) can then be avoided. It is expected that expanding the rightmost factor further for $\nu^2 + p_n^2 \gg 1$ leads to considerable cancellations when combined with $\zeta'_{\text{tot,as}}(0)$ [14].

We split the asymptotic terms in (3-18) into those strictly needed to make the sums convergent and those that ultimately will not contribute. For example, we expand according to

$$\ln(1+z^2)\Big|_{\sqrt{p_n^2+m^2}/\nu}$$
$$= \ln\left(1 + \frac{p_n^2+m^2}{\nu^2}\right) = \ln\left(1 + \frac{p_n^2}{\nu^2}\right) + \ln\left(1 + \frac{m^2}{\nu^2+p_n^2}\right)$$
$$= \ln\left(1 + \frac{p_n^2}{\nu^2}\right) + \frac{m^2}{\nu^2+p_n^2} + \left[\ln\left(1 + \frac{m^2}{\nu^2+p_n^2}\right) - \frac{m^2}{\nu^2+p_n^2}\right].$$

The first two terms have to be subtracted in (3-18) in order to make the summations convergent. The terms in brackets are of the order $\mathcal{O}(1/(v^2+p_n^2)^2)$ and even after performing the summations in (3-18) a finite result follows. Thus the first two terms represent a minimal set of terms to be subtracted in (3-18) in order to make the sums finite. This minimal set of necessary terms will be called $\ln f_\ell^{\mathrm{asym},(1)}(i\sqrt{p_n^2+m^2})$. The last two terms can be summed separately yielding a finite answer; they are summarized under $\ln f_\ell^{\mathrm{asym},(2)}(i\sqrt{p_n^2+m^2})$. One can proceed along the same lines for all other terms. With the definition

$$\ln f_\ell^{\mathrm{asym}}(i\sqrt{p_n^2+m^2})$$
$$= -\ln(2v) - \tfrac{1}{2}\ln(1+z^2) + 2v^{-2}D_2(t)\Big|_{z=\sqrt{p_n^2+m^2}/v}$$
$$= \ln f_\ell^{\mathrm{asym},(1)}(i\sqrt{p_n^2+m^2}) + \ln f_\ell^{\mathrm{asym},(2)}(i\sqrt{p_n^2+m^2}) \quad (3\text{-}20)$$

the splitting is

$$\ln f_\ell^{\mathrm{asym},(1)}(i\sqrt{p_n^2+m^2}) = -\ln(2v) - \frac{1}{2}\ln\left(1+\frac{p_n^2}{v^2}\right) - \frac{1}{2}\frac{m^2}{v^2+p_n^2}$$
$$+ \frac{2}{v^2}\left[\frac{1}{16}\left(1+\frac{p_n^2}{v^2}\right)^{-1} - \frac{3}{8}\left(1+\frac{p_n^2}{v^2}\right)^{-2} + \frac{5}{16}\left(1+\frac{p_n^2}{v^2}\right)^{-3}\right], \quad (3\text{-}21)$$

$$\ln f_\ell^{\mathrm{asym},(2)}(i\sqrt{p_n^2+m^2}) = -\frac{1}{2}\ln\left(1+\frac{m^2}{v^2+p_n^2}\right) + \frac{1}{2}\frac{m^2}{v^2+p_n^2}$$
$$+ \frac{2}{v^2}\bigg[\frac{1}{16}\left(1+\frac{p_n^2}{v^2}\right)^{-1}\left(\left(1+\frac{m^2}{v^2+p_n^2}\right)^{-1}-1\right)$$
$$- \frac{3}{8}\left(1+\frac{p_n^2}{v^2}\right)^{-2}\left(\left(1+\frac{m^2}{v^2+p_n^2}\right)^{-2}-1\right)$$
$$+ \frac{5}{16}\left(1+\frac{p_n^2}{v^2}\right)^{-3}\left(\left(1+\frac{m^2}{v^2+p_n^2}\right)^{-3}-1\right)\bigg]. \quad (3\text{-}22)$$

We have used the given notation for the asymptotics to make a comparison with [14] as easy as possible. With these asymptotic quantities we rewrite $\zeta'_{\mathrm{tot},f}(0)$ as

$$\zeta'_{\mathrm{tot},f}(0) = -\sum_{n=-\infty}^{\infty}\sum_{\ell=0}^{\infty}(2\ell+1)\Big(\ln(I_v(\sqrt{p_n^2+m^2})K_v(\sqrt{p_n^2+m^2}))$$
$$- \ln f_\ell^{\mathrm{asym},(1)}(i\sqrt{p_n^2+m^2})\Big)$$
$$+ \sum_{n=-\infty}^{\infty}\sum_{\ell=0}^{\infty}(2\ell+1)\ln f_\ell^{\mathrm{asym},(2)}(i\sqrt{p_n^2+m^2}). \quad (3\text{-}23)$$

Let us next analyze $\zeta'_{\text{tot,as}}(0)$. To further analyze $\zeta_{\text{tot,as}}(s)$, equation (3-17), we use the integrals

$$\int_{\sqrt{p_n^2+m^2}}^{\infty} \frac{du}{\sqrt{p_n^2+m^2}} (u^2 - p_n^2 - m^2)^{-s} \frac{d}{du} \ln\left(1 + \frac{u^2}{v^2}\right) = \frac{\pi}{\sin \pi s}(m^2 + v^2 + p_n^2)^{-s},$$

$$\int_{\sqrt{p_n^2+m^2}}^{\infty} \frac{du}{\sqrt{p_n^2+m^2}} (u^2 - p_n^2 - m^2)^{-s} \frac{d}{du}\left(1 + \frac{u^2}{v^2}\right)^{-N/2}$$
$$= -\frac{\pi \Gamma\left(s + \frac{N}{2}\right)}{\sin(\pi s) \Gamma\left(\frac{N}{2}\right) \Gamma(s)} v^{-2s}\left(1 + \frac{p_n^2 + m^2}{v^2}\right)^{-s - \frac{N}{2}},$$

which are the relevant ones after substituting $zv = u$. This shows that

$$\zeta_{\text{tot,as}}(s) = -\sum_{n=-\infty}^{\infty} \sum_{\ell=0}^{\infty} v^{1-2s}\left(1 + \frac{p_n^2+m^2}{v^2}\right)^{-s}$$
$$- \frac{1}{4}s \sum_{n=-\infty}^{\infty} \sum_{\ell=0}^{\infty} v^{-2s-1}\left(1 + \frac{p_n^2+m^2}{v^2}\right)^{-s-1}$$
$$+ \frac{3}{2}s(s+1) \sum_{n=-\infty}^{\infty} \sum_{\ell=0}^{\infty} v^{-2s-1}\left(1 + \frac{p_n^2+m^2}{v^2}\right)^{-s-2}$$
$$- \frac{5}{8}s(s+1)(s+2) \sum_{n=-\infty}^{\infty} \sum_{\ell=0}^{\infty} v^{-2s-1}\left(1 + \frac{p_n^2+m^2}{v^2}\right)^{-s-3}. \quad (3\text{-}24)$$

To each of these terms we apply the rewriting (3-19). Intermediate expressions are relatively lengthy and we explain details only for the first term. We proceed as for the splitting in (3-21) and (3-22) and write

$$-\sum_{n=-\infty}^{\infty} \sum_{\ell=0}^{\infty} v^{1-2s}\left(1 + \frac{p_n^2+m^2}{v^2}\right)^{-s}$$
$$= -\sum_{n=-\infty}^{\infty} \sum_{\ell=0}^{\infty} v^{1-2s}\left(1 + \frac{p_n^2}{v^2}\right)^{-s}\left(1 + \frac{m^2}{v^2 + p_n^2}\right)^{-s}$$
$$= -\sum_{n=-\infty}^{\infty} \sum_{\ell=0}^{\infty} \frac{v}{(v^2 + p_n^2)^s}\left[\left(1 + \frac{m^2}{v^2 + p_n^2}\right)^{-s} - 1 + s\frac{m^2}{v^2 + p_n^2} + 1 - s\frac{m^2}{v^2 + p_n^2}\right]$$
$$= -\sum_{n=-\infty}^{\infty} \sum_{\ell=0}^{\infty} \frac{v}{(v^2 + p_n^2)^s}\left[\left(1 + \frac{m^2}{v^2 + p_n^2}\right)^{-s} - 1 + s\frac{m^2}{v^2 + p_n^2}\right]$$
$$- \frac{1}{2}\sum_{n=-\infty}^{\infty} \sum_{\ell=0}^{\infty} \frac{2v}{(v^2 + p_n^2)^s} + \frac{sm^2}{2}\sum_{n=-\infty}^{\infty} \sum_{\ell=0}^{\infty} \frac{2v}{(v^2 + p_n^2)^{s+1}}. \quad (3\text{-}25)$$

The first line is seen to be analytic about $s = 0$. We have subtracted the minimal number of terms to make the sums convergent. The remaining terms represent zeta functions of Epstein type,

$$E^{(k)}(s,a) = \sum_{n=-\infty}^{\infty} \sum_{\ell=0}^{\infty} 2v \frac{v^k}{(v^2+a^2n^2)^s}, \qquad (3\text{-}26)$$

the analytical continuation of which is well understood. Performing a Poisson resummation on the n-summation [2; 15; 30] yields

$$E^{(k)}(s,a) = \frac{2\sqrt{\pi}}{a} \frac{\Gamma(s-\frac{1}{2})}{\Gamma(s)} \zeta_H(2s-k-2, \tfrac{1}{2})$$

$$+ \frac{8\pi^s}{\Gamma(s)a^{s+1/2}} \sum_{\ell=0}^{\infty} v^{k+(3/2)-s} \sum_{n=1}^{\infty} n^{s-1/2} K_{1/2-s}\left(2\pi v \frac{n}{a}\right). \qquad (3\text{-}27)$$

The first line has poles at $s = \frac{1}{2} - j$, $j \in \mathbb{N}_0$, and for $s = \frac{1}{2}(k+3)$, the second line is analytic for $s \in \mathbb{C}$.

In terms of these Epstein functions, in equation (3-25) we have shown that

$$-\sum_{n=-\infty}^{\infty} \sum_{\ell=0}^{\infty} v^{1-2s}\left(1+\frac{p_n^2+m^2}{v^2}\right)^{-s} = -\tfrac{1}{2} E^{(0)}\left(s, \frac{2\pi}{\beta}\right) + \frac{sm^2}{2} E^{(0)}\left(s+1, \frac{2\pi}{\beta}\right)$$

$$- \sum_{n=-\infty}^{\infty} \sum_{\ell=0}^{\infty} \frac{v}{(v^2+p_n^2)^s}\left[\left(1+\frac{m^2}{v^2+p_n^2}\right)^{-s} - 1 + s\frac{m^2}{v^2+p_n^2}\right]. \qquad (3\text{-}28)$$

Noting from equation (3-27) that $E^{(0)}(s,a)$ and $E^{(0)}(s+1,a)$ are analytic about $s = 0$, we get

$$-\frac{d}{ds} \sum_{n=-\infty}^{\infty} \sum_{\ell=0}^{\infty} v^{1-2s}\left(1+\frac{p_n^2+m^2}{v^2}\right)^{-s} = -\tfrac{1}{2} E^{(0)\prime}\left(0, \frac{2\pi}{\beta}\right) + \tfrac{1}{2} m^2 E^{(0)\prime}\left(1, \frac{2\pi}{\beta}\right)$$

$$- \sum_{n=-\infty}^{\infty} \sum_{\ell=0}^{\infty} v\left[-\ln\left(1+\frac{m^2}{v^2+p_n^2}\right) + \frac{m^2}{v^2+p_n^2}\right]. \qquad (3\text{-}29)$$

The last term on the right cancels the first line from $\ln f_\ell^{\mathrm{asym},(2)}(i\sqrt{p_n^2+m^2})$ in equation (3-22), the remaining terms are easily found from (3-27).

One can proceed in exactly this way for the other terms in $\zeta_{\mathrm{tot,as}}(s)$; there are always terms that cancel with terms from $\ln f_\ell^{\mathrm{asym},(2)}(i\sqrt{p_n^2+m^2})$ in (3-22) and terms expressible using the Epstein type zeta functions given in (3-26). Adding up all contributions, the second line in (3-23) completely cancels and we obtain the following closed form for the finite-T zeta function:

$$\zeta'_{tot}(0) = -\sum_{n=-\infty}^{\infty}\sum_{\ell=0}^{\infty} 2\nu\bigg[\ln\big(I_\nu(\sqrt{p_n^2+m^2})K_\nu(\sqrt{p_n^2+m^2})\big) + \ln(\nu^2+p_n^2)$$
$$+ \frac{m^2}{\nu^2+p_n^2} - \frac{1}{8}\frac{1}{\nu^2+p_n^2}\bigg(1 - \frac{6\nu^2}{(\nu^2+p_n^2)^2} + \frac{5\nu^4}{(\nu^2+p_n^2)^3}\bigg)\bigg]$$
$$- \tfrac{1}{2}E^{(0)\prime}\Big(0,\tfrac{2\pi}{\beta}\Big) + \tfrac{1}{2}m^2 E^{(0)}\Big(1,\tfrac{2\pi}{\beta}\Big)$$
$$+ \tfrac{3}{4}E^{(2)}\Big(2,\tfrac{2\pi}{\beta}\Big) - \tfrac{5}{8}E^{(4)}\Big(3,\tfrac{2\pi}{\beta}\Big). \tag{3-30}$$

From (3-27) it is clear that the Epstein type zeta functions contain zero temperature contributions to the Casimir energy (first line in (3-27)) and exponentially damped contributions for small temperature described by the Bessel functions (second line in (3-27)). As it turns out, the zero temperature contributions from the Epstein type zeta functions in (3-30) all vanish. The remaining zero temperature contributions in (3-30) are found replacing the Riemann sum over n by an integral,

$$\sum_{n=-\infty}^{\infty} f(n) \implies \frac{\beta}{2\pi}\int_{-\infty}^{\infty} dp\, f(p).$$

As $\beta \to 0$ this shows that

$$\frac{1}{\beta}\zeta'_{tot}(0)$$
$$= -\frac{1}{2\pi}\int_{-\infty}^{\infty} dp \sum_{\ell=0}^{\infty} 2\nu\bigg[\ln\big(I_\nu(\sqrt{p^2+m^2})K_\nu(\sqrt{p^2+m^2})\big) + \ln(\nu^2+p^2)$$
$$+ \frac{m^2}{\nu^2+p^2} - \frac{1}{8}\frac{1}{\nu^2+p^2}\bigg(1 - \frac{6\nu^2}{(\nu^2+p^2)^2} + \frac{5\nu^4}{(\nu^2+p^2)^3}\bigg)\bigg], \tag{3-31}$$

from which the Casimir energy (3-3) is trivially obtained. The result is much simpler than previous results given [24; 5] and a numerical evaluation could easily be performed.

4. Functional determinant on a two dimensional torus

As our next example let us consider a two dimensional torus $S^1 \times S^1$. For convenience we choose the perimeter of the circles to be 1. The relevant eigenvalue problem to be considered then is

$$P\phi_\lambda(x,y) := \bigg(-\frac{\partial^2}{\partial x^2} - \frac{\partial^2}{\partial y^2}\bigg)\phi_\lambda(x,y) = \lambda^2 \phi_\lambda(x,y),$$

and we choose periodic boundary conditions

$$\phi_\lambda(x, y) = \phi_\lambda(x+1, y), \quad \frac{\partial \phi_\lambda(x, y)}{\partial x} = \frac{\partial \phi_\lambda(x+1, y)}{\partial x},$$

$$\phi_\lambda(x, y) = \phi_\lambda(x, y+1), \quad \frac{\partial \phi_\lambda(x, y)}{\partial y} = \frac{\partial \phi_\lambda(x, y+1)}{\partial y}.$$

The eigenfunctions and eigenvalues clearly are

$$\phi_{m,n}(x, y) = e^{-2\pi i m x} e^{-2\pi i n y}, \quad \lambda^2 = (2\pi)^2 (m^2 + n^2), \quad n, m \in \mathbb{Z}.$$

The related zeta function then reads

$$\zeta_P(s) = (2\pi)^{-2s} \sum_{(m,n) \in \mathbb{Z}^2 \setminus \{(0,0)\}} (m^2 + n^2)^{-s}; \qquad (4\text{-}1)$$

note that the zero mode $m = n = 0$ has to be omitted in the summation to make $\zeta_P(s)$ well defined. The zeta function in equation (4-1) is an Epstein zeta function and $\zeta_P'(0)$ can be evaluated using the Kronecker limit formula [18; 19]. Here, we apply the contour approach previously outlined which simplifies the calculation.

Instead of using the fact that the eigenvalues can be given in closed form, we proceed differently. We say that

$$\lambda^2 = (2\pi)^2 (n^2 + k^2), \qquad n \in \mathbb{Z},$$

where k is determined as a solution to the equation

$$e^{\pi i k} - e^{-\pi i k} = 0. \qquad (4\text{-}2)$$

Of course, solutions are given by $k \in \mathbb{Z}$ and the correct eigenvalues follow. Using equation (4-2) determining the eigenvalues in the way we have used equations (2-3) and (3-4), the zeta function can be represented as the contour integral

$$\zeta_P(s) = 4 \sum_{n=1}^{\infty} \int_\gamma \frac{dk}{2\pi i} (2\pi)^{-2s} (n^2 + k^2)^{-s} \frac{d}{dk} \ln\left(\frac{e^{\pi i k} - e^{-\pi i k}}{2\pi i k} \right)$$
$$+ 4(2\pi)^{-2s} \zeta_R(2s). \qquad (4\text{-}3)$$

The last term represents the part where one of the two indices m or n is zero in equation (4-1). The first line represents the remaining contributions. The factor of 4 is a result of summing over positive n only and because the contour γ is supposed to enclose positive integers only. The reason that we have used

$$\frac{e^{\pi i k} - e^{-\pi i k}}{2\pi i k}$$

instead of equation (4-2) is that

$$\lim_{k \to 0} \frac{e^{\pi i k} - e^{-\pi i k}}{2\pi i k} = 1,$$

which will allow us to shift the contour in a way as to include the origin; see the discussion below equation (3-7). Let us evaluate the contour integral

$$\zeta_n(s) = \int_\gamma \frac{dk}{2\pi i} (2\pi)^{-2s} (n^2 + k^2)^{-s} \frac{d}{dk} \ln \frac{e^{\pi i k} - e^{-\pi i k}}{2\pi i k}.$$

Substituting $k = \sqrt{z}$ and deforming the contour to the negative real axis along the lines described previously, an intermediate result is

$$\zeta_n(s) = (2\pi)^{-2s} \frac{\sin \pi s}{\pi} \int_{n^2}^\infty dz (z - n^2)^{-s} \frac{d}{dz} \ln \frac{e^{\pi \sqrt{z}} - e^{-\pi \sqrt{z}}}{2\pi \sqrt{z}}. \quad (4\text{-}4)$$

From the behavior of the integrand as $z \to \infty$ and $z \to n^2$ this representation is seen to be valid for $\frac{1}{2} < \operatorname{Re} s < 1$. In order to construct the analytical continuation to a neighborhood of $s = 0$ we note that

$$\frac{e^{\pi \sqrt{z}} - e^{-\pi \sqrt{z}}}{2\pi \sqrt{z}} = \frac{e^{\pi \sqrt{z}}}{2\pi \sqrt{z}} (1 - e^{-2\pi \sqrt{z}}).$$

We therefore write

$$\zeta_n(s) = (2\pi)^{-2s} \frac{\sin \pi s}{\pi} \int_{n^2}^\infty dz (z - n^2)^{-s} \frac{d}{dz} \ln \frac{e^{\pi \sqrt{z}}}{2\pi \sqrt{z}}$$

$$+ (2\pi)^{-2s} \frac{\sin \pi s}{\pi} \int_{n^2}^\infty dz (z - n^2)^{-s} \frac{d}{dz} \ln(1 - e^{-2\pi \sqrt{z}}).$$

The first line is evaluated using

$$\int_{n^2}^\infty dz \frac{(z - n^2)^{-s}}{\sqrt{z}} = \frac{n^{1-2s}}{\sqrt{\pi}} \Gamma(1 - s) \Gamma(-\tfrac{1}{2} + s),$$

$$\int_{n^2}^\infty dz \frac{(z - n^2)^{-s}}{z} = \frac{\pi n^{-2s}}{\sin \pi s}.$$

With the identity [28] $\dfrac{\sin \pi s}{\pi} \Gamma(1 - s) = \dfrac{1}{\Gamma(s)}$, this produces

$$\zeta_n(s) = \tfrac{1}{4}(2\pi)^{-2s+1} \frac{n^{1-2s} \Gamma(-\tfrac{1}{2} + s)}{\sqrt{\pi} \Gamma(s)} - \tfrac{1}{2}(2\pi)^{-2s} n^{-2s}$$

$$+ (2\pi)^{-2s} \frac{\sin \pi s}{\pi} \int_{n^2}^\infty dz (z - n^2)^{-s} \frac{d}{dz} \ln(1 - e^{-2\pi \sqrt{z}}).$$

This is the form that allows the sum over n to be (partly) performed and it shows that

$$\zeta_P(s) = 4(2\pi)^{-2s}\frac{\sin \pi s}{\pi}\sum_{n=1}^{\infty}\int_{n^2}^{\infty}dz(z-n^2)^{-s}\frac{d}{dz}\ln\left(1-e^{-2\pi\sqrt{z}}\right)$$
$$-2(2\pi)^{-2s}\zeta_R(2s) + (2\pi)^{-2s+1}\frac{\Gamma(-\frac{1}{2}+s)}{\sqrt{\pi}\Gamma(s)}\zeta_R(2s-1). \quad (4\text{-}5)$$

This form allows for the evaluation of $\zeta_P'(0)$. From known elementary properties of the Γ-function and the zeta function of Riemann [28] we obtain

$$\frac{d}{ds}\bigg|_{s=0}\left[(2\pi)^{-2s}\zeta_R(2s)\right] = -2\ln(2\pi)\zeta_R(0) + 2\zeta_R'(0) = 0,$$

$$\frac{d}{ds}\bigg|_{s=0}\left[(2\pi)^{-2s+1}\frac{\Gamma(-\frac{1}{2}+s)}{\sqrt{\pi}\Gamma(s)}\zeta_R(2s-1)\right] = \frac{\pi}{3}.$$

The first line in (4-5) is also easily evaluated because

$$\frac{d}{ds}\bigg|_{s=0}4(2\pi)^{-2s}\frac{\sin \pi s}{\pi}\sum_{n=1}^{\infty}\int_{n^2}^{\infty}dz(z-n^2)^{-s}\frac{d}{dz}\ln\left(1-e^{-2\pi\sqrt{z}}\right)$$
$$= 4\sum_{n=1}^{\infty}\int_{n^2}^{\infty}dz\frac{d}{dz}\ln\left(1-e^{-2\pi\sqrt{z}}\right) = -4\sum_{n=1}^{\infty}\ln\left(1-e^{-2\pi n}\right).$$

This can be reexpressed using the Dedekind eta function

$$\eta(\tau) := e^{i\pi\tau/12}\prod_{n=1}^{\infty}\left(1-e^{2\pi i n \tau}\right)$$

for $\tau \in \mathbb{C}$, $\operatorname{Re}\tau > 0$. The relation relevant for us follows by setting $\tau = i$:

$$\ln|\eta(i)|^4 = -\frac{\pi}{3} + 4\sum_{n=1}^{\infty}\ln\left(1-e^{-2\pi n}\right).$$

Adding up all contributions for $\zeta_P'(0)$, the final answer reads

$$\zeta_P'(0) = \frac{\pi}{3} - 4\sum_{n=1}^{\infty}\ln\left(1-e^{-2\pi n}\right) = -\ln|\eta(i)|^4, \quad (4\text{-}6)$$

in agreement with known answers; see [46; 50], for example.

5. Conclusions

We have shown that contour integrals are very useful and effective tools for the evaluation of determinants of differential operators. Although the results look very simple only in one dimension — see equation (2-10) — , for particular configurations also in higher dimensions closed answers can be found suitable for numerical evaluation, as in equations (3-31) and (4-6). Here we have provided answers only for the torus and a spherically symmetric situation. But the same ideas should apply when separability of the partial differential equations in other coordinate systems is possible. Results in this direction will be presented elsewhere.

Acknowledgement

Kirsten acknowledges support by the NSF through grant PHY-0757791. Part of the work was done while the author enjoyed the hospitality and partial support of the Department of Physics and Astronomy of the University of Oklahoma. Thanks go in particular to Kimball Milton and his group who made this very pleasant and exciting visit possible.

References

[1] M. Abramowitz and I. A. Stegun. *Handbook of mathematical functions*. Dover, New York, 1970.

[2] J. Ambjorn and S. Wolfram. Properties of the vacuum. 1. Mechanical and thermodynamic. *Ann. Phys.*, 147:1–32, 1983.

[3] A. O. Barvinsky, A. Yu. Kamenshchik, and I. P. Karmazin. One loop quantum cosmology: Zeta function technique for the Hartle–Hawking wave function of the universe. *Ann. Phys.*, 219:201–242, 1992.

[4] M. Bordag, E. Elizalde, and K. Kirsten. Heat kernel coefficients of the Laplace operator on the D-dimensional ball. *J. Math. Phys.*, 37:895–916, 1996.

[5] M. Bordag, E. Elizalde, K. Kirsten, and S. Leseduarte. Casimir energies for massive fields in a spherical geometry. *Phys. Rev.*, D56:4896–4904, 1997.

[6] M. Bordag, B. Geyer, K. Kirsten, and E. Elizalde. Zeta function determinant of the Laplace operator on the D-dimensional ball. *Commun. Math. Phys.*, 179:215–234, 1996.

[7] M. Bordag, K. Kirsten, and J. S. Dowker. Heat kernels and functional determinants on the generalized cone. *Commun. Math. Phys.*, 182:371–394, 1996.

[8] D. Burghelea, L. Friedlander, and T. Kappeler. On the determinant of elliptic differential and finite difference operators in vector bundles over S^1. *Commun. Math. Phys.*, 138:1–18, 1991.

[9] A. A. Bytsenko, G. Cognola, L. Vanzo, and S. Zerbini. Quantum fields and extended objects in space-times with constant curvature spatial section. *Phys. Rept.*, 266:1–126, 1996.

[10] S. Coleman. *Aspects of symmetry: Selected lectures of Sidney Coleman*. Cambridge University Press, Cambridge, 1985.

[11] J. S. Dowker and R. Critchley. Effective Lagrangian and energy momentum tensor in de Sitter space. *Phys. Rev.*, D13:3224–3232, 1976.

[12] T. Dreyfuss and H. Dym. Product formulas for the eigenvalues of a class of boundary value problems. *Duke Math. J.*, 45:15–37, 1978.

[13] G. V. Dunne and K. Kirsten. Functional determinants for radial operators. *J. Phys.*, A39:11915–11928, 2006.

[14] G. V. Dunne and K. Kirsten. Simplified vacuum energy expressions for radial backgrounds and Domain Walls. *J. Phys. A*, 42:075402, 2009.

[15] E. Elizalde. On the zeta-function regularization of a two-dimensional series of Epstein-Hurwitz type. *J. Math. Phys.*, 31:170–174, 1990.

[16] E. Elizalde. *Ten physical applications of spectral zeta functions*. Lecture Notes in Physics m35, Springer, Berlin, 1995.

[17] E. Elizalde, S. D. Odintsov, A. Romeo, A. A. Bytsenko, and S. Zerbini. *Zeta regularization techniques with applications*. World Scientific, Singapore, 1994.

[18] P. Epstein. Zur Theorie allgemeiner Zetafunctionen. *Math. Ann.*, 56:615–644, 1903.

[19] P. Epstein. Zur Theorie allgemeiner Zetafunctionen II. *Math. Ann.*, 63:205–216, 1907.

[20] A. Erdélyi, W. Magnus, F. Oberhettinger, and F. G. Tricomi. *Higher transcendental functions*. Based on the notes of Harry Bateman, McGraw-Hill Book Company, New York, 1955.

[21] R. P. Feynman and A. R. Hibbs. *Quantum mechanics and path integrals*. McGraw-Hill, New York, 1965.

[22] R. Forman. Functional determinants and geometry. *Invent. Math.*, 88:447–493, 1987; erratum in 108:453-454, 1992.

[23] R. Forman. Determinants, finite-difference operators and boundary value problems. *Commun. Math. Phys.*, 147:485–526, 1992.

[24] M. De Francia. Free energy for massless confined fields. *Phys. Rev.*, D50:2908–2919, 1994.

[25] M. De Francia, H. Falomir, and M. Loewe. Massless fermions in a bag at finite density and temperature. *Phys. Rev.*, D55:2477–2485, 1997.

[26] I. M. Gelfand and A. M. Yaglom. Integration in functional spaces and its applications in quantum physics. *J. Math. Phys.*, 1:48–69, 1960.

[27] P. B. Gilkey, K. Kirsten, and D. V. Vassilevich. Heat trace asymptotics with transmittal boundary conditions and quantum brane world scenario. *Nucl. Phys.*, B601:125–148, 2001.

[28] I. S. Gradshteyn and I. M. Ryzhik. *Table of integrals, series and products*. Academic Press, New York, 1965.

[29] S. W. Hawking. Zeta function regularization of path integrals in curved space-time. *Commun. Math. Phys.*, 55:133–148, 1977.

[30] K. Kirsten. Topological gauge field mass generation by toroidal space-time. *J. Phys. A*, 26:2421–2435, 1993.

[31] K. Kirsten. *Spectral functions in mathematics and physics*. Chapman&Hall/CRC, Boca Raton, FL, 2001.

[32] K. Kirsten. Heat kernel asymptotics: more special case calculations. *Nucl. Phys. B (Proc. Suppl.)*, 104:119–126, 2002.

[33] K. Kirsten and P. Loya. Computation of determinants using contour integrals. *Am. J. Phys.*, 76:60–64, 2008.

[34] K. Kirsten and A. J. McKane. Functional determinants by contour integration methods. *Ann. Phys.*, 308:502–527, 2003.

[35] K. Kirsten and A. J. McKane. Functional determinants for general Sturm–Liouville problems. *J. Phys. A*, 37:4649–4670, 2004.

[36] H. Kleinert. *Path integrals in quantum mechanics, statistics, polymer physics, and financial markets*. World Scientific, Singapore, 2006.

[37] H. Kleinert and A. Chervyakov. Simple explicit formulas for Gaussian path integrals with time-dependent frequencies. *Phys. Lett. A*, 245:345–357, 1998.

[38] H. Kleinert and A. Chervyakov. Functional determinants from Wronskian Green functions. *J. Math. Phys.*, 40:6044–6051, 1999.

[39] M. Lesch. Determinants of regular singular Sturm–Liouville operators. *Math. Nachr.*, 194:139–170, 1998.

[40] M. Lesch and J. Tolksdorf. On the determinant of one-dimensional elliptic boundary value problems. *Commun. Math. Phys.*, 193:643–660, 1998.

[41] B. M. Levitan and I. S. Sargsjan. *Introduction to spectral theory: Selfadjoint ordinary differential operators*. Translations of Mathematical Monographs 39. AMS, Providence, R. I., 1975.

[42] A. J. McKane and M. B. Tarlie. Regularisation of functional determinants using boundary perturbations. *J. Phys. A*, 28:6931–6942, 1995.

[43] P. Moon and D. E. Spencer. *Field theory handbook*. Springer, Berlin, 1961.

[44] F. W. Olver. The asymptotic expansion of Bessel functions of large order. *Philos. Trans. Roy. Soc. London*, A247:328–368, 1954.

[45] F. W. J. Olver. *Asymptotics and special functions*. Academic Press, New York, 1974.

[46] J. Polchinski. *String theory, 1: An introduction to the bosonic string.* Cambridge University Press, Cambridge, 1998.

[47] D. B. Ray and I. M. Singer. R-torsion and the Laplacian on Riemannian manifolds. *Advances in Math.*, 7:145–210, 1971.

[48] L. S. Schulman. *Techniques and applications of path integration.* Wiley, New York, 1981.

[49] H. Weyl. Das asymptotische Verteilungsgesetz der Eigenwerte linearer partieller Differentialgleichungen. *Math. Ann.*, 71:441–479, 1912.

[50] F. Williams. *Topics in quantum mechanics.* Birkhäuser, New York, 2003.

KLAUS KIRSTEN
DEPARTMENT OF MATHEMATICS
BAYLOR UNIVERSITY
WACO, TX 76798
UNITED STATES
Klaus_Kirsten@baylor.edu

The role of the Patterson–Selberg zeta function of a hyperbolic cylinder in three-dimensional gravity with a negative cosmological constant

FLOYD L. WILLIAMS

To the memory of Kenneth Hoffman

1. Introduction

A few years ago, the author took note of certain sums that appeared in the physics literature in connection with thermodynamic of the BTZ black hole — a three-dimensional solution discovered by M. Bañados, C. Teitelboim, and J. Zanelli [1] of the Einstein gravitational field equations

$$R_{ij} - \tfrac{1}{2} R g_{ij} - \Lambda g_{ij} = 0 \tag{1.1}$$

with *negative cosmological constant* Λ. Here $R_{ij} = R_{ij}(g)$, $R = R(g)$ are the Ricci tensor and Ricci scalar curvature, respectively, of the solution metric $g = [g_{ij}]$. We describe the BTZ metric in equation (2.1) below. These sums were used to express, for example, the nondivergent part of the effective BTZ action, or corrections to classical Bekenstein–Hawking entropy [4; 13; 15] — sums that physicists evidently did not realize were related to the *Patterson–Selberg zeta function* $Z_\Gamma(s)$ of a hyperbolic cylinder. The paper [21], for example, was written to point out this relation and thus to establish a thermodynamics-zeta function connection. Another such connection appears in my Lecture 6 of this volume.

In [23; 25; 26], for example, we see that the Mann–Solodukhin quantum correction to black hole entropy [15] is expressed, in fact, in terms of a suitable "deformation" of $Z_\Gamma(s)$. It is also possible to keep track of a corresponding deformation of the black hole topology. We review the deformation of zeta, and of the BTZ topology, in Section 4 below where we use it to set up a one-loop determinant formula (or an effective action formula) in the presence of conical singularities.

In Section 3 we express the one-loop quantum field partition function, the one-loop gravity partition function, and the full gravity partition function all in terms of the zeta function $Z_\Gamma(s)$. Using the holomorphic sector of the one-loop gravity partition function and the classical elliptic modular function $j(\tau)$, one can build up (with the help of Hecke operators) modular invariant partition functions of proposed holomorphic conformal field theories with central charge $24k$, where k is a positive integer — theories first defined by G. Höhn [12] and proposed by E. Witten [28] as the holographic dual of pure $2+1$ gravity. That is, these partition functions exist even if the theories do not (although for $k=1$ existence has been established by I. Frenkel, J. Lepowsky, and A. Meurman in [8]), and in Section 5 we take a close look (in Theorem 5.16, for example) at their Fourier coefficients — the asymptotics of which provide for quantum corrections to holomorphic sector black hole entropy.

The lecture, after this introduction, consists of four sections and an appendix:

- The BTZ black hole
- Patterson–Selberg zeta function and a one-loop determinant formula
- Determinant formula in the presence of conical singularities
- Extremal partition functions of conformal field theories with central charge $24k$
- Appendix to Section 5: Computation of $Z_k(\tau)$ for $k=2,3$
- References

The author dedicates this lecture to the memory of Professor Kenneth Hoffman. His kind support and friendliness to me, as a young MIT postdoc, remains most highly appreciated these many years later.

2. The BTZ black hole

The BTZ metric that solves the vacuum Einstein equations (1.1) in three dimensions with $\Lambda < 0$ is given (in Euclidean form) by

$$ds^2_{\text{BTZ}} = \left(N_1(r)^2 + r^2 N_2(r)^2\right) d\tau^2 + N_1(r)^{-2} dr^2 + 2r^2 N_2(r)\, d\phi\, d\tau + r^2\, d\phi^2, \quad (2.1)$$

in coordinates (r, ϕ, τ) on a region of anti-deSitter space where for mass and angular momentum parameters $M > 0$, $J \geq 0$, respectively,

$$N_1(r)^2 \overset{\text{def.}}{=} -M - \Lambda r^2 - J^2/4r^2, \quad N_2(r) = -J/2r^2. \quad (2.2)$$

In equation (2.1), one has periodicity of the Schwarzschild variable ϕ; i.e. there is the identification $\phi \sim \phi + 2\pi n$ for $n \in \mathbb{Z}$, the ring of integers. We return to this important point shortly. Not all scalar curvatures are created equal. So in equation (1.1), our sign convention is such that $R = -6\Lambda > 0$; in particular ds_{BTZ}^2 is a constant curvature solution. The metric ds_{BTZ}^2 is also a black hole solution with outer and inner event radii r_+, r_- given by

$$r_+^2 = \frac{M\sigma^2}{2}\left[1 + \left(1 + \frac{J^2}{M^2\sigma^2}\right)^{1/2}\right], \quad r_- = -\frac{\sigma J i}{2r_+}, \tag{2.3}$$

where $i^2 = -1$ and $\sigma \stackrel{\text{def.}}{=} 1/\sqrt{-\Lambda} > 0$. Here $r_+ > 0$, but $r_- \in i\mathbb{R}$ is pure imaginary (since we are working with the Euclidean form of BTZ). Note that

$$r_-^2 = \frac{M\sigma^2}{2}\left[1 - \left(1 + \frac{J^2}{M^2\sigma^2}\right)^{1/2}\right], \quad |r_-| = \frac{\sigma J}{2r_+} = i r_-. \tag{2.4}$$

Of course $r_- = 0$ is equivalent to $J = 0$, which is the case of the nonspinning black hole, in which case there is a single event horizon.

Given the periodicity $\phi \sim \phi + 2\pi n$, $n \in \mathbb{Z}$, of the Schwarzschild variable ϕ, as mentioned earlier, one can describe the topology of the space-time (where ds_{BTZ}^2 lives, with τ regarded as a time variable) as a quotient space

$$B_\Gamma = \Gamma \backslash \mathbb{H}^3 \tag{2.5}$$

where $\mathbb{H}^3 \stackrel{\text{def.}}{=} \{(x, y, z) \in \mathbb{R}^3 \mid z > 0\}$ is hyperbolic 3-space, and where

$$\Gamma = \Gamma_{(a,b)} \stackrel{\text{def.}}{=} \{\gamma^n \mid n \in \mathbb{Z}\} \quad \text{for } \gamma = \begin{bmatrix} e^{a+ib} & 0 \\ 0 & e^{-(a+ib)} \end{bmatrix}, \tag{2.6}$$

with $a \stackrel{\text{def.}}{=} \pi r_+/\sigma > 0$, and $b \stackrel{\text{def.}}{=} \pi |r_-|/\sigma = \pi J/2r_+ \geq 0$; see equations (2.3). Thus $\Gamma \subset \mathrm{SL}(2, \mathbb{C})$ is the cyclic subgroup with generator $\gamma \in \mathrm{SL}(2, \mathbb{C})$. The action of Γ on $\mathbb{H}^3$ is given by $\gamma^n \cdot (x, y, z) = (x', y', z')$ for

$$\begin{aligned} x' &= e^{2an}(x \cos 2bn - y \sin 2bn), \\ y' &= e^{2an}(x \sin 2bn + y \cos 2bn), \\ z' &= e^{2an} z. \end{aligned} \tag{2.7}$$

A fundamental domain F for this action is given by

$$F = \{(x, y, z) \in \mathbb{H}^3 \mid 1 < \sqrt{x^2 + y^2 + z^2} < e^{2a}\}, \tag{2.8}$$

a proof of which is given in Appendix A3 of [27], for example. In particular $\Gamma \subset \mathrm{SL}(2, \mathbb{C})$ is a *Kleinian* subgroup; that is, F has *infinite* hyperbolic volume:

$$\operatorname{vol} F \stackrel{\text{def.}}{=} \int_F dx\, dy\, dz/z^3 = \infty. \tag{2.9}$$

The description (2.5) is derived by way of a suitable change of variables $(r, \phi, \tau) \to (x, y, z)$, $z > 0$, whereby (remarkably) the BTZ metric ds_{BTZ}^2 in equation (2.1) is transformed, in fact, to a multiple ds^2 of the standard hyperbolic metric $(dx^2 + dy^2 + dz^2)/z^2$ on $\mathbb{H}^3$:

$$ds^2 = \sigma^2(dx^2 + dy^2 + dz^2)/z^2 \tag{2.10}$$

where $\sigma^2 = 1/(-\Lambda)$, by definition (2.3); see [7; 18], for example.

3. Patterson–Selberg zeta function and a one-loop determinant formula

Going back to the fact that Γ is Kleinian, we can assign to the black hole $B_\Gamma = \Gamma \backslash \mathbb{H}^3$ a natural zeta function (an Euler product)

$$Z_\Gamma(s) \stackrel{\text{def.}}{=} \prod_{0 \leq k_1, k_2 \in \mathbb{Z}}^{\infty} \left(1 - (e^{2bi})^{k_1}(e^{-2bi})^{k_2} e^{-(k_1+k_2+s)2a}\right), \tag{3.1}$$

which is the *Patterson–Selberg zeta function* attached to the hyperbolic cylinder $\Gamma \backslash \mathbb{H}^3$; see [16; 21]. $Z_\Gamma(s)$ is an entire function whose zeros are the numbers

$$N_{k_1,k_2,n} \stackrel{\text{def.}}{=} -(k_1+k_2) + (k_1-k_2)\frac{2bi}{2a} + \frac{2\pi n i}{2a}$$

for $k_1, k_2, n \in \mathbb{Z}$, $k_1, k_2 \geq 0$, that come from the zeros of its factors. In particular, $Z_\Gamma(s) \neq 0$ for $\operatorname{Re} s > 0$. In fact, for $\operatorname{Re} s > 0$, $Z_\Gamma(s) = e^{\log Z_\Gamma(s)}$, where

$$\log Z_\Gamma(s) \stackrel{\text{def.}}{=} -\sum_{n=1}^{\infty} \frac{e^{-(s-1)2an}}{4n(\sinh^2(an) + \sin^2(bn))}$$

$$= -\sum_{n=1}^{\infty} \frac{e^{-(s-1)2an}}{2n(\cosh(2an) - \cos(2bn))}. \tag{3.2}$$

In [10], the authors study the one-loop partition function of a free quantum field ϕ propagating in a locally anti-de Sitter background. The results they obtain cover not only the BTZ case, but higher genus generalizations of it, as well as the case of nonscalar fields ϕ (say gauge and graviton excitations). In the special BTZ case with ϕ a scalar field, for example, the *one-loop determinant formula* (equation (4.9) of [10])

$$-\log \det \Delta = \sum_{n=1}^{\infty} \frac{e^{-2\pi n \sqrt{1+m^2}\, \operatorname{Im}\tau}}{2n|\sin \pi n \tau|^2} \tag{3.3}$$

is derived, where now $\tau \stackrel{\text{def.}}{=} \frac{1}{2\pi}(\theta + i\beta)$ denotes the modular parameter corresponding to the anti-de Sitter temperature β^{-1} and angular potential θ, where

$$K(t,r) = \frac{e^{-(m^2+1)t - r^2/4t}}{(4\pi t)^{3/2}} \frac{r}{\sinh r} \tag{3.4}$$

is the scalar heat kernel on $\mathbb{H}^3$, and where in (3.3) the divergent contribution proportional to vol F (see equation (2.9)) is disregarded. We indicate how formula (3.3) also follows quite quickly from our result in [3], and we point out that the right-hand side of (3.3) in fact coincides with the special value $-2\log Z_\Gamma(1+\sqrt{1+m^2})$ (where we identify $\beta/2$ with a and $\theta/2$ with b) — this observation being an example of our initial remarks regarding sums appearing in the physics literature that are expressible in terms of the zeta function $Z_\Gamma(s)$ — a point unnoticed by physicists.

We start with the reminder that for $z = x + iy \in \mathbb{C}$, $\sin z = \sin x \cosh y + i\cos x \sinh y$. In particular $|\sin z|^2 = \sin^2 x \cosh^2 y + \cos^2 x \sinh^2 y = \sin^2 x \cdot \cosh^2 y + (1-\sin^2 x)\sinh^2 y = \sin^2 x(\cosh^2 y - \sinh^2 y) + \sinh^2 y = \sin^2 x + \sinh^2 y \Rightarrow |\sin \pi n \tau|^2 \stackrel{\text{def.}}{=} \sin^2\left(\frac{\theta n}{2}\right) + \sinh^2\left(\frac{\beta n}{2}\right) \Rightarrow$ the right-hand side of formula (3.3) is (since $\operatorname{Im}\tau \stackrel{\text{def.}}{=} \frac{\beta}{2\pi}$)

$$2\sum_{n=1}^{\infty} \frac{e^{-\sqrt{1+m^2}\beta n}}{4n\left[\sinh^2\left(\frac{\beta n}{2}\right) + \sin^2\left(\frac{\theta n}{2}\right)\right]} \stackrel{\cdot}{=} -2\log Z_\Gamma(1+\sqrt{1+m^2}), \quad (3.5)$$

by definition (3.2).

On the other hand, we have considered in [3; 22] a *truncated heat kernel*

$$K_t^{*\Gamma}(\widetilde{p}_1, \widetilde{p}_2) \stackrel{\text{def.}}{=} \sum_{n \in \mathbb{Z}-\{0\}} K_t(p_1, \gamma^n \cdot p_2) \quad (3.6)$$

for $B_\Gamma = \Gamma\backslash\mathbb{H}^3$, $t > 0$, where $\widetilde{p}_j \in B_\Gamma$ denotes the Γ-orbit of $p_j \in \mathbb{H}^3$, $j = 1, 2$, $\gamma^n \cdot p_2$ is given by definition (2.7), and where

$$K_t(p_1, p_2) \stackrel{\text{def.}}{=} \frac{e^{-t-d(p_1,p_2)^2/4t}}{(4\pi t)^{3/2}} \frac{d(p_1, p_2)}{\sinh(d(p_1, p_2))} \quad (3.7)$$

(compare equation (3.4)) for $d(p_1, p_2)$ the hyperbolic distance between p_1 and p_2, given by

$$\cosh(d(p_1, p_2)) = 1 + \frac{(x_1-x_2)^2 + (y_1-y_2)^2 + (z_1-z_2)^2}{2z_1 z_2} \quad (3.8)$$

for $p_j = (x_j, y_j, z_j)$. The expression $K_t^{*\Gamma}(\widetilde{p}_1, \widetilde{p}_2)$ gives rise to the theta function (or heat trace)

$$\theta_\Gamma(t) = \operatorname{trace} K_t^{*\Gamma} \stackrel{\text{def.}}{=} \iiint_F K_t^{*\Gamma}(\widetilde{p}, \widetilde{p})\, dv(p) \quad (3.9)$$

where $dv = dx\, dy\, dz/z^3$ is the hyperbolic volume element; see (2.8) and (2.9). We regard the integral

$$I(m) \stackrel{\text{def.}}{=} \int_0^\infty e^{-m^2 t} \text{ trace } K_t^{*\Gamma} \frac{dt}{t} \qquad (3.10)$$

as the meaning of the expression $-\log \det \Delta$ in the left-hand side of (3.3), with the understanding that by restricting the summation to $n \neq 0$ in (3.6), we disregard the divergent term

$$\int_0^\infty e^{-m^2 t} \left(\iiint_F \frac{e^{-t}}{(4\pi t)^{3/2}} dv \right) \frac{dt}{t} = \frac{\text{vol } F}{(4\pi)^{3/2}} \int_0^\infty e^{-(1+m^2)t} e^{-3/2 - 1} dt. \qquad (3.11)$$

Namely, $n = 0$ implies

$$K_t(p, \gamma^n \cdot p) = K_t(p, p) = e^{-t} (4\pi t)^{-3/2},$$

by (3.7) and (3.8), and this expression is independent of $p \in \mathbb{H}^3$. Thus if we were to include the term for $n = 0$ in (3.6), there would be a manifest contribution $\iiint_F e^{-t} (4\pi t)^{-3/2} dv$ to (3.9), which in turn would lead to the contribution $\int_0^\infty e^{-m^2 t} \left(\iiint_F \frac{e^{-t}}{(4\pi t)^{-3/2}} dv \right) dt/t$ to (3.10). This explains the divergent term mentioned in (3.11), where one notes not only the "infrared" divergence vol F (by (2.9)), but also the "ultraviolet" divergence reflected by the *negative* $-3/2$ in the integral $J(m) \stackrel{\text{def.}}{=} \int_0^\infty e^{-(1+m^2)t} t^{-3/2 - 1} dt$. Given the formula

$$\int_0^\infty e^{-(1+m^2)t} t^{v-1} dt = \frac{\Gamma(v)}{(1+m^2)^v} \qquad (3.12)$$

for *positive* v, the authors in [10] (also compare [5]) remove the ultraviolet divergence by assigning to $J(m)$ the value

$$\frac{\Gamma(-3/2)}{(1+m^2)^{-3/2}} = \frac{4\sqrt{\pi}}{3} (1+m^2)^{3/2}.$$

Thus, in summary, the divergent term being disregarded is (by (3.11)) equal to

$$\frac{\text{vol } F}{(4\pi)^{3/2}} \frac{4\sqrt{\pi}}{3} (1+m^2)^{3/2} = \frac{(1+m^2)^{3/2}}{6\pi} \text{vol } F,$$

and we regard the one-loop determinant formula (3.3) as the statement that

$$I(m) \stackrel{(3.10)}{=} \int_0^\infty e^{-m^2 t} \text{ trace } K_t^{*\Gamma} \frac{dt}{t} = -2 \log Z_\Gamma (1 + \sqrt{1+m^2}), \qquad (3.13)$$

since we have noted that the right-hand side of (3.3) is the right-hand side of equation (3.5).

Now formula (3.13) is easy to prove since the theta function $\theta_\Gamma(t) = \text{trace } K_t^{*\Gamma}$ was computed in [3]; also compare [4; 15; 17]. Namely

$$\theta_\Gamma(t) = \frac{a}{\sqrt{4\pi t}} \sum_{n=1}^{\infty} \frac{e^{-(t+n^2 a^2/t)}}{\sinh^2(na) + \sin^2(nb)}. \qquad (3.14)$$

Also by formula 32. on page 1145 of [11]

$$\int_0^\infty t^{-3/2} e^{-A/4t} e^{-Bt}\, dt = 2\sqrt{\frac{\pi}{A}} e^{-(AB)^{1/2}} \qquad (3.15)$$

for $A > 0$, $B \geq 0$. Commutation of the integration in (3.13) with the summation in (3.14) is okay:

$$I(m) = \frac{a}{\sqrt{4\pi}} \sum_{n=1}^{\infty} \frac{1}{\sinh^2(na) + \sin^2(nb)} \int_0^\infty t^{-3/2} e^{-4n^2 a^2/(4t)} e^{-(1+m^2)t}\, dt$$

$$\stackrel{(3.15)}{=} \frac{a}{\sqrt{4\pi}} \sum_{n=1}^{\infty} \frac{1}{\sinh^2(na) + \sin^2(nb)} 2\sqrt{\frac{\pi}{4n^2 a^2}} e^{-[4n^2 a^2 (1+m^2)]^{1/2}}$$

$$= 2 \sum_{n=1}^{\infty} \frac{e^{-\sqrt{1+m^2}\, 2an}}{4n(\sinh^2(na) + \sin^2(nb))}$$

$$= -2 \log Z_\Gamma(1 + \sqrt{1+m^2}), \qquad (3.16)$$

as desired (again by (3.2)).

Given formula (3.13), we can go a step further and obtain in terms of $Z_\Gamma(s)$ the *one-loop partition function* denoted by $Z_{\text{scalar}}^{\text{1-loop}}(\tau, \overline{\tau})$ in [10]. By definition, it equals $(\det \Delta)^{-1/2}$, which we take to mean $(e^{-I(m)})^{-1/2}$. Thus, by (3.13),

$$Z_{\text{scalar}}^{\text{1-loop}}(\tau, \overline{\tau}) = \frac{1}{Z_\Gamma(1 + \sqrt{1+m^2})}. \qquad (3.17)$$

Let $q \stackrel{\text{def.}}{=} e^{2\pi i \tau} = e^{2bi - 2a}$, $\overline{q} = e^{-2bi - 2a}$, $h \stackrel{\text{def.}}{=} (1 + \sqrt{1+m^2})/2$, and note that for $0 \leq k_1, k_2 \in \mathbb{Z}$ one has

$$q^{k_1 + h}(\overline{q})^{k_2 + h} = e^{(2bi - 2a)(k_1 + h)} e^{(-2bi - 2a)(k_2 + h)}$$

$$= (e^{2bi})^{k_1} (e^{-2bi})^{k_2} e^{-(k_1 + k_2 + 2h) 2a}.$$

Therefore by definition (3.1) we can also write

$$\frac{1}{Z_\Gamma(1 + \sqrt{1+m^2})} = \prod_{0 \leq k_1, k_2 \in \mathbb{Z}} \frac{1}{1 - q^{k_1 + h}(\overline{q})^{k_2 + h}} \qquad (3.18)$$

where, as noted in [10], the right-hand side has the form trace $q^{L_0}\bar{q}^{\bar{L}_0}$ for Virasoro operators $L_0, \bar{L}_0$ that generate scale transformations (in the language of boundary conformal field theory).

The one-loop gravity partition function $Z_{\text{gravity}}^{\text{1-loop}}(\tau)$ is also computed in [10]. The result is

$$Z_{\text{gravity}}^{\text{1-loop}}(\tau) = \prod_{m=2}^{\infty} \frac{1}{|1-q^m|^2} = \prod_{m=2}^{\infty} \frac{1}{|1-e^{2mbi}e^{-2am}|^2}, \qquad (3.19)$$

from which one obtains the full gravity partition function

$$Z_{\text{gravity}}(\tau) = |q|^{-2k} Z_{\text{gravity}}^{\text{1-loop}}(\tau). \qquad (3.20)$$

for the Chern–Simon coupling constant $k = \sigma/16G$ (see (2.3)), G being the Newton constant; see [10; 14; 28]. We claim that $Z_{\text{gravity}}^{\text{1-loop}}(\tau)$ can also be expressed in terms of the zeta function $Z_\Gamma(s)$. We have a factorization

$$Z_{\text{gravity}}^{\text{1-loop}}(\tau) = Z_{\text{hol}}(\tau)\overline{Z}_{\text{hol}}(\tau) \qquad (3.21)$$

for

$$Z_{\text{hol}}(\tau) \stackrel{\text{def.}}{=} \prod_{m=2}^{\infty} \frac{1}{1-q^m} \qquad (3.22)$$

its holomorphic sector. Since $a > 0$, we have $|q^m| = e^{-2am} < 1$ for $m > 0$; hence $\log(1-q^m) = -\sum_{n=1}^{\infty} q^{mn}/n$. That is,

$$\log Z_{\text{hol}}(\tau) = -\sum_{m=2}^{\infty} \log(1-q^m) = \sum_{n=1}^{\infty} \frac{1}{n} \sum_{m=2}^{\infty} (q^n)^m$$

$$= \sum_{n=1}^{\infty} \frac{1}{n} \frac{q^{2n}}{(1-q^n)} \frac{1-\bar{q}^n}{1-\bar{q}^n} = \sum_{n=1}^{\infty} \frac{e^{4bni}e^{-4an}(1-e^{-2bni}e^{-2an})}{n|1-q^n|^2}$$

$$= \sum_{n=1}^{\infty} \frac{e^{4bni}e^{-4an} - e^{2bni}e^{-4an}e^{-2an}}{n|1-q^n|^2}. \qquad (3.23)$$

On the other hand, $\sin^2(bn) + \sinh^2(an) = \frac{1}{2}(\cosh(2an) - \cos(2bn))$ — this identity was used in (3.2) and will be used later in (4.8). Thus

$$\frac{|1-q^n|^2}{4|q|^n} = \frac{1-q^n-\bar{q}^n+|q|^{2n}}{4|q|^n} = \frac{1-e^{2bni}e^{-2an}-e^{-2bni}e^{-2an}+e^{-4an}}{4e^{-2an}}$$

$$= \tfrac{1}{4}(e^{2an} - 2\cos(2bn) + e^{-2an})$$

$$= \tfrac{1}{2}(\cosh(2an) - \cos(2bn)) = \sin^2(bn) + \sinh^2(an). \qquad (3.24)$$

That is, $\dfrac{1}{|1-q^n|^2} = \dfrac{1}{e^{-2an}4(\sinh^2(an)+\sin^2(bn))}$, which by (3.23) lets us write

$$\log Z_{\text{hol}}(\tau) = \sum_{n=1}^{\infty} \frac{e^{4bni}e^{-2an} - e^{2bni}e^{-4an}}{4n(\sinh^2(an)+\sin^2(bn))}$$

$$= \sum_{n=1}^{\infty} \frac{e^{-(-\frac{2b}{a}i+1)2an} - e^{-(-\frac{b}{a}i+2)2an}}{4n(\sinh^2(an)+\sin^2(bn))}$$

$$= -\log Z_\Gamma\left(2 - \frac{2b}{a}i\right) + \log Z_\Gamma\left(3 - \frac{b}{a}i\right) \quad (3.25)$$

by definition (3.2). By definition (3.1), $\overline{Z_\Gamma(s)} = Z_\Gamma(\bar{s})$. By equation (3.25) we have therefore established:

THEOREM 3.26. *For* $\tau = \dfrac{b}{\pi} + \dfrac{ai}{\pi}$, $Z_{\text{hol}}(\tau) = Z_\Gamma\left(3-\dfrac{b}{a}i\right) \Big/ Z_\Gamma\left(2-\dfrac{2b}{a}i\right)$. *In particular, by* (3.21),

$$Z_{\text{gravity}}^{\text{1-loop}}(\tau) = \frac{Z_\Gamma\left(3-\dfrac{b}{a}i\right) Z_\Gamma\left(3+\dfrac{b}{a}i\right)}{Z_\Gamma\left(2-\dfrac{2b}{a}i\right) Z_\Gamma\left(2+\dfrac{2b}{a}i\right)}, \quad (3.27)$$

and thus by equation (3.20), $Z_{\text{gravity}}(\tau)$ also has the explicit expression e^{4ak} (the right-hand side of equation (3.27)) *in terms of the Patterson–Selberg zeta function* $Z_\Gamma(s)$.

4. Determinant formula in the presence of conical singularities

We will now extend the one-loop determinant formula (3.13) to the BTZ black holes $B_{\Gamma^{(\alpha)}}$ with *conical singularities*. Here we fix $0 < \alpha \le 1$ and for

$$\gamma_\alpha \stackrel{\text{def.}}{=} \begin{bmatrix} e^{i\pi\alpha} & 0 \\ 0 & e^{-i\pi\alpha} \end{bmatrix}$$

we define $\Gamma^{(\alpha)}$ to be the subgroup of $\text{SL}(2,\mathbb{C})$ generated by γ and γ_α, for γ as in (2.6):

$$\Gamma^{(\alpha)} \stackrel{\text{def.}}{=} \{\gamma^n \gamma_\alpha^m \mid n, m \in \mathbb{Z}\}. \quad (4.1)$$

$\Gamma^{(\alpha)}$ acts on $\mathbb{H}^3$ by $(\gamma^n \gamma_\alpha^m) \cdot (x, y, z) = (x', y', z')$, where

$$\begin{aligned} x' &= e^{2an}(x\cos 2(bn+\pi\alpha m) - y\sin 2(bn+\pi\alpha m)), \\ y' &= e^{2an}(x\sin 2(bn+\pi\alpha m) + y\cos 2(bn+\pi\alpha m)), \\ z' &= e^{2an}z. \end{aligned} \quad (4.2)$$

This action, like that defined in equation (2.7), is the restriction of the standard action of $SL(2, \mathbb{C})$ on $\mathbb{H}^3$. We take

$$B_{\Gamma^{(\alpha)}} \stackrel{\text{def.}}{=} \Gamma^{(\alpha)} \backslash \mathbb{H}^3.$$

If $\alpha = 1$, then $\Gamma^{(\alpha)} = \Gamma$ and $B_{\Gamma^{(\alpha)}} = B_\Gamma$. However, in general, $B_{\Gamma^{(\alpha)}}$ is *not* a smooth manifold since the action of $\Gamma^{(\alpha)}$ is not free. For example each point $(0, 0, z)$, $z > 0$, on the positive z-axis is a fixed point of γ_α^m, by definition (4.2).

To understand the topology of $B_{\Gamma^{(\alpha)}}$ a little better, consider the action of $\mathbb{Z}$ on $\mathbb{R}^2$ given by

$$m \cdot \begin{bmatrix} x \\ y \end{bmatrix} \stackrel{\text{def.}}{=} \begin{bmatrix} x\cos(2\pi m\alpha) - y\sin(2\pi m\alpha) \\ x\sin(2\pi m\alpha) + y\cos(2\pi m\alpha) \end{bmatrix}$$

$$= \begin{bmatrix} \cos(2\pi m\alpha) & -\sin(2\pi m\alpha) \\ \sin(2\pi m\alpha) & \cos(2\pi m\alpha) \end{bmatrix} \begin{bmatrix} x \\ y \end{bmatrix} \quad (4.3)$$

for $m \in \mathbb{Z}$ and $x, y \in \mathbb{R}$.

Thus the action is a rotation, the angle of rotation being $2\pi m\alpha$. Let $(\mathbb{Z}\backslash\mathbb{R}^2)^{(\alpha)}$ denote the corresponding quotient space, and let $S^1 = \{z \in \mathbb{C} \mid |z| = 1\}$ denote the unit circle. In [24] we construct a well-defined surjective homeomorphism $\psi_\alpha : B_{\Gamma^{(\alpha)}} \to (\mathbb{Z}\backslash\mathbb{R}^2)^{(\alpha)} \times S^1$. In fact, given $(x, y, z) \in \mathbb{H}^3$ define

$$r = r(x, y, z) \stackrel{\text{def.}}{=} \frac{\pi}{a} \log z \quad (\text{since } z > 0),$$

$$u = u(x, y, z) \stackrel{\text{def.}}{=} \frac{x}{z}\cos\frac{rb}{\pi} + \frac{y}{z}\sin\frac{rb}{\pi}, \quad (4.4)$$

$$v = v(x, y, z) \stackrel{\text{def.}}{=} -\frac{x}{z}\sin\frac{rb}{\pi} + \frac{y}{z}\cos\frac{rb}{\pi}.$$

If $\widetilde{(x, y, z)} \in B_{\Gamma^{(\alpha)}}$ denotes the $\Gamma^{(\alpha)}$-orbit of $(x, y, z) \in \mathbb{H}^3$, and $\widetilde{(u, v)} \in (\mathbb{Z}\backslash\mathbb{R}^2)^{(\alpha)}$ denotes the $\mathbb{Z}$-orbit of $(u, v) \in \mathbb{R}^2$, then

$$\psi_\alpha(\widetilde{(x, y, z)}) \stackrel{\text{def.}}{=} (\widetilde{(u, v)}, e^{ir} = e^{i\frac{\pi}{a}\log z}). \quad (4.5)$$

Similarly, the inverse function $\psi_\alpha^{-1} : (\mathbb{Z}\backslash\mathbb{R}^2)^{(\alpha)} \times S^1 \to B_{\Gamma^{(\alpha)}}$ is explicated in [24]. If $\alpha = 1/l$ with $2 \leq l \in \mathbb{Z}$, for example, then one computes that a fundamental domain for the $\mathbb{Z}$ action in (4.3) is given by a cone in $\mathbb{R}^2$ with vertex at $(0, 0)$, and with opening angle $2\pi/l = 2\pi\alpha$. Given that the black holes $B_{\Gamma^{(\alpha)}}$ have the topology $(\mathbb{Z}\backslash\mathbb{R}^2)^{(\alpha)} \times S^1$, as just indicated, we see that they have conical singularities. In particular B_Γ has the topology $\mathbb{R}^2 \times S^1$, as is well-known.

The family $\{B_{\Gamma^{(\alpha)}}\}_{0<\alpha\leq 1}$ of topological spaces is a "deformation" of B_Γ in the sense that $B_{\Gamma^{(1)}} = B_\Gamma$, as we have noted. Similarly, as indicated in the introductory remarks, we have constructed a family $\{Z_{\Gamma^{(\alpha)}}\}_{0<\alpha\leq 1}$ of zeta functions such that $Z_{\Gamma^{(1)}} = Z_\Gamma$ in (3.1). We review this construction as it is the key to the extension of formula (3.13). For convenience we will also write $Z(s; \alpha)$ for $Z_{\Gamma^{(\alpha)}}(s)$.

$$Z_{\Gamma^{(\alpha)}}(s) = Z(s; \alpha) \overset{\text{def.}}{=} \tag{4.6}$$

$$\prod_{0 \le k_1, k_2 \in \mathbb{Z}} \left(1 - (e^{2bi/\alpha})^{k_1}(e^{-2bi/\alpha})^{k_2} e^{-(k_1+k_2+\alpha s)2a/\alpha}\right)$$

$$\times \prod_{0 \le k_1, k_2, k_3 \in \mathbb{Z}} \frac{1 - e^{-2(k_3+1)2a}(e^{2bi/\alpha})^{k_1}(e^{-2bi/\alpha})^{k_2} e^{-(k_1+k_2+\alpha s)2a/\alpha}}{1 - e^{-2(k_3+1/\alpha)2a}(e^{2bi/\alpha})^{k_1}(e^{-2bi/\alpha})^{k_2} e^{-(k_1+k_2+\alpha s)2a/\alpha}}.$$

For $\alpha = 1$, the product over $k_1, k_2, k_3 \ge 0$ here is 1. So clearly $Z_{\Gamma^{(1)}} = Z_\Gamma(s)$ by definition (3.1). On the other hand, it is proved in [27] that we can also write

$$Z(s; \alpha) = \prod_{0 \le k_1, k_2 \in \mathbb{Z}} \left(1 - e^{(ib-a)2k_1/\alpha} e^{-4ak_2} e^{-2as}\right) \cdot$$

$$\prod_{0 \le k_1, k_2 \in \mathbb{Z}} \left(1 - e^{-(ib+a)(2/\alpha)(k_1+1)} e^{-4ak_2} e^{-2as}\right), \tag{4.7}$$

which shows that $Z(s; \alpha)$ is an entire function. A third expression for $Z(s; \alpha)$ is also proved in [27]. The definition (4.6) might appear to be a bit opaque, but there are physical motivations for it. Namely, the author's interest was to find a zeta function meaning of results in [15]. By deforming $Z_\Gamma(s)$, we wished to construct a statistical mechanics type function $\log Z(s; \alpha)$ such that the evaluation $(\alpha \, \partial/\partial \alpha - 1) \log Z(s; \alpha)\big|_{\alpha=1}$, for a special value of s, would capture the quantum correction of R. Mann and S. Solodukhin to BTZ black hole entropy; see [23; 25; 26]. From definition (4.6) one can obtain for Re $s > 0$ the equality $Z(s; \alpha) = e^{\log Z(s;\alpha)}$, where

$$\log Z(s; \alpha) \overset{\text{def.}}{=} -\sum_{n=1}^\infty \frac{\sinh(2an/\alpha) \, e^{-(s-1)2an}}{4n \sinh(2an) \left(\sinh^2(an/\alpha) + \sin^2(bn/\alpha)\right)}$$

$$= -\sum_{n=1}^\infty \frac{\sinh(2an/\alpha) \, e^{-(s-1)2an}}{2n \sinh(2an) \left(\cosh(2an/\alpha) - \cos(2bn/\alpha)\right)}. \tag{4.8}$$

Of course the formulas in (4.8) reduce to those in (3.2) in case $\alpha = 1$.

A final ingredient needed is an extension of the trace formula (3.14). Fortunately, this is available from [27] in case $\alpha = 1/l$ (again) for $1 \le l \in \mathbb{Z}$, which we therefore assume. By averaging the heat kernel $K_t^{*\Gamma}(\tilde{p}_1, \tilde{p}_2)$ in (3.6) over the finite group $l\mathbb{Z}\backslash\mathbb{Z}$ we obtain the truncated heat kernel (for $t > 0$)

$$K_t^{*\Gamma^{(\alpha)}}(\tilde{p}_1, \tilde{p}_2) \overset{\text{def.}}{=} \sum_{m=0}^{l-1} K_t^{*\Gamma}(\tilde{p}_1, \widetilde{\gamma_\alpha^m \cdot p_2}), \tag{4.9}$$

which equals

$$\frac{1}{(4\pi t)^{3/2}} \sum_{m=0}^{l-1} \sum_{n \in \mathbb{Z}-\{0\}} e^{-t-d(p_1,(\gamma^n\gamma_\alpha^m)\cdot p_2)^2/4t} \frac{d(p_1,(\gamma^n\gamma_\alpha^m)\cdot p_2)}{\sinh d(p_1,(\gamma^n\gamma_\alpha^m)\cdot p_2)}$$

for $B_{\Gamma^{(\alpha)}}$, from which we can define (compare definition (3.9)) the theta function (for $t > 0$)

$$\theta_{\Gamma^{(\alpha)}}(t) = \text{trace } K_t^{*\Gamma^{(\alpha)}} \stackrel{\text{def.}}{=} \iiint_{F^{(\alpha)}} K_t^{*(\alpha)}(\tilde{p},\tilde{p}) \, dv(p), \qquad (4.10)$$

where $F^{(\alpha)} \subset \mathbb{H}^3$ is defined in terms of spherical coordinates $x = \rho \sin\phi \cos\theta$, $y = \rho \sin\phi \sin\theta$, $z = \rho \cos\phi$, with $\rho \geq 0$, $0 \leq \theta < 2\pi$, $0 \leq \phi < \pi/2$:

$$F^{(\alpha)} \stackrel{\text{def.}}{=} \{(x,y,z) \in \mathbb{H}^3 \mid \begin{matrix} 1 < \rho < e^{2a} \\ 2\pi(1-\alpha) \leq \theta \leq 2\pi \end{matrix}\}. \qquad (4.11)$$

Thus $F^{(\alpha)}$ is the upper hemispherical region in $\mathbb{R}^3$ between the spheres of radii 1 and e^{2a}, but with θ at least $2\pi(1-\alpha)$, called the *defect angle*. If we choose $\theta_1 = 2\pi(1-\alpha)$ and $\theta_2 = 2\pi$ in formula (4.8) of [27] we obtain, for $t > 0$:

THEOREM 4.12. *For $\alpha = 1/l$, with $1 \leq l \in \mathbb{Z}$, one has*

$$\theta_{\Gamma^{(\alpha)}}(t) = \frac{\alpha a}{2\sqrt{4\pi t}} \sum_{\substack{n \in \mathbb{Z}-\{0\} \\ m \in \mathbb{Z} \\ 0 \leq m \leq l-1}} \frac{e^{-t-a^2n^2/t}}{\sinh^2(an) + \sin^2(bn+\pi\alpha m)}.$$

This theorem generalizes the trace formula (3.14). Similarly the following theorem generalizes the one-loop determinant formula (3.13):

THEOREM 4.13. *For $\alpha = 1/l$, with $1 \leq l \in \mathbb{Z}$, one has*

$$\int_0^\infty e^{-m^2 t} \text{ trace } K_t^{*\Gamma^{(\alpha)}} \frac{dt}{t} = -2 \log Z_{\Gamma^{(\alpha)}}(1 + \sqrt{1+m^2}). \qquad (4.14)$$

PROOF. We follow the argument above in the proof of (3.13), given Theorem 4.12.

$$\int_0^\infty e^{-m^2 t} \text{ trace } K_t^{*\Gamma^{(\alpha)}} \frac{dt}{t} =$$

$$\frac{\alpha a}{2\sqrt{4\pi}} \sum_{\substack{n \neq 0 \\ 0 \leq m \leq l-1}} \frac{1}{\sinh^2(an) + \sin^2(bn+\pi\alpha m)} \int_0^\infty t^{-3/2} e^{-4n^2 a^2/4t} e^{-(1+m^2)t} \, dt$$

$$\stackrel{(3.15)}{=} \frac{\alpha a}{2\sqrt{4\pi}} \sum_{\substack{n \neq 0 \\ 0 \leq m \leq l-1}} \frac{1}{\sinh^2(an) + \sin^2(bn+\pi\alpha m)} \cdot 2\sqrt{\frac{\pi}{4n^2 a^2}} e^{-\sqrt{4n^2 a^2(1+m^2)}}$$

$$= \frac{\alpha}{2} \sum_{\substack{n \neq 0 \\ 0 \leq m \leq l-1}} \frac{e^{-2|n|a\sqrt{1+m^2}}}{|n|\, 2\left(\sinh^2(an) + \sin^2(bn + \pi\alpha m)\right)}$$

$$= \frac{\alpha}{2} \sum_{\substack{n \neq 0 \\ 0 \leq m \leq l-1}} \frac{e^{-2|n|a\sqrt{1+m^2}}}{|n|\left(\cosh(2an) - \cos 2(bn + \pi\alpha m)\right)}. \quad (4.15)$$

To proceed further we employ the identity[1]

$$\sum_{m=0}^{l-1} \frac{1}{\cosh u - \cos(v - 2\pi m/l)} = \frac{l \sinh(lu)}{\sinh u \left(\cosh(lu) - \cos(lv)\right)}. \quad (4.16)$$

We apply it with $u \overset{\text{def.}}{=} 2an$ and $v \overset{\text{def.}}{=} -2bn$ to rewrite the last sum in (4.15) as

$$\frac{\alpha}{2} \sum_{n \in \mathbb{Z} - \{0\}} \frac{e^{-2|n|a\sqrt{1+m^2}} l \sinh(l2an)}{|n| \sinh(2an) \left(\cosh(l2an) - \cos(l2bn)\right)}$$

$$= 2 \sum_{n=1}^{\infty} \frac{\sinh(2an/\alpha)\, e^{-2an\sqrt{1+m^2}}}{2n \sinh(2an)\left(\cosh(2an/\alpha) - \cos(2bn/\alpha)\right)} \quad (4.17)$$

$$= -2 \log Z_{\Gamma(\alpha)}(1 + \sqrt{1+m^2}),$$

by definition (4.8), which concludes the proof of Theorem 4.13. □

In the effective action formula (4.14) we have assumed that $\alpha^{-1} \in \mathbb{Z}$. This assumption can be removed and thus a more general formula can be presented if we appeal to an old contour integral formula that goes back to A. Sommerfeld in 1897, in his amazing diffraction studies. The reader can consult the references [13; 15], for example, on this point — references which of course do not employ the zeta function $Z_{\Gamma(\alpha)}(s)$.

5. Extremal partition functions of conformal field theories with central charge $24k$

In this section we consider the modular invariant partition function $Z_k(\tau)$ of a holomorphic conformal field theory (CFT) with central charge $c = 24k$, $k = 1, 2, 3, 4, \ldots$. Such a theory was introduced by G. Höhn [12], and is called an *extremal* CFT (ECFT) — which according to a bold proposal of E. Witten [28] is the dual to 3-dimensional pure gravity with a negative cosmological constant;

---

[1] Formula (4.16) corrects a misprint in [27]. Namely, the expression $\sin(lu)$ in formula (4.10) of [27] should read $\sinh(lu)$. Also in equations (4.5) and (4.6) of [27] the often occurring expression $\gamma_m^{(\alpha)}$ should read γ_α^m.

also compare [14]. Apart from the case $k = 1$, however, there is uncertainty regarding the existence of ECFT's. I. Frenkel, J. Lepowsky, and A. Meurman (FLM) [8] have indeed constructed a holomorphic CFT with central charge $c = 24$ (i.e., with $k = 1$) and with $Z_1(\tau) = j(\tau) - 744$, where $j(\tau)$ is the classical elliptic modular invariant (see (5.6) below).

An important point regarding the FLM construction is *monster symmetry*: the states of the theory transform as a representation of the finite, simple, Fischer–Griess group M, of order $|M| = 2^{46} \cdot 3^{20} \cdot 5^9 \cdot 7^6 \cdot 11^2 \cdot 13^2 \cdot 17 \cdot 19 \cdot 23 \cdot 29 \cdot 31 \cdot 41 \cdot 47 \cdot 59 \cdot 71 \approx 10^{54}$, called the *monster* (or the *friendly giant*). However for $k = 2$, D. Gaiotto [9] has slain the "two-headed monster": there exists no self-dual ECFT for $c = 48$ with monster symmetry.

We begin by indicating how $Z_k(\tau)$ (defined for $\mathrm{Im}\,\tau > 0$) can be explicitly constructed from the FLM $Z_1(\tau)$ and the one-loop partition function $Z_{\mathrm{hol}}(\tau)$ of definition (3.22), with help of Hecke operators.

Fix $k = 1, 2, 3, 4, \ldots$, and for $\tau \in \mathbb{C}$ with $\mathrm{Im}\,\tau > 0$ set $q = q(\tau) \stackrel{\mathrm{def.}}{=} e^{2\pi i \tau}$, so $|q| < 1$. Define

$$Z_0(\tau) \stackrel{\mathrm{def.}}{=} q^{-k} Z_{\mathrm{hol}}(\tau) \stackrel{\mathrm{def.}}{=} q^{-k} \prod_{n=2}^{\infty} \frac{1}{1-q^n}; \qquad (5.1)$$

compare definition (3.22). The full gravity partition function of definition (3.20) therefore admits the factorization

$$Z_{\mathrm{gravity}}(\tau) = Z_0(\tau) \overline{Z}_0(\tau), \qquad (5.2)$$

by equation (3.21).

Let p denote the partition function on $\mathbb{Z}^+$; that is, $p(n)$ is the number of ways of writing a positive integer n as a sum of positive integers, without regard to order. Euler's formula (equation (9.10) of my introductory lectures, page 72) says that

$$\frac{1}{\prod_{n=1}^{\infty}(1-z^n)} = \sum_{n=0}^{\infty} p(n) z^n \qquad (5.3)$$

for $|z| < 1$, where $p(0) \stackrel{\mathrm{def.}}{=} 1$. Therefore we can write

$$Z_0(\tau) = q^{-k}(1-q) \prod_{n=1}^{\infty} \frac{1}{1-q^n} = q^{-k}(1-q) \sum_{n=0}^{\infty} p(n) q^n$$
$$= \sum_{n=0}^{\infty} p(n) q^{n-k} - \sum_{n=0}^{\infty} p(n) q^{n+1-k}.$$

Collecting coefficients here we see that

$$Z_0(\tau) = \sum_{r=-k}^{\infty} a_r(k) q^r \qquad (5.4)$$

for
$$a_r(k) \stackrel{\text{def.}}{=} p(r+k) - p(r+k-1), \quad r \geq -k, \tag{5.5}$$
where we set $p(-1) \stackrel{\text{def.}}{=} 0$.

As mentioned in my introductory lectures (equations (4.44) and (4.45) on page 42), the modular j-invariant has a q-expansion with all Fourier coefficients $c_n \in \mathbb{Z}$. That is, defining

$$j(\tau) \stackrel{\text{def.}}{=} \frac{1728(60 G_4(\tau))^3}{(60 G_4(\tau))^3 - 27(140 G_6(\tau))^2}, \tag{5.6}$$

we have

$$j(\tau) = \frac{1}{q} + \sum_{n=0}^{\infty} c_n q^n \quad \text{with}$$

$$c_0 = 744,$$
$$c_1 = 196{,}884,$$
$$c_2 = 21{,}493{,}760, \tag{5.7}$$
$$c_3 = 864{,}299{,}970,$$
$$c_4 = 20{,}245{,}856{,}256,$$
$$c_5 = 333{,}202{,}640{,}600,$$
$$c_6 = 4{,}252{,}023{,}300{,}096.$$

The denominator in (5.6) is the Dedekind–Klein discriminant form $\Delta(\tau)$, and $G_l(z)$ is the holomorphic Eisenstein series of weight l given in definition (4.4) of the introductory lectures (page 31). *Here we depart from convention in using $J(\tau)$ not in the classical sense but to denote the function $j(\tau) - c_0$:*

$$J(\tau) \stackrel{\text{def.}}{=} j(\tau) - 744 = \frac{1}{q} + 196{,}884\, q + 21{,}493{,}760\, q^2 +$$
$$864{,}299{,}970\, q^3 + 20{,}245{,}856{,}256\, q^4 + \cdots . \tag{5.8}$$

Now recall the n-th Hecke operator $T(n)$ of weight w acting on a function $f(\tau)$, $\operatorname{Im} \tau > 0$, where $n, w \in \mathbb{Z}$, $n \geq 1$, $w \geq 0$. As seen in (3.22) of the introductory lectures (page 28), it is given by

$$(T(n)f)(\tau) \stackrel{\text{def.}}{=} n^{w-1} \sum_{\substack{d>0 \\ d\mid n}} \sum_{a=0}^{d-1} d^{-w} f\left(\frac{n\tau + da}{d^2}\right). \tag{5.9}$$

In particular

$$n(T(n)f)(\tau) \stackrel{\text{def.}}{=} \sum_{\substack{d>0 \\ d\mid n}} \sum_{a=0}^{d-1} f\left(\frac{n\tau + da}{d^2}\right) \tag{5.10}$$

for $w = 0$, which is the only case we will need, since $J(\tau)$ in (5.8) has weight zero. Of course $T(1)f = f$ for any weight w. We can now define the main object, where we have fixed an integer $k > 0$:

$$Z_k(\tau) \stackrel{\text{def.}}{=} a_0(k) + \sum_{r=1}^{k} a_{-r}(k) r(T(r)J)(\tau)$$

$$= p(k) - p(k-1) + \sum_{r=1}^{k} (p(k-r) - p(k-r-1)) r(T(r)J)(\tau) \quad (5.11)$$

by definition (5.5), where $r(T(r)J)(\tau)$ is given by equation (5.10) applied to $f = J$. Since $p(0) = p(1) = 1$, $p(-1) = 0$, and $T(1)f = f$ we see that $Z_1(\tau) \equiv J(\tau)$, which (as remarked on earlier) is the partition function of the FLM holomorphic CFT of central charge $c = 24$, with monster symmetry. Frenkel, Lepowsky, and Meurman also conjecture that this ECFT is unique — a result that remains unproved at the present time.

To be a bit more precise, these authors construct a graded, infinite-dimensional M-module $V^\natural = V_0 \oplus V_1 \oplus V_2 \oplus V_3 \oplus V_4 \oplus \cdots$ (the *moonshine module*), where V_0 is the trivial representation π of M, $V_1 = \{0\}$, $V_2 = \pi_1 \oplus \pi_{196,833}$, $V_3 = \pi_1 \oplus \pi_{196,833} \oplus \pi_{21,296,876}$, and so on; π_d is the irreducible representation of M of degree d, for $d \geq 1$. A remarkable observation, first made by John McKay in 1978, is that the early Fourier coefficients c_n in (5.7) are integral linear combinations of the degrees d; thus $c_1 = 196,884 = 1 + 196,883$, $c_2 = 21,493,760 = 1 + 196,883 + 21,296,876$, and

$$c_3 = 864,299,970 = 2 \times 1 + 2 \times (196,883) + 21,296,876 + 842,609,326.$$

$V^\natural$ has the structure, in fact, of a *vertex operator algebra* (VOA), a subject thoroughly discussed by G. Mason and M. Tuite in their lectures in this book. The submodule V_2 is actually an algebra (which is commutative but not associative), the *Griess algebra*, which has the monster M as its full symmetry group (i.e., as its automorphism group).

By equations (5.7), (5.8) we have the Fourier expansion $J(\tau) = \sum_{n \geq -1} c_n q^n$, where $c_{-1} = 1, c_0 = 0$. Accordingly, $(T(n)J)(\tau)$ has Fourier expansion

$$(T(n)J)(\tau) = \sum_{m \geq -n} c_m^{(n)} q^m,$$

where

$$c_m^{(n)} = \sum_{\substack{d > 0 \\ d \mid n, d \mid m}} \frac{c_{mn/d^2}}{d}, \ m \geq 1; \quad c_0^{(n)} = c_0 \sum_{\substack{d > 0 \\ d \mid n}} \frac{1}{d} \stackrel{.}{=} 0 = c_{-m}^{(n)}, \ 1 \leq m < n;$$

$$c_{-n}^{(n)} = \frac{1}{n}. \quad (5.12)$$

That is,
$$(T(n)J)(\tau) = \frac{q^{-n}}{n} + \sum_{m=1}^{\infty} c_m^{(n)} q^m \qquad (5.13)$$

which we use in definition (5.11):
$$Z_k(\tau) = a_0(k) + \sum_{r=1}^{k} a_{-r}(k)\left(q^{-r} + r\sum_{n=1}^{\infty} c_n^{(r)} q^n\right).$$

Collecting coefficients here, we see that
$$Z_k(\tau) = a_{-k}(k)q^{-k} + \cdots + a_{-2}(k)q^{-2} + a_{-1}(k)q^{-1}$$
$$+ a_0(k) + \sum_{n=1}^{\infty} b_{k,n} q^n, \qquad (5.14)$$

where
$$a_{-k}(k) = 1, \quad b_{k,n} \stackrel{\text{def.}}{=} \sum_{r=1}^{k} r a_{-r}(k) c_n^{(r)}, \quad n \geq 1, \qquad (5.15)$$

with $a_{-r}(k) = p(k-r) - p(k-r-1)$ for $1 \leq r \leq k$ given by (5.5), and $c_n^{(r)} = \sum_d c_{rn/d^2}/d$ given by (5.12).

Before commenting on the important physical significance of the coefficients $b_{k,n}$ in (5.15), we state the following result:

THEOREM 5.16. *For* $k, n \in \mathbb{Z}, k, n \geq 1$, *let*
$$b_{k,n}^{\infty} \stackrel{\text{def.}}{=} k e^{4\pi\sqrt{kn}}/\sqrt{2}(kn)^{3/4}. \qquad (5.17)$$

Then $b_{k,n}$ *equals*
$$b_{k,n}^{\infty}\left(1 - \frac{3}{32\pi\sqrt{kn}} + \varepsilon_{kn} + T(k,n)\right.$$
$$\left.+ \frac{1}{k^{1/4}} \sum_{r=1}^{k-1} \frac{r^{1/4} a_{-r}(k)}{e^{4\pi\sqrt{n}(\sqrt{k}-\sqrt{r})}}\left(1 - \frac{3}{32\sqrt{rn}} + \varepsilon_{rn} + T(r,n)\right)\right), \qquad (5.18)$$

where $|\varepsilon_m| \leq .055/m$ *for integer* $m \geq 1$, *and* $0 \leq T(r,n)$ *is bounded above by both* $r^{3/2}\zeta(\frac{3}{2})/(2e^{2\pi\sqrt{rn}})$ *and* $n^{3/2}\zeta(\frac{3}{2})/(2e^{2\pi\sqrt{rn}})$ *for* $1 \leq r \leq k$, *where* $\zeta(s)$ *is the Riemann zeta function.*

In setting up the proof of Theorem 5.16, the author relied heavily on the following result of N. Brisebarre and G. Philibert [2] (as mentioned in equation (9.32) on page 76 of my introductory lectures): For $m \geq 1$
$$c_m = \frac{e^{4\pi\sqrt{m}}}{\sqrt{2}m^{3/4}}\left(1 - \frac{3}{32\pi\sqrt{m}} + \varepsilon_m\right), \qquad (5.19)$$

where again $|\varepsilon_m| \leq .055/m$. Equation (5.19) immediately implies the weaker asymptotic result (see equation (9.31) on page 76)

$$c_m \sim e^{4\pi\sqrt{m}}/\sqrt{2}m^{3/4} \text{ as } m \to \infty, \tag{5.20}$$

due to H. Petersson in 1932 and H. Rademacher in 1938, who was unaware of Petersson's proof. Similarly, Theorem 5.16 immediately implies the weaker asymptotic result

$$b_{k,n} \sim b_{k,n}^\infty \stackrel{\text{def.}}{=} k e^{4\pi\sqrt{kn}}/\sqrt{2}(kn)^{3/4} \text{ as } n \to \infty \tag{5.21}$$

for every fixed k, as observed by E. Witten in Section 3 of [28]. Actually Witten assumes that k is large with n/k fixed, but we see that this assumption is unnecessary for the statement (5.21).

Now from $\log b_{k,n}$, say for n sufficiently large, one obtains both the classical, holomorphic sector Bekenstein–Hawking black hole entropy $S_{\text{hol}} = 4\pi\sqrt{kn}$ (the leading asymptotic term) and corrections (subleading asymptotic terms) to that entropy:

$$\log b_{k,n}^\infty = 4\pi\sqrt{kn} + \left(\tfrac{1}{4}\log k - \tfrac{3}{4}\log n - \tfrac{1}{2}\log 2\right), \tag{5.22}$$

by (5.21).

We offer further explanation regarding equation (5.22). In particular we explain why the leading term $S_{\text{hol}} = 4\pi\sqrt{kn}$ there was referred to as the holomorphic sector entropy. In formulas (2.3), (2.4), the outer and inner black hole radii for the BTZ metric in *Euclidean* form (2.1) are given by

$$r_\pm^2 = \frac{M\sigma^2}{2}\left(1 \pm \sqrt{1 + \left(\frac{J}{M\sigma}\right)^2}\right).$$

For convenience, we also consider the *Lorentzian form* of the metric

$$ds_L^2 = \left(-N_1(r)^2 + r^2 N_2(r)^2\right)dt^2 + N_1(r)^{-2}dr^2 \\ + 2r^2 N_2(r)\,d\phi\,dt + r^2\,d\phi^2, \tag{5.23}$$

where now

$$N_1(r) = \left(-8GM + \frac{r^2}{\sigma^2} + \frac{16G^2}{r^2}J^2\right)^{1/2}, \tag{5.24}$$

with the gravitational constant G also included for generality. We omit the definition of $N_2(r)$, which will not be needed. The corresponding radii, which we again denote by $r_\pm$, are (by definition) solutions of the quartic equation $N_1(r) = 0$:

$$r_\pm^2 = 4GM\sigma^2\left(1 \pm \sqrt{1 - \left(\frac{J}{M\sigma}\right)^2}\right). \tag{5.25}$$

Here we assume (so ds_L^2 can have a black hole structure) that $1-(J/\sigma M)^2 \geq 0$, which is to say $|J| \leq \sigma M$; then $r_\pm \geq 0$. The key point here is that the classical Bekenstein–Hawking entropy, given by

$$S_{BH} = \frac{\pi r^+}{2G}, \tag{5.26}$$

is also given by the formula of J. Cardy [6] (see equation (9.3) on page 70)

$$S_{BH} = 2\pi\sqrt{\frac{cL_0}{6}} + 2\pi\sqrt{\frac{c\bar{L}_0}{6}}, \tag{5.27}$$

where L_0 and $\bar{L}_0$ are eigenvalues of the holomorphic and antiholomorphic Virasoro generators, respectively, and where the central charge c equals $3\sigma/2G$. To check equation (5.27) we use the equalities

$$L_0 = (\sigma M + J)/2, \qquad \bar{L}_0 = (\sigma M - J)/2. \tag{5.28}$$

By definition (5.25) we can write $r_+^2 + r_-^2 = 8GM\sigma^2$ and $r_+ r_- = 4G\sigma J$; therefore $(r_+ \pm r_-)^2 = 8GM\sigma^2 \pm 8G\sigma J = 16G\sigma(\sigma M \pm J)/2$, which yields

$$L_0 = (r_+ + r_-)^2/16G\sigma, \qquad \bar{L}_0 = (r_+ - r_-)^2/16G\sigma. \tag{5.29}$$

By (5.26), the right-hand side of equation (5.27) is then

$$2\pi\sqrt{\frac{c}{6}}\left(\frac{r_+ + r_- + r_+ - r_-}{4\sqrt{G\sigma}}\right) = \pi r^+/2G = S_{BH}$$

for $c = 3\sigma/2G$, which verifies equation (5.27).

Recall the Chern–Simon coupling constant $k = \sigma/16G$ following equation (3.20). Since $c = 3\sigma/2G$, we see that $c = 24k$. Thus our ongoing assumption $c \in 24\mathbb{Z}^+$ amounts now to the "quantization" of k; that is, $k = \sigma/16G$ is a positive integer. If we call the first term $2\pi\sqrt{cL_0/6}$ in equation (5.27) the *holomorphic sector entropy* S_{hol} (for obvious reasons), then for $c = 24k$ we have $S_{hol} = 2\pi\sqrt{24kL_0/6} = 4\pi\sqrt{kL_0}$, which is the leading asymptotic term in equation (5.22), where n there is identified with the Virasoro eigenvalue L_0. This is a justification for referring to that leading term as holomorphic sector entropy.

Appendix to Section 5: Computation of $Z_k(\tau)$ for $k = 2, 3$

The explicit formulas (5.14) and (5.15) are sufficient for the direct computation of the initial terms of $Z_k(\tau)$, say for small values of k. One could employ a computer program to deal with larger values of k. For example, take $k = 2$. Then, by (5.5) and (5.15),

$$a_{-1}(2) = p(2-1) - p(2-2) = 1 - 1 = 0,$$
$$a_{-2}(2) = 1,$$
$$a_0(2) = p(2) - p(1) = 1.$$

Also (5.15) gives, for $n \geq 1$,

$$b_{2,n} = \sum_{r=1}^{2} r a_{-r}(2) \sum_{\substack{d>0 \\ d|r, d|n}} \frac{1}{d} c_{rn/d^2} \doteq 2 \sum_{\substack{d>0 \\ d|2, d|n}} \frac{1}{d} c_{2n/d^2} = \begin{cases} 2c_{2n} & \text{if } 2 \nmid n, \\ 2c_{2n} + c_{n/2} & \text{if } 2 \mid n, \end{cases}$$

leading to

$$b_{2,1} = 2c_2 = 42{,}987{,}520,$$
$$b_{2,2} = 2c_4 + c_1 = 40{,}491{,}909{,}396,$$
$$b_{2,3} = 2c_6 = 8{,}540{,}046{,}600{,}192,$$

by (5.7). Therefore by (5.14)

$$Z_2(\tau) =$$
$$q^{-2} + 1 + 42{,}987{,}520q + 40{,}491{,}909{,}396q^2 + 8{,}504{,}046{,}600{,}192q^3 + \cdots.$$

Of course,

$$Z_1(\tau) \stackrel{\text{def.}}{=} J(\tau) \stackrel{\text{def.}}{=} j(\tau) - 744 =$$
$$q^{-1} + 196{,}884q + 21{,}493{,}760q^2 + 864{,}299{,}970q^3 + 20{,}245{,}856{,}256q^4 + \cdots,$$

by equation (5.8).

Similarly for $k = 3$ we have $a_0(3) = 1$, $a_{-1}(3) = 1$, $a_{-2}(3) = 0$, $a_{-3}(3) = 1$, so that

$$b_{3,n} = \sum_{r=1}^{3} r a_{-r}(3) \sum_{\substack{d>0 \\ d|r, d|n}} \frac{1}{d} c_{rn/d^2} = c_n + 3 \sum_{\substack{d>0 \\ d|3, d|n}} \frac{1}{d} c_{3n/d^2},$$

which leads to

$$b_{3,1} = c_1 + 3c_3 = 2{,}593{,}096{,}794,$$
$$b_{3,2} = c_2 + 3c_6 = 12{,}756{,}091{,}394{,}048,$$

and hence

$$Z_3(\tau) = q^{-3} + q^{-1} + 1 + 2{,}593{,}096{,}794q + 12{,}756{,}091{,}394{,}048q^2 + \cdots.$$

References

[1] Máximo Bañados, Claudio Teitelboim, and Jorge Zanelli, *Black hole in three-dimensional spacetime*, Phys. Rev. Lett. **69** (1992), no. 13, 1849–1851.

[2] Nicolas Brisebarre and Georges Philibert, *Effective lower and upper bounds for the Fourier coefficients of powers of the modular invariant j*, J. Ramanujan Math. Soc. **20** (2005), no. 4, 255–282.

[3] A. A. Bytsenko, M. E. X. Guimarães, and F. L. Williams, *Remarks on the spectrum and truncated heat kernel of the BTZ black hole*, Lett. Math. Phys. **79** (2007), no. 2, 203–211.

[4] Andrei A. Bytsenko, Luciano Vanzo, and Sergio Zerbini, *Quantum correction to the entropy of the $(2+1)$-dimensional black hole*, Phys. Rev. D (3) **57** (1998), no. 8, 4917–4924.

[5] Roberto Camporesi, *Harmonic analysis and propagators on homogeneous spaces*, Phys. Rep. **196** (1990), no. 1-2, 1–134.

[6] John L. Cardy, *Operator content of two-dimensional conformally invariant theories*, Nuclear Phys. B **270** (1986), no. 2, 186–204.

[7] Steven Carlip and Claudio Teitelboim, *Aspects of black hole quantum mechanics and thermodynamics in $2+1$ dimensions*, Phys. Rev. D (3) **51** (1995), no. 2, 622–631.

[8] Igor B. Frenkel, James Lepowsky, and Arne Meurman, *A moonshine module for the Monster*, Vertex operators in mathematics and physics (Berkeley, Calif., 1983), Math. Sci. Res. Inst. Publ., vol. 3, Springer, New York, 1985, pp. 231–273.

[9] D. Gaiotto, *Monster symmetry and extremal CFTs*, preprint, 2008, arXiv:0801.0988.

[10] Simone Giombi, Alexander Maloney, and Xi Yin, *One-loop partition functions of 3D gravity*, J. High Energy Phys. (2008), no. 8, 007, 25.

[11] I. Gradshteyn and I. Ryzhik, *Table of integrals, series, and products*, corrected and enlarged ed., Academic Press, 1980.

[12] Gerald Höhn, *Selbstduale Vertexoperatorsuperalgebren und das Babymonster*, Bonner Mathematische Schriften, no. 286, Universität Bonn Mathematisches Institut, Bonn, 1996, Dissertation, Rheinische Friedrich-Wilhelms-Universität Bonn, 1995.

[13] Ikuo Ichinose and Yuji Satoh, *Entropies of scalar fields on three-dimensional black holes*, Nuclear Phys. B **447** (1995), no. 2-3, 340–370.

[14] A. Maloney and E. Witten, *Quantum gravity partition functions in three dimensions*, preprint, 2007, arXiv:0712.0155.

[15] Robert B. Mann and Sergey N. Solodukhin, *Quantum scalar field on a three-dimensional (BTZ) black hole instanton: heat kernel, effective action, and thermodynamics*, Phys. Rev. D (3) **55** (1997), no. 6, 3622–3632.

[16] S. J. Patterson, *The Selberg zeta-function of a Kleinian group*, Number theory, trace formulas and discrete groups (Oslo, 1987), Academic Press, Boston, MA, 1989, pp. 409–441.

[17] P. Perry, *Heat trace and zeta function for the hyperbolic cylinder in three dimensions*, four page fax based in part on notes of F. Williams, 2001.

[18] P. Perry and F. Williams, *Selberg zeta function and trace formula for the BTZ black hole*, Int. J. Pure Appl. Math. **9** (2003), no. 1, 1–21.

[19] Hans Petersson, *Über die Entwicklungskoeffizienten der automorphen Formen*, Acta Math. **58** (1932), no. 1, 169–215.

[20] Hans Rademacher, *The Fourier Coefficients of the Modular Invariant $J(\tau)$*, Amer. J. Math. **60** (1938), no. 2, 501–512.

[21] F. Williams, *A zeta function for the BTZ black hole*, Internat. J. Modern Phys. A **18** (2003), no. 12, 2205–2209.

[22] _____, *BTZ black hole and Jacobi inversion for fundamental domains of infinite volume*, Algebraic structures and their representations, Contemp. Math., vol. 376, Amer. Math. Soc., Providence, RI, 2005, pp. 385–391.

[23] _____, *Conical defect zeta function for the BTZ black hole*, One hundred years of relativity: Proceedings of the Einstein Symposium (Iais, Romania, 2005), 2005, Scientific Annals of Alexandru Ioan Cuza Univ., pp. 54–58.

[24] _____, *Topology of the BTZ black hole with a conical singularity*, unpublished lecture notes, 2005.

[25] _____, *Note on quantum correction to BTZ instanton entropy*, Proceedings of Science (IC 2006) (2006), no. 006, Available at http://pos.sissa.it/archive/conferences/031/006/IC2006˙006.pdf.

[26] _____, *A deformation of the Patterson–Selberg zeta function*, Actas del XVI Coloquio Latinoamericano de Álgebra (Colonia del Sacramento, Uruguay, August 2005) (W. Santos, G. González-Sprinberg, A. Rittatore, and A. Solotar, eds.), Rev. Mat. Iberoamericana, Madrid, 2007, pp. 109–114.

[27] _____, *A resolvent trace formula for the BTZ black hole with conical singularity*, Council for African American Researchers in the Mathematical Sciences. Vol. V, Contemp. Math., vol. 467, Amer. Math. Soc., Providence, RI, 2008, pp. 49–62.

[28] E. Witten, *Three-dimensional gravity revisited*, preprint, 2007, arXiv:0706.3359.

Added in proof

The author has discovered that a (slightly incorrect) version of formula (3.27), page 337, has been obtained, independently, by A. Bytsenko and M. Guimarães, (see formula (4.13) in their *Truncated heat kernel and one-loop determinants for the BTZ geometry*, Eur. Phys. J. C **58** (2008), pp. 511–516).

The following reference provides for further connections of the Patterson–Selberg zeta function to BTZ physics: D. Diaz, *Holographic formula for the determinant of the scattering operator in thermal AdS*, preprint, arXiv:0812.2158v3 (2009).

FLOYD L. WILLIAMS
DEPARTMENT OF MATHEMATICS AND STATISTICS
LEDERLE GRADUATE RESEARCH TOWER
710 NORTH PLEASANT STREET
UNIVERSITY OF MASSACHUSETTS
AMHERST, MA 01003-9305
UNITED STATES
   williams@math.umass.edu

For EU product safety concerns, contact us at Calle de José Abascal, 56–1º,
28003 Madrid, Spain or eugpsr@cambridge.org.

www.ingramcontent.com/pod-product-compliance
Lightning Source LLC
LaVergne TN
LVHW021942060526
838200LV00042B/1891